D1243386

SILICON CHEMISTRY

INTERNATIONAL UNION OF PURE AND APPLIED CHEMISTRY

Proceedings of the Eighth International Symposium on Organosilicon Chemistry, St. Louis, Missouri USA, June 7-12, 1987

Based on the invited and special lectures presented at the 8th International Symposium on Organosilicon Chemistry, St Louis, Missouri, USA. June 7–12, 1987.

SILICON CHEMISTRY

Editors:

E. R. COREY
Department of Chemistry
University of Missouri–St Louis, USA

J. Y. COREY
Department of Chemistry
University of Missouri–St Louis, USA

and

P. P. GASPAR
Department of Chemistry
Washington University, St Louis, USA

Published for the
INTERNATIONAL UNION OF PURE AND APPLIED CHEMISTRY

ELLIS HORWOOD LIMITED
Publishers · Chichester

Halsted Press: a division of
JOHN WILEY & SONS
New York · Chichester · Brisbane · Toronto

First published in 1988 by
ELLIS HORWOOD LIMITED
Market Cross House, Cooper Street,
Chichester, West Sussex, PO19 1EB, England
The publisher's colophon is reproduced from James Gillison's drawing of the ancient Market Cross, Chichester.

Distributors:

Australia and New Zealand:
JACARANDA WILEY LIMITED
GPO Box 859, Brisbane, Queensland 4001, Australia

Canada:
JOHN WILEY & SONS CANADA LIMITED
22 Worcester Road, Rexdale, Ontario, Canada

Europe and Africa:
JOHN WILEY & SONS LIMITED
Baffins Lane, Chichester, West Sussex, England

North and South America and the rest of the world:
Halsted Press: a division of
JOHN WILEY & SONS
605 Third Avenue, New York, NY 10158, USA

South-East Asia
JOHN WILEY & SONS (SEA) PTE LIMITED
37 Jalan Pemimpin # 05-04
Block B, Union Industrial Building, Singapore 2057

Indian Subcontinent
WILEY EASTERN LIMITED
4835/24 Ansari Road
Daryaganj, New Delhi 110002, India

© 1988 International Union of Pure and Applied Chemistry IUPAC Secretariat:
Bank Court Chambers, 2-3 Pound Way, Cowley Centre, Oxford OX4 3YF, UK

British Library Cataloguing in Publication Data
Silicon Chemistry.
1. Silicon compounds. Chemical analysis
I. Corey, Eugene R.   II. Corey, Joyce Y.   III. Gaspar, Peter P.
IV. International Union of Pure and Applied Chemistry
546'.6836
Library of Congress CIP data available

ISBN 0–7458–0528–0 (Ellis Horwood Limited)
ISBN 0–470–21081–8 (Halsted Press)

Printed in Great Britain by Hartnolls, Bodmin.

TABLE OF CONTENTS

Foreword

Eugene R. Corey and Joyce Y. Corey – Department of Chemistry, University of Missouri-St Louis, St. Louis MO 63121, USA.

Peter P. Gaspar – Department of Chemistry, Washington University, St. Louis, MO 63130, USA.

The Eighth International Symposium on Organosilicon Chemistry was held the week of June 7, 1987 on the Washington University campus in St.Louis, Missouri. A goal of the symposium was coverage of all areas of current interest in silicon chemistry. Three plenary lecturers and forty-nine invited lecturers delivered oral papers in the areas of silicon-assisted organic synthesis, the organic chemistry of silicon, silicon in living systems, silicon reactive intermediates, silicon-silicon chemistry, silicon-oxygen polymers and materials, the inorganic chemistry of silicon, silicon in solid state technology; and physical chemistry, theoretical studies, and spectroscopy of silicon compounds. Silicon Chemistry: Current Vistas is a collection of papers written by the majority of the fifty-two invited speakers. Chapters 1, 16, and 27 are the reports of the plenary lecturers and the remaining 46 chapters were written by the invited lecturers. Abstracts for more than 150 oral presentations and over 110 poster presentations were included in the technical program for the Eighth International Organosilicon Symposium. The range of topics is broad and includes several in addition to the traditional organosilicon subjects. We hope this volume will provide background information for specialists and newcomers to the rapidly changing fields of silicon chemistry.

PART I

SILICON-ASSISTED
ORGANIC SYNTHESIS

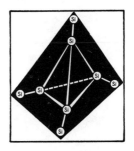

Chapter 1

Stereochemical and reactivity patterns in silyl-substituted cycloalkanes and acyclic analogues

Leo A. Paquette — Evans Chemical Laboratories, The Ohio State University, Columbus, Ohio 43210

ABSTRACT

Silyl-substituted cyclopropanes are becoming increasingly available and are developing into serviceable reagents for a relatively wide variety of synthetic transformations. A major portion of the lecture is dedicated to a survey of this rapidly growing field of organosilicon chemistry. The configurational stability of α-silyl carbanions is next examined. When such reactive intermediates are generated by Haller-Bauer cleavage of the corresponding phenyl ketones, retention of configuration is invariably seen. Consequently, chiral cyclopropyl, cyclic, and open-chain silanes can easily be generated in optically active condition from appropriate precursors, implying of course that effective planarization does not materialize during the C-C bond cleavage and subsequent protonation.

NEW DEVELOPMENTS IN SILYLCYCLOPROPANE CHEMISTRY

The versatility of silyl-substituted cyclopropanes as synthetic reagents has recently been underscored [1]. Since late 1985 when this review was written, a number of interesting new developments pertinent to this field have appeared. These relevant findings, to be addressed first, are presented in parallel to the format utilized in reference 1.

Shiori and his coworkers have reported a convenient and efficient preparation of (trimethylsilyl)diazomethane invol-

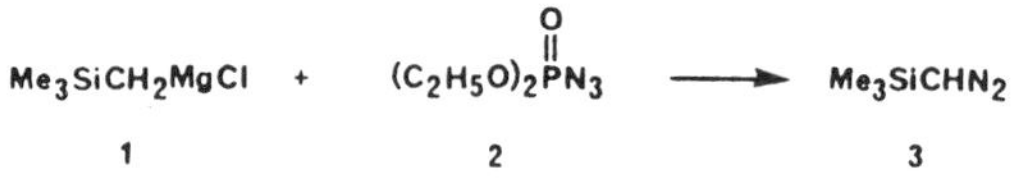

$$Me_3SiCH_2MgCl \quad + \quad (C_2H_5O)_2\overset{\overset{O}{\|}}{P}N_3 \quad \longrightarrow \quad Me_3SiCHN_2$$

$$\quad\quad 1 \quad\quad\quad\quad\quad\quad\quad 2 \quad\quad\quad\quad\quad\quad\quad\quad 3$$

ving a diazo transfer reaction. Exposure of Grignard reagent
1 to diphenyl phosphorazidate (2) in diethyl or di-_n_-butyl
ether at room temperature provides 3 in > 80% yield [2]. The
photochemical decomposition of 3 in the presence of alkenes
leads stereospecifically via the singlet carbene to silyl-
substituted cyclopropanes [3]. Catalysis by cuprous chloride
is less stereocontrolled [4].

Quenching of the dianions of the three- to six-membered
carboxylic acids 4 with chlorodiphenylmethylsilane followed by
aqueous hydrolysis leads to 5 in 60-85% yield.[5] Although the
direct C-silylation of the cyclopropyl system had been earlier

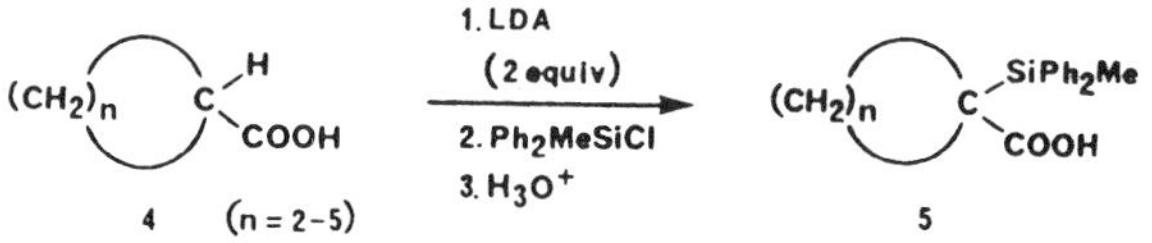

demonstrated [6], ketene acetal production (i.e., O-silylation)
normally operates when the ring is larger and other chlorosilanes are utilized.

Halogen-metal exchange within cyclopropylalkyne 6 and
silylation of the lithium derivative permit conversion to 7 [7]
and subsequently to 8 and 9 [8].

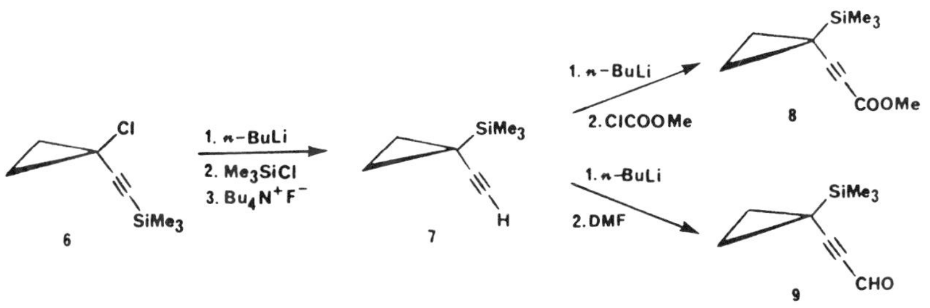

The ease with which benzocyclopropene (10) undergoes me-
talation lends itself particularly well to bissilylation of the
cyclopropyl methylene site [9]. Whereas 11 is unreactive
toward molecular bromine, regioselective nitration to give 12
occurs in the presence of 67% nitric acid. Reduction of the
nitro group in 12 takes place smoothly and the availability of
functionalized aniline 13 allows access to azo compound 14.
Only when 12 was treated with zinc in strongly alkaline solution was cleavage of the bissilylated cyclopropane ring noted

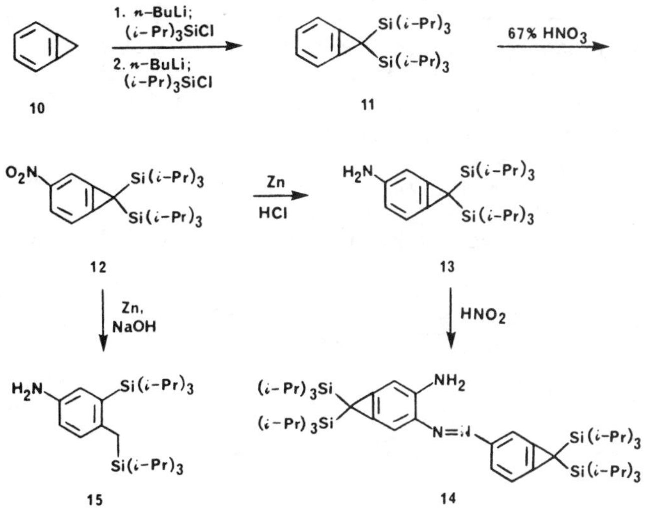

as in **15** [9].

Nakajima and coworkers have successfully prepared a series of previously unknown cyclopropyl trimethylsilyl ketones (20) by reaction of two equivalents of the corresponding 1-(trimethylsilyl)cyclopropyllithium compounds (16) with dichloromethyl methyl ether [10]. The facility with which the intramolecular 1,2-silyl shift occurs has been rationalized in terms of the trapping of _in situ_-generated chloromethoxycarbene and passage through intermediates **17-19** as shown.

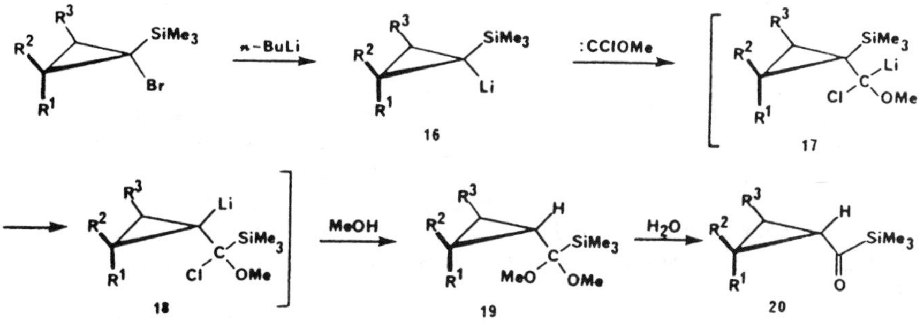

The trans-disubstituted cyclopropane **21** happens to be unreactive toward $ClSO_3SiMe_3$ at ambient temperature. In refluxing cyclohexane solution, however, conversion to **22** occurs in 70% yield [11]. The hydrolysis of **22** in water proceeds in an exothermic manner to deliver **23**, only the second known cyclopropanesulfonic acid.

In contrast to the above, cis isomer **24** undergoes instantaneous reaction with $ClSO_3SiMe_3$ at 20°C. The product is not
one of sulfonation, however, but of isomerization (i.e., **21**).
Knowledge of the precise mechanistic course of this process
would be of interest. Friedel-Crafts acylation of either **21** or
24 effectively provides ketone **25** [11].

Trimethylsilylation of **10** and 1H-cyclopropa[b]naphthalene,
deprotonation with strong base, and reaction of the resulting
anion with an aldehyde or ketone provides via Peterson olefina-

tion alkylidene cycloproparenes (**26**) [12], benzocalicenes (**27**)
[13], and benzotriaheptafulvenes (**28**) [13].

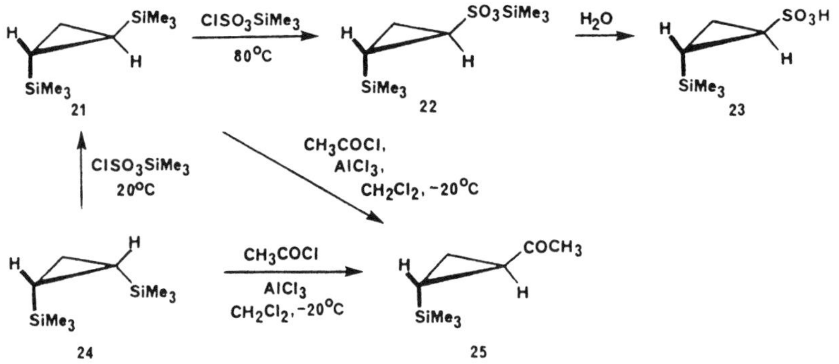

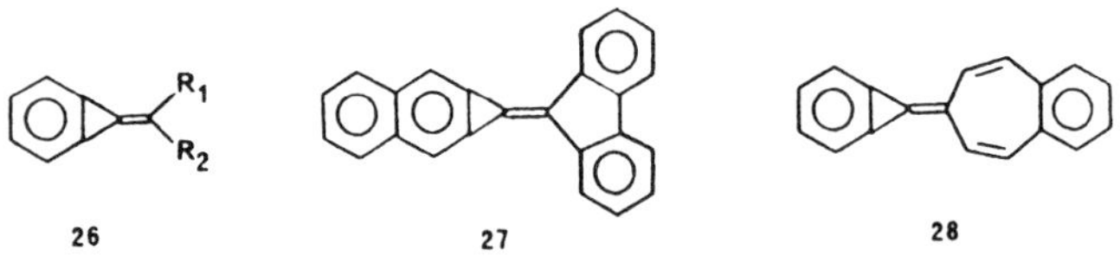

β-Halocyclopropylsilanes such as **29** and **31** can be converted
in high yield to the corresponding cyclopropenes in the gas
phase through use of tetra-n-butylammonium fluoride deposited
on glass helices as the solid phase [14,15]. The highly
strained products **30** and **32** are prone to spontaneous dimerization, the structures of which have recently been elucidated
[15].

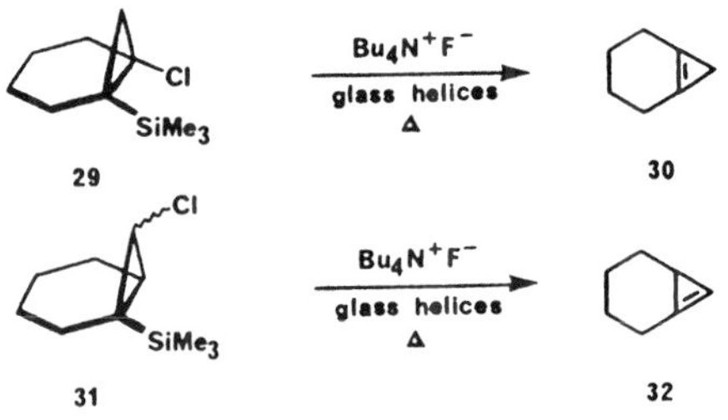

The reaction of cyclopropene **33** with thiophenoxide ion has been examined [16]. In the presence of excess base, **36** proved to be the exclusive isolated product. At shorter reaction times and in the presence of an equivalent amount of base, **34** and **35** were formed efficiently in a 2:1 ratio. Both of these products were readily desilylated to **36**, removal of the α-trimethylsilyl group undoubtedly occuring by carbanion generation. Significantly, both desilylation reactions generate only the trans isomer of **36**. This has been construed as a manifestation of an anomeric effect induced by the β-sulfonyl group which facilitates carbanion interconversion.

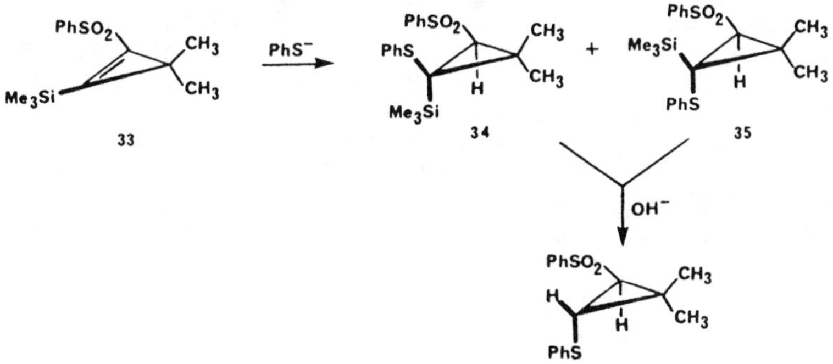

An interesting metal insertion reaction has been observed upon reaction of **37** with (η³-allyl)(η⁵-cyclopentadienyl)palladium (**38**) in the presence of trimethylphosphine [17]. The product cyclobutapalladium **39**, which is stable at room temperature but decomposes in air, represents the first well defined palladium complex to be obtained by ring opening of a benzocyclopropene.

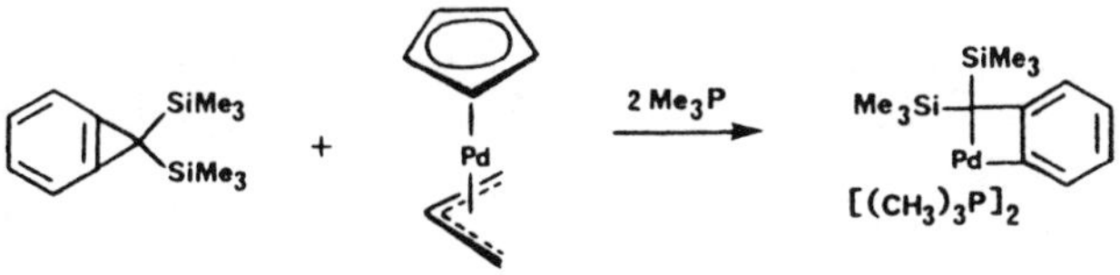

The electrophilic opening of **40** occurs in a highly regioselective manner with stannic chloride [18]. The specific homoallyltrichlorostannane (**41**) arises as a consequence of combined α and β activating contributions from the pair of silicon substituents.

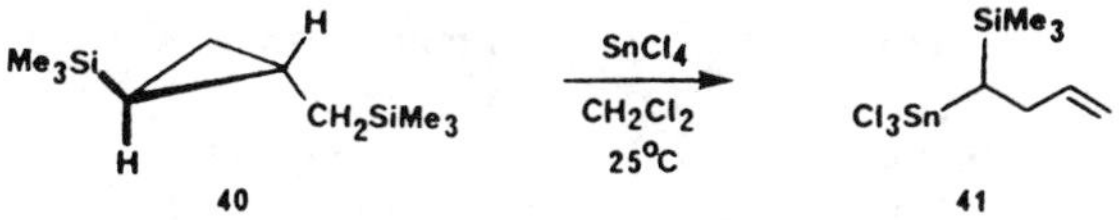

The gas-phase thermal decomposition of (trimethylsilyl)-
cyclopropane to allyltrimethylsilane (91.9%) and (E)- and (Z)-
1-propenyltrimethylsilane (8.1%) has been investigated at 416-
478°C and 14 torr [19]. The difference in activation energies
has been attributed to a β effect.

CONFIGURATIONAL STABILITY OF CYCLIC 1-TRIMETHYLSILYL CARBANIONS

In 1984, the details of a study aimed at elucidating the ability of an optically active α-silyl cyclopropyl carbanion were
reported [20]. The levorotatory carboxylic acid 42 was obtained by resolution of its l-cinchonidine salt and shown to
possess the R configuration by conversion to 43, a molecule
alternatively prepared from the known (R)-(-)-bromide 44. Following conversion of 42 to the (R)-(-)-carboxaldehyde 45, condensation with phenyllithium was effected. Oxidation of the
resulting carbinol with manganese dioxide provided the optically active ketone 46, Haller-Bauer cleavage of which led to
(R)-(-)-47 with complete retention of stereochemistry. The
latter conclusion was based on an independent stereodefined
synthesis of the product silylcyclopropane 47 [20].

Previously, Walborsky had made particularly elegant use of
the Haller-Bauer reaction [21] in demonstrating that anionic
centers on three-membered rings are configurationally stable
[22]. The high stereochemical retention associated with the
(R)-(+)-46 to (R)-(-)-47 conversion is an indicator that a
trimethylsilyl group present at the carbanion site does not
lower the racemization barrier.

However, the inability of cyclopropyl anions to undergo inversion of configuration is intrinsic to these systems [23] and
sheds no light per se on the generalized stereochemical course
of the Haller-Bauer process. Consequently, we were led to

examine the cleavage of phenyl ketones by amide ion in several
cyclic and open-chain silicon-containing systems [24].

Attention was initially given to the cyclopentyl example
54, whose structural features were designed to eliminate any
element of serious steric compression at the reaction site.
The remote isopropylidene group is, of course, necessary for
chirality. Following diisobutylaluminum hydride reduction of
48, the alcohol was transformed into its O-acetylmandelate
ester (50) and the diastereomers were partially separated by
chromatography. After reductive cleavage of one of the esters,
the proper absolute configurational assignment to the resultant
(-)-**49** (30% ee) was formalized by sequential highly stereoselective catalytic hydrogenation to (-)-**51**, Peterson olefina-

tion, and ozonolytic cleavage of the exo-methylene functionality so unmasked. That the product ketone is **52** rests on
direct comparison with the known (S)-(-)-3-isopropylcyclopenta-
none [25].

Once the enantiomeric (+)-**49** (40% ee) had been transformed into phenyl ketone **54** as before, cleavage to (-)-**55** was
accomplished with two amide bases in refluxing anhydrous benzene solution. While $NaNH_2$ furnished the silane with 95% retention of stereochemistry, recourse to KNH_2 proved almost as
effective (91% retention) [24]. Direct information bearing on
the S configuration of (-)-**55** and its enantiomeric purity was
gained by hydroboration-oxidation [26a] of the known (S)-(-)-3-
(trimethylsilyl)cyclopentene (**56**) [26b] to give (S)-(-)-**57** to
which (-)-**55** was directly corrolated by ozonolysis.

Thus, we see that bond scission in **54** also gives rise to a non-racemic α-silyl carbanion that is subject to electrophilic protonation with almost complete retention of configuration.

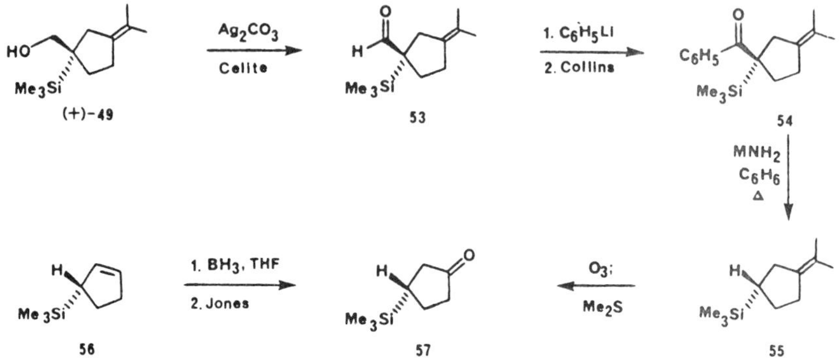

GENERATION AND TRAPPING OF CHIRAL OPEN-CHAIN α-TRIMETHYLSILYL CARBANIONS

The next stage of our investigation involving acyclic counterparts of **46** and **54** was initiated by preparation of the

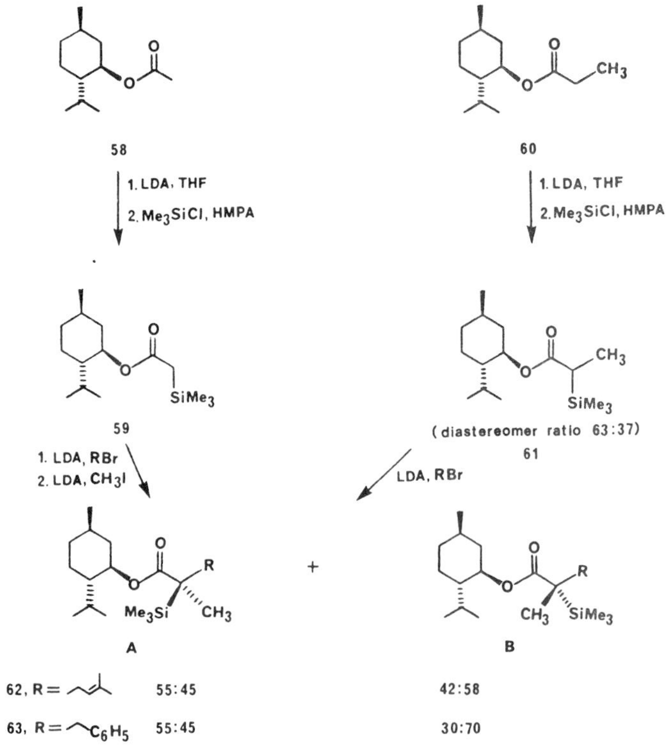

	A		B
62, R =	55:45		42:58
63, R = C_6H_5	55:45		30:70

diastereomeric pairs of menthyl esters **62** and **63**. When the
pathway to **62/63** involved C-silylation of menthyl acetate **(58)**
and reiterative alkylation of **59** first with prenyl or benzyl
bromide and then with methyl iodide, the diastereomers labelled
A were found to predominate somewhat. Conversely, the route
from menthyl propionate **(60)** via **61** led predominantly to
diastereomers **B**. To our great convenience, **63A** could be obtained diastereomerically pure by simple recrystallization,
thus enabling elucidation of its stereochemistry by X-ray
crystallography (Figure 1) [27]. Structural assignment to **62a**
and **62B** is based on the expectedly related stereochemical
events adopted during alkylation of both **59** and **61**. The de of
62A could be improved to 46% by medium pressure liquid chromatography.

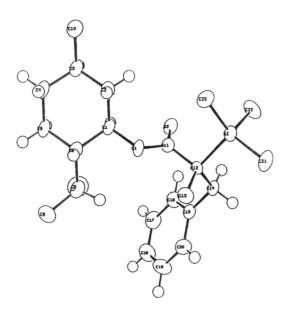

Fig. 1. Computer-generated three-dimensional view of 63A showing the absolute
stereochemistry.

Following arrival at (+)-**64** (46% ee) and (+)-**65** (>98% ee),
cleavage was again effected with sodium and potassium amides in
hot benzene. The results are summarized in Table I. The
constancy of the dextrorotatory nature of **66** and **67** signaled
immediately that retention of configuration had also materi-
alized in these examples. The structurally related secondary
silanes **68-70** have previously been prepared by Hayashi and
coworkers [28] who demonstrated that their levorotatory character corresponded to the S̲ absolute configuration. Although the
extent of retention in **66** and **67** has not been established at
the time of this writing, it is noteworthy that the differences

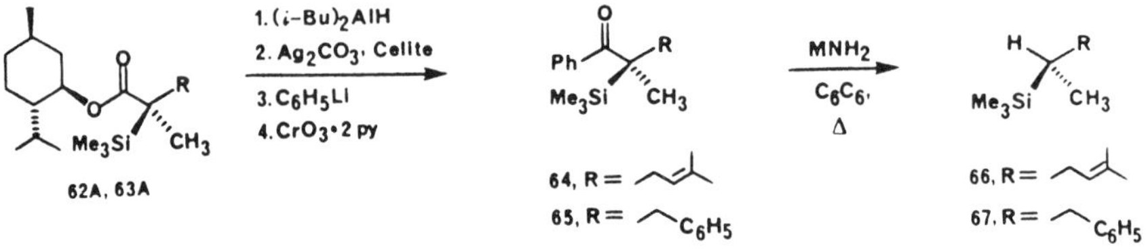

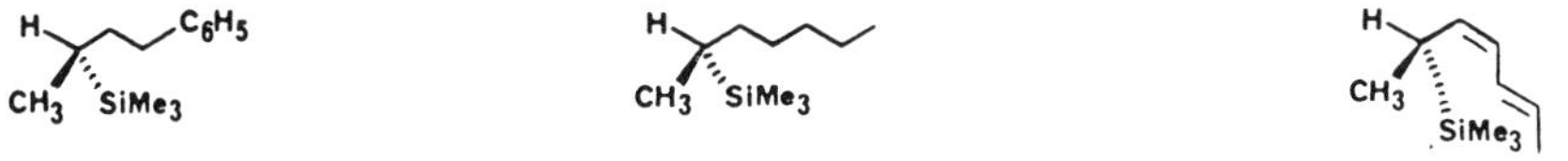

seen with sodium and potassium amide appear to be strikingly
constant for (−)-**54**, and (+)-**64**, and (+)-**65**. In all three
examples, the use of potassium as counterion results in a 7-13%
drop in stereochemical retention relative to the sodium base.
The reverse relative ordering has been observed during comparable generation of benzylic carbanions [29]. Importantly,
such changes do not affect the obviously strong preference for
replacing benzoyl by hydrogen from the front side of the
stereogenic center.

Table 1. Haller-Bauer Cleavage of (+)−64 and (+)−65.

compd	base	solvent	$[\alpha]^{25\,a}$	optical course
(+)−64	NaNH$_2$	C$_6$H$_6$, Δ	+9.6°	retention
	KNH$_2$	C$_6$H$_6$, Δ	+8.5°	retention
(+)−65	NaNH$_2$	C$_6$H$_6$, Δ	+26.7°	retention
	KNH$_2$	C$_6$H$_6$, Δ	+23.9°	retention

[a]The optical rotations for 64 and 65 were recorded at 365 nm and
the sodium D line, respectively.

What lies behind the tendency of the Haller-Bauer process
to preserve configuration? Phenomena closely related to those
discussed by Cram [30] are certainly at play. From a preparative viewpoint, the present work establishes that the Haller-
Bauer reaction, by virtue of one's ability to obtain α-silyl
carboxylic acids in optically active (and ideally homochiral)
condition, holds generality for the preparation of optically
active organosilanes of varied type.

ACKNOWLEDGEMENT

We are grateful to the National Science Foundation for
financial support.

REFERENCES

(1)   Paquette, L. A. Chem. Rev. 1986, 86, 733; Isr. J. Chem. 1981, 21, 128.
(2)   Mori, S.; Sakai, I.; Aoyama, T.; Shiori, T. Chem. Pharm. Bull. 1982, 30, 3380.
(3)   Hazeldine, R. N.; Scott, D. L.; Tipping, A. E. J. Chem. Soc., Perkin Trans. 1 1974, 1440.
(4)   (a) Seyferth, D.; Dow, A. W.; Menzel, H.; Flood, T. C. J. Am. Chem. Soc. 1969, 90, 1080.
(5)   Larson, G. L.; de Maldonado, V. C.; Berrios, R. R. Synth. Commun. 1986, 16, 1347.
(6)   Ainsworth, C.; Kuo, Y. J. Organometal. Chem. 1972, 46, 73.
(7)   Liese, T.; de Meijere, A. Chem. Ber. 1986, 119, 2995.
(8)   Bengtson, G.; Keyaniyan, S.; de Meijere, A. Chem. Ber. 1986, 119, 3607.
(9)   Neidlein, R.; Christen, D. Helv. Chim. Acta 1986, 69, 1623.
(10)   Nakajima, T.; Tanabe, M.; Ohno, K.; Segi, M.; Suga, S. Chem. Lett. 1986, 177.
(11)   Grignon-Dubois, M.; Dunoguès, J. J. Organomet. Chem. 1986, 309, 35.
(12)   Halton, B.; Randall, C. J.; Gainsford, G. J.; Stang, P. J. J. Am. Chem. Soc. 1986, 108, 5949.
(13)   Halton, B.; Buckland, S. J.; Mei, Q.; Stang, P. J. Tetrahedron Lett. 1986, 5159.
(14)   Billups, W. E.; Lin, L.-J. Tetrahedron 1986, 42, 1575.
(15)   Billups, W. E.; Arney, B. E., Jr. private communication.
(16)   Padwa, A.; Wannamaker, M. W. Tetrahedron Lett. 1986, 2555.
(17)   Schwager, H.; Benn, R.; Wilke, G. Angew. Chem., Int. Ed. Engl. 1987, 26, 67.
(18)   Ryu, I.; Suzuki, H.; Mural, S.; Sonoda, N. Organometallics 1987 6, 212.
(19)   Conlin, R. T.; Kwak, Y.-W. Organometallics 1986, 5, 1205
(20)   Paquette, L. A.; Uchida, T.; Gallucci, J. C. J. Am. Chem. Soc. 1984, 106, 335.
(21)   (a) Hamlin, K. E.; Weston, A. W. Org. React. 1957, 9, 1. (b) Kaiser, E. M.; Warner, C. D. Synthesis 1975, 395.
(22)   (a) Walborsky, H. M.; Impastato, F. J. Chem. Ind. (London) 1958, 1690. (b) Walborsky, H. M. Rec. Chem. Progr. 1962, 23, 75. (c) Impastato, F. J.; Walborsky, H. M. J. Am. Chem. Soc. 1962, 84, 4348. (d) Walborsky, H. M.; Allen, L. E.; Traenckner, H.-J.; Powers, E. J. J. Org. Chem. 1971, 36, 2937.
(23)   (a) Walborsky, H. M.; Impastato, F. J. J. Am. Chem. Soc. 1959, 81, 5835. (b) Walborsky, H. M.; Young, A. E. Ibid. 1961, 83, 2595; Ibid. 1964, 86, 3288. (c) Pierce, J. B.; Walborsky, H. M. J. Org. Chem. 1968, 33, 1962. (d) Walborsky, H. M.; Johnson, F. P.; Pierce, J. B. J. Am. Chem. Soc. 1968, 90, 5222. (e) Webb, J. L.; Mann, C. K.; Periasamy, M. P. Ibid. 1974, 96, 3711. (f) Hoell, D.; Schnieders, C.; Müllen, K. Angew. Chem., Int. Ed. Engl. 1983, 22, 243.
(24)   Paquette, L. A.; Ra, C. S.; Gilday, J. P.; Hoppe, M. manuscript in preparation.

(25) (a) Naves, Y.-R. Bull. Soc. Chim. France **1958**, 1372. (b) Nakazaki, M. Bull. Chem. Soc. Japan **1962**, <u>35</u>, 1904. (c) Posner, G.; Frye, L. L.; Hulce, M. Tetrahedron **1984**, <u>40</u>, 1401.

(26) (a) De Jesus, M.; Rosario, O.; Larson, G. L. J. Organometal. Chem. **1977**, <u>132</u>, 301. (b) Hayashi, T.; Kabeta, K.; Yamamoto, T.; Tameo, K.; Kumada, M. Tetrahedron Lett. **1983**, 5661.

(27) This study was performed at -110°C by Dr. J. C. Gallucci (Ohio State Crystallographic Service) whom we thank.

(28) Hayashi, T.; Konishi, M.; Okamoto, Y.; Kabeta, K.; Kumada, M. J. Org. Chem. **1986**, <u>51</u>, 3772.

(29) Paquette, L. A.; Gilday, J. P.; Ra, C. S. submitted for publication.

(30) Cram, D. J. "Fundamentals of Carbanion Chemistry", Academic Press, New York, 1965, Chapter IV.

Chapter 2

Trialkylsilyl protecting groups in the synthesis of α-hydroperoxy carbonyl compounds

Waldemar Adam, Luiz H. Catalani and Axel Griesbeck – Institute of Organic Chemistry, University of

Würzburg, am Hubland, D-8700 Würzburg, West-Germany.

ABSTRACT

α-Hydroperoxy carbonyl compounds, which are postulated as transient species in the horse radish peroxi-
dase-catalyzed oxygenation of aldehydes, ketones and
carboxylic acids leading to triplet excited states
via the corresponding hydroxy-1,2-dioxetanes, can be
prepared readily in the trialkylsilyl protected form
by employing the photooxygenation-silylation-ozonoly-
sis sequence. Attempts to release the trialkylsilyl
protecting groups leads to decomposition with chemiluminescence, providing evidence that the postulated
biological oxygenation proceeds via such α-hydroperoxy carbonyl products as intermediates. The alternative approach to prepare hydroxy-1,2-dioxetane precursors via photooxygenation of tiglic aldehyde
gives instead the hydroxy-1,2-dioxolanes involving
regioselective ene reaction at the α-methyl group.
Protecting the aldehydic functionality in the form
of trialkylsilyl cyanohydrins does promote ene reaction at the desired ß-methyl substituent; but more
significantly, both regioisomers are formed in high
and identical diastereomeric ratios. Similar but still
more pronounced diastereomeric control was observed
for the trialkylsilyl cyanohydrins of senecic aldehyde. These novel results implicate perepoxide-like
ene reactions of singlet oxygen with these substrates.

Enzymatic α-hydroperoxylation of carbonyl compounds, leading to α-hydroperoxy aldehydes, ketones, carboxylic acids and esters, constitutes an important

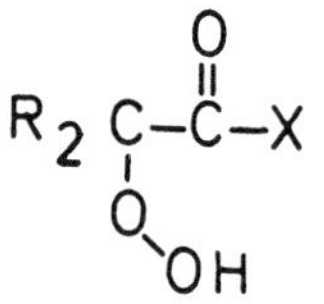

$$R_2 \underset{\underset{OH}{\overset{|}{O}}}{\overset{O}{\underset{||}{C}}}-\overset{O}{\underset{||}{C}}-X \qquad\qquad X = R, Ar, H, OH, OR, OAr$$

process in biological autoxidation, most significantly in the phenomenon of bioluminescence [1] and more generally in the enzymatic generation of electronically excited carbonyl products [2]. For example, in firefly bioluminescence, one of the best studied cases (Eq.1), the luciferin is autoxidative-

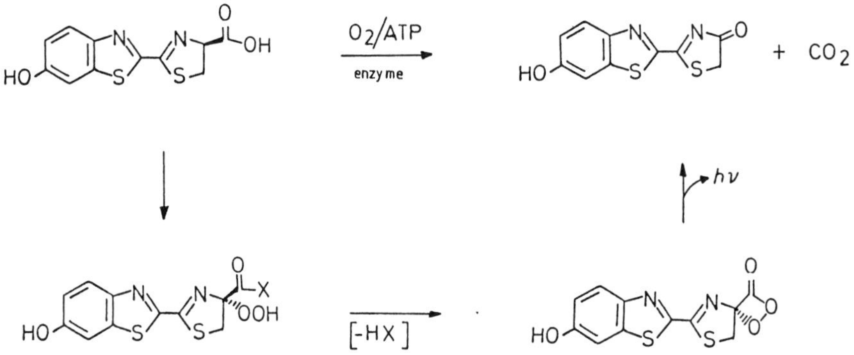

(1)

ly decarboxylated into the oxyluciferin with light emission. The precursor for the electronically excited oxyluciferin is the α-peroxy lactone, which in turn is derived from the α-hydroperoxy luciferin by dehydrative cyclization.

The key step in this enzymatic transformation is the introduction of the α-hydroperoxy functionality. Particularly the α-hydroperoxy carboxylic acids are rather prone to suffer decarboxylative decomposition either in acidic or basic media via the respective transition states below, so that gentle synthetic

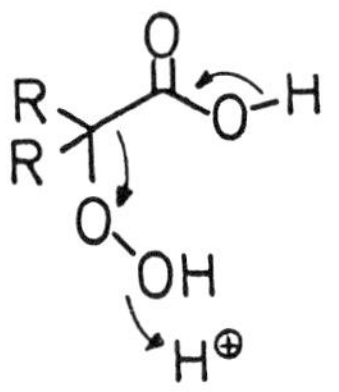

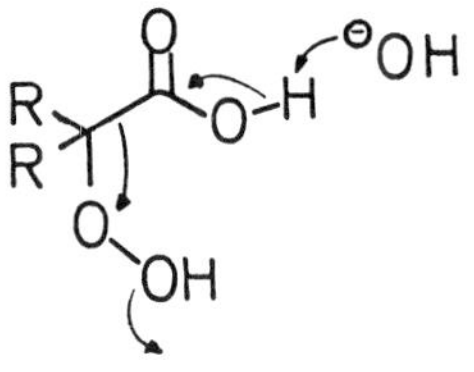

methods are essential to prepare these sensitive
compounds. For this purpose trialkylsilyl protecting
groups proved to be especially helpful, as witnessed
the synthesis of the first α-peroxylactone (Eq.2),

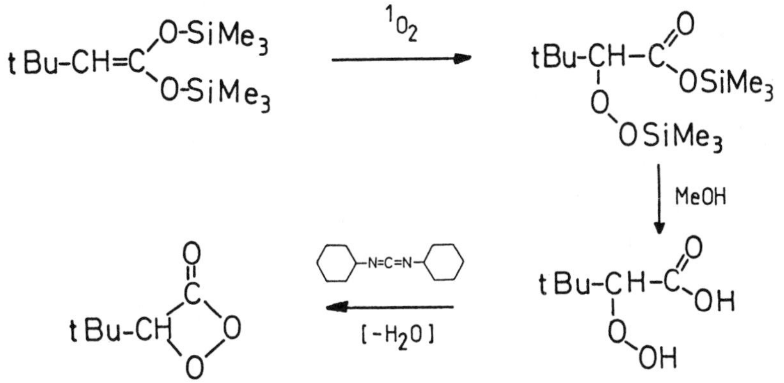

$$tBu-CH=C\overset{O-SiMe_3}{\underset{O-SiMe_3}{}} \xrightarrow{{}^1O_2} tBu-CH-C\overset{O}{\underset{OSiMe_3}{}}$$

$$\downarrow MeOH$$

$$tBu-CH\overset{O}{\underset{O}{}}C=O \xleftarrow[{[-H_2O]}]{\bigcirc-N=C=N-\bigcirc} tBu-CH-C\overset{O}{\underset{OH}{}}$$

(2)

employing the photooxygenation of silyl ketone
acetals [3]. The advantage of the silylated α-hydroperoxy acid, readily obtained via the silatropic ene
reaction with singlet oxygen, is that the trimethylsilyl groups protect these labile compounds towards
decarboxylation, provide sufficient volatility to
permit purification by vacuum distillation, and are
gently removed by subambient treatment with methanol. This synthetic break-through opened the way to
chemical model studies on the biological generation
of electronically excited states [4].

    In this context, the most established enzymatic
source of triplet state acetone is the horse radish
peroxidase (HRP)-catalyzed oxygenation of isobutyr-
aldehyde (Eq.3), postulated to be derived from the

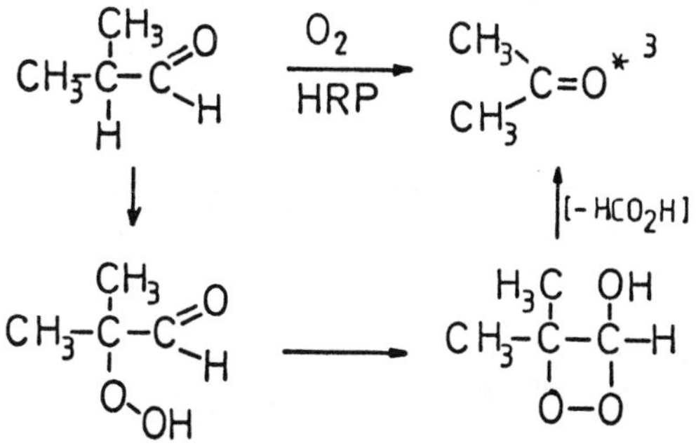

$$CH_3-\overset{CH_3}{\underset{H}{C}}-C\overset{O}{\underset{H}{}} \xrightarrow[HRP]{O_2} \overset{CH_3}{\underset{CH_3}{}}C=O^* {}^3$$

$$\downarrow \qquad\qquad \uparrow [-HCO_2H]$$

$$CH_3-\overset{CH_3}{\underset{O-OH}{C}}-C\overset{O}{\underset{H}{}} \longrightarrow CH_3-\overset{H_3C}{\underset{O-O}{C}}-\overset{OH}{\underset{}{C}}-H$$

(3)

hydroxydioxetane, which in turn is the cyclic tautomer of the essential α-hydroperoxy aldehyde [5]. To
this date no stable examples of these extremely
labile hydroperoxides have been isolated, except for
the trialkylsilyl derivative (Eq.4). In this novel

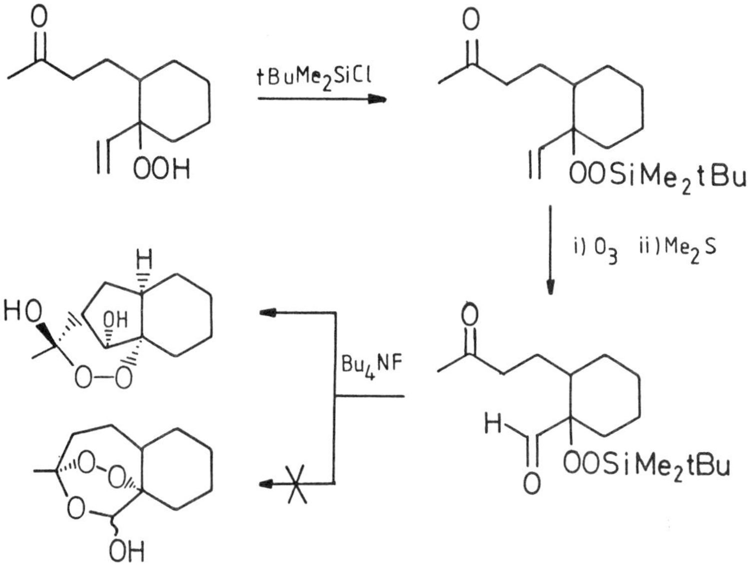

(4)

sequence [6], the hydroperoxy function is first protected by trialkylsilylation, which permits ozonolysis of the vinyl group without risking the loss of
the formyl substituent. However, on attempted
release of the hydroperoxy function by desilylation
with fluoride ion, aldolization with subsequent
cyclotautomerization led to the tricyclic peroxide
instead of the desired partial Qinghaosu structure.

We tested the scope of this novel synthetic tool
for the preparation of α-hydroperoxy aldehydes,
which are implicated in the HRP-catalyzed oxygenation (Eq.3). Our particular interest was to provide
evidence for the hitherto elusive hydroxydioxetanes, the cyclic four-membered ring tautomers of
the α-hydroperoxy aldehydes.

To become familiarized with this methodology, we
tried out the sequence of Eq.4 on the simpler

3-hydroperoxy-3-methyl-2-butanone (Eq.5), a known

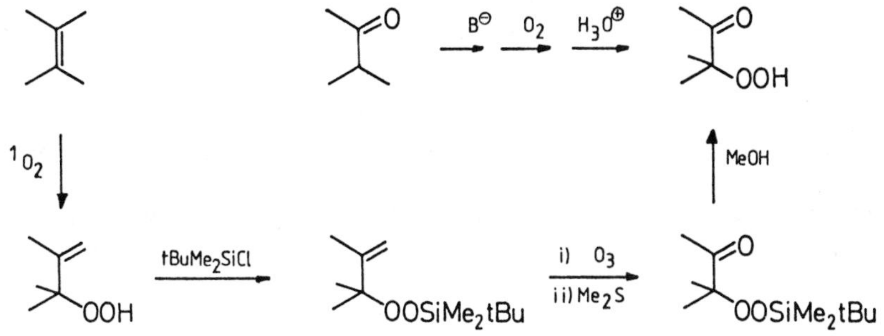

(5)

[7] and isolable stable α-hydroperoxy ketone. Indeed, no difficulties were encountered with this
sequence, and it is preferred with respect to yield
and purity of product to the direct, base-catalyzed
oxygenation of 2-methyl-2-butanone, the previous
source to the authentic α-hydroperoxy ketone [7].

The way was now open to use this trialkylsilyl
protection route to make the considerably more labile
and thus elusive α-hydroperoxy aldehydes. This sequence proved successful (Eq.6) for the preparation

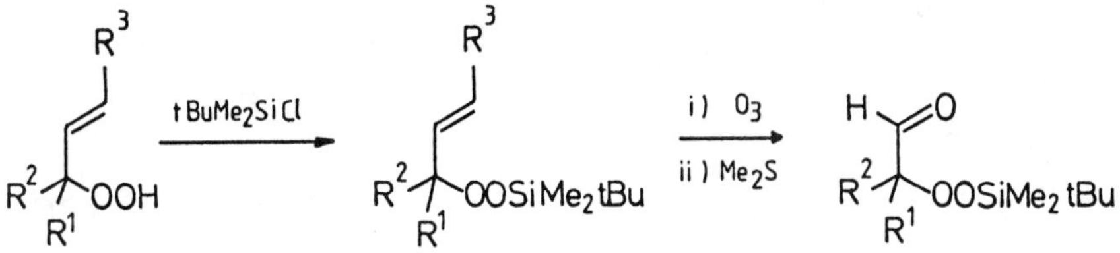

$\underline{1}$ $\underline{2}$ $\underline{3}$

a: $R^1 = R^2 = Me$, $R^3 = H$; b: $R^1 = R^3 = H$, $R^2 = Me$; c: $R^1 = H$, $R^2 = Et$, $R^3 = Me$

(6)

of the silylated aldehydes 3a-c, starting from the
respective hydroperoxides 1a-c, readily available
via photooxygenation of the appropriate alkenes [8].

Attempts to remove the trialkylsilyl protecting
group to release the free α-hydroperoxy aldehydes 3
proved difficult. Desilylation in methanol led to the
hemiacetals (Eq.7), which could not be demethanolated

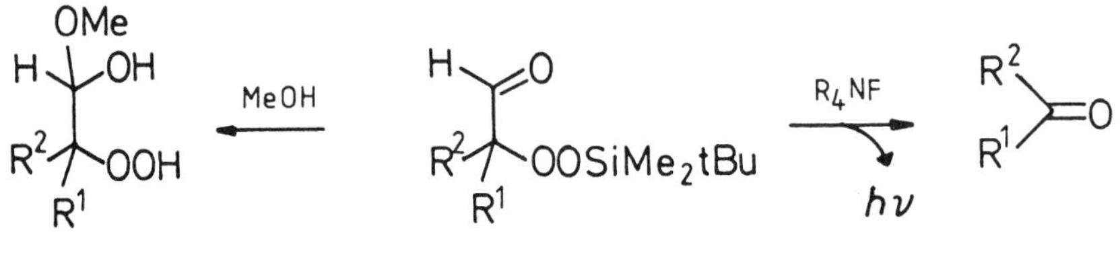

(7)

without risking decomposition. The aldehyde function
is sufficiently more electrophilic than normal ketones and add protic nucleophiles directly without
acid or base catalysis. On the other hand, desilylation with fluoride ion gave the corresponding excited
carbonyl product (Eq.7), as manifested by chemiluminescence. Although the free α-hydroperoxy aldehyde
could not be isolated as stable substance, it is gra-
tifying to confirm that these serve as precursors to
electronically excited carbonyl products via the
transient hydroxydioxetanes. This provides experimental confirmation of the mechanism postulated [5] in
the HRP-catalyzed autoxidation of aliphatic aldehydes (Eq.3).

As an alternative approach, the photooxygenation
of α,β-unsaturated aldehydes, e.g. tiglic aldehyde,
was undertaken (Eq.8), with the expectation that the

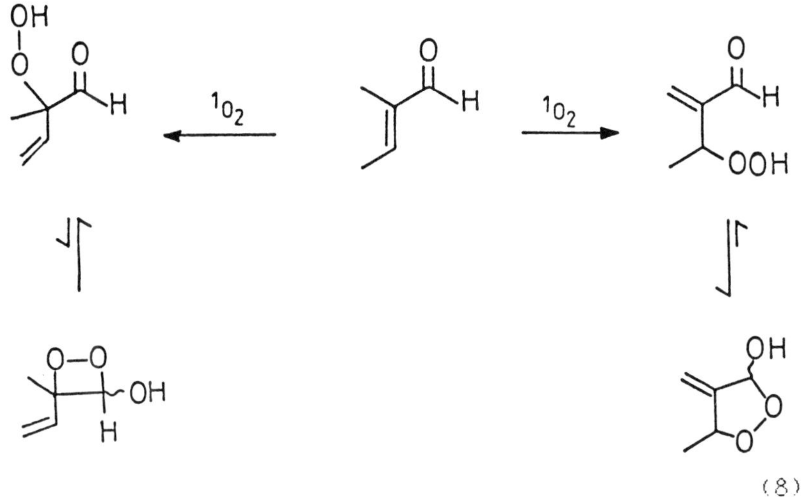

(8)

α-hydroperoxy aldehyde would be formed, which should serve as precursor to the corresponding hydroxydio-xetane, its cyclic tautomer. Much to our surprise [9], exclusively the ß-hydroperoxy aldehyde was formed in form of its cyclic hydroxy-1,2-dioxolane tautomer (Eq.8). In fact, this interesting regioche-mical preference appears to be quite general [10] in the photooxygenation of methyl-substituted α,ß-unsa-turated carbonyl compounds (Eq.9). The mechanistic

X =OH, OR, NR₂

(9)

details of the directing effect of the carbonyl sub-stituent in this ene reaction with singlet oxygen is still poorly understood.

To counteract this regioselectivity, the aldehyde function was protected in form of its trialkylsilyl cyanohydrin derivative (Eq.10). As anticipated, in-deed, both possible regioisomers were now produced, with the desired α-hydroperoxy derivative predomi-nating (56%) for both the trimethyl- and t-butyldi-methylsilyl cases [11]. Still more interesting me-chanistically speaking, is the appreciable diaste-

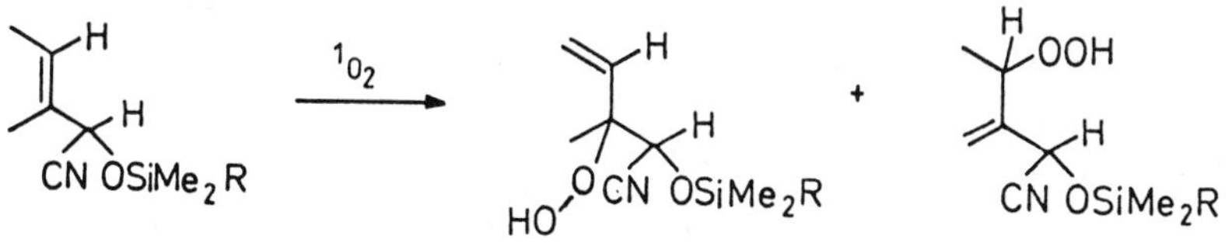

56%   44%

R=Me   d.r.=75:25   d.r.=75:25
R=tBu   d.r.=73:27   d.r.=73:27

(10)

reoselectivity, i.e. a diastereomeric ratio d.r.=75:25 (R=Me) and d.r.=73:27 (R=tBu) in both regioisomers.

This diastereomeric preference in the ene reaction with singlet oxygen is still more pronounced in the case of the trialkylsilyl cyanohydrin of senecic aldehyde. While the senecic aldehyde itself is completely inert towards singlet oxygen, its trialkylsilyl cyanohydrin derivative affords regioselectively (68%) the α-hydroperoxy isomer (Eq.11). More

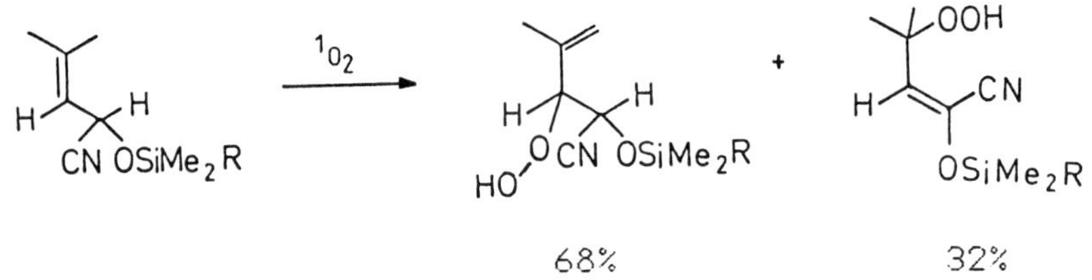

$$^1O_2$$

68%   32%

R=Me   d.r.=96:4   d.r.=96:4
R=tBu   d.r.>99:1   d.r.=99:1

(11)

significant is the diastereoselectivity, the highest yet observed in the ene reaction of singlet oxygen with acyclic substrates, with a diastereomeric ratio d.r.=94:4 (R=Me) and d.r. > 99:1 (R=tBu) again for both regioisomers.

This high diastereoselectivity is astonishing, but mechanistically more revealing is that the same diastereomeric ratio (d.r.) is observed for both regioisomers of the cyanohydrins of tiglic aldehyde (Eq.10) and of senecic aldehyde (Eq.11). Thus, not only has the trialkylsilyl protecting group favored the α-hydroperoxy regioisomer for both substrates, but more significantly it provided diastereomeric control.

Unsymmetric pathways for the singlet oxygen ene reaction via the concerted transition state A or the dipolar and diradical intermediates B and C, respec-

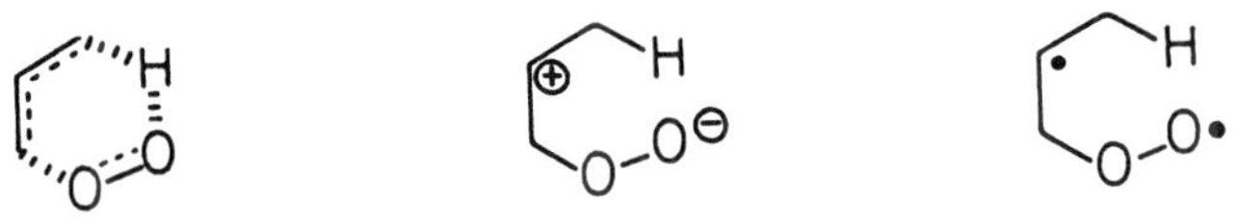

A   B   C

tively, cannot account for this stereochemical results. Instead, a common symmetric species is required to rationalize these unprecedented diastereo-
selectivities in the regioisomers. We postulate the
perepoxide-like transition states D and E for the

D   E

silyl cyanohydrins of tiglic and senecic aldehydes,
respectively. The left-hand transition state in each
set leads to the major diastereomer and the right-
hand one to the minor. Besides the cis effect [12],
which provides an anchor for the pendant oxygen atom
and thus the necessary rigidity for stereochemical
control, conformational preference is dictated by
the steric interactions with the bulky trialkylsilyl
substituent. Thus, besides serving as protecting
group, the trialkylsilyl moiety exercises stereochemical control, which should be synthetically
useful especially in the oxygen functionalization
with singlet oxygen.

In regard to our synthetic goal of preparing isolable, stable α-hydroperoxy aldehydes as precursors
to the desired hydroxydioxetanes, also the silyl
cyanohydrins proved of little use. Attempted release
of the aldehyde function by desilylcyanation led to
decomposition. Clearly, still more gentle and synthetic methods must be sought to make these labile
and elusive compounds accessible.

ACKNOWLEDGEMENTS

We thank the Deutsche Forschungsgemeinschaft
(Sonderforschungsbereich 172, "Molekulare Mechanis-
men Canzerogener Primärveränderungen"), the Fonds
der Chemischen Industrie, the Stiftung Volkswagen-
werk, the A. von Humboldt-Stiftung (postdoctoral
fellowship for L.H.C.) and the Fritz Thyssen
Stiftung for generous funding.

REFERENCES

[1] Adam, W.; Cilento, G. (Editors) in "Chemical and Biological Generation of Excited States", Academic Press, New York (1982).

[2] Adam, W.; Cilento, G. Angew. Chem. Int. Ed. Engl. 1983 22, 529.

[3] Adam, W.; Liu, J.-C. J. Am. Chem. Soc. 1972, 94, 2894.

[4] Adam, W. in "Small Ring Heterocycles", Hassner, A. (Ed.) John Wiley and Sons, New York (1985), Vol. 42, Part 3, Chap. 4, p. 351.

[5] Cilento, G. Pure Appl. Chem. 1984, 54, 1179.

[6] Clark, G. R.; Nikaido, M. M.; Fair, C. K.; Lin, J. J. Org. Chem. 1985, 50, 1994.

[7] Cubbon, R. C. P.; Hewlett, C. J. Chem. Soc. 1968, 2978. Richardson, W. H.; Hodge, W. F.; Stigall, D. L.; Yelvington, M. B.; Montgomery, F. C. J. Am. Chem. Soc. 1974, 96, 6652. Sawaki, Y.; Ogata, Y. J. Am. Chem. Soc. 1975, 97, 6983.

[8] Frimer, A.A. (Ed.) "Singlet Oxygenation", CRC Press, Boca Raton, Florida (1985).

[9] Adam, W.; Griesbeck, A. Synthesis 1986, 1050.

[10] Ensley, H. E.; Carr, R. V. C.; Martin, R. S.; Pierre, T. E. J. Am. Schem. Soc. 1980, 102, 2838. Orfanopoulos, M.; Foote, C. S. Tetrahedron Lett. 1985, 26, 5991. Adam, W.; Griesbeck, A. Angew. Chem. Int. Ed. Engl. 1985, 24, 1070.

[11] Adam, W.; Catalani, L. H.; Griesbeck, A. J. Org. Chem. 1986, 51, 5496.

[12] Orfanopoulos, M.; Girdina, M. B.; Stephenson, L. M. J. Am. Chem. Soc. 1979, 101, 275. Schulte-Elte, K. H.; Müller, B. L.; Rautenstrauch, V. J. Am. Chem. Soc. 1980, 102, 1738.

Chapter 3

Applications of organosilicon chemistry to the synthesis of polyoxygenated natural products

S.J. Danishefsky, D.C. Myles, S.L. DeNinno, F.E. Wincott and D.M. Armistead – Department of Chemistry Yale University, New Haven, Connecticut 06511.

INTRODUCTION

The field of natural products total synthesis serves a variety of important interests [1]. Historically, synthesis was an important confirmatory component in the proof of structure argument. With the advent of modern spectroscopic resources, not to mention the compelling power of X-ray crystallographic techniques, the centrality of the role of synthesis in structure deduction has been diminished.

The possibility that synthesis might provide access to natural products which are difficultly available from natural sources, continues to be an exciting feature of the field. Moreover, synthesis often provides the best opportunity to evaluate analog structures with respect to biological or other properties. Furthermore, synthetic studies can offer a unique vantage point from which the chemistry and even the mode of biological activity of the natural product itself can be examined [2].

Still another opportunity arising from complex target-oriented synthesis, and the one most germane to this Symposium, is that of examining new synthetic methodology. The symbiosis here is very clear. Natural products provide methodology with important as well as challenging targets. The new chemistry might in turn simplify the synthesis. The total synthesis exercise can provide a framework for evaluating the scope and limitations of reactions in a critical way.

Typically, new synthetic methods are developed with relatively simple substrates which are chosen to highlight a particular functional group conversion. However, extrapolations about the success of a new method from such simple settings are frequently disappointing. Demonstrations of transformations in the context of multifunctional substrates are therefore particularly valuable.

In this report we relate some recent applications of organosilicon chemistry to the synthesis of natural products. The field of organosilicon chemistry has been the beneficiary of a great deal of creative exploration. Silicon based reagents play a significant role in the armamentarium of organic synthesis, and that role is surely growing [3].

Our research dealing with applications of silicon based reagents to the synthesis of formidable target natural products has focused on two types of systems - dienoxysilanes **1**, and allylic silanes **2**. In our early investigations silyl enol ethers were incorporated as substructural units of functionalized dienes in the all-carbon Diels Alder reaction [4]. A number of natural products were synthesized via such silyl based dienes.

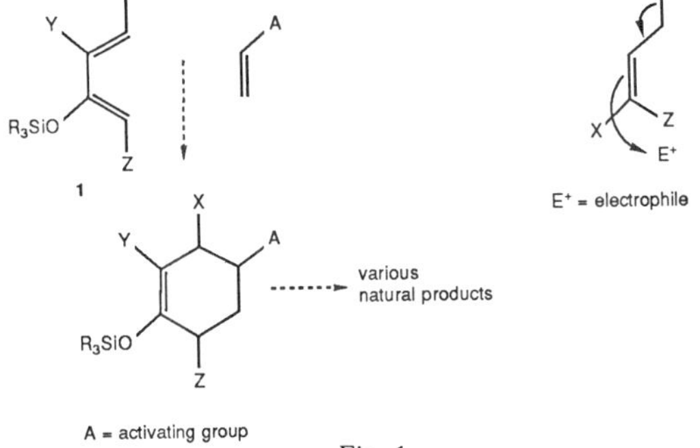

Fig. 1.

More recently, we have explored the potentialities inherent in the Lewis acid catalyzed cyclocondensation of dienoxylsilanes with a variety of aldehydes [5,6]. The reaction may produce the immediate cycloaddition product **3**, a cycloaddition-hydrolysis product **4**, or an unravelling product, **5**. The choice of Lewis acids and the conditions of workup play no small role in disposing the outcome. Several favorable circumstances converge to ensure that the cyclocondensation reaction will enjoy wide usage. Of course, it benefits from accessibility of its required constituents. There is no dearth of aldehydes! In a similar vein the technology to prepare a variety of α,β-unsaturated aldehydes, ketones, and esters is really quite powerful. Given the proficiency which has been attained in the enol silylation of such compounds, including those bearing extensive functionality, excellent access to the silyloxydienes is available [7].

Fortunately, the cyclocondensation reaction can indeed tolerate a remarkably broad range of functionality in both the diene (see X, Y and Z in structure **1**) and in the aldehyde components. Finally, we note that a great deal of stereochemical control at both the topographic and diastereofacial levels can be realized through judicious and informed selection of substrates and catalysts [5,6].

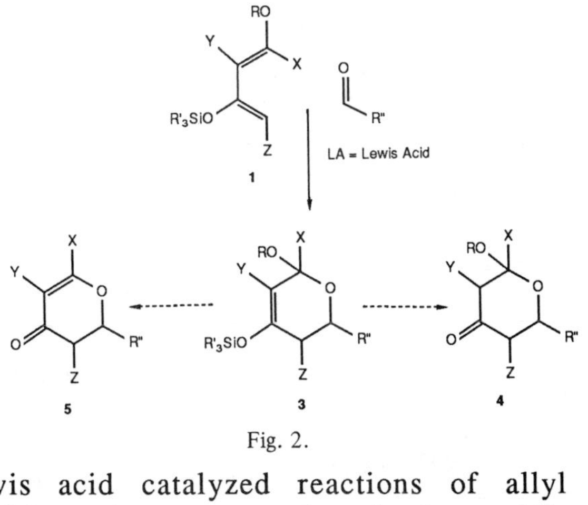

Fig. 2.

Lewis acid catalyzed reactions of allyl and crotyl silanes with various oxygen-based electrophiles have also found application in several of our total synthesis

endeavors.   Two of the prototype transformations which
have been particularly valuable have been the reactions of
allyl and crotyl silanes with glycal analogs (cf. **6**) and with
aldoseuloses systems (cf.**8**).

The reactions of allyl silanes with glycals mediated by
BF_3·etherate have proven to be a valuable resource in the
synthesis of C-glycosides [8].   The reaction manifests high
facial selectivity in that the allyl group attacks anti to the
substituent at C_5 (pyranose numbering).   If it be assumed
that the C_5 substituent adopts an equatorial disposition in
the reacting conformer, the result corresponds to axial
attack by the carbon-Ferrier type nucleophile [9].

The reaction of allyltrimethylsilane with aldehydes
such as **8** is a valuable device for extending the chiral bases
of furanose and pyranose rings to emerging stereogenic
centers on a side chain [10].   As previously reported, with
important aldehyde building blocks in the ribose and
galactose series, it is possible through appropriate
manipulation of the catalytic systems to obtain either the
apparent Cram-Felkin (C-F-**9**) or controlled (C-C-**10**)
allylation product [11].

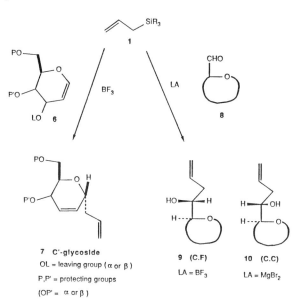

Fig. 3.

More recently, we have investigated the reactions of geometrically defined crotyl silanes in both of these generic transformations [12,13]. In the carbon-Ferrier-like series, E and Z crotyl silanes of the type **12E** and **12Z** give rise to major products **13** and **14** with good to excellent margins of selectivity. Very influential in controlling the selectivity is the nature of X and L of the glycal, and R of the crotyl silanes. We have not broadly investigated the corresponding reactions of crotyl silanes with aldoseulose systems. However, an important application of such a reaction will be shown in connection with our avermectin program (vide infra).

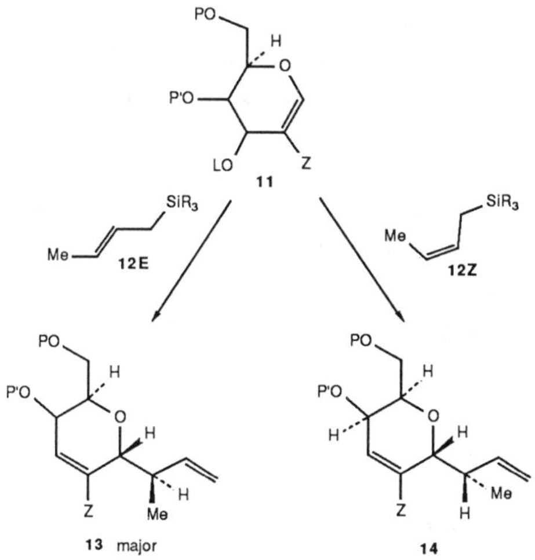

Fig. 4.

APPLICATIONS

(i)   6a-Deoxyerythronolide B. The route which we are following with the expectation of synthesizing 6a-deoxyerythronolide B (**15**) is appropriate for illustrating the reiterative features of the cyclocondensation strategy. The subgoal of the synthesis is a structure such as **16**, in which all of the stereochemical information is encoded. This subgoal system contains several interesting variations in its pattern of stereogenic centers relative to those encountered in our previous work [14].

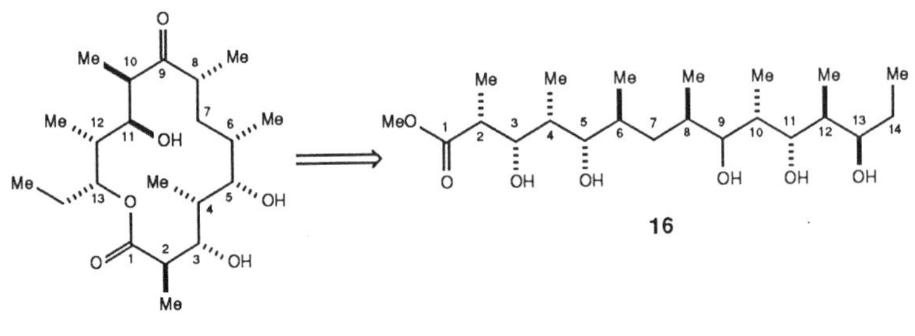

Fig. 5.

Since our primary concerns were the testing of the outer limits of the reiterative cyclocondensation strategy, and the evaluation of new designs for stereochemical communication, there was no need to operate with single enantiomers [15]. Accordingly, we started with the cyclocondensation of diene **17** with formaldehyde under catalysis by anhydrous zinc chloride. A 70-80% yield of pure dihydropyrone **18** was obtained. Reduction (LAH) was followed by Ferrier rearrangement leading to **19** (80% yield). However, ordinarily the crude material (ca. 100%) is carried forward. Hydroboration-oxidation and benzylation (NaH; BnBr: Bu4NI-THF) provided the methyl glycoside, **20**.

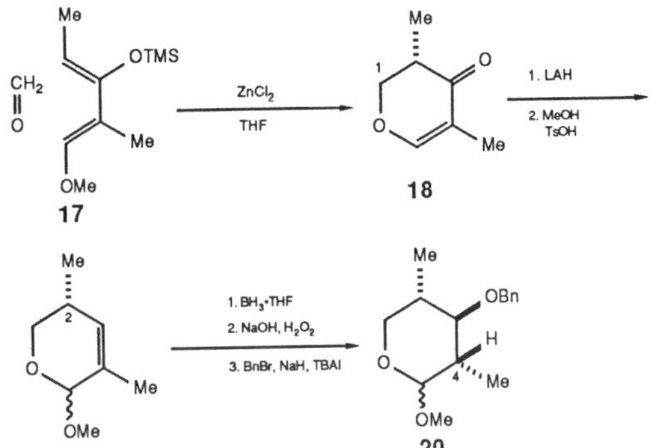

Fig. 6.

Reaction of this compound with 1,3-propane dithiol (TiCl4) gave the desired ring opening. The primary alcohol, destined to become the C_1 carboxyl was protected as its t-butyldiphenyl silyl ether, leading to compound **21** (85% yield from **20**). Unveiling of the aldehyde function was accomplished through the action of N-bromosuccinimide in aqueous acetone leading to compound **22** in 75% yield.

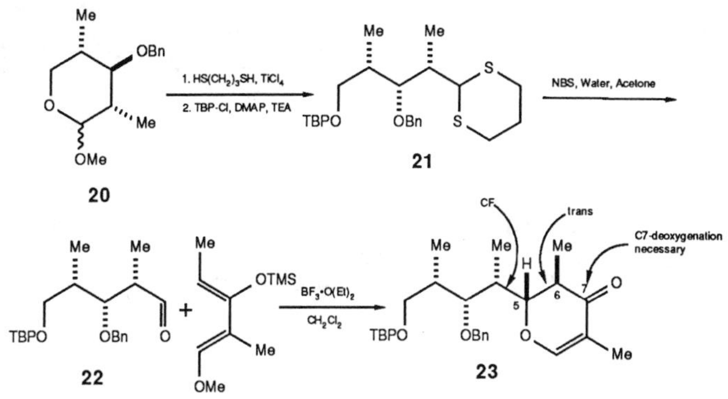

Fig. 7.

Compound **22** served as the aldehyde component of the second cyclcocondensation, again utilizing diene **17**. Unlike the first case, where formaldehyde was the substrate, this reaction has, *a priori,* both topographic and diastereofacial uncertainties. In the event, under BF_3 catalysis in methylene chloride at $-78^\circ C$, a 63% yield of compound **23** was obtained with excellent selectivity [16]. A different set of maneuvers was now necessary to establish the required stereochemistry and substitution pattern within the second pyranoid matrix. Reduction of the ketone with sodium borohydride-$CeCl_3$ afforded glycal **24**, containing an equatorial alcohol. A Ferrier rearrangement was triggered, using 2-propanol as the nucleophile in the presence of p-TsOH. Glycoside **25** was obtained in 90% yield from **23**.

The combined effects of the β-isopropoxy and methyl groups in this compound provided a high degree of facial bias for the next step. Thus, catalytic reduction ($H_2/10\%$ $Pd-Al_2O_3$) provided an 80% yield of **26**. A three-step sequence: (i) propane 1,3-dithiol, (ii) benzylation as above (cf. formation of **20**) and (iii) unveiling of the aldehyde (cf. formation of compound **22**) afforded aldehyde **27** in 40% overall yield.

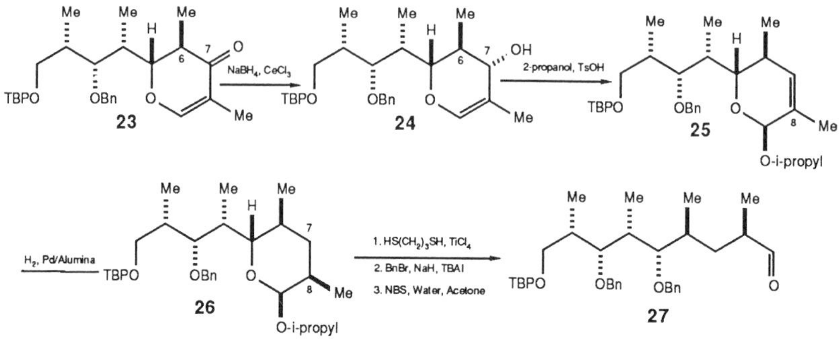

Fig. 8.

It will be recognized that the aldehyde function of **27** is destined to become the achiral C_9 ketone in the target system **15**. Nonetheless, the stereochemistry of the cyclocondensation of **27** with diene **17** is very crucial. The nature of the product will establish the agenda for installing the stereogenic centers at carbons 10, 11, 12, and 13. The reader will assume for the moment the capacity to generate a trans dihydropyrone stereochemistry in the cyclocondensation reaction. To exploit this capability it is necessary that diastereofacial sense of the reaction be in the C-F mode [11]. Actually, a significant amount of information accumulating in our laboratory indicates that high margins of selectivity can be achieved in the cyclocondensation reaction even when using aldehydes with seemingly weakly directing dissymmetry elements.

For the present, we simply report that cyclocondensation of compound **27** with diene **17** under mediation by $BF_3 \cdot$etherate afforded a 57% yield of dihydropyrone **28** with excellent stereoselectivity [17]. Reduction of the keto group under standard ($NaBH_4$-$CeCl_3$) conditions provided compound **29**.

It is seen that compound **29** carries the stereochemical imprint from carbon 1 through 11 of subgoal system **16**. To reach the required configuration at C_{12} in the context of the pyrone matrix would require installation of an axial methyl group. Current research is addressed to solving the problem at C_{12} either in the pyran ring or in a different fashion.

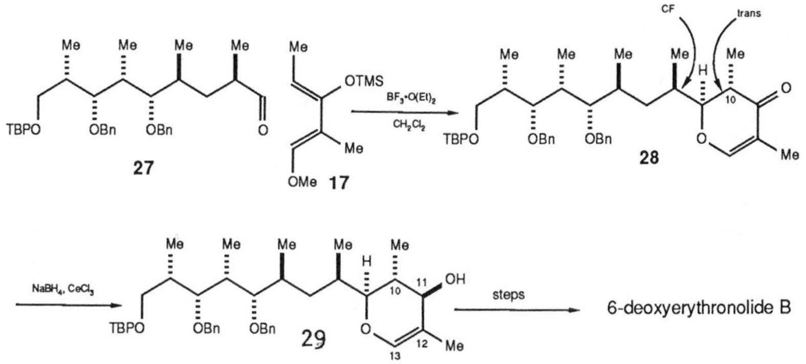

Fig. 9.

(ii) **Peracetyltunicaminyluracil, 30**   Recently we reported the synthesis of compound **30**, which is a precursor of any of the members of the tunicamycin family **31**[18].   The latter compounds are potentially valuable research resources in that they inhibit the enzyme U.D.P.-galactose transferase which is involved in the bioassembly of oligosaccharides.   Given the importance of **31**, we have persisted in an effort to achieve a more processable total synthesis of **30** and eventually **31**.   Several significant advances in this regard are described below.   We start with a uridine derivative such as **32**.   In our earlier venture the uracil residue was installed at a late stage.   This delay necessitated a difficult and inefficient differentiation of the ribosyl and aminogalactosyl residues.   The ability to start in the uridine series is a major advance.

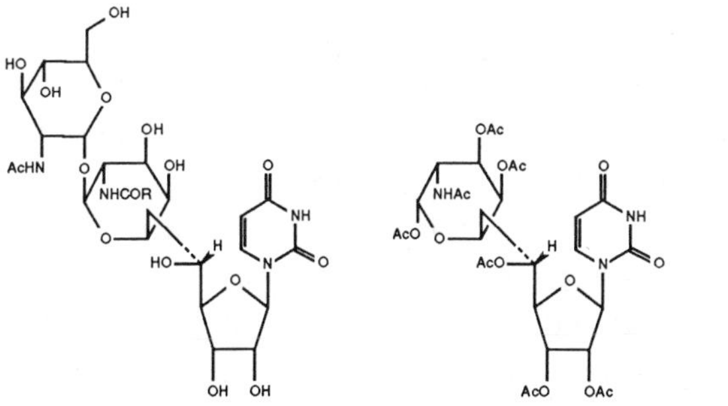

31

The tunicamycins

R= ~CH=CH-(CH$_2$)$_n$-CH(CH$_3$)$_2$
n=7,8,9,10,11
R= ~CH=CH-(CH$_2$)$_n$-CH$_3$
n=10,11,12,13

30

heptaacetyltunicaminyluracil

Fig. 10.

The optimal protecting group for the uracil moiety appears to be a p-methoxybenzyl group because of its ease of removal. We report our findings in this series. In this research we rely upon sequential applications of allylsilane and dienoxysilane methodology.

Reaction of **32** with allyltrimethyl silane afforded, with high selectivity (>10:1), carbinol **33**, which was further protected as its O-benzyloxymethyl (BOM) ether **34** (84% yield from **32**). After oxidative cleavage of the terminal methylene group [(i) OsO4, N-methylmorpholine-N-oxide; (ii) Pb(OAc)4], the aldehyde **35** was in hand in 74% yield.

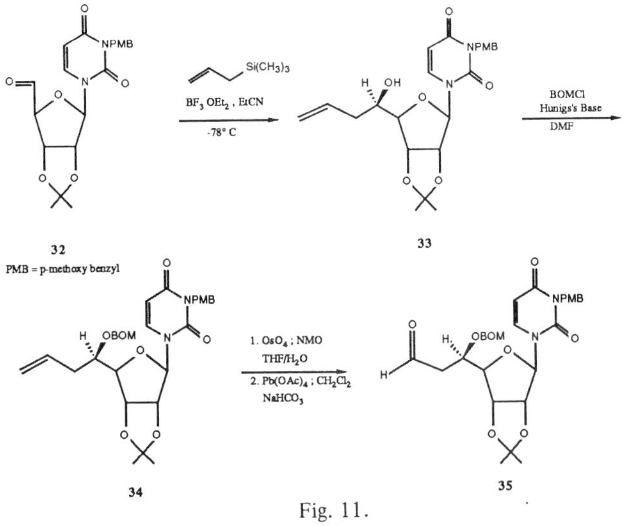

Fig. 11.

Cyclocondensation of **35** with parent diene **36**, using SnCl4 as the catalyst afforded dihydropyrone **37** (45% with 10% recovered **35**). It was hoped that the overall sense of the reaction would be consistent with a scenario wherein a chelated version of **35** (met = metal) would be attacked from its less hindered α-face, as shown in Figure 12. Such a result had been realized on an analog of **35** wherein there was a β-methoxy group at the anomeric center rather than a protected uracil [18]. Surprisingly, in the case of **35** the selectivity of the reaction was totally eroded and a 1:1 mixture of pyrones was obtained. Studies are in progress to determine the optimal protecting group for the secondary alcohol to obtain the best chelation results. Work with a MOM group at that position appears promising.

Reduction of **37** under standard conditions afforded **38** and thence, after cleavage of the benzoate and acetalization, the isopropylidene derivative **39** in 67% overall yield.

At this stage we were able to take advantage of a reaction recently developed by Fitzsimmons and co-workers [19]. Thus, photolysis of **39** in the presence of dibenzylazodicarboxylate (DBAD) afforded cycloadduct **40**. Subsequent methanolysis, followed by hydrogenolysis and acetylation, afforded compound **41** in 30% yield, overall. It is recognized that this compound contains all of the necessary stereochemical information to reach goal system **30** [20].

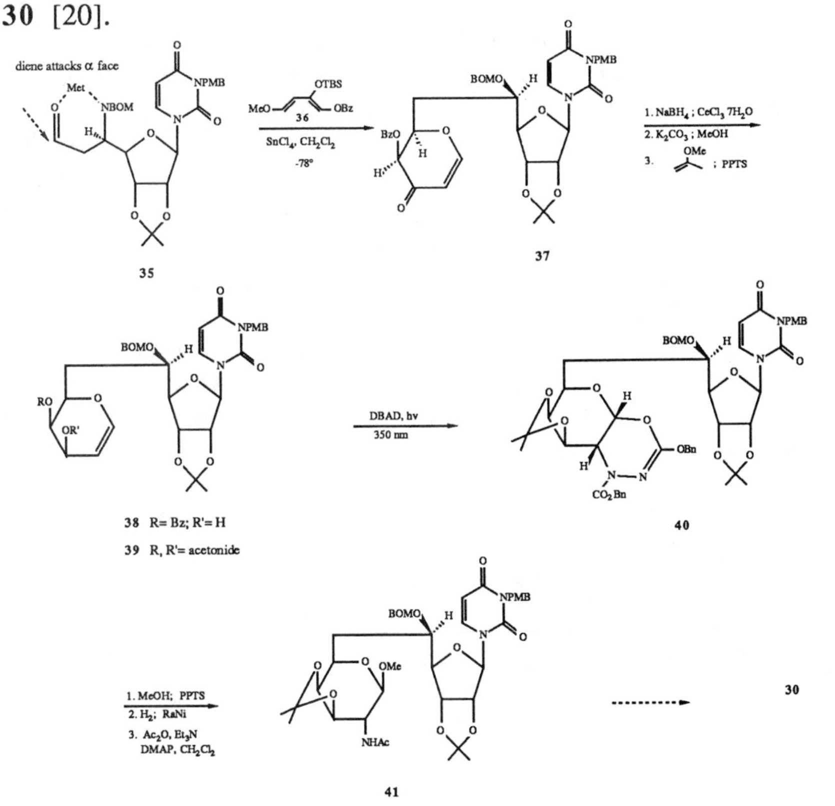

Fig. 12.

(iii) <u>Avermectins</u>   We have been seeking to develop a fully synthetic route to the avermectin family of macrolide-like antibiotics (see avermectin A_{1a}) [21]. Pursuant to this goal we have had occasion to exploit various permutations of allyl (and crotyl) silane methodology in league with cyclocondensation technology.   In previous studies directed toward zincophorin and indanomycin we made use of reactions of E-crotylsilanes (**12E**) with activated glycals to reach "anti" type structures, **13** [12,13].   With a view toward reaching the avermectins, the reaction of Z-crotyltriphenylsilane with D-glucal tripivalate was examined.   A high yield of a 4-5:1 ratio of **42** : **43** was obtained.   While efforts to improve upon this ratio are in progress, even this result allowed the effort to move forward.   Catalytic reduction of the side chain double bond followed by allylic displacement of the pivalate through the action of lithium dimethylcuprate afforded **44** (ca. 60% overall).   Shortly thereafter the aldehyde **45** became available.   Cyclocondensation of **45** with **46** under the influence of anhydrous magnesium bromide afforded **47**. Once again, reduction of the ketone ($NaBH_4$-$CeCl_3$) afforded the equatorial alcohol which was protected as its tert butyl dimethylsilyl ether **48**.   Using standard carbohydrate chemistry [(i) hydroxybromination with N-bromosuccinimide, (ii) debromination with tri-n-butyl tin hydride, (iii) reductive cleavage with lithium borohydride, (iv) di-pivaloylation, and (v) de-silylation] compound **49** was in hand.   Happily, oxidative cyclization could be achieved in ca. 70% yield through the action of HgO-I_2 [22] leading to subunit **50**.

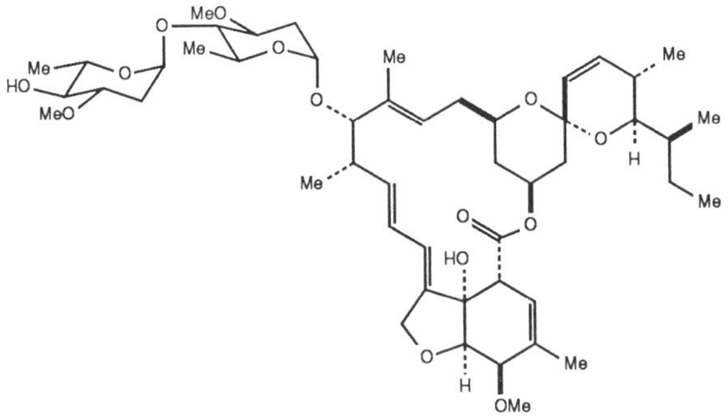

Avermectin A$_{1a}$

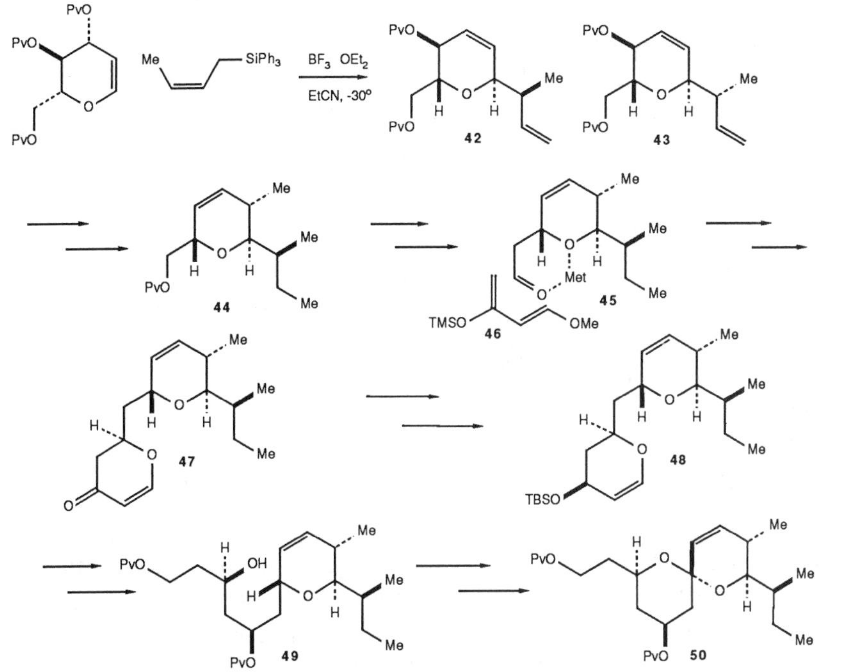

Fig. 13.

Our program for a stereoselective synthesis of the oxahydrindane benefitted from crotylsilane chemistry. Thus reaction of aldoseulose 51 with Z-crotyltrimethylsilane afforded a three component mixture of homoallylic alcohols after preliminary separation followed by treatment with sodium hydride-methyl iodide. The major product was obtained in 78% overall yield as its methyl ether 52. The shape of our synthetic plan becomes clearer as we relate the results of a critical model process on compound 53 obtained from 52 by straightforward methodology. Reaction of 53 with the aluminum "ate" species shown in Figure 14 afforded a 94% yield of the octahydrobenzofuran 54 as a single diastereomer. This same logic, applied to a rather more elaborate enal, helped us to achieve a total synthesis of avermectin A_{1a} [23].

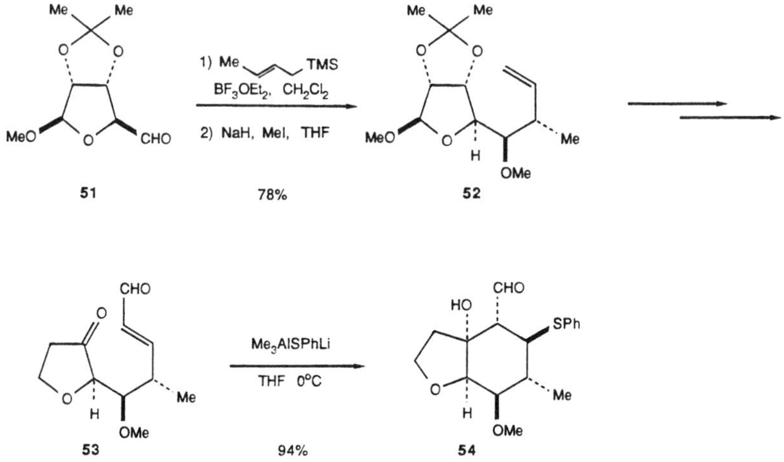

Fig. 14.

CONCLUSION   It is clear that organosilicon chemistry has great potentialities in the solution of a variety of complex total synthesis problems.   The applications shown here in the erythronolide, tunicamycin and avermectin areas are but illustrative of the opportunities for discovery.   We conclude with the expectation that continuing interactions between that community of chemists which is primarily concerned with natural products, and the one which focuses on organosilicon chemistry will serve the interests of both of these worthy pursuits.

ACKNOWLEDGMENT   This research was supported by PHS Grants AI16943 and HL25848.   A Dox Fellowship to D.C.M. and a PHS Postdoctoral Fellowship (Grant 5 F32 GM11051) to D.M.A. are gratefully acknowledged.   Additional support from the Merck Laboratories was very helpful.   NMR spectra were obtained through the auspices of the Northeast Regional NSF/NMR Facility at Yale University, which was supported by NSF Chemistry Division Grant CHE7916210.

REFERENCES

1.   For a discussion of some of our recent thoughts in the field of total synthesis, see: Danishefsky, S. J. *Adrichimica Acta*, **1986**, *19*, 59.

2.   For an example of this type of interactivity see: Egbertson, M; Danishefsky, S. J. *J. Am. Chem. Soc.* 1987, *109*, 2204.

3.   Colvin, E. W., "Silicon in Organic Synthesis", Butterworth & Co. Ltd., London, 1981.

4.   Danishefsky, S. *Accts. Chem. Res.* **1981**, *14*, 400.

5.   Danishefsky, S. J.; Larson, E.; Askin, D.; Kato, N. *J. Am. Chem. Soc.* **1985**, *107*, 1246.

6.   Danishefsky, S. J.; Pearson, W. H.; Harvey, D. F.; Maring, C. J.; Springer, J. P. *J. Am. Chem. Soc.* **1985**, *107*, 1256.

7.   Emde, H.; Domsch, D.; Feger, H.; Frick, U.; Gotz, A.; Hergott, H. H.; Hofmann, K.; Kober, W.; Krageloh, K.; Oesterle, T.; Steppan, W.; West. W.; Simchen, G. *Synthesis* **1982**, 1.

8   Danishefsky, S.; Kerwin, J. F., Jr. *J. Org. Chem.* **1982**, *47*, 3803.

9.   For the original Ferrier rearrangement with a hetero atom nucleophile see: Ferrier, R. *J. Chem. Soc.* **1964**, 5443.

10. Danishefsky, S. J.; DeNinno, M. P.; Phillips, G. B.; Zelle, R. E.; Lartey, P. A. *Tetrahedron* **1986**, *42*, 2809.

11. For a discussion of these descriptors see Reference 1, footnotes 13 and 15.

12. Danishefsky, S. J.; Selnick, H. G.; DeNinno, M. P.; Zelle, R. E. *J. Am. Chem. Soc.* **1987**, *109*, 1572.

13. Danishefsky, S. J.; DeNinno, S.; Lartey, P. *J. Am. Chem. Soc.* **1987**, *109*, 2082.

14. Danishefsky, S. J.; Myles, D. C.; Harvey, D. F. *J. Am. Chem. Soc.* **1987**, *109*, 862.

15. The strategy implemented here is one of continuous stereochemical communication. No coupling of chiral fragments is required. As such, there is no need for the use of homochiral intermediates. The distinction between stereochemical communication and stereochemical correlation is discussed in reference 1.

16. A small amount (ca. 9%) of the C_6 epimer of **23** was also obtained. These compounds can be separated by silica gel chromatography.

17. Another pyrone was isolated from this experiment. This minor component (ca. 7%) has not been fully characterized.

18. Danishefsky, S.; Barbachyn, M. *J. Am. Chem. Soc.* **1985**, *107*, 7761.

19. Fitzsimmons, B. J.; Le Blanc, Y.; Rokach, J. *J. Am. Chem. Soc.* **1987**, *109*, 285.

20. In the interim, this goal has been achieved, thus confirming the stereochemical assignments shown herein.

21. Fisher, M. H.; Mrozik, H. "Macrolide Antibiotics", Academic Press: New York, NY, 1984; p.553.

22. (a) Cf. Kay, I. T.; Williams, E. G. *Tetrahedron Lett.* **1983**, *24*, 5915. (b) Kay, I. T.; Bartholomew, D. *Tetrahedron Lett.* **1984**, *25*, 2035.

23. D. Armistead, F. Wincott, and H. Selnick, Yale University, unpublished results.

Chapter 4

Silicon-mediated or group transfer polymerization

Owen W. Webster and Dotsevi Y. Sogah – E.I. Du Pont de Nemours & Co., Inc., CR&D Department, Experimental Station, Wilmington, Delaware, USA.

INTRODUCTION

Group Transfer Polymerization (GTP) is a fundamentally new method for the living synthesis of acrylic polymers [1–8]. It involves the repeated addition of monomer to a growing polymer chain end bearing a silyl ketene acetal group. During the addition, the silyl group transfers to incoming monomer, hence the name group transfer polymerization. Equation 1 illustrates the method for polymerization of methyl methacrylate.

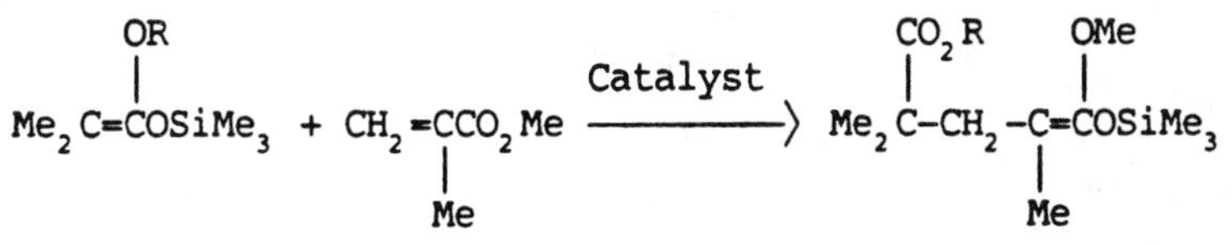

$$\underset{\mathbf{1}}{Me_2C=\overset{\displaystyle OR}{\overset{|}{C}}OSiMe_3} \; + \; CH_2=\underset{\underset{\displaystyle Me}{|}}{C}CO_2Me \; \xrightarrow{\text{Catalyst}} \; \underset{\mathbf{2}}{Me_2\overset{\displaystyle CO_2R}{\overset{|}{C}}-CH_2-\underset{\underset{\displaystyle Me}{|}}{C}=\overset{\displaystyle OMe}{\overset{|}{C}}OSiMe_3} \qquad (1)$$

$$\mathbf{2} \; + \; n \; CH_2=\underset{\underset{\displaystyle Me}{|}}{C}CO_2Me \; \longrightarrow \; Me_2\overset{\displaystyle CO_2R}{\overset{|}{C}}(CH_2\underset{\underset{\displaystyle Me}{|}}{\overset{\displaystyle CO_2Me}{\overset{|}{C}}})_n CH_2\underset{\underset{\displaystyle Me}{|}}{C}=\overset{\displaystyle OMe}{\overset{|}{C}}OSiMe_3$$

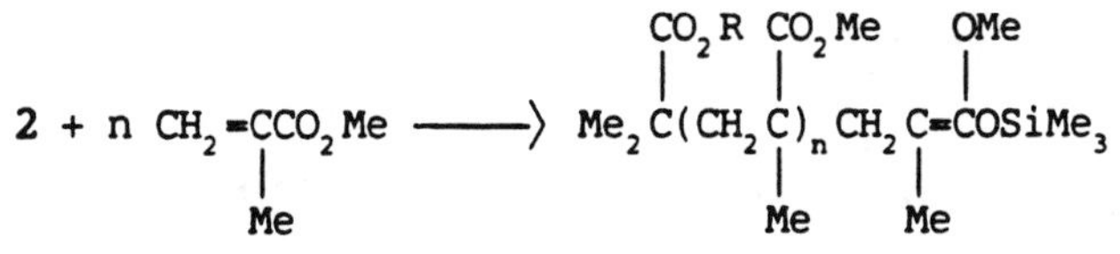

3

A catalyst, for example a soluble bifluoride, is necessary
for GTP to proceed. The chief characteristic of GTP is that
it is a "living" polymerization and thus, molecular weight
and polymer architecture can be controlled. It is
industrially attractive, since it operates at room
temperature or above.

THE GTP PROCESS

Group-Transfer polymerization works best with methacrylates,
to which most of the published data refer. Functional groups
sensitive to free radicals, such as allyl and sorbyl, can be
present on the ester groups of the monomer and remain
unreacted in the polymer. Active hydrogen compounds
interfere with GTP. Thus, the polymerization of methacrylic
acid or hydroxyethyl methacrylate requires the use of
protective groups that can be subsequently removed. Carboxy
groups can be introduced by using trimethylsilyl methacrylate
as the monomer for GTP[7,9] followed by mild hydrolysis.
Pendent hydroxy groups can be introduced by
using 2-(trimethylsilyloxy)ethyl methacrylate, as the
monomer[4,7]. The trimethylsilyl group is removed from this
polymer with methanolic tetrabutylammonium fluoride.
Glycidyl methacrylate can be polymerized without involving
the epoxy groups provided the temperature is held below 0°C.

Acrylates[3,7,10], acrylonitrile[6,7],
methacrylonitrile[6,7], N,N-dimethylacrylamide[6,7],
pentadienoates[7] and α-methylene-γ-butyrolactones[6,11] have
also been polymerized by GTP. Obtaining high molecular
weights with acrylates is difficult. Possibly because the
silyl function on the growing polymer end migrates to other
positions on the chain[6]. Acrylonitrile is extremely active
in GTP and gives high molecular weight polymer but control of
molecular weight is difficult.

For initiation of methacrylates, a silyl ketene acetal,1,
with an alkyl group in the 2 position is ideal. Large alkyl
groups on the silicon reduce the polymerization rate[6].
α-Trimethylsilyl, -stanyl and germyl esters operate as GTP
initiators, possibly by first rearranging to the ketene
acetals[6]; but they tend to give poorer molecular weight
control than do the silyl ketene acetals. Silyl derivatives
that add to methacrylates to produce ketene acetals, for
example, $Me_3SiCN[1,7,12,13]$ (Equation 2), $Me_3SiSMe[6,7,10]$,
$Me_3SiSPh[6,7]$, $Me_3SiCR_2CN[6,7]$ or $R_2PO SiMe_3[6,7]$ also
initiate GTP.

$$Me_3SiCN + CH_2 = \overset{\overset{\displaystyle Me}{|}}{C}CO_2Me \longrightarrow NCCH_2\underset{\underset{\displaystyle Me}{|}}{\overset{\overset{\displaystyle OMe}{|}}{C}} = COSiMe_3 \qquad (2)$$

Catalysts for GTP fall into two classes, anionic and Lewis acid. Among the anionic catalysts are fluoride[1,7], azide[1,7], cyanide[1,7], carboxylates[8], phenolates[8], sulfinates[8], phosphinates[8], nitrite[8], and cyanate[8]. Even though GTP does not proceed when protonic solvents or monomers are used, some of the better anionic catalysts are derived from the combination of an anion with its conjugate acid e.g. bifluoride[1,7] or biacetate[8]. The anionic catalysts are usually used as their tetraalkyl ammonium or tris(dimethylamino)sulfonium (TAS) salts. They catalyze GTP by coordination with the silicon atom[5]. GTP works best with low levels of nucleophilic catalyst[12] as little as 0.01% based on initator.

Examples of Lewis acid catalysts for GTP are zinc halides, diakylaluminum halides and tetraalkyl aluminoxanes. These catalysts operate by coordination to monomer[3]. Relatively large amounts of catalyst are required, 10% or more based on initiator. Anionic catalysts work best for methacrylates, Lewis acid catalysts for acrylates.

CONTROL OF POLYMER STRUCTURE

The molecular weight of polymer obtained by GTP is determined by the molar ratio of initiator to monomer. Although this property of GTP allows one to control molecular weight accurately in the 1,000 to 60,000 range, obtaining high molecular weight polymer is more difficult since the amount of initiator needed is so low that it begins to match interfering impurity levels. By using very pure monomers, solvents, catalysts, and initiators, however, one can make polymer in the 100,000 to 200,000 molecular weight range.

The molecular weight distribution, Mw/Mn, for GTP polymers under the best conditions is close to one, as it should be for a living polymerization. Dispersities higher than one could be due to chain termination by cyclization of the end group with the penultimate ester group[14], reaction of the end group with impurities, an initiation rate which is slower than the propagation rate or a catalyst transfer rate which is slower than the propagation rate.

The principle parameters influencing tacticity of GTP PMMA are catalyst and temperature. The precise variation of tactic composition of GTP-PMMA with polymerization temperature was determined by performing isothermal polymerizations of MMA with TASHF$_2$ as catalyst and THF as solvent[6,9,15]. The isotactic component remains small under all conditions, while the syndiotactic component increases from 50 to 80% as the temperature is lowered from 60° to -90°C. The dyad statistics are essentially Bernouillian.

These data are almost identical with those obtained for free
radical polymerization of MMA[16]. Lewis acid catalysis of
GTP of MMA generally provides a much more syndiotactic PPMA
than does anion catalysis, but detailed temperature studies
have not been carried out.

Polymers with reactive functionality on one end can be
readily made by GTP by merely starting with a functionalized
initiator. Thus, terminal hydroxyl or carboxyl groups are
readily prepared by using the functionalized initiators 4a or
4b respectively, (Equation 3) and then hydrolyzing the
resulting polymer with refluxing methanolic
tetrabutylammonium fluoride or with dilute methanolic HCl at
ambient temperature to give PMMA—OH (6a) or PMMA—COOH(6b),
respectively. On treatment of hydroxyl ended polymethacrylates
with isocyanatoethyl methacrylates one obtains macromonomers
useful for formation of comb polymers[17]. Initiation of GTP
with other functionalized silyl compounds has given –CN,
–SME, or –PO(OSiMe$_3$)$_2$, ended polymer.

Another way to functionalize polymer chain ends is to react
the living silyl ketene end group with electophilic reagents.
Examples are benzaldehyde to give a masked hydroxyl group[7],
and bromine to give an α–bromoester[2],

$$
\begin{array}{ccc}
\mathbf{4} & \mathbf{5} & \mathbf{6} \qquad (3)
\end{array}
$$

a. R = CH$_2$CH$_2$OSiMe$_3$ a. R' = CH$_2$CH$_2$OH

b. R = SiMe$_3$ b. R' = H

α,ω–Difunctional (telechelic) polymers are useful for
construction of block, as well as, chain extended polymers.
They can be readily synthesized by coupling living GTP chelic
polymers,5. Suitable coupling agents are p–xylylene
dibromide or Br$_2$ (Equation 4). With low molecular weight
PMMA (in the 2000 range) one can obtain over 99% pure
telechelic polymer[2]. However, it is difficult to obtain
polymer in the 10,000 to 20,000 molecular weight range of
greater than 95% telechelic purity.

$$5 + BrCH_2\langle\bigcirc\rangle CH_2Br \xrightarrow[H_2O]{F^-} R'OC-C-PMMA-CH_2\langle\bigcirc\rangle CH_2-PMMA-C-COR' \quad (4)$$

$$R' = CH_2CH_2OH, \text{ or } H$$

As long as one uses monomers belonging to the same family (all methacrylate or all acrylate, for example), one can make random copolymers simply by adding a mixture of the monomers to the initiator and catalyst. The large difference in reactivity between the various acrylic monomer types (e.g., acrylonitrile, methacrylonitrile, acrylates and methacrylates) prevents random copolymer formation by GTP of their mixtures. At this point, reactivity ratios are not available.

A useful attribute of "living" polymer systems is the ability to prepare block copolymers of predetermined block length and sequence. GTP provides a particularly facile methodology for the sequencing of both homopolymer blocks and random copolymer blocks. One simply adds a new monomer or mixture of monomers when the first batch is depleted[6,7].

Superior pigment dispersants for acrylic finishes have been made by forming AB block polymer with PMMA or PBMA blocks coupled with short glycidyl methacrylate blocks. The epoxy functions are then converted to polar absorbing groups by reaction with carboxylic acids or amines[17].

Multiarmed star polymers,8, can be made by addition of a difunctional methacrylate, for example ethylene glycol dimethacrylate,7, to a living GTP polymer,3, (Equation 5). One use for these stars is in toughening plastics and coatings. The addition of hydroxy functional stars to other polyols and then co-crosslinking produces two phase films which are clear, flexible and hard[17]. Another use for stars is in rheology control for application of coatings.

$$3 + (CH_2=C-CO_2-CH_2\text{—})_2 \longrightarrow \qquad\qquad (5)$$

7   8

MECHANISM

In GTP, the silyl group does not dissociate from the living ends before transfer to monomer[5]. Labeling studies have shown that the trialkylsilyl fluoride does not exchange with living ends under reaction conditions. Furthermore, the living ends do not exchange with other living ends. The

exact nature of the intermediates for associative transfer
has not been established, but the sequence of reactions shown
in Equation 6 is emerging as a likely pathway. Based on
stopped-flow kinetics, an enthalpy of activation of 6.4 Kcal

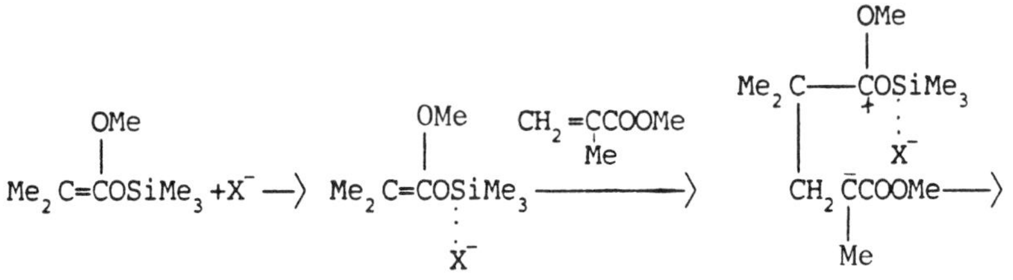

$$Me_2C=COSiMe_3 + X^- \longrightarrow Me_2C=COSiMe_3 \xrightarrow{\substack{CH_2=CCOOMe \\ Me}} \begin{array}{c} OMe \\ | \\ Me_2C—COSiMe_3 \\ | \qquad + \\ X^- \\ CH_2\bar{C}COOMe \\ | \\ Me \end{array} \longrightarrow$$

X⁻ = Nucleophilic Anion   (6)

$$\begin{array}{c} OMe \\ | \\ Me_2C-C=O \quad OMe \\ | \qquad\qquad | \\ CH_2-C=COSiMe_3 \\ | \quad : \\ Me \quad X^- \end{array} \longrightarrow \begin{array}{c} OMe \\ | \\ Me_2C-C=O \\ | \qquad\qquad OMe \\ | \qquad\qquad | \\ CH_2-C=COSiMe_3 + X^- \\ | \\ Me \end{array}$$

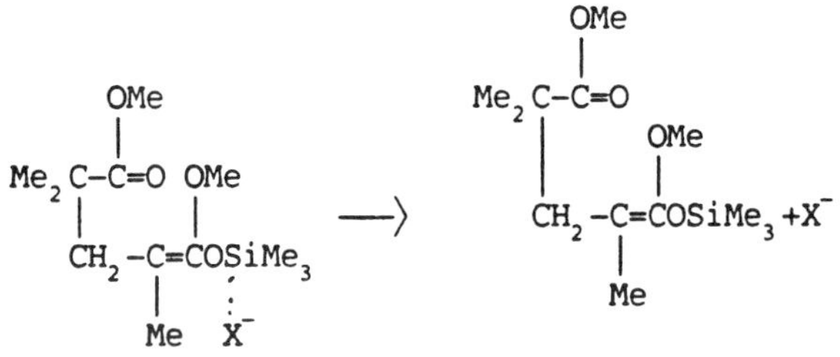

per mole and an entropy of −41.6 units were calculated [14].
Such a high entropy of activation agrees with the associative
mechanism depicted in Equation 6 in which monomer bonds to
initiator and the trimethylsily group transfers to monomer in
one step.

SUMMARY

The large number of functional methacrylates available,
coupled with the numerous structural variations possible,
makes GTP an extremely useful tool for polymer research. In
addition, even though GTP is a complex procedure, requiring
very pure reagents, dry conditions and much fine turning, it
appears that many industrial applications will be
forthcoming.

REFERENCES

1. O. W. Webster, W. R. Hertler, D. Y. Sogah,
   W. B. Farnham, and T. V. RajanBabu, J. Amer. Chem. Soc.,
   105, 5706 (1983).

2. D. Y. Sogah and O. W. Webster, J. Polym. Soc. Lett. Ed.
   21, 927 (1983).

3. W. R. Hertler, D. Y. Sogah, O. W. Webster, and
   B. M. Trost, Macromolecules, 17, 1417 (1984).

4. O. W. Webster, W. R. Hertler, D. Y. Sogah,
   W. B. Farnham, and T. V. RajanBabu, J. Macromol.
   Sci-Chem., A21(8&9), 943-960 (1984).

5. D. Y. Sogah and W. B. Farnham, "Organosilicon and
   Bioorganosilicon Chemistry: Structures, Bonding,
   Reactivity and Synthetic Application", H. Sakruai, Ed.,
   John Wiley & Sons, N.Y. 1985, Chapter 20.

6. D. Y. Sogah, W. R. Hertler, O. W. Webster and
   G. M. Cohen, Macromolecules, in press.

7. O. W. Webster, (Du Pont) U.S. Pat. 4,417,034 (1983)
   (Chem. Abstr., 1984, 100, 86327A); 4,508,880 (1985)
   (Chem. Abstr., 1985, 102, 221336) W. B. Farnham,
   D. Y. Sogah, (Du Pont) U.S. Pat. 4,414,372 (1983); (Chem
   Abstr., 1984, 100, 68964) 4,581,428 (1986) (Chem.
   Abstr., 1986, 105, 98133).

8. I. B. Dicker, W. B. Farnham, W. R. Hertler,
   E. D. Laganis, D. Y. Sogah, T. W. Del Pesco and
   P. H. Fitzgerald and I. B. Dicker, (Du Pont Inc.)
   U.S. Pat. 4,588,795 (1986) (Chem. Abstr., 1986, 105,
   98135).

9. Y. Wei and G. Wnek, Polym. Prepr., Am. Chem. Soc., Div.
   Polym. Chem., 28, 252 (1987).

10. M. T. Reetz, R. Ostarek, K.-E. Piejko, P. Arlt, and
    B. Bömer, Angew. Chem. Ent. Ed. Engl., 1986, 25, 1108.

11. J. Suenaga, D. M. Sutherlin, and J. K. Stille,
    Macromolecules, 17, 2913 (1984).

12. F. Bandermann, H. P. Sitz and H. D. Speikamp, Polym.
    Prepr. Am. Chem. Soc., Div. Polym. Chem., 27, 169
    (1986).

13. F. Bandermann and H. D. Speikamp, Makromol. Chem., Rapid Commun., **6**, 336 (1985).

14. W. J. Brittain and D. Y. Sogah, unpublished results.

15. M. A. Muller and M. Stickler, Makromol. Chem., Rapid Commun., **7**, 575 (1986).

16. T. G. Fox and H. W. Schnecko, Polymer, **3**, 575 (1962).

17. J. A. Simms and H. J. Spinelli, J. Coatings Tech., 1987, **00**, 000. (Du Pont) PCT Int. Pat. Appl., WO86/626 (1986) (Chem. Abstr., 1986, **105**, 24811).

Chapter 5

Effect of substituent on reactions remote from silicon — application in organic synthesis

T.H. Chan*, **K. Koumaglo, R. Horvath** – Department of Chemistry, McGill University, Montreal, Quebec, Canada H3A 2K6.
D. Wang, Z.Y. Wei, G.L. Yi and J.S. Li – Institute of Chemistry, Academia Sinica, Beijing, People's Republic of China.

ABSTRACT

By changing the substituent on a silyl group, reactions remote from silicon can be modified. Examples in organic synthesis include: (1) regioselective reaction of α-silylallyl anion with alkyl halides in the synthesis of insect sex pheromones and (2) enantioselective reaction of chiral allylsilanes with carbonyl compounds to give homoallylic alcohols.

In the last fifteen years, organic chemists have learned to exploit the properties of organosilicon compounds for organic synthesis. In most cases, the substituents on silicon are alkyl, typically trimethyl, groups. It is possible to change the substituents on silicon to control the reactions occurring at silicon. Such a substituent effect is well recognised and has been used to advantage in organic synthesis. An example is the use of t-butyldimethylsilyl- [1] in contrast to trimethylsilyl- as protective group [2] for the hydroxy function. The bulky t-butyl substituent confers reasonable stability to the silyl ether moiety towards hydrolysis (Scheme 1).

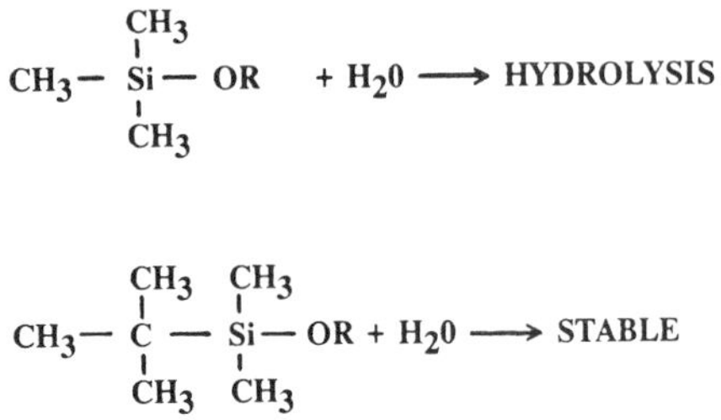

$$
CH_3 - \underset{\underset{CH_3}{|}}{\overset{\overset{CH_3}{|}}{Si}} - OR \quad + H_2O \longrightarrow HYDROLYSIS
$$

$$
CH_3 - \underset{\underset{CH_3}{|}}{\overset{\overset{CH_3}{|}}{C}} - \underset{\underset{CH_3}{|}}{\overset{\overset{CH_3}{|}}{Si}} - OR + H_2O \longrightarrow STABLE
$$

Scheme 1

Less explored is the effect of substituent on reactions remote from silicon. The general concept is expressed in Scheme 2. Compound 1 reacts with reagent A under a set of conditions to give product 2. The reaction site (X → Y) is remote from, or at least not directly at, silicon. By modifying the substituent R to R' as in compound 3, the course of the reaction with reagent A under identical conditions is changed (i.e., X → Z).

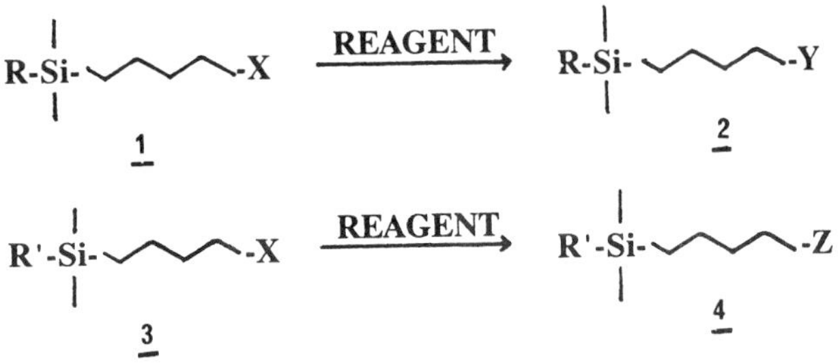

$$
R\text{-Si-}\diagdown\diagup\diagdown\diagup\text{-X} \quad \xrightarrow{\textbf{REAGENT}} \quad R\text{-Si-}\diagdown\diagup\diagdown\diagup\text{-Y}
$$

$$
\underline{\mathbf{1}} \qquad\qquad\qquad\qquad \underline{\mathbf{2}}
$$

$$
R'\text{-Si-}\diagdown\diagup\diagdown\diagup\text{-X} \quad \xrightarrow{\textbf{REAGENT}} \quad R'\text{-Si-}\diagdown\diagup\diagdown\diagup\text{-Z}
$$

Scheme 2

$$
\underline{\mathbf{3}} \qquad\qquad\qquad\qquad \underline{\mathbf{4}}
$$

In our laboratories, we have examined several reactions of allylsilanes by systematically varying the substituents on silicon. It is found that the substituent on silicon can affect the chemical reactivity, the regioselectivity and the enantioselectivity of reactions remote from silicon.

STERIC EFFECT IN DIRECTING γ-ALKYLATION OF ANIONS

In 1975, Corriu et al. reported the generation of α-silylallyl anion 5 and its reaction with a number of electrophiles [3]. Reaction of 5 with carbonyl compounds gave regioselectively the γ-adduct. On the other hand, we found that when alkyl halides were used as electrophile, the reaction was far less regioselective. A mixture of α- and γ-alkylated products was obtained (Scheme 3) [4]. The ratio of α- to γ-products did not vary significantly with external factors, such as the solvent system used (ether or THF-HMPA), by addition of DABCO or 12-

crown-4, or by addition of metal salts [e.g., MgBr$_2$, CuI, ZnCl$_2$, CsF, NiCl$_2$, or Al(iBu)$_3$].

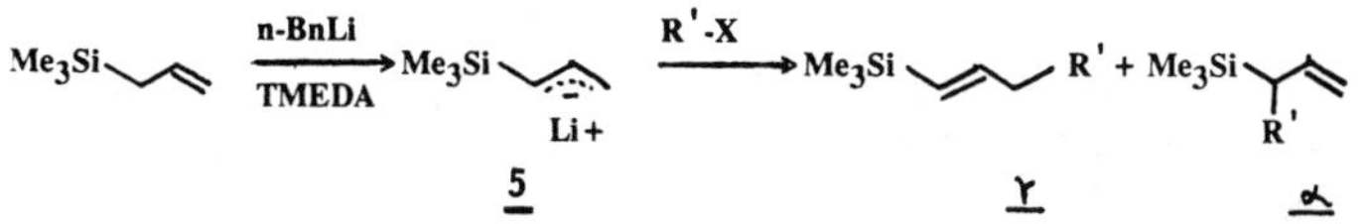

R'X =	n-Bu-I	65%	33%
	n-C$_{10}$H$_{21}$I	65	35
	n-C$_{10}$H$_{21}$-Br	57	42

Scheme 3

It was therefore somewhat surprising to find that when n-BuLi and K-O-t-Bu in hexane, the so-called "Schlosser's" base [5], was used to generate the silylallyl anion, the regioselectivity of the reaction was much improved (Scheme 4) [4].

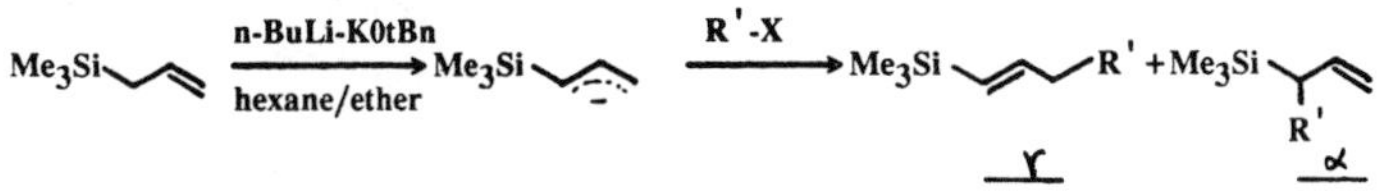

R'X =	n-C$_3$H$_7$Br	83%	17%
	n-C$_7$H$_{15}$Br	80	18
	n-C$_8$H$_{17}$Cl	80	16
	n-C$_9$H$_{19}$I	86	14
	n-C$_{12}$H$_{25}$I	85	15

Scheme 4

In trying to understand the origin of this regioselectivity, a series of silylallyl anions, in which the substituents on silicon were varied in a systematic manner, from methyl to ethyl to propyl, were prepared. The results in Scheme 5 clearly show that, as the size of the substituent on silicon increases, the regioselectivity, as measured by the ratio of γ to α isomer (γ/α), increases.

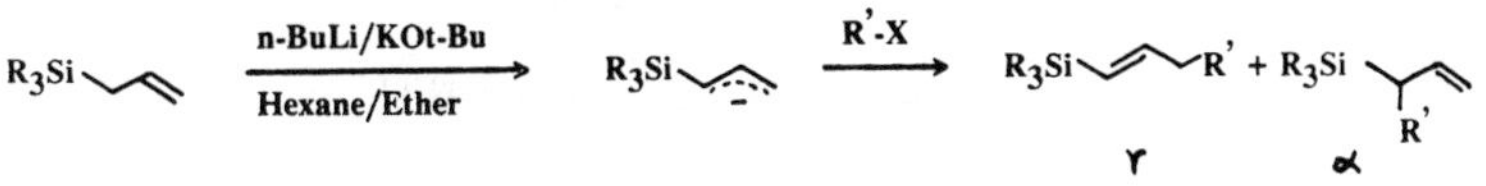

		γ/α Ratio
Me$_3$Si -		4/1 to 9/1
Et$_3$Si -	n-C$_3$H$_7$I	16/1
	n-C$_{12}$H$_{25}$Br	18/1
Ph$_3$Si -	n-C$_3$H$_7$-Br	16/1
Pr$_3$Si -	n-C$_3$H$_7$Br	46/1
	THP-O-(CH$_2$)$_6$-Br	36/1

Scheme 5   Steric effect in controlling regioselection

This is, therefore, an example of the effect of substituent on a reaction occurring remote from silicon.

It should be noted that the γ-isomer formed in this reaction is nearly exclusively the E-isomer. The alkylation is not only regioselective, but stereoselective as well. Furthermore, the minor amount of the α-isomer can be removed readily from the mixture by selective proto-desilylation followed by distillation (Scheme 6). The pure E-vinylsilanes thus obtained can be further transformed, using iodine monochloride/potassium fluoride [6] conditions, to give stereoselectively with inversion the Z-vinyliodides (Scheme 7).

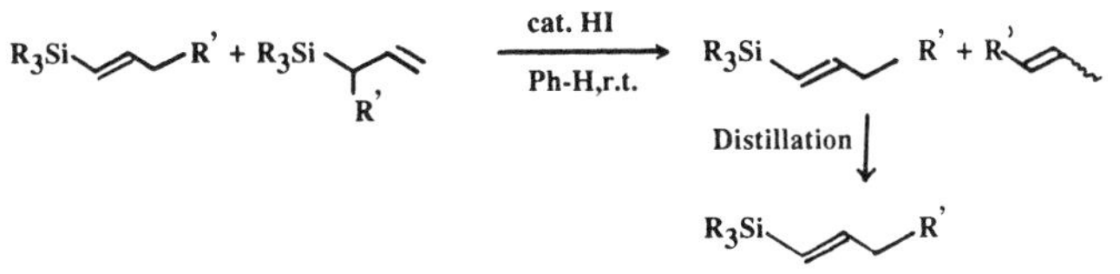

Scheme 6

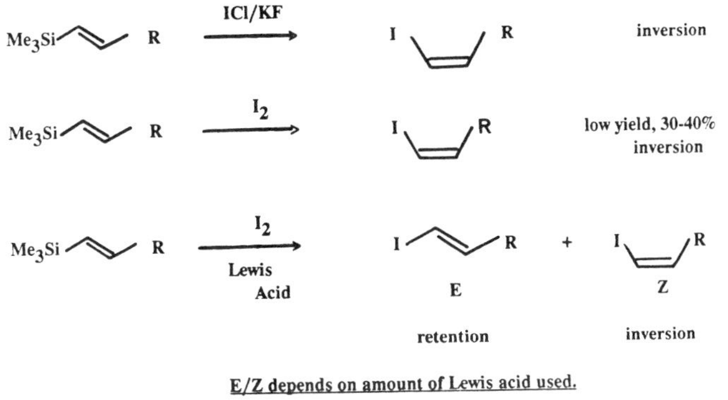

Scheme 7: Stereoselectivity in iododesilylation of terminal vinylsilanes.

We have used this sequence of reactions as a general method for the stereoselective synthesis of insect sex pheromones [7]. As illustrated in Scheme 8, the sex pheromones of the oriental fruit moth 6 (n=6, m=2) and the cabbage looper moth 7 (n=5, m=3) have been synthesized via the silicon method. One of the key steps is the stereospecific coupling of the Z-vinyliodides (8) with alkylzinc chloride using Pd(0) as catalyst [8]. Equally effective is the alkylation

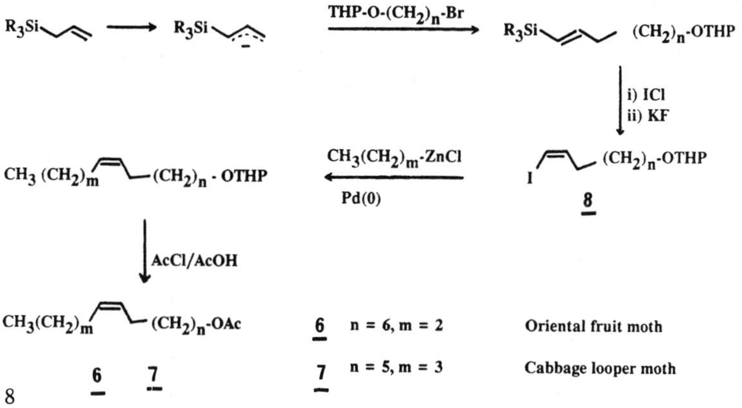

Scheme 8

$6 \quad 7$

| $\underline{6}$ | $n = 6, m = 2$ | Oriental fruit moth |
| $\underline{7}$ | $n = 5, m = 3$ | Cabbage looper moth |

of the silylallyl anion 5 with an alkyl halide containing an ethylene ketal function. This has led to an efficient synthesis of the sex pheromone of the Douglas Fir Lepidoptera [9] (Scheme 9).

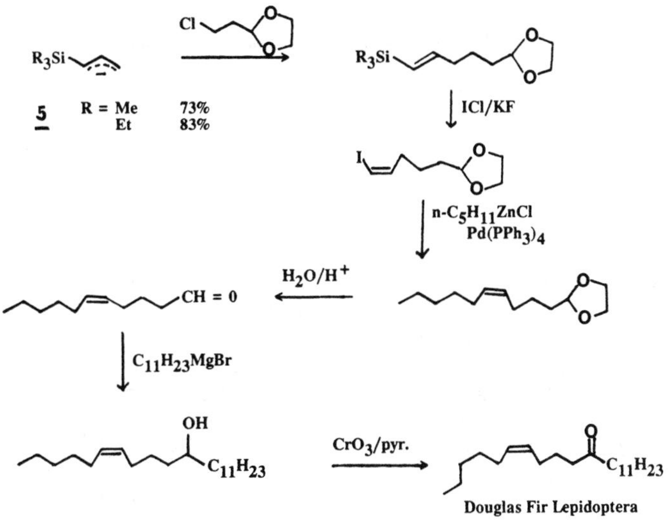

Scheme 9

Up to this point, the approach is limited to the synthesis of Z alkenes. In trying to extend the method to the synthesis of the E-isomer, we examined the stereochemistry of the reaction of E-vinylsilanes with iodine. The reaction was previously reported to give iododesilylation with retention [10]. To our surprise the reaction of these terminal E-vinyl-silanes with iodine gave a poor yield of vinyl-iodides, and with inversion of stereochemistry. We found that, by adding some Lewis acid together with the iodine, the yield of the vinyliodides is much improved. In addition, the stereochemical outcome of the reaction depends very much on the amount of Lewis acid used. Using 0.1 eq of $AlCl_3$, the E-vinylsilane is converted with 90% inversion to the Z-vinyliodide (Figure 1). As the amount of $AlCl_3$ increases to 1

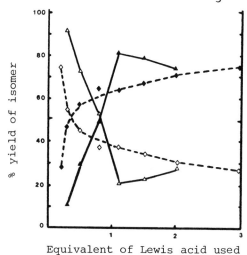

Fig. 1. Variation in the E/Z isomeric ration of 13-iodotetradec-12-enyl acetate in the iododesilylation of the precursor E-vinylsilane. ------$AlCl_3$, △ Z, ▲ E; ------ $SnCl_4$, ◇ Z, ◆ E.

equivalent, the product gave predominantly retention, 80% E-isomer. The relative ratio of the two isomers, E and Z, in the product mixture can therefore be controlled by using a controlled amount of the Lewis acid. We proposed the term a "tunable" stereoselective synthesis to describe such a reaction [11]. It is known that the sex pheromones of a number of insects are often specific blends of E and Z geometrical isomers of alkenes. For example, the Asian corn borer sex pheromone is found to be a 50:50 mixture of E- and Z-12-tetradecenyl acetate (9) [12]. The most common approach to the synthesis of alkenes with a specific E/Z ratio is done by stereospecific semihydrogenation of the alkyne precursors to give the E- and Z-isomers separately, followed by blending the two geometrically pure isomers in the required ratio. We have applied the "tunable" concept to the synthesis of the Asian corn borer sex pheromone (Scheme 10). The E-vinylsilane 10 was treated with one equivalent of iodine and 0.4 equivalent of $SnCl_4$ to give the vinyliodide 11 with a E/Z ratio of 1.

Coupling of <u>11</u> with methylmagnesium bromide followed by acetylation gave the sex pheromone <u>9</u> in the desired E/Z ratio.

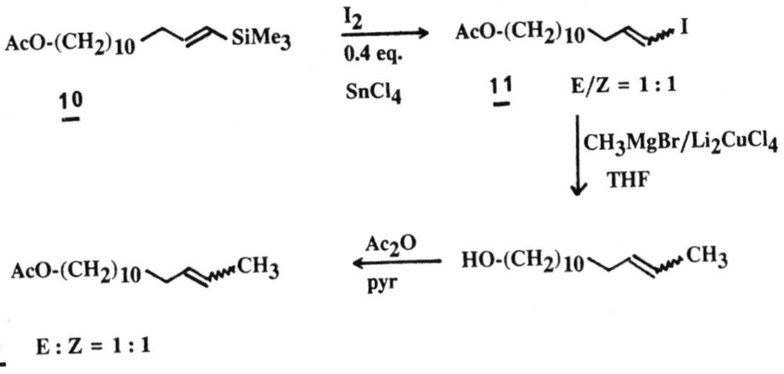

Scheme 10: A "tunable" stereoselective synthesis of the sex pheromone of the Asian corn borer.

POLAR AND COMPLEXATION EFFECT IN DIRECTING α-ALKYLATION OF α-SILYLALLYL ANIONS

In the alkylation reaction of α-silylallyl anion <u>5</u>, we have shown that the steric effect of the substituent on silicon can direct the alkylation to the γ-position. A more demanding question is whether any effect of substituent could have directed the alkylation to the α-site. Because of the chemical reactivity of the Si-H bond, we cannot replace the methyl group on silicon with the sterically smaller hydrogen.

We examined first the effect of replacing the alkyl group with a more electron-withdrawing alkoxy group. Firstly, it was noted that proton abstraction of triethoxyallylsilane (<u>12</u>) can now be performed with lithium diisopropylamide (LDA), a much weaker base than n-BuLi. This is presumably due to the electron-withdrawing effect of the ethoxy group in stabilising the anion <u>13</u> [13]. More surprisingly, alkylation of <u>13</u> with methyl iodide gave a predominance of the α-isomer ($\alpha/\gamma_t = 4.5$) (Scheme 11). Replacing the ethoxy group by a methoxyethoxy group (<u>14</u>) did not improve the regioselectivity (Table 1). On the other hand, methoxyethoxydimethylsilylallyl anion <u>15</u> reacted with methyliodide to give an even better α-selection ($\alpha/\gamma_t = 7$). Similarly, dimethylaminoethoxydimethylsilylallyl anion <u>16</u> and ethoxyethoxyethoxydimethylsilylallyl anion <u>17</u> showed similar regioselectivity.

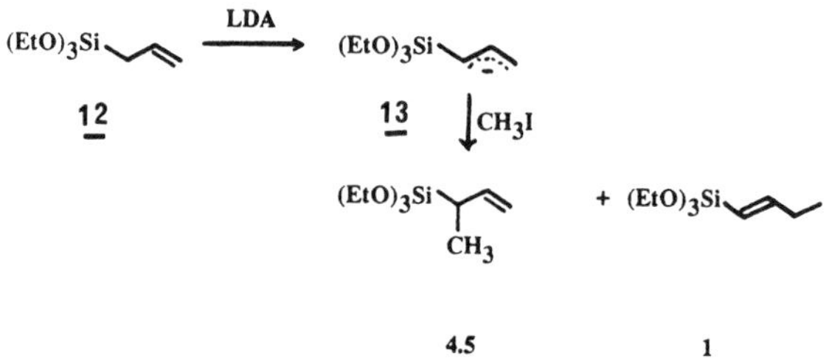

Scheme 11   4.5   1

Table 1. Regioselection of the alkylation of alkoxysilylallyl anion with methyl iodide.

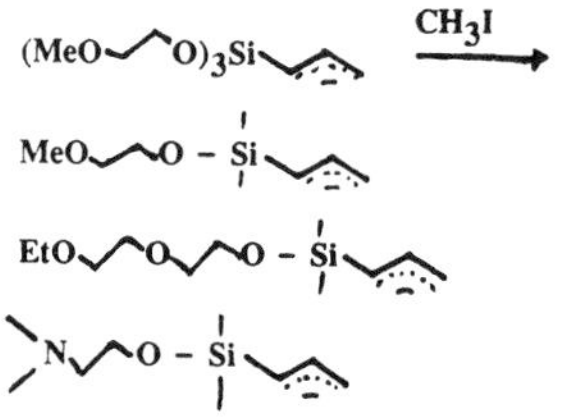

		α	γ_t
13	$(EtO)_3Si$	4.5	1
14	$(MeO O)_3Si$	3	1
15	$MeO O - Si$	7	1
16	$EtO O O - Si$	7	1
17	$N O - Si$	6	1

Using the alkoxysilylallyl anion **17** as the common nucleophile, we probed the regioselectivity with a number of alkyl halides. It is clear that other than methyl iodide, the regioselection (α/γ_t) is not high (Table 2). It is interesting to note however that γ,γ-dimethylallyl bromide gave a better α-selection than the less bulky allyl bromide. The origin of this improved regioselectivity is not clear at the moment.

Table 2. Regioselection in the alkylation of ethoxyethoxyethoxydimethylsilylallyl anion.

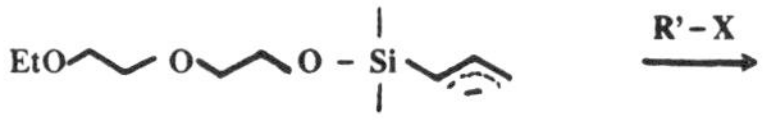

	α	γ_t
$R'-X = MeI$	7	1
$n-C_6H_{13}I$	1.5	1
 $\diagup\!\!\diagdown Br$	0.67	1
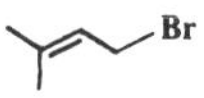 $\diagup\!\!\diagdown Br$	3	1

Also unresolved is the origin of the effect of the
ethoxyethoxyethoxy group. Is it simply a matter of
the electron-withdrawing effect of the oxygen, or is
it due to the complexation of the lithium counter ion
by the three oxygen atoms which control the regioselection? In order to distinguish between the two
possibilities, we prepared a series of heteroatom
substituted methylsilylallyl systems. Reaction of
diethylamine with chloromethyldimethylallylsilane 18
gave diethylaminomethyldimethylallylsilane 19
(Scheme 12).

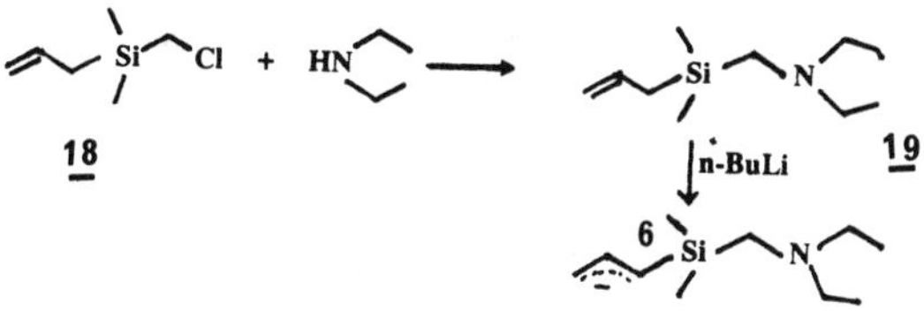

Scheme 12

Similar amino-substituted compounds can be prepared
readily. Reaction of 19 with n-butyllithium
generated the corresponding allyl anion 20 in good
yield. Quenching the reaction mixture with MeI gave
the three alkylation products in a ratio of
2.6:1:0.25 for $\alpha:\gamma_t:\gamma_c$. For the first time, the γ_c
product was observed. We attributed the higher
α-regioselection to the complexing ability of the
amino nitrogen with lithium ion. Replacing the
diethylamino group by the pyrrolidine group led to a
poorer α-selection. On the other hand, the better
complexing ligands, both bis(ethoxyethyl)amino and
dimethylaminoethylmethylamino groups, gave much
enhanced α-selection with α/γ_t = 9.6 to 13.5,
respectively (Table 3). Again, using the bis-
(ethoxyethyl)aminomethyldimethylsilylallyl anion as

Table 3. Effect of intramolecular metal ion complexation on regioselection.

	α	γ_t	γ_c
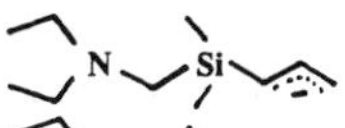	2.6	1	0.25
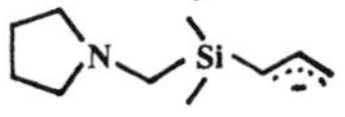	1.6	1	0.11
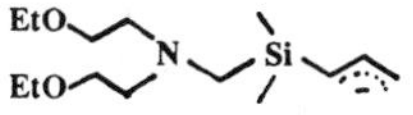	9.6	1	
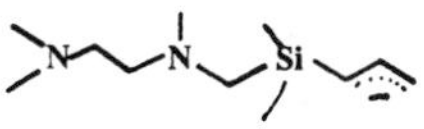	13.5	1	

the common nucleophile, we can see that with a number
of alkyl halides the α-selection remains fairly high,
at a synthetically useful level (Table 4).

Table 4.

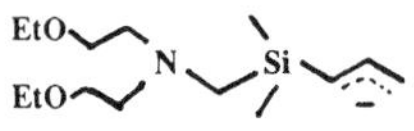

		$\underline{\alpha}$	$\underline{r_x}$	$\underline{r_c}$
	MeI	9.6	1	
	$n\text{-}C_6H_{13}I$	2.5	1	0.22
	I	4.1	1	1.85
	Br	7.1	1	0.81

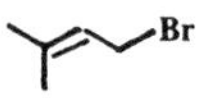

In summary, it is clear that by manipulation of the
substituent on silicon, it is possible to control the
reaction at a site removed from silicon.

ENANTIOSELECTIVE SYNTHESIS USING CHIRAL ORGANOSILICON COMPOUNDS

Another topic of interest to us is the possibility of
using chiral organosilicon compounds for enantioselective synthesis. Optically active organosilicon
compounds with chirality at silicon (e.g., 21) are
well known since the pioneering study of Sommer [14].

$$21 \qquad \begin{array}{c} \alpha\text{-Np} \\ | \\ Ph - Si - \\ \backslash \\ CH_3 \end{array}$$

Since the advent of the use of organosilicon compounds for organic synthesis, the use of chiral
organosilicon compounds of the type 21 for asymmetric
induction has not escaped the attention of the
organic chemists. However, the few attempts reported
in the literature have not been successful [15].
Right from the beginning, we have rejected the
possibility of using such chiral silicon compounds as
the chiral auxiliary. This is based on the perceived
disadvantages that: (1) the preparation of such
chiral auxiliary compounds requires tedious optical
resolution, (2) the structural feature available on
such silicon compounds is somewhat limited, and (3)
since silicon compounds can undergo inversion,
retention or racemization in many reactions, the
recovery of such chiral auxiliary with the optical
activity intact is somewhat uncertain. We have opted

for using optically active organosilicon compounds
with chirality remote from silicon [16]. We felt
that such compounds can perhaps be synthesized
readily with available optically active natural
products. Secondly, since the chiral centre is
located remote from silicon, the stereochemical
integrity of the chiral auxiliary would not be com-
promised in the reaction and the optically active
compound can perhaps be recovered and recycled again.

As a first effort, we prepared the optically active
α-pinanyldimethylallylsilane 22 from the precursor
chloro compound 23 [17]. The chlorosilane 23 was in
turn derived from the hydrosilylation of α-pinene
with dimethylchlorosilane [16]. The absolute configuration of 22 is deduced to be 1S, 2S, 5S (Scheme
13). Reaction of 22 with aldehydes under the

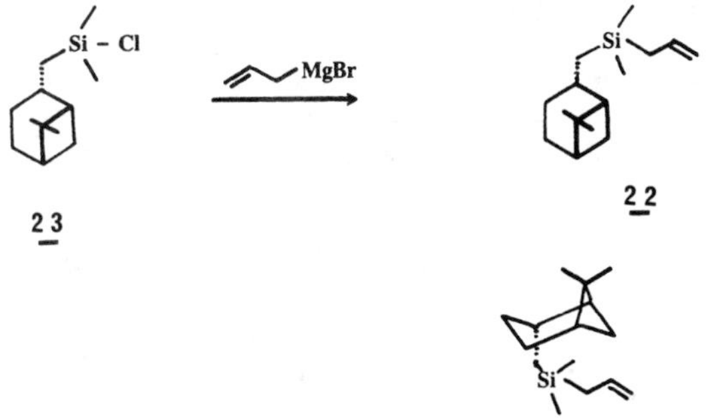

Scheme 13

standard conditions using Lewis acids gave the homoallylic alcohols 24 in good yields. Several
features of the reaction are of interest. The first
is the fact that α-pinanyldimethylsilanol 25 and its
disiloxane can be recovered readily from the reaction
and can therefore be recycled. The second is that
the homoallylic alcohols 24 displayed definite,
albeit low, enantiomeric excess. The third point is
that the predominant enantiomer obtained depends on
the Lewis acid used. With R = n-C$_3$H$_7$-, TiCl$_4$ gave
the S- alcohol 24, but SnCl$_4$ or BF$_3$ gave the R-
enantiomer (Scheme 14).

The general stereochemistry in the reaction of allylsilanes with electrophiles in an S$_E$2' reaction has
been deduced to be anti- from the work of Kumada [18]
and Fleming [19]. It is difficult to reconcile the
stereochemical results observed in our reaction with
an antiperiplanar transition state A as postulated by
Kumada [20]. On the other hand, Denmark reported
that for intramolecular reaction of an allylsilane
with an aldehyde moiety using Lewis acid, there is a

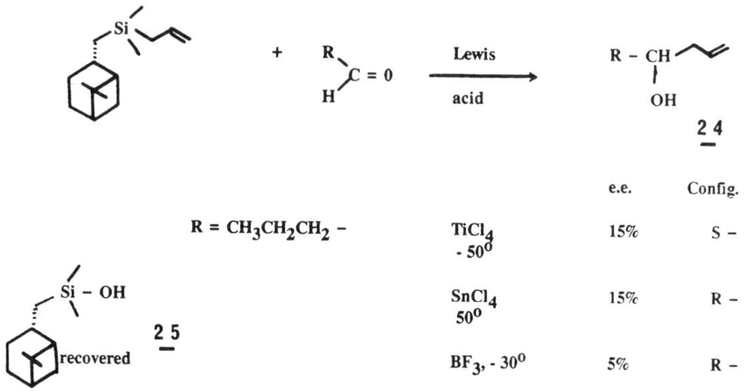

Scheme 14

substantial component of syn- stereochemistry in the
approach of the two components [21] (Scheme 15). It
seems to us that in order for substituents on silicon
to have an effect on the enantiomeric yield of the
reaction, a syn- transition state B would be required. A corollary of this argument is that any

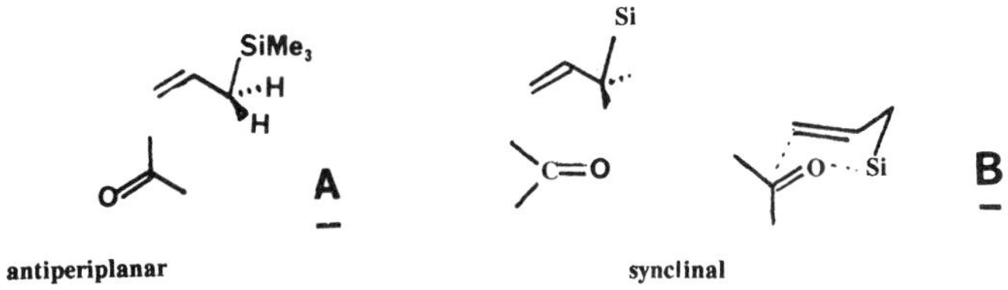

Scheme 15

structural feature which enhances the predominance of
the syn- approach will likely enhance the enantiomeric yield of the reaction. Indeed the recent work
of Taddei, using the methoxy-substituted pinanyl-
silane 26 as the chiral auxiliary, reacted with
carbonyl compounds under Lewis acid conditions to
give homoallylic alcohols with greater enantiomeric
excess (21 to 56%) (Scheme 16) [22]. However, the
synthesis of the chiral organosilicon compound 26
required five steps and is therefore less convenient.

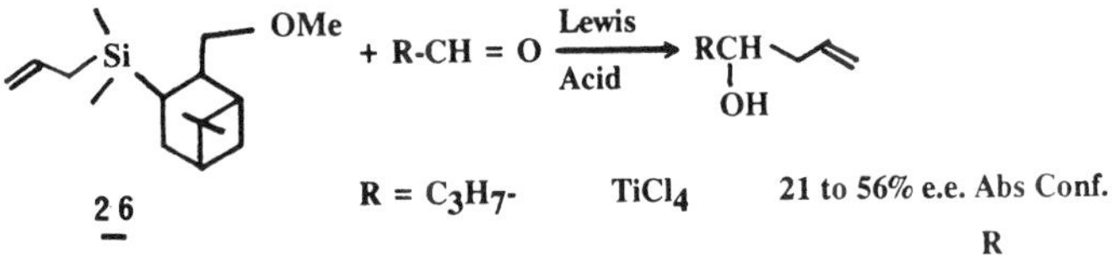

Scheme 16

Our own approach centered on the use of alkoxy-dimethylallylsilanes __27__ where the alkoxy group can be any naturally occurring optically active alcohol (e.g., ℓ-menthol). Compounds __27__ can be readily prepared by one of the two routes in Scheme 17.

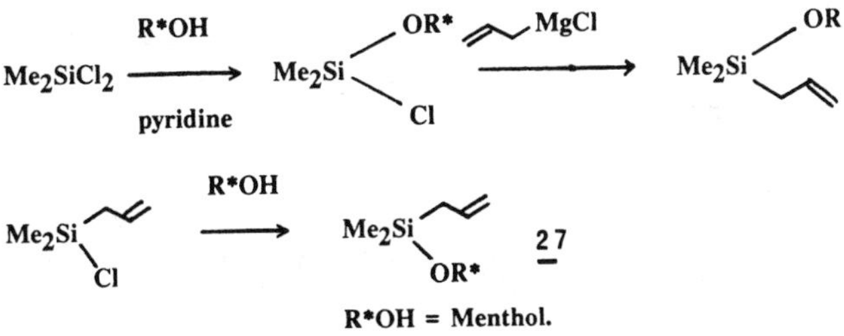

Scheme 17

R*OH = Menthol.

Reaction of ℓ-menthoxydimethylallylsilane __27__ with aldehydes using a limited amount of Lewis acid (less than 1/2 equivalent) gave the homoallylic alcohol __24__ in good chemical yields. The optical yield is now in the range of 20-25% e.e. (Scheme 18), an improvement over the pinanyl system __22__. A particularly striking example is the increase in e.e. for the reaction with n-nonanal [23]. We have applied this reaction for

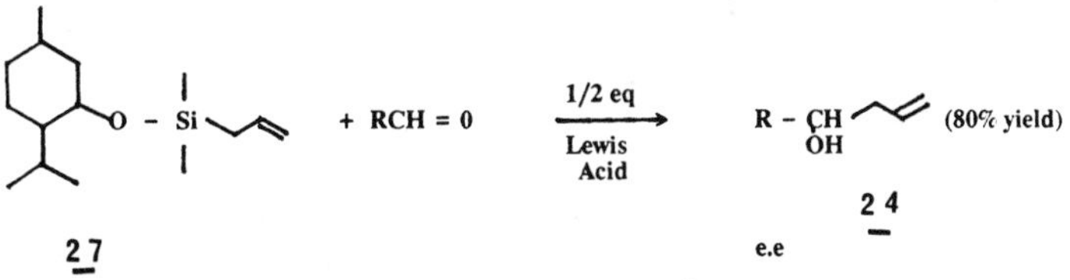

R =	n-C$_3$H$_7$-	TiCl$_4$, -78°	23%	S -
n-C$_6$H$_{13}$-	BF$_3$, -78°	(~25%)	S -
n-C$_8$H$_{17}$-	TiCl$_4$, -78°	21%	S -

Scheme 18

the synthesis of the γ-lactone __28__, a component of the sex pheromone of the Rove Beetle according to Scheme 19 [17].

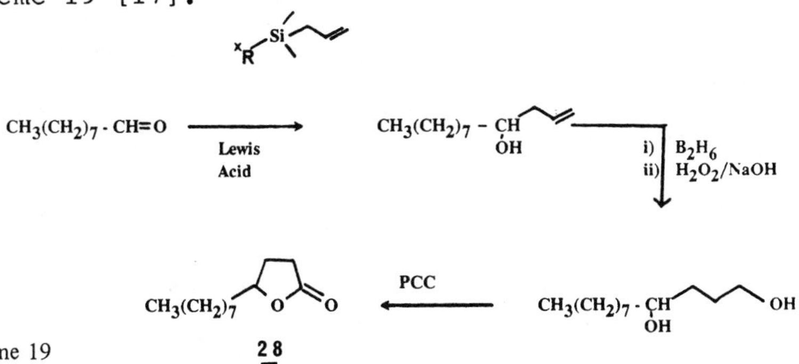

Scheme 19   __28__

One would expect that the optical yield of the reaction can be further enhanced if the chiral auxiliary has more than one functional group for complexation. We are actively exploring this avenue.

ACKNOWLEDGEMENTS

We thank the Natural Sciences and Engineering Research Council of Canada, the Ministry of Education of Quebec, and the United Nations Development Program for financial support.

REFERENCES

1. E.J. Corey and A. Venkateswarlu, J. Amer. Chem. Soc., 94, 6190 (1972).
2. M. Lalonde and T.H. Chan, Synthesis, 817 (1985).
3. R.J.P. Corriu, J. Maase and D. Samate, J. Organomet. Chem., 93, 71 (1975).
4. T.H. Chan and K. Koumaglo, Tetrahedron Letters, 25, 717 (1984).
5. M. Schlosser and J. Hartmann, J. Amer. Chem. Soc., 98, 4674 (1976).
6. R.B. Miller and G. McGarvey, Synth. Commun., 8, 291 (1978).
7. T.H. Chan and K. Koumaglo, J. Organomet. Chem., 281, 109 (1985).
8. E.I. Negishi, Acc. Chem. Res., 15, 340 (1982).
9. T.H. Chan, Z.Y. Wei, J.S. Li and D. Wang, to be published.
10. T.H. Chan, P.W.K. Lau, M. Mychajlowskij, Tetrahedron Letters, 3317 (1977).
11. T.H. Chan and K. Koumaglo, Tetrahedron Letters, 27, 883 (1986).
12. T.C. Cheng, C.M. Hsiao, Y.S. Ho, S.J. Huang, C.C. Lo, C.M. Yeng and C.H. Yang, K'o Hsueh T'ung Pao, 25, 658 (1980).
13. We acknowledge gratefully the suggestion by Professor R.R. Fraser of using an alkoxy group on silicon to stabilize α-silylcarbanions.
14. L.H. Sommer, "Stereochemistry, Mechanism and Silicon", McGraw Hill, New York, 1965.
15. (a) R.G. Daniels and L.A. Paquette, Organometallics, 1, 1449 (1982);   (b) G.L. Larson and E. Torres, J. Organomet. Chem., 293, 19 (1985).
16. D. Wang and T.H. Chan, Tetrahedron Letters, 24, 1573 (1983).
17. D. Wang, G.L. Yi and T.H. Chan, in press.
18. T. Hayashi, M. Konishi, H. Ito and M. Kumada, J. Amer. Chem. Soc., 104, 4962 (1982).
19. I. Fleming and N.K. Terrett, Tetrahedron Letters, 24, 4153 (1983).

20. T. Hayashi, K. Kateba, I. Hamachi and M. Kumada,
     Tetrahedron Letters, 28, 2865 (1983).
21. S.E. Denmark and E.J. Weber, Helv. Chim. Acta,
     66, 1655 (1983).
22. L. Coppi, A. Mordini, M. Taddei, Tetrahedron
     Letters, 28, 969 (1987).
23. D. Wang, Z.Y. Wei and T.H. Chan, to be
     published.

Chapter 6

Silicon — assisted synthesis of β-lactams

Ernest W. Colvin, Daniel McGarry and Mark Nugent – Department of Chemistry, University of Glasgow, Glasgow G12 8QQ.

N-Unsubstituted azetidin-2-ones offer considerable synthetic opportunity in the synthesis of β-lactam antibiotics such as the carbapenems and penams and the monobactams [1], and synthetic analogues. Routes to such potentially valuable intermediates have hitherto involved either creation of an N-functionalised β-lactam followed by N-liberation, or degradation of a naturally-occurring bicyclic β-lactam [2].

N-Silyl imines, in particular those derived formally from diaryl ketones [3], have been known for some time [4]; enolisable carbonyl compounds do not lead to pure silyl imines because of tautomeric equilibrium with enamines. The N-trimethylsilyl imine derived from benzaldehyde has been reported [5] to react with diphenylketene to give 3,3,4-triphenylazetidin-2-one in modest yield. More recently, the first direct route [6] to N-unsubstituted azetidin-2-ones was described; reaction of certain ester lithium enolates with several N-trimethylsilyl imines provided the target β-lactams directly.

This paper [7] describes an efficient procedure for the assembly of a variety of N-unsubstituted azetidin-2-ones, based on Lewis acid catalysed reactions between N-trialkylsilyl aldimines and ketene silyl acetals.

Two routes to the required aldimines are detailed: the first, a modification of a known procedure, converts non-enolisable aldehydes into N-trimethylsilyl aldimines, whereas the second provides a novel route from primary amines to otherwise unobtainable N-t-butyldimethylsilyl aldimines.

N-TRIMETHYLSILYL ALDIMINES

Treatment of a solution of lithium hexamethyldisilazide in THF with one equivalent of a non-enolisable aldehyde at 0°C, followed by one equivalent of Me_3SiCl [to convert the produced Me_3SiO^- ion into $(Me_3Si)_2O$] gave, after non-aqueous isolation and distillation (Kugelrohr), the range of aldimines shown in Scheme 1.

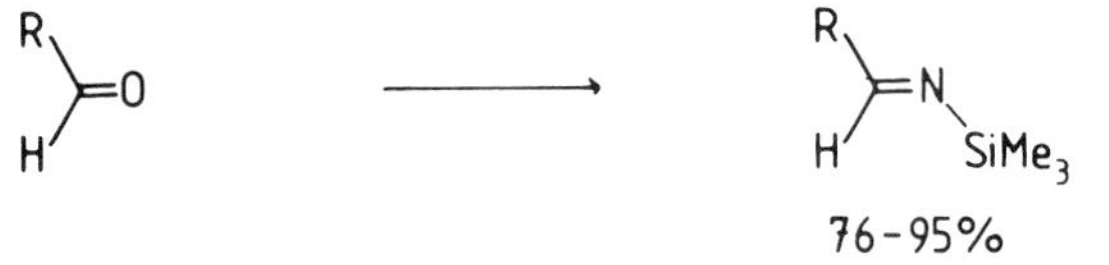

76-95%

(1), R = Ph
(2),   = 2-furyl
(3),   = PhCH=CH
(4),   = PhC≡C
(5),   = $Me_3SiC≡C$
(6),   = $Me_3SiCH=CH$

Scheme 1

The probable mechanism of this reaction is shown in Scheme 2. N→O Silyl migration does not appear to proceed at a significant rate at -78°C; under these conditions, the aminoacetal adduct can be trapped and isolated as its tris(trimethylsilyl) derivative.

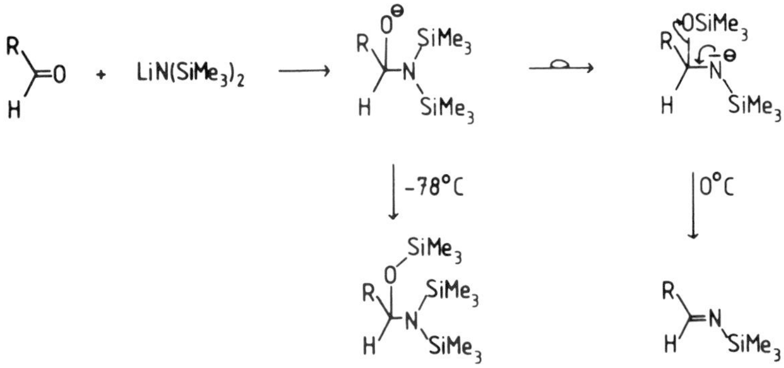

Scheme 2

N-t-BUTYLDIMETHYLSILYL ALDIMINES

N-t-Butyldimethylsilyl aldimines are of greater potential value
than the corresponding N-trimethylsilyl analogues: the silyl
protecting group is much more stable to hydrolytic cleavage,
and it survives the β-lactam forming conditions about to be
discussed.   However, different methodology has to be employed
– methodology that also gives access to the hitherto elusive
N-silyl imine of ethyl glyoxylate.

The primary amine was converted into its N-t-butyldimethylsilyl
derivative, which on reaction with t-butyl hypochlorite gave
the N-t-butyldimethylsilyl-N-chloro species.   This, on
elimination of HCl, produced the desired N-t-butyldimethylsilyl
imines (Scheme 3) in good to excellent yield.

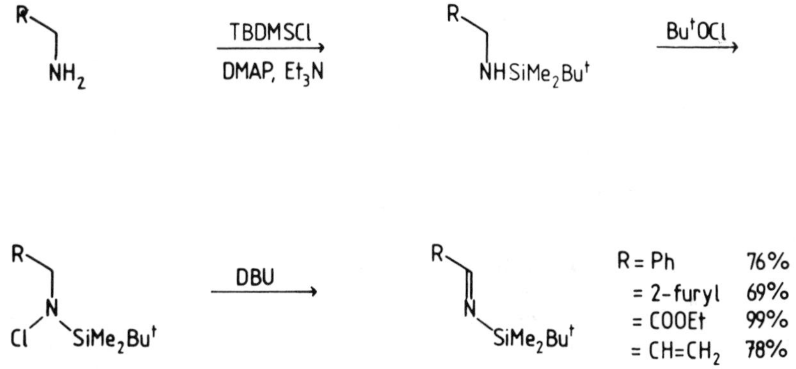

Scheme 3

R = Ph   76%
  = 2-furyl   69%
  = COOEt   99%
  = CH=CH$_2$ 78%

KETENE SILYL ACETALS

A range of simple ketene trimethylsilyl acetals was prepared
following literature procedures [8], by deprotonation using
LiN(i-Pr)$_2$ followed by O-silylation.   Deprotonation under
these conditions normally leads selectively to the Z-enolate
and thence the E-ketene acetal.   The separate Z- and E-ketene
t-butyldimethylsilyl acetals of methyl propionate were prepared
using Ireland's procedure [9], and the vinyl-substituted
ketene acetal was prepared from crotonyl chloride [10] (Scheme
4).

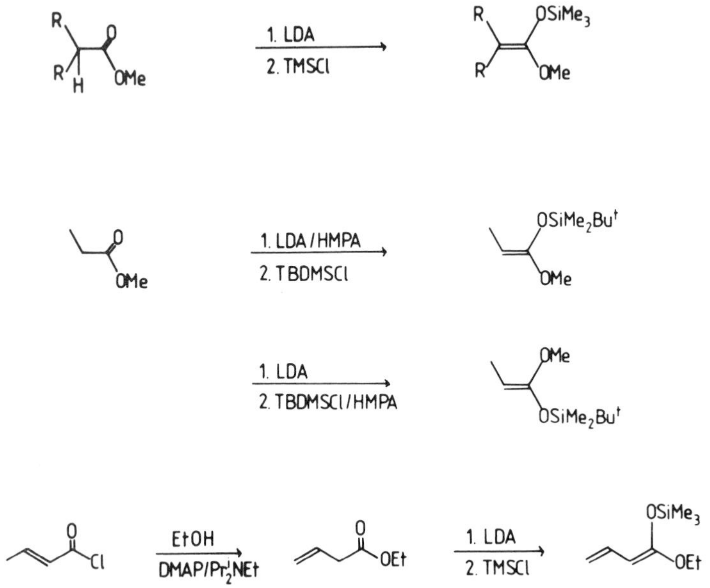

Scheme 4

REACTION OF SILYL ALDIMINES WITH KETENE SILYL ACETALS

The first catalyst examined for this reaction was fluoride ion. Although successful for the direct production of certain of the target β-lactams, the sequence proved to be irreproducible, possibly because of the vagaries associated with obtaining catalytically active 'anhydrous' tetrabutylammonium fluoride [11].

Activation of the aldimine component with a Lewis acid was examined first using $TiCl_4$, a reagent employed successfully in a related reaction [12] with N-aryl and -alkyl imines. This and a variety of other Lewis acids proved to be of limited utility, but it was observed that use of ZnI_2 in ether produced an imino-ester in high yield (Scheme 5). On the assumption that this had been formed by trans-amination between unreacted imine and an intermediate metallo-amide, the same reaction was performed in the presence of a weak proton source, t-butanol. This yielded β-aminoester as sole isolated product. Indeed, treatment of the reaction mixture prior to work-up with MeMgBr [13] gave, on isolation, β-lactams directly. Employing this general technique of complexation of the imine with ZnI_2 in ether, followed by immediate and sequential addition of the ketene acetal and t-butanol, and, after 2-3 h, of MeMgBr, the range of β-lactams listed in Table 1 was obtained.

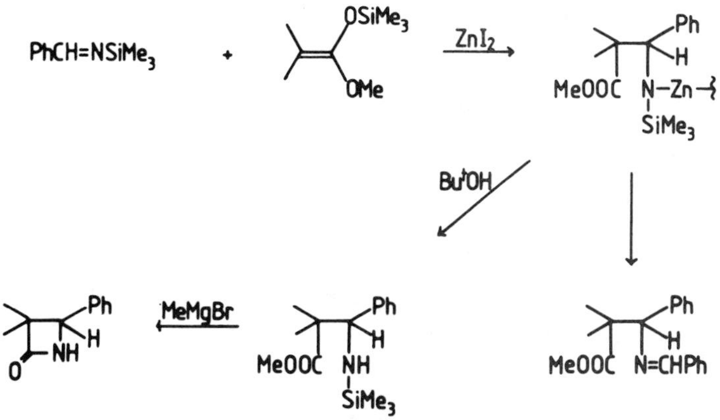

Scheme 5

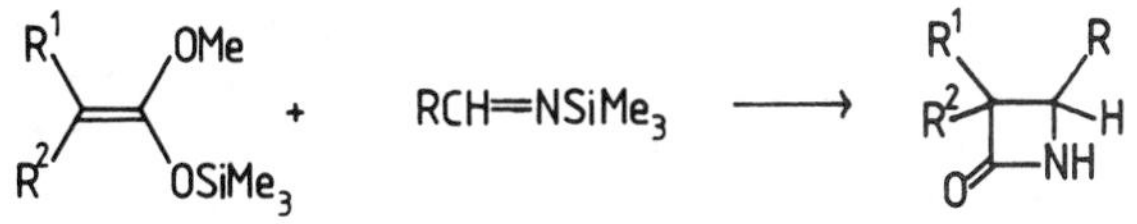

R^1	R^2	Imine	% Yield	cis:trans ratio
Me	Me	(1)	75	
		(2)	76	
		(4)	78	
		(6)	75	
H	H	(1)	27[1]	
H	PhO	(2)	58	1:0.77
Et	H	(1)	61	1:9[2]
Me	H	(4)	82	1:3
		(5)	62	1:1.5
Ph	H	(5)	53	1:12[2]

[1] Using t-butyldimethylsilyl ketene acetal

[2] Initial additions at -78 °C

Table 1.

STEREOCHEMISTRY AND REGIOCHEMISTRY

A degree of trans-stereoselectivity, which could be enhanced using lower temperatures, was observed, in contrast to the cis-stereoselectivity which prevails [6] when ester lithium enolates are employed. This threo-diastereoselectivity in the initial bond-forming step can be explained by involving transition states such as those shown in Scheme 6. In this interpretation, the ketene acetal eclipses the imine, and the steric interactions so produced favour the transition state leading to the threo-aminoester.

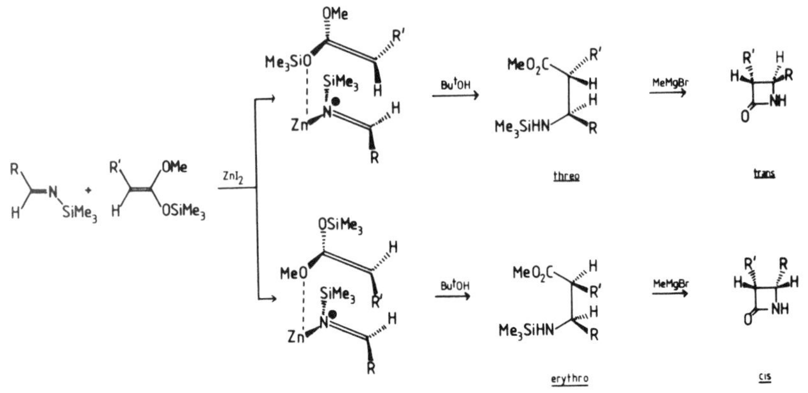

Scheme 6

The effect of ketene acetal geometry on diastereoselectivity
was examined by reacting separately the Z- and E-O-t-butyl-
dimethylsilyl ketene acetals of methyl propionate with N-
trimethylsilylbenzaldimine.   These results showed very
little dependence on the ketene acetal geometry, and suggest
that the controlling interactions are between the phenyl group
of the imine and the alkyl substituent on the ketene acetal.
As expected, the crotonate-derived unsaturated ketene acetal
reacted with a high degree of γ-regioselectivity; this could be
reduced somewhat by operating at lower temperatures.

TMSOTf-INDUCED REACTIONS

The aldimine derived from ethyl glycinate did not react
satisfactorily under the above ZnI$_2$ conditions.   However,
catalytic use of TMSOTf did allow smooth reaction to take place.
Selective desilylation and Grignard-mediated ring closure
afforded the desired β-lactam (Scheme 7) in an overall yield
of 60%.

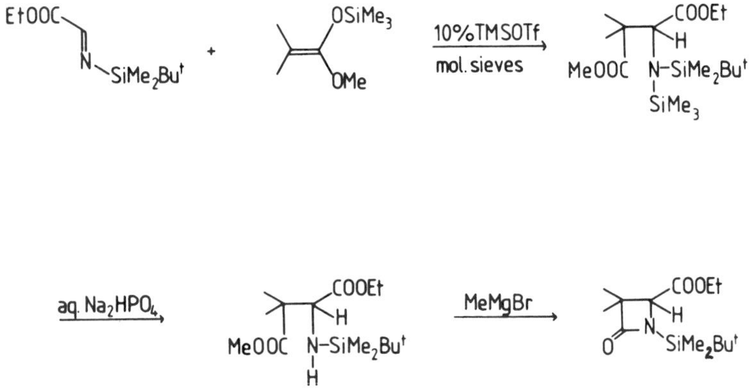

Scheme 7

The amine acetals obtained by low temperature silylation (Scheme 8) could also be employed as precursors of the reactive iminium ion species.   For example, pre-complexation of the furfural-derived amine acetal with a full equivalent of TMSOTf, followed by addition of ketene acetal and aqueous isolation gave the β-aminoester and thence the β-lactam, again with a preference for the <u>trans</u>-isomer.

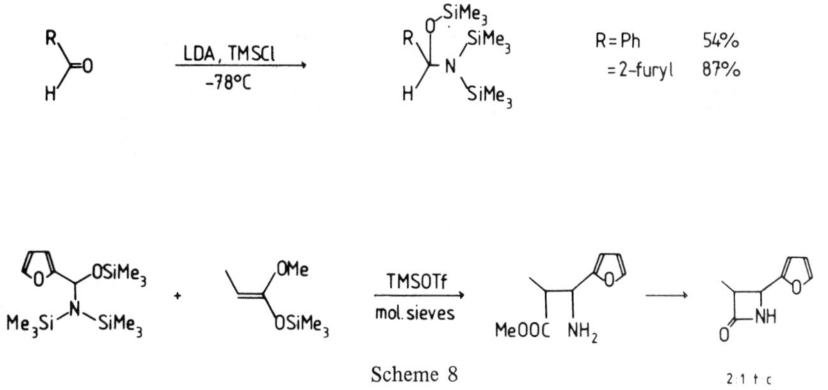

Scheme 8

We thank the S.E.R.C. for financial support of this work.

REFERENCES

1.   Southgate, R. and Elson, S., "Naturally Occurring β-Lactams", <u>Prog. Chem. Org. Nat. Prods.</u>, <u>47</u>, 1 (1985).

2.   "Chemistry and Biology of β-Lactam Antibiotics", vol. 2, eds. Morin, R.B. and Gorman, M., Academic Press, New York and London, 1982.

3.   Chan, L.-H. and Rochow, E.G., <u>J. Organomet. Chem.</u>, <u>9</u>, 231 (1967).

4.   Krüger, C., Rochow, E.G., and Wannagat, U., <u>Chem. Ber.</u>, <u>96</u>, 2132 (1963).

5.   Birkofer, L. and Schramm, J., <u>Liebigs Ann. Chem.</u>, 760 (1977).

6.   Ha, D.-C., Hart, D.J., and Yang, T.-K., <u>J. Am. Chem. Soc.</u>, <u>106</u>, 4819 (1984) and references therein.

7.   For a preliminary account of some of this work, see Colvin, E.W. and McGarry, D.G., <u>J. Chem. Soc., Chem. Commun.</u>, 539 (1985).

8. Ainsworth, C., Chen, F., and Kuo, Y.N., J. Organomet.
    Chem., 46, 59 (1972).

9. Ireland, R.E., Mueller, R.H., and Willard, A.K.,
    J. Am. Chem. Soc., 98, 2868 (1976).

10. Lombardo, L., Tetrahedron Lett., 26, 381 (1985).

11. Sharma, R.K. and Fry, J.L., J. Org. Chem., 48, 2112 (1983).

12. Ojima, I. and Inaba, S., Tetrahedron Lett., 21, 2077,
    2081 (1980) and references therein; see also Okano, K.,
    Morimoto, T., and Sekiya, M., J. Chem. Soc., Chem. Commun.,
    883 (1984).

13. Birkofer, L. and Schramm, J., Liebigs Ann. Chem., 2195
    (1975).

Chapter 7

Directing effects of a silyl group on cationic carbon skeleton rearrangements

I. Kuwajima – Department of Chemistry, Tokyo Institute of Technology.

Cation-stabilizing effects of silyl group have been utilized for selective carbon-chain homologation or introduction of functional groups as shown in allyl-, vinyl-, or homoallylsilanes[1]. These effects may also be expected to play important roles in several cationic carbon skeleton rearrangement reactions[2]. We have attempted to utilize the well documented β- and γ-cation stabilizing effects of a silyl group for controlling the direction of cationic rearrangement reactions, and have found some interesting features which seems to be very useful in organic synthesis.

REARRANGEMENT REACTIONS OF 3-(TRIMETHYLSILYL)–METHYLCYCLOHEXYL MESYLATES [3]

Treatment of a 3-substituted 3-(trimethyl-silyl)methylcyclohexyl mesylate 1a or 1b with dimethylaluminium triflate in hexane led to the formation of two types of rearrangement products, a methylenecyclohexane 2 and a vinylcyclopentane 3, along with a small amount of an acyclic diene 4 resulting from a fragmentation reaction (Eq 1). Other stronger Lewis acids such as Ti(IV) or Sn(IV) tetrachloride gave cyclohexenes through deprotonation from initially formed cationic species.

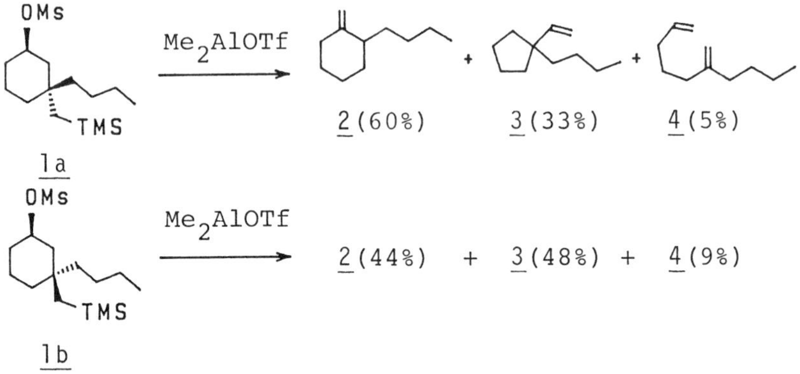

$$\text{(1)}$$

These results indicate that the silyl group has an effect to attract a cationic site from δ-position to γ- and finally β-position, which induces a selective hydride shift followed by rearrangement of an alkyl group.

However, since 5, which generates a common cationic intermediate A, gave 2 exclusively under similar conditions (Eq 2), steric factor of TMS-methyl group and/or the leaving group appear to play also important roles for determining the reaction course.

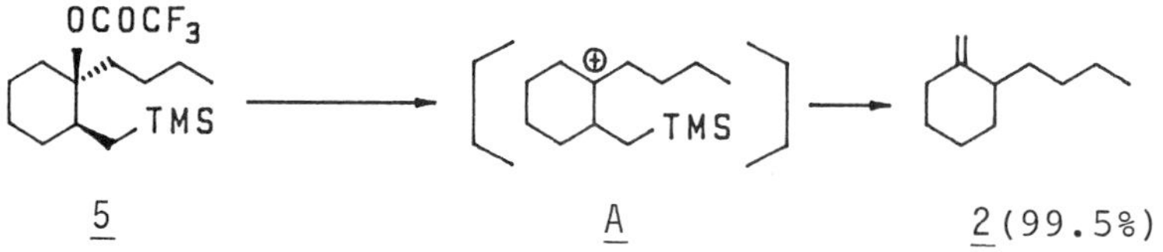

$$\text{(2)}$$

Results of the reaction with conformationally rigid substrates 1c-1f have disclosed the following features (see Eq 3). Thus, for determining the reaction course, the stereochemistry of TMS-methyl group is much more important than that of the leaving group: An axial TMS-methyl group on 1c and 1d exhibits an essentially similar effect which favor the ring contraction irrespective of stereochemistry of the leaving group except the formation of a fragmentation product 8. On the other hand, in the reaction of substrates bearing an equatorial TMS-methyl (1e and 1f), the stereochemistry of the leaving group has also a great influence on the direction of rearrangement, and the formation of ring contraction product 6 was completely excluded. Further, three types of methylenecyclohexanes are formed in these reactions, but their structural features are strictly dependent on the stereochemistry of TMS-methyl group of the starting materials.

In a similar manner with usual cationic rear-

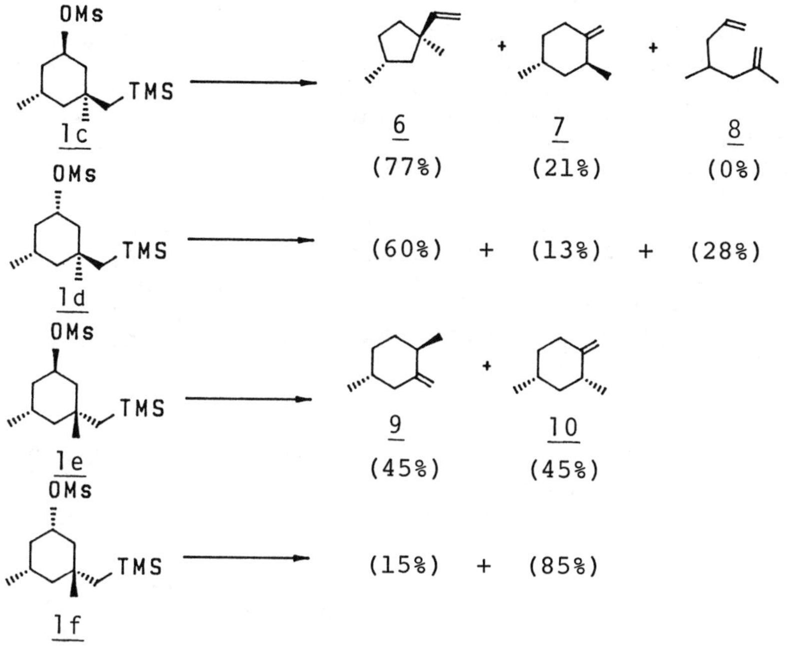
OMs
TMS
1c
6
(77%)
7
(21%)
8
(0%)
OMs
TMS
1d
(60%) + (13%) + (28%)
OMs
TMS
1e
9
(45%)
10
(45%)
OMs
TMS
1f
(15%) + (85%)

(3)

rangement processes, successive migration of two antiperiplanar substituents on axial positions may be greatly favored in this silicon-directed rearrangement too. However, it appears to be necessary to take additional factor into account to explain these puzzling results. We would like to propose such an assumption that two successive 1,2-rearrangement steps may involve conformational transformation of a cyclohexane ring, and most of sp^3 character may be still retained [4] in the cationic species generated through such processes. Thus, in the reaction of 1c or 1d, preferential rearrangements of an axial hydrogen and TMS-methyl accompany conformational transformation to give the intermediate B bearing a vacant orbital antiperiplanar to a ring carbon, which leads to the formation of

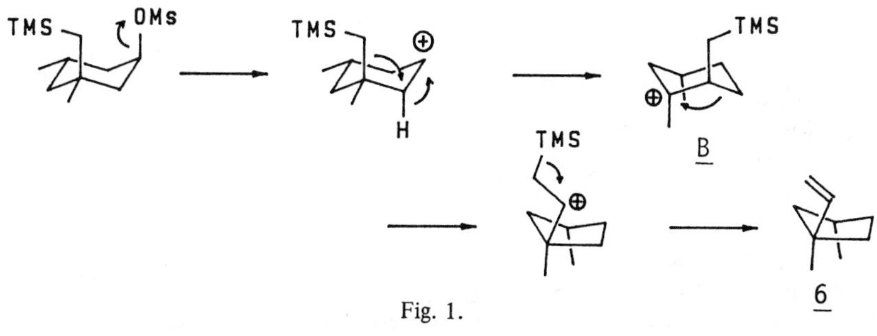
TMS
OMs
TMS
H
TMS
B
TMS
6

Fig. 1.

the ring contraction product 6 via a carbon skeleton rearrangement so as to yield a stable cationic
species (Fig 1). Formation of 7 may be attributable
to a less favored synclinal hydride shift from B.

Similarly, successive hydride and methyl rearrangements from the substrates bearing an equatorial TMS-methyl group produce a β-silicon stabilized intermediate C, which gives 9 exclusively.
However, TMS-methyl seems to have much greater
tendency for migration over methyl, and its 1,2-
synperiplanar rearrangement may form an intermediate D bearing an axial vacant orbital, which
undergoes an antiperiplanar axial hydrogen shift to
afford 10 (Fig 2).

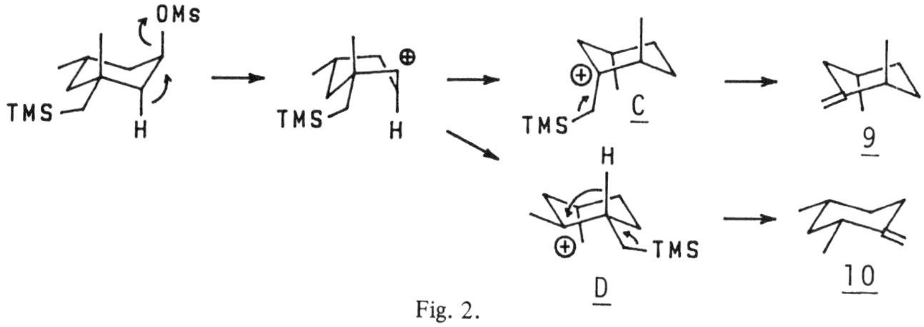

Fig. 2.

In contrast to such larger migration aptitude
of TMS-methyl, a ring carbon or methyl migration
follows an initial hydride shift in the reaction of
2-substituted ones 1g or 1h to afford a mixture of
11 and 12 (Eq 4). In these cases, such a second
alkyl group migration as above allows a direct
generation of a stable β-silyl tertiary cationic
intermediate, whereas TMS-methyl rearrangement apparently brings little stabilization to the resulting species.

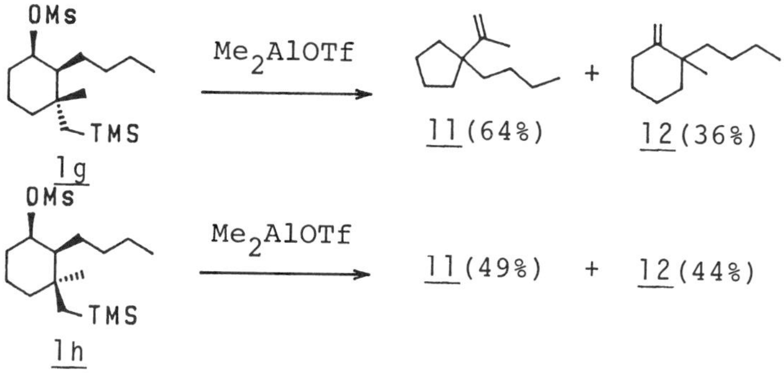

(4)

Similar effects have also worked on bicyclic
compounds. As predicted from the above results, the
substrates 1i gave methylenedecalin 13 through

rearrangement of a hydride and TMS-methyl, whereas
1j selectively afforded *trans*-hydroazulene 14 via a
ring carbon migration (Eq 5).

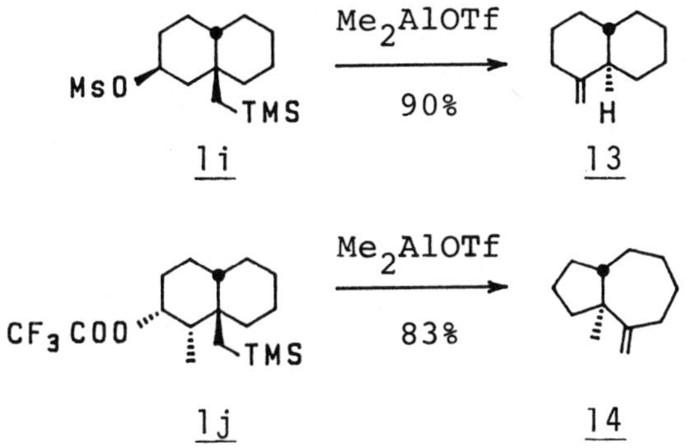

$$(5)$$

RING ENLARGEMENT REACTIONS OF 1-(TRIMETHYLSILYL)–METHYLCYCLOALKANECARBALDEHYDES AND THEIR ACETALS

In order to make use of β-cation-stabilizing
effect of a silyl group for carbon skeleton ring
enlargement, we have studied the generation of
several cationic species such as E and the course
of their reactions.

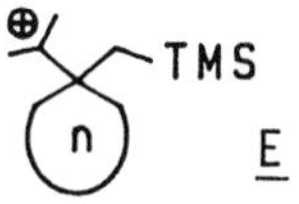

Under the influence of dimethylaluminium triflate, several types of 1-(trimethylsilylmethyl)-
cyclohexylmethyl mesylates 15 gave the corresponding vinylcyclohexane 16 exclusively without
any formation of the desired ring enlargement products (Eq 6).

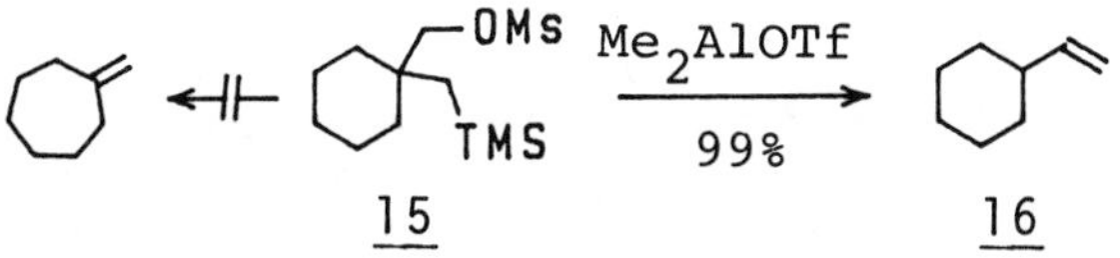

$$(6)$$

Thus, on the cationic species generated from
mesylates, a greater migration tendency of TMS-
methyl predominates over the β-cation-stabilizing
effect to determine the reaction course (Fig 3).

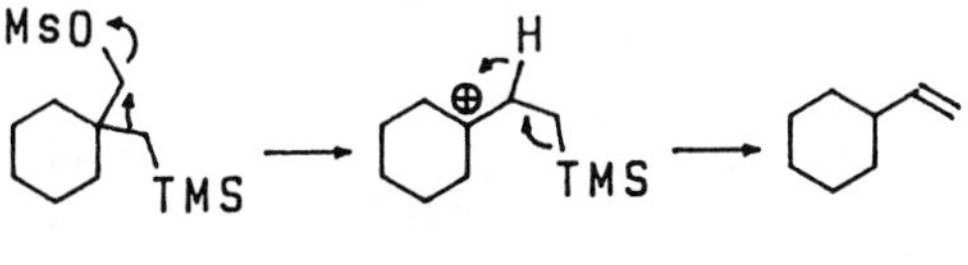

(Fig 3)

Use of acyl cation also failed to induce the expected ring enlargement reaction: Acyl chloride 17 gave 18 as a sole product on treating with Ti(IV) tetrachloride. In this case, the silyl group apparently accelerates the decarbonylation of an initially formed acyl cation to yield methylene-cyclohexane which undergoes Friedel-Crafts like acylation under the present reaction conditions (Fig 4).

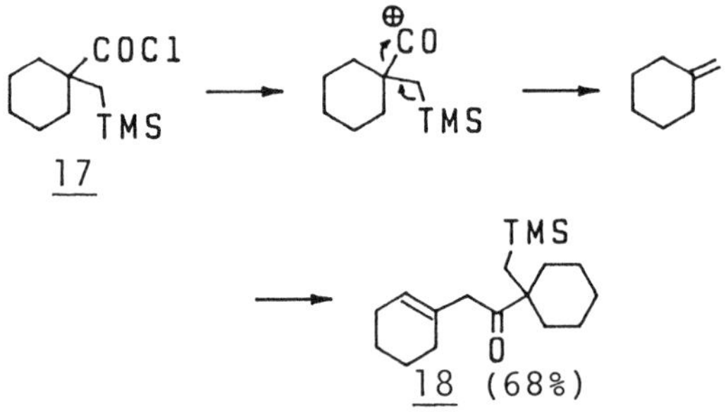

(Fig 4)

On exposure to aluminium chloride, the reac-tion courses of oxiranes were different, depending on the substitution patterns. As shown in Eq 7, a mono-substituted oxirane underwent TMS-methyl group migration exclusively to yield 2-TMS-methyl alcohol, whereas a disubstituted oxirane gave the ring enlargement product along with the aldehyde resulting from isomerization of an oxirane.

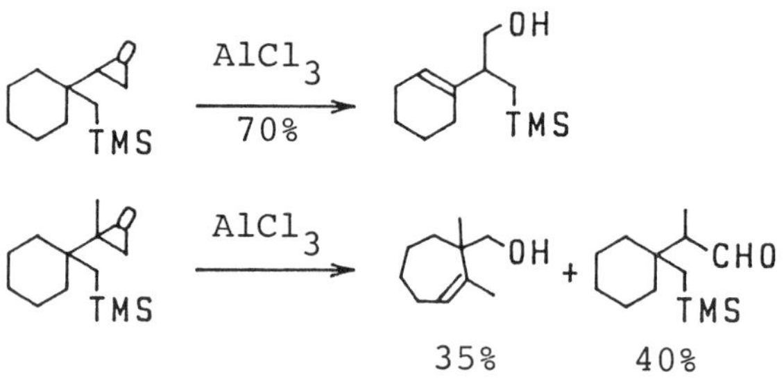

(7)

Use of 1-(trimethylsilylmethyl)cycloalkane-carbaldehydes 19 has improved the situation to induce the desired selective one-carbon ring en-largement reactions. Thus, on treating with alumi-nium chloride or methylaluminum dichloride in methylene chloride, cyclohexanecarbaldehyde 19a afforded the two types of ring enlargement pro-ducts, 2-(trimethylsilyl)methylcycloheptanone 20a as the major product and 2-methylenecycloheptanol 21a as the minor one (Eq 8) [5].
    As shown in Fig 5, the silyl group exerts a strong directing influence on an initial step of a sequence of this rearrangement reaction. From the

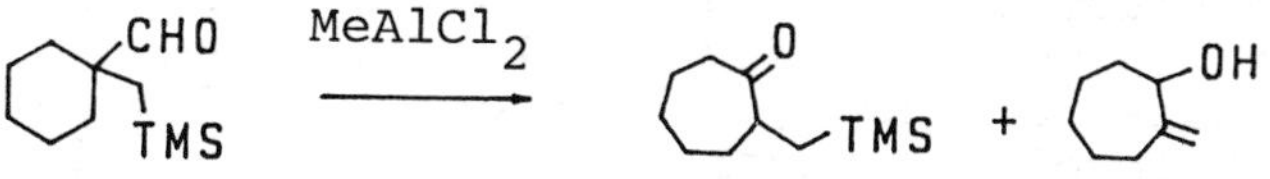

$$(8)$$

20a (77%)   21a (Trace)

resulting cationic intermediate **F**, a pinacol rearrangement-like 1,2-hydride shift takes place predominantly, which makes a good contrast with a well precedented behavior of β-silicon substituted cationic species, e.g. removal of silyl group, and may attract much attention on mechanistic as well as synthetic point of view [6].

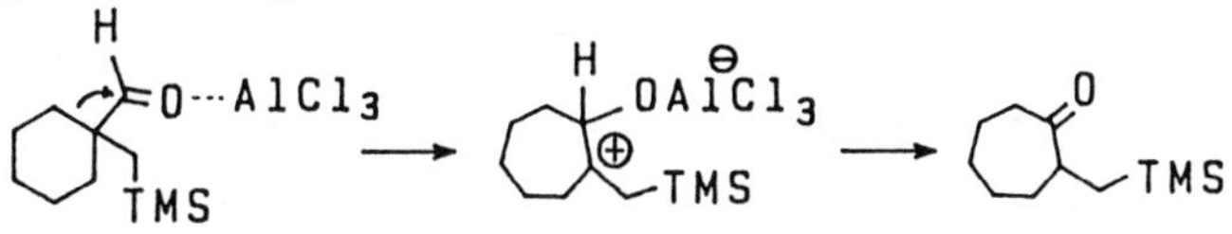

(Fig 5)

F

Aldehydes **19b-c** bearing various sized carbon rings also undergo similar rearrangement to give the corresponding ring-enlarged ketones **20b-c** and alcohols **21b-c** (Eq 9) [5].

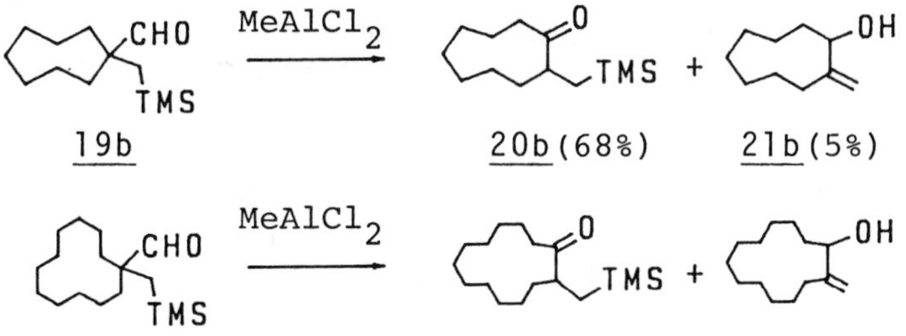

19b   20b (68%)   21b (5%)   (9)

19c   20c (57%)   21c (18%)

In addition to such direction control, a silyl group appears to have also an accelerating effect. In the absence of a silyl subtituent, this type of reaction did not occur under similar reaction conditions, and under more forcing conditions, 1-methylcyclohexanecarbaldehyde underwent an exclusive rearrangement of methyl group [7] very slowly to yield a small amount of cyclohexyl methyl ketone, accompanied with major recovery of the starting material (Eq 10).

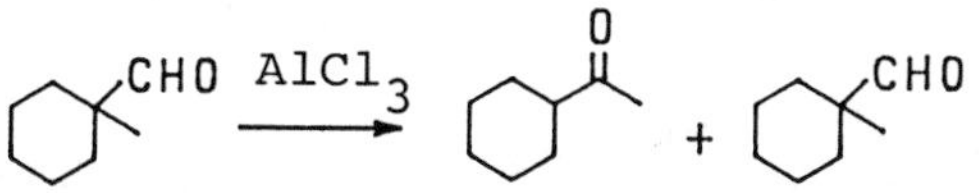

$$(10)$$

20%   80%

In the first step of this rearrangement process,   migration aptitudes of primary vs. seconda-

ry or tertiary alkyl groups seem to be different
enough to induce a preferential rearrangement of a
more substituted group; the reaction of 2-methyl-1-
silylmethylcyclohexanecarbaldehyde 19d gave 2-
methyl-7-(silylmethyl)cycloheptanone 20d as a mixture of *cis*- and *trans*-isomers (Eq 11), whereas two
types of rearrangement products were formed in ca.
2:1 ratio from 3-methyl derivative (Eq 12).

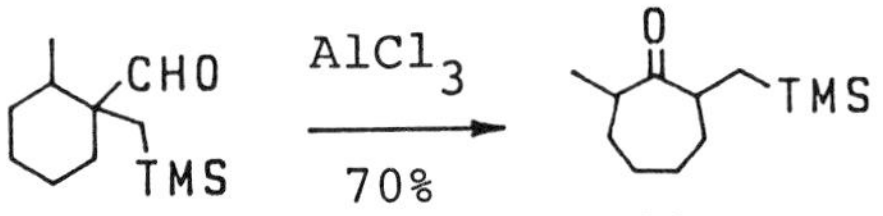

$$\text{19d} \xrightarrow[\text{70\%}]{\text{AlCl}_3} \text{20d} \tag{11}$$

$$\xrightarrow[\text{84\%}]{\text{MeAlCl}_2} \tag{12}$$

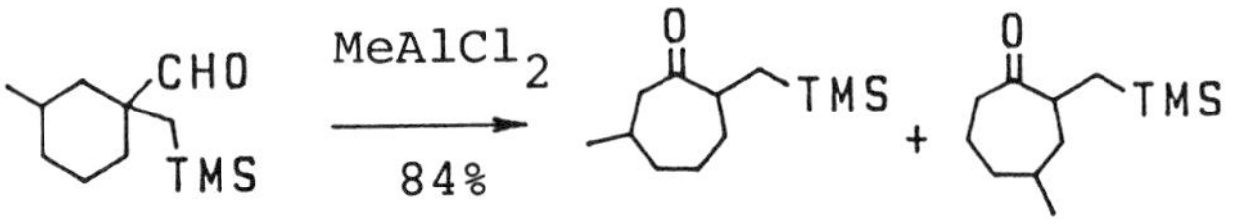

In contrast, a substrate 22 with an aldehyde
function on the allylic position failed to afford
one-carbon ring enlargement products, but gave the
corresponding methyl ketone 23 (Eq 13). A cation-
stabilizing effect of an allylic group as well as a
larger migration aptitude of TMS-methyl may
account for the difference of rearrangement direction.

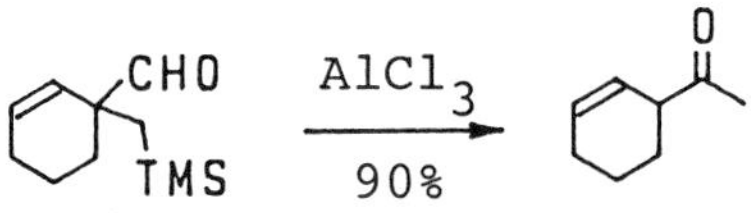

$$\text{22} \xrightarrow[\text{90\%}]{\text{AlCl}_3} \text{23} \tag{13}$$

Acetals 24 derived from the above aldehydes
also undergo a selective one-carbon ring enlargement reaction. In this case, use of milder Lewis
acid is usually preferable. Thus, on heating with
zinc bromide in methylene chloride for 30 min, an
acetal 24 can be converted to the corresponding

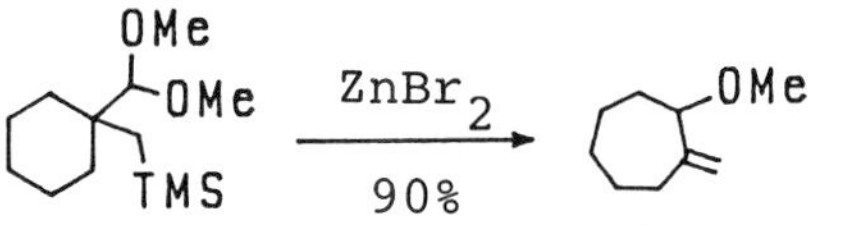

$$\text{24a} \xrightarrow[\text{90\%}]{\text{ZnBr}_2} \text{25a} \tag{14}$$

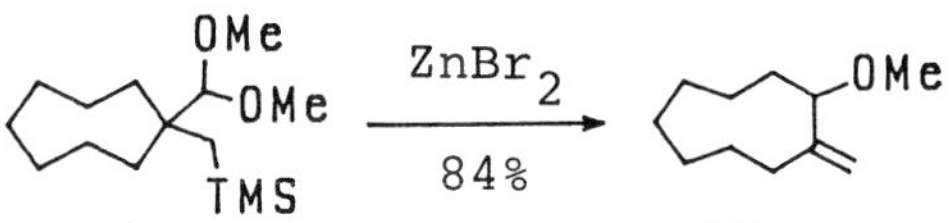

$$\text{24b} \xrightarrow[\text{84\%}]{\text{ZnBr}_2} \text{25b}$$

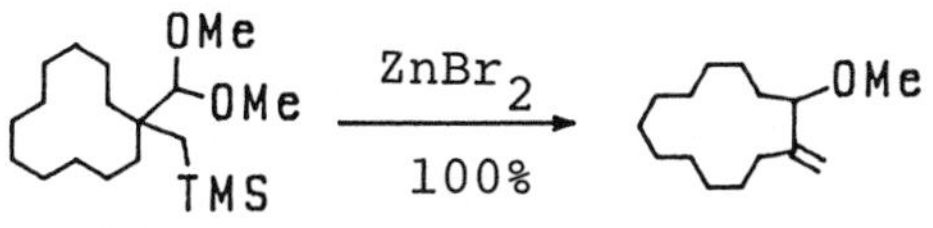

24c   25c

product <u>25</u> almost quantitatively (Eq 14) [8].
    Further, regio- and stereochemical outcomes
appear to be controlled rigorously through this
ring enlargement process; a more-substituted group
rearranges preferentially with retention of its
configuration (Eq 15).

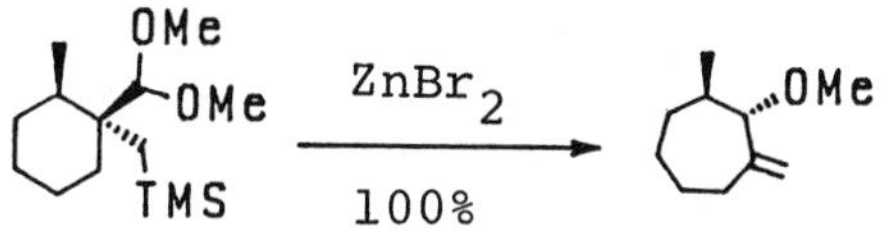

$$(15)$$

    Use of TMS-based nucleophiles such as TMS-OMe
(2 equiv) in the presence of TMS-OTf (1 mol%) has
allowed a direct conversion of aldehydes to the
corresponding ring enlarged ethers in excellent
yields without any formation of TMS-methyl ketones
(Eq 16).

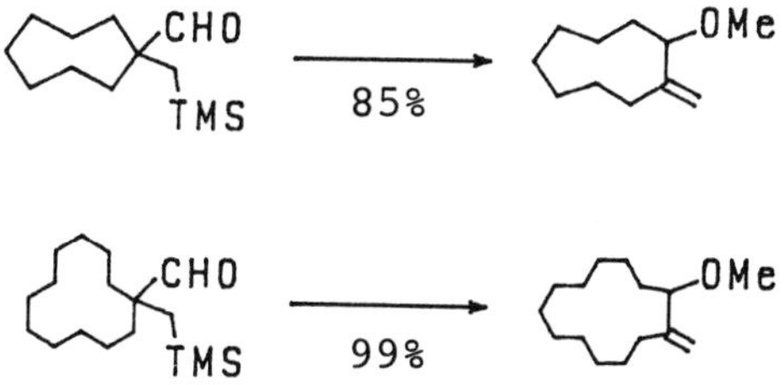

$$(16)$$

    In conclusion, three types of effects of a
silyl group; (i) an attraction of a cationic site
from remote position to β-position, (ii) a greater
migration tendency of TMS-methyl group, and (iii)
control of the direction of rearrangement due to
β-cation stabilizing effect, are important in
cationic carbon skeleton rearrangement reaction.
    Further, as described in the second part, a
preference between the effect (ii) and (iii) is
critically dependent on the nature of the cationic
species; the former is predominant with non-stabi-
lized ones such as primary carbocations, whereas
relatively stabilized cations undergo rearrangement
controlled by the latter effect. Thus, an appropriate choice of leaving and cation-stabilizing
substituents may further lead to the development of
selective carbon skeleton rearrangement reactions.

REFERENCES

1) E. W. Colvin, "Silicon in Organic Synthesis", Butterworth, London (1981), chapter 3 and 9. W. P. Weber, "Silicon Reagents for Organic Synthesis", Springer-Verlag, Berlin (1983), chapter 11. H. Sakurai, Pure Appld. Chem., $\underline{54}$, 1 (1982).

2) I. Fleming and I. P. Michael, J. Chem. Soc., Chem. Commun., 245 (1978); I. Fleming, and S. K. Petel, Tetrahedron Lett., $\underline{22}$, 2321 (1981).

3) K. Tanino, Y. Hatanaka, and I. Kuwajima, Chem. Lett., 385 (1987).

4) V. D. Shiner, Jr., and M. W. Ensinger, J. Am. Chem. Soc., $\underline{108}$, 842 (1986). E. R. Davidson and V. J. Shiner, Jr., ibid., $\underline{108}$, 3135 (1986).

5) K. Tanino, T. Katoh, and I. Kuwajima, submitted for publication.

6) For conversion of this type of ketones to α-methylene ketones, see, I. Fleming and J. Goldhill, J. Chem. Soc., Chem. Commun., 176 (1978).

7) H. Hopff, C. D. Nenitzescu, D. A. Isacescu, and I. P. Cantuniari, Ber., $\underline{69B}$, 2244 (1936).

8) T. Katoh, K. Tanino, and I. Kuwajima, to be published.

Chapter 8

Silicon mediated transformations in organic synthesis

Philip Magnus*, Daniel P. Becker, Peter M. Cairns and John Moursounidis — Department of Chemistry, Indiana University, Bloomington, Indiana 47405, USA.

INTRODUCTION

Increasingly, organic chemists are finding new and useful roles for the electronic and stereochemical properties inherent in the carbon-silicon σ-bond. These stereoelectronic attributes can be used to great effect to control the regio- and stereochemical outcome of reactions that would otherwise lack specificity.[1] In this lecture some examples of the positive role that the trimethylsilyl ($-SiMe_3$) group can play in natural product synthesis are described, with particular reference to the so-called β-effect, and the steric bulk of the $-SiMe_3$ group.[2]

The retro-Diels-Alder extrusion of cyclopentadiene, as depicted in Scheme 1, is part of the strategy we have developed for the synthesis of *aspidosperma*-type indole alkaloids.[3] Heating 1 at 180-190°C for 120h resulted in a clean extrusion of cyclopentadiene to give 3 (67%); similarly, heating 4 at 190-200°C for 24h gave 6 (>95%). While these reactions proceed under relatively mild conditions, (at least, compared with most cyclopentadiene extrusion processes), the reaction times are inconveniently lengthy.[4] An intriguing solution to this problem, and one of general interest, is to test the following hypothesis. A trimethylsilyl group *trans*-coplanar to the C-C bond, which is being broken in the retro-Diels-Alder reaction, could lower the activation energy of the extrusion process by virtue

of charge stabilization β- to the -SiMe₃ group,
Scheme 2.

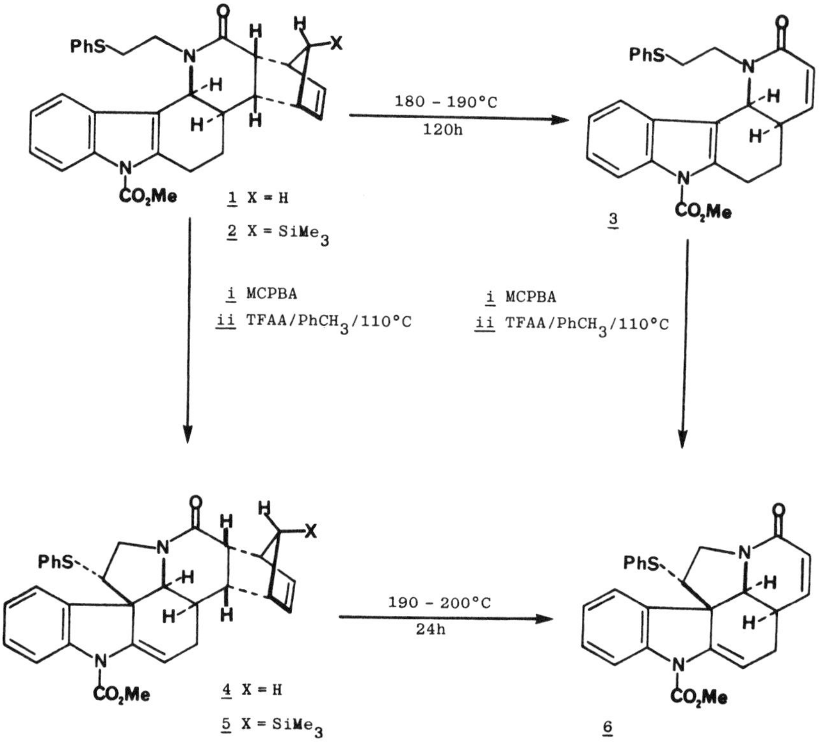

Scheme 1

5-(Trimethylsilyl)cyclopentadiene 7⁵ reacted with 5-
hydroxybutenolide 8 at 20°C to give an 8:1 mixture
of Diels-Alder adducts (SiMe₃ regioisomers), which
on recrystallization gave the pure isomer 9 (78%).
It should be noted, that in all probability 8 reacts
with 7 in the aldehydic form. The lactol 9 was converted into the acid 10 (94%), and thence to the
acid chloride 11, by standard methods.

When the imine 12 was treated with the acid
chloride 11 in toluene/i-Pr₂NEt/110°C, (identical
conditions to those used to give 1, in the 7-H
series), 2 (71%) and the α,β-unsaturated lactam 3
(3%) were isolated. Heating 2 in toluene 180-190°C/
7h gave 3 (ca.100%). The contrast with the 7-H

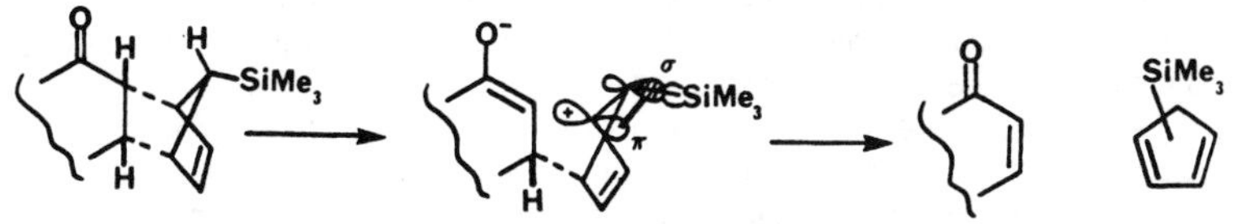

Scheme 2

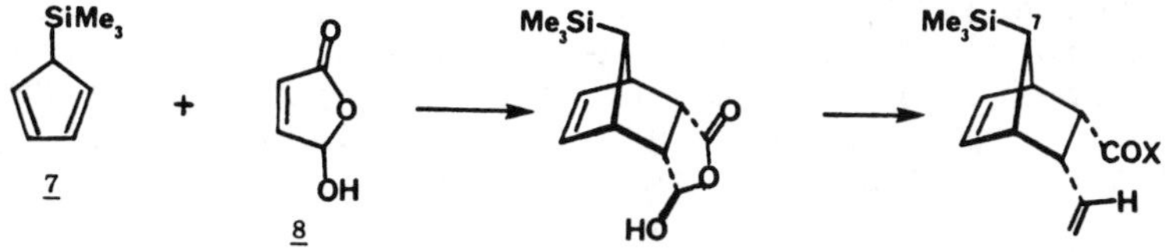

7   8   9   10 X = OH

11 X = Cl

series (180-190°C for 120h) is dramatic.   Conversion
of 2 into 5 using the Pummerer reaction (MCPBA/CH$_2$Cl$_2$
oxidation to the derived sulfoxide, followed by
TFAA/110°C/1.5h) gave 5 (15%), and the retro-Diels-
Alder product 6 (67%).   Clearly, the 7-trimethylsilyl
group has a substantial accelerating effect.   To put

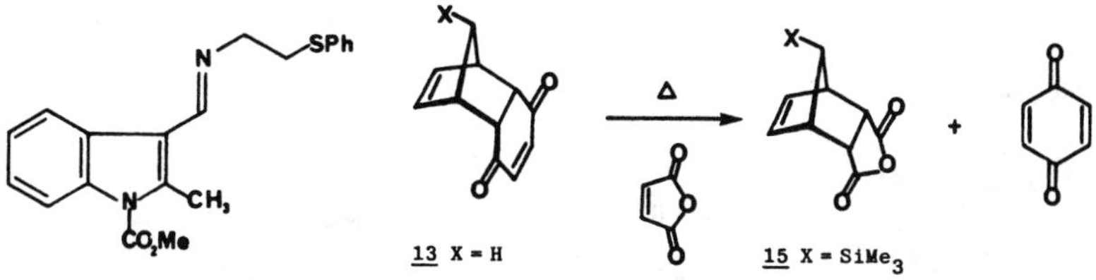

12   13 X = H   15 X = SiMe$_3$

14 X = SiMe$_3$

this effect on a more quantitative footing we have
examined the system 14.   This was chosen because
Wasserman[6] studied the retro-Diels-Alder reaction of
13, and determined the activation energy in benzene
as 29±1.5 kcal. mol.$^{-1}$ The substrate 14 is known,
and was made from benzoquinone and 5-(trimethylsilyl)-
cyclopentadiene 7.[5]   The adduct 14 was heated in
chlorobenzene at temperatures ranging from 58.5±0.2°C

to 78.0±0.2°C in the presence of maleic anhydride, and the increase in absorbance at 435nm (benzoquinone) was measured. The adduct 15 was isolated in these kinetic runs, demonstrating that 5-(trimethylsilyl)-cyclopentadiene is extruded in the retro [$4\pi + 2\pi$] process, and desilylation is not taking place. The adduct 14 undergoes the retro-Diels-Alder reaction approximately 95 times as fast as 13 under comparable reaction conditions. When the rate data are used in an Arrhenius plot, the following thermodynamic parameters were obtained: Ea = 24.1±1.2 kcal. mol^{-1}; $\Delta S^{\ddagger}$ = -5.8±0.5eu; $\Delta H^{\ddagger}$ = 24.1±1.2 kcal. mol$^-$; $\Delta G^{\ddagger}$ = 26.0±1.4 kcal. mol^{-1}. Thus, the 7-trimethylsilyl substituent has lowered the Arrhenius activation energy in the extrusion process by approximately 4.2 kcal. mol^{-1}, and $\Delta\Delta G^{\ddagger}$ is 3.3 kcal. mol^{-1} at 25°C.

Force field calculations that accurately reproduce the A value of -SiMe$_3$ indicate that there are no steric effects responsible for the observed acceleration.[7] Therefore, we attribute the lowering of activation energy to a polarized transition state, Scheme 2, where the -SiMe$_3$ group is able to stablize through hyperconjugation (p-σ)π the build-up of cationic character β to it. The ΔEa value of 4.2 kcal. mol^{-1} is a measure of the β-effect.

To provide further examples of the use of a 7-trimethylsilyl substituent to activate the retro-Diels-Alder process we have carried out the reactions shown below. The conditions needed to convert 16 into 17, and 19 into 20, roughly parallel those associated with the corresponding fulvene-type adducts.[8]

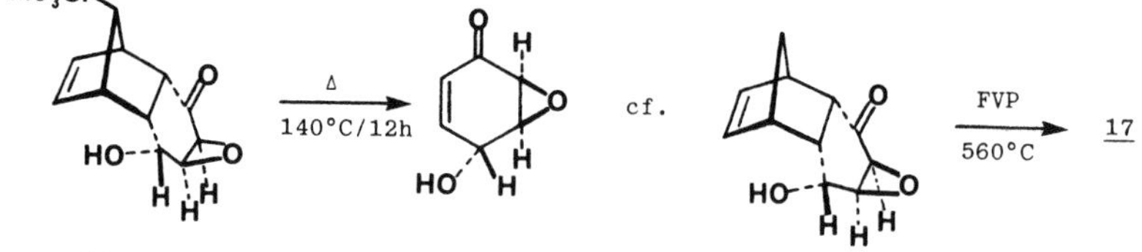

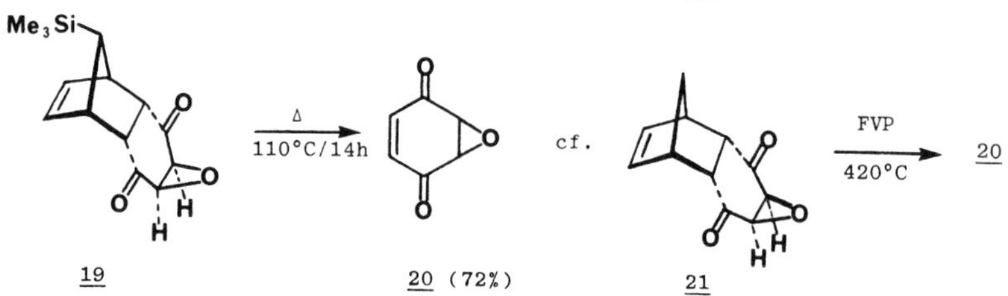

The Pauson-Khand reaction lends itself to an
exceptionally concise retrosynthetic representation
of the synthesis of 6a-carboprostaglandin, Scheme 3.[9]

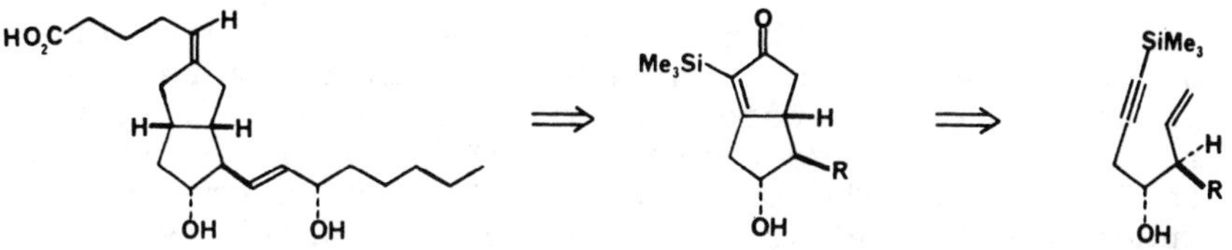

Scheme 3

The question one might ask is, "What is the function
of the -SiMe₃ group in the substrates 23 and 24?" It
serves *two* crucial and invaluable functions. The
first is to ensure that the conversion of 24 into 23,
mediated by dicobaltoctacarbonyl [Co₂(CO)₈], takes
place in a highly stereoselective manner; and the
second, to prevent the competitive trimerization of
24 to benzenoid products. We first observed the
influence of the terminal -SiMe₃ group when examining the Co₂(CO)₈ enyne-mediated cyclization on the
substrates 25 and 26 to give the bicyclo[3.3.0]-
octenones 27/28 and 29/30 respectively.[10]

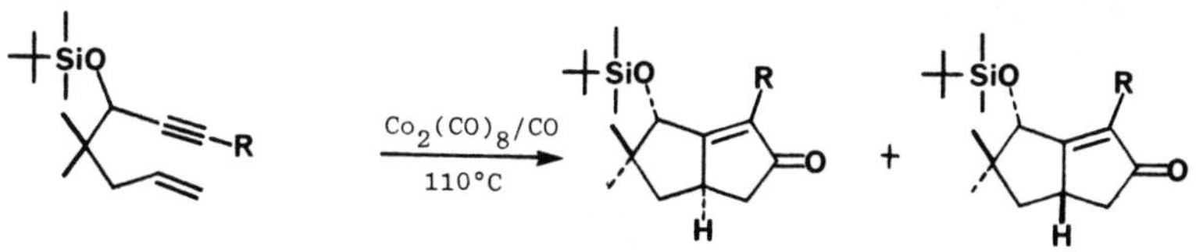

25 R = SiMe₃   27 R = SiMe₃   28 R = SiMe₃

26 R = Me   29 R = Me   30 R = Me

For R = SiMe₃ the ratio of 27 to 28 was 26:1 in favor
of the *cis*-bicyclo[3.3.0]octenone; whereas, for R =
Me the ratio of 29 to 30 was 3:1. Clearly, the size

of the group on the terminus of the acetylene has a
controlling influence on the 1,3-stereoselectivity.
The isolable complex 31 can form two cobalt metallo-
cycles, 32 and/or 33, upon alkene insertion into the
internal C-Co bond. The newly formed five-membered
ring Co-metallocycle is presumably *cis*-fused, since
the *trans*-fused arrangement is unacceptably strained.
The metallocycle 32 minimizes the steric interactions
between R^{1}O- and R-; whereas, 33 has a severe 1,3-
pseudo-diaxial interaction on the *endo*-face. Consequently, a large R-group (-SiMe$_3$) would be expected
to favor 32. The metallocycle 32 can undergo CO-
insertion to the acyl-Co complex 34, which is set up
to migrate the C-Co bond to the adjacent electrophilic carbonyl group to give 35. Reductive elimination of the cobalt carbonyl residue in 35 establishes the cyclopentenone double bond.

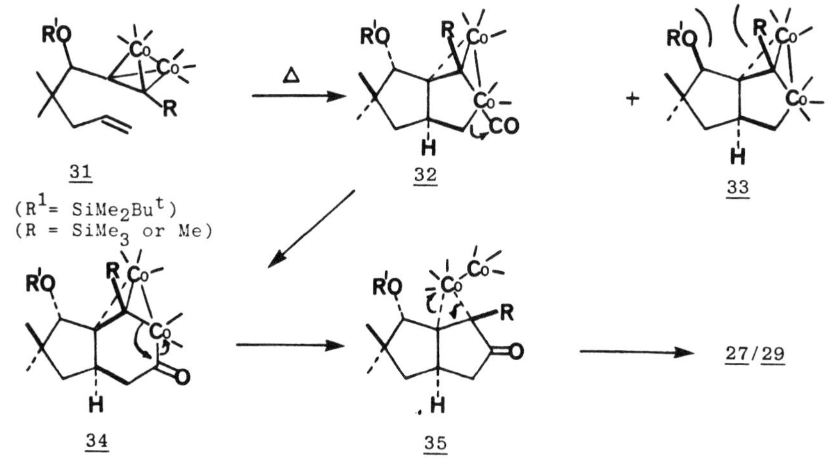

$(R^1 = SiMe_2Bu^t)$
$(R = SiMe_3 \ or \ Me)$

Scheme 4

As a generalization, this hypothesis predicts that
both allylic and propargylic substituents in the
resulting [3.3.0]bicyclooctenone appear on the *exo*-
face, which usually corresponds to the more stable
thermodynamic situation. If we apply this analysis
to the carbocycline precursor 24, as shown in Scheme
5, then the prediction is that the desired stereo-
isomer 23 will be the major product.

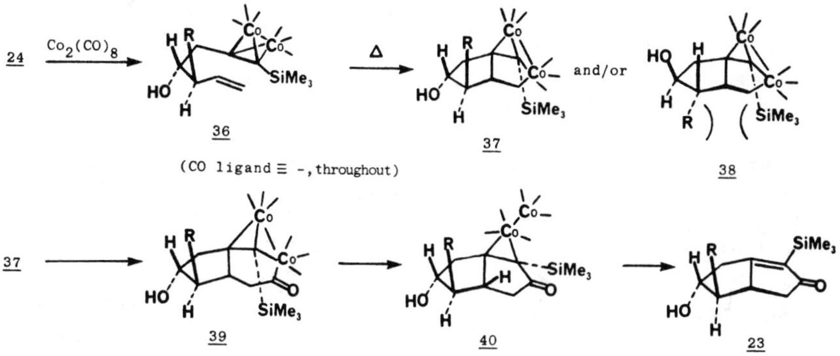

Scheme 5

D-(+)-Ribonolactone 41 was converted into the ortho-ester 42, which was pyrolyzed at 200°C/40mm Hg to give the butenolide 43 (48%).[11] Treatment of 43 with trisylbromide/pyridine gave 44 (77%), which underwent the required 1,4-conjugate addition with $Li_2(\!\!=\!\!\!\searrow)_2 CuCN$ to give 45 (64%). When 45 was exposed to $MeOH/K_2CO_3$ the epoxide 46 (94%) was isolated. Addition of the epoxy methyl ester 46 to a solution of lithium trimethylsilylacetylide and $BF_3.OEt_2$ gave the lactone 47 (73%). The lactone 47 was reduced with $LiAlH_4$ to give the diol 48 (93%), which on treat-

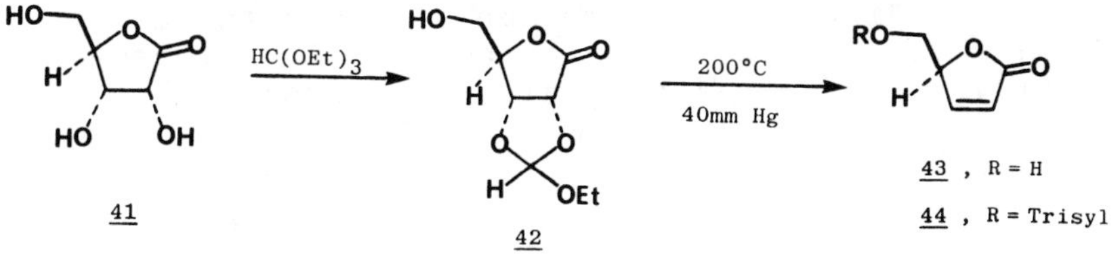

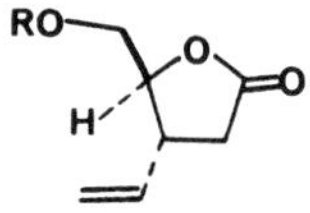

45 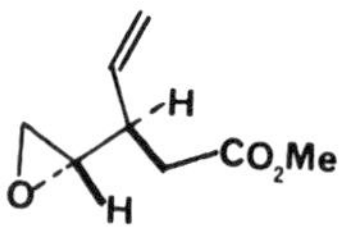   46 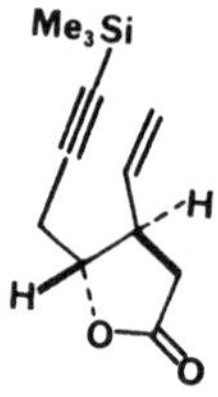   47

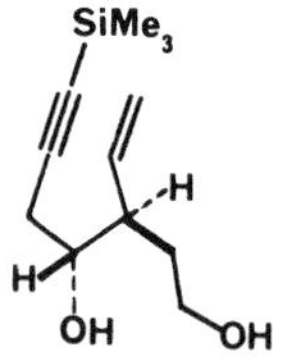

48

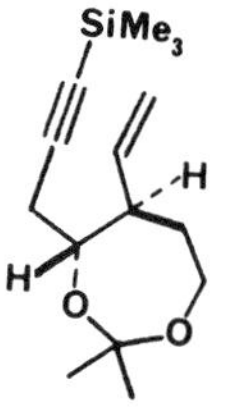

49 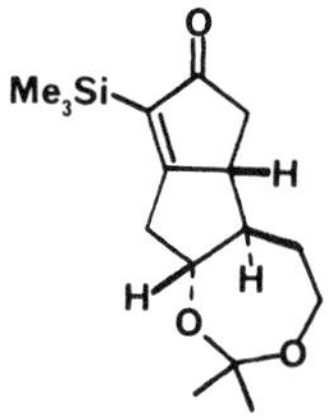   50

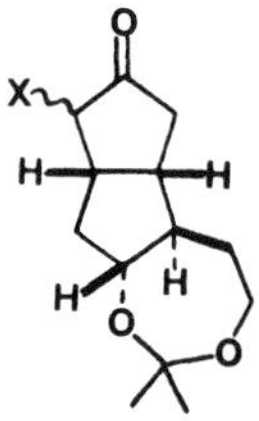

51 , X = SiMe₃

52 , X = H

ment with acetone/H⁺ gave the acetonide 49 (92%).
When 49 was treated with Co₂(CO)₈/P(O)Bu₃/85°C/3 days
the enone 50 (45%) was formed. We could not detect
any other stereoisomers. The structure and stereochemistry of 50 was confirmed by hydrogenation over
5% Pd/C₊to give 51 (94%), which was desilylated
using NBu₄F⁻/THF to give 52. The structure of 52
was established by a single crystal X-ray crystallographic determination of the derived bis-(4-bromo-
benzylidene) derivative 53, thus confirming the
absolute stereochemistry, as shown. This provides
unambiguous confirmation that the crucial Co₂(CO)₈
mediated cyclization proceeded with the stereoselectivity predicted by the mechanistic hypothesis,
which uses a –SiMe₃ group to control the stereochemistry at the newly created ring fusion.

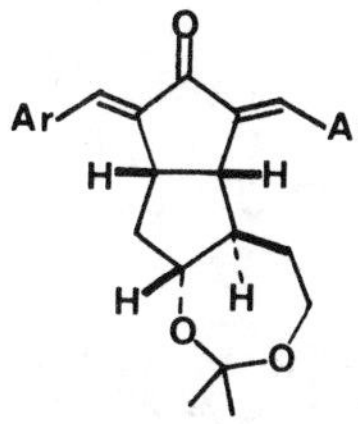

53, Ar = C_6H_4Br-p

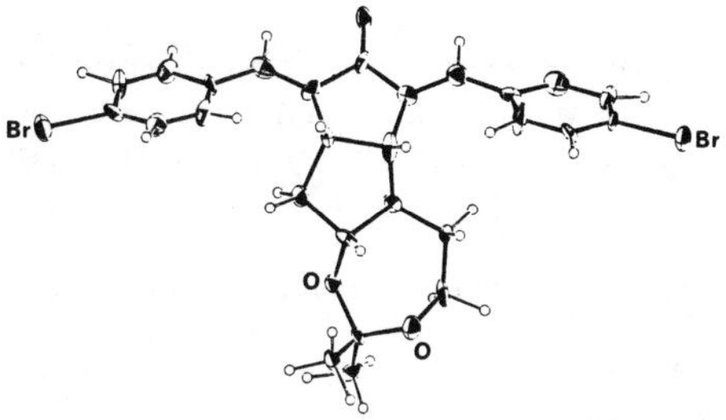

ORTEP of 53

ACKNOWLEDGEMENTS   The National Science Foundation, and The National Institutes of Health are thanked for their financial support of this work.

REFERENCES

1. P. Magnus, T. Sarkar and S. Djuric,"Comprehensive Organometallic Chemistry", Vol. 5, Chapter 40. Pergamon Press, 1982.

2. T.G. Traylor, W. Hanstein, H.J. Berwin, N.A. Clinton and R.S. Brown, *J. Am. Chem. Soc.*, 1971, *93*, 5715; J.B. Lambert and R.B. Finzel, *Ibid.*, 1982, *104*, 2020. See also reference 3.

3. P. Magnus, P.M. Cairns and J. Moursounidis, *J. Am. Chem. Soc.*, 1987, *109*, 2469. For other quantitative determinations of the β-effects see:- R.T. Conlin and Y-W. Kwak, *Organometallics*, 1986, *5*, 1205.

4. P. Magnus and P.M. Cairns, *J. Am. Chem. Soc.*, 1986, *108*, 217. For a general review of the indole-2,3-quinodimethane strategy: P. Magnus, P. Brown, P. Pappalardo, *Acc. Chem. Res.*, 1984, *17*, 35.

5. G.A. Tolstikov, B.M. Lerman, F.Z. Galin, N.A. Danilova, *Zh. Obsh. Khim.*, 1977, *47*, 1656.

6. B.S. Khambata and A.J. Wasserman, *J. Chem. Soc.*, 1939, 375.

7. W. Kitching, H.A. Olszowy, G.M. Drew and W. Adcock

J. Org. Chem., 1982, *47*, 5153.

8. For reviews of the Retro-Diels-Alder Reaction:
   H. Kwart, K. King, *Chem. Rev.*, 1968, *68*, 415;
   A. Ichihara, *Synthesis*, 1987, 207; M-C. Lasne,
   J-L. Ripoll, *Ibid.*, 1985, 121.

9. I.U. Khand, G.R. Knox, P.L. Pauson and W.E. Watts
   J. Chem. Soc., Perkin Trans. 1, 1973, 975; P.L.
   Pauson, *Tetrahedron*, 1985, *41*, 5855, and refer-
   ences cited therein.

10. P. Magnus and C. Exon, *J. Am. Chem. Soc.*, 1983,
    105, 2477; P. Magnus, C. Exon and P. Albaugh-
    Robertson, *Tetrahedron*, 1985, *41*, 5861;
    P. Magnus and L.M. Principe, *Tetrahedron Letters*,
    1985, *26*, 4851; P. Magnus, M. Slater and L.M.
    Principe, *J. Org. Chem.*, 1987, *52*, 1483.

11. P. Camps, J. Cardellach, J. Font, R.M. Ortuno
    and O. Ponsati, *Tetrahedron*, 1982, *38*, 2395.

PART II

ORGANIC CHEMISTRY OF SILICON

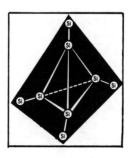

Chapter 9

Elimination reactions of β-functional organosilicon compounds

Paul F. Hudrlik* and Edwin L.O. Agwaramgbo – Department of Chemistry, Howard University, Washington, D.C. 20059.

Olefin-forming β-elimination reactions of β-halosilanes and β-hydroxysilanes were first reported more than 40 years ago [1]. Although these elimination reactions were the subject of several mechanistic studies over the next two decades [2], most organic chemists were not aware of their potential synthetic utility until 1968, when Donald Peterson showed the elimination reactions of β-hydroxysilanes could be used in a useful alternative to the Wittig reaction [3]. He found that a variety of silicon-stabilized carbanions react with aldehydes and ketones to yield olefinic compounds as shown in Scheme 1. Although mixtures of cis and trans olefins are generally obtained, this type of reaction, which has become known as the Peterson olefination reaction, has become a useful and widely-used method for the olefination of carbonyl compounds [4].

Scheme 1

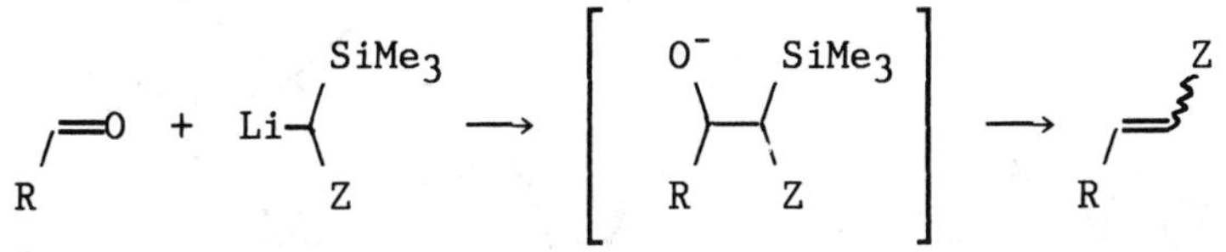

The Peterson olefination reaction has generally been assumed [4,5] to take place via a β-oxidosilane as shown above (although recent work in our laboratory, discussed below, has led us to question this assumption). When carbanion-stabilizing groups (-Z) are present on the carbon bearing silicon, β-elimination apparently occurs <u>in situ</u>, and it is often not possible to isolate the β-hydroxysilane. However, when -Z = -H or alkyl (or methoxy), the lithium alkoxide is stable, and the β-hydroxysilane can be isolated, and converted to olefin by treatment with either acid (e.g. H_2SO_4 or $BF_3 \cdot Et_2O$) or base (e.g. KH).

Some time ago (David) Peterson and I began to investigate alternative approaches for the preparation of β-hydroxysilanes, in order to study the stereochemistry of their β-elimination reactions, and with the hope that methods for the stereospecific synthesis of olefins could be found [6]. We developed several new methods for the synthesis of diastereomerically-pure β-hydroxysilanes, and found that the acid- and base-induced elimination reactions were highly stereospecific [6b,c], and that the acid-catalyzed eliminations were anti, while the base-induced reactions were syn [6d]. Thus, either cis or trans olefin can be prepared from a single β-hydroxy-silane.

Among the most useful methods for the preparation of β-hydroxysilanes in high diastereomeric purity are the hydride

Scheme 2

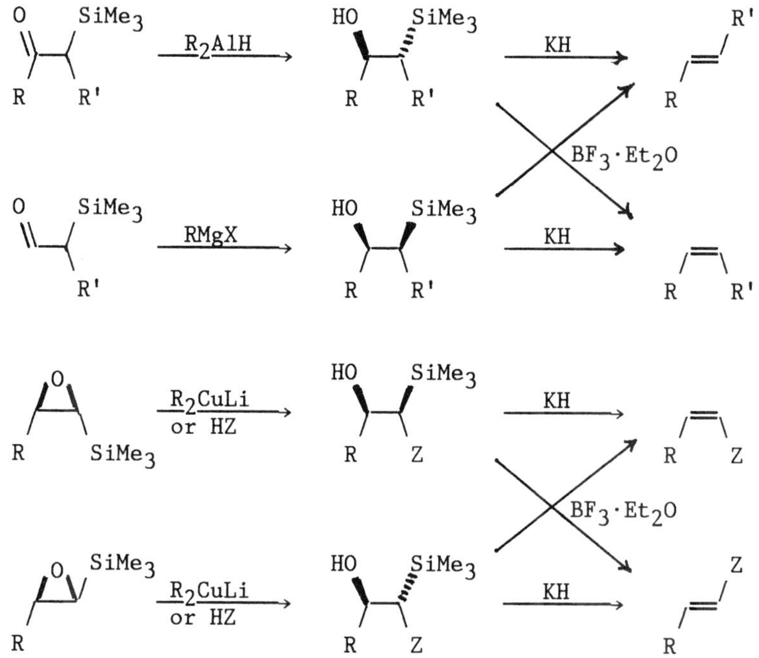

with KH: -Z = -R, -NR$_2$, -NHAc, -OMe, -SPh

with $BF_3 \cdot Et_2O$:   -Z = -R, -OAc, -OSO$_2$Me, -SPh, -F, -Br

reductions of α-silyl ketones (β-ketosilanes) [6b,c], the reactions of α-silyl ketones [6a,7] and aldehydes [8] with organometallic reagents, and the ring-opening reactions of α,β-epoxysilanes [6d,9] (see Scheme 2). The acid-catalyzed reactions of α,β-epoxysilanes with a variety of heteroatom nucleophiles [9] are particularly valuable for the stereo-specific synthesis of α-substituted β-hydroxysilanes, and enable α,β-epoxysilanes to serve as "stereospecific vinyl cation equivalents" for the synthesis of a variety of hetero-atom-substituted olefins. β-Hydroxysilanes can also be prepared from β-silyl ketones via Baeyer-Villiger oxidation [10], and from the reactions of epoxides with silylmetallic reagents [11].

In addition to the reactions discussed above, several other types of β-elimination reactions of β-functional organo-silicon compounds exist. From a mechanistic point of view, β-elimination reactions of organosilicon compounds can be divided into five groups.

1. E1-Type Elimination Reactions. There is considerable evidence to indicate that cations β to silicon are considerably stabilized by the silicon, particularly when the cationic p orbital can be parallel to the adjacent C-Si bond [2,12]. Many solvolytic and acid-catalyzed elimination reactions of β-functional organosilicon compounds are facilitated by silicon, and such cations may be involved as intermediates. Solvolytic elimination reactions of β-bromosilanes [13], and acid cata-lyzed elimination reactions of β-hydroxysilanes [6d] and β-alkoxysilanes [14] have been shown to take place with **anti** stereochemistry, as have related elimination reactions invol-ving other elements, e.g. Sn [15], Hg [16], and B [17].

2. E2-Type Elimination Reactions. Elimination reactions of β-functional organosilicon compounds carried out under strongly basic or nucleophilic conditions may be occurring by an E2 mechanism (or possibly by a mixture of E1 and E2 mechan-isms). The fluoride-induced elimination reactions of β-halosilanes, introduced by Cunico [18] and extensively devel-oped by Chan [4b], have been shown to be particularly useful for the preparation or generation of highly strained multiple bonds such as benzyne, allenes, allene oxides, and cyclo-propenes. In unstrained systems, there is ample evidence to indicate that these reactions take place with **anti** stereo-chemistry [10b,19,20].

Elimination reactions of α-oxidosilanes having a leaving group β to silicon are mechanistically very similar to the E2-type eliminations discussed above. Such reactions were first postulated in 1963 [21], but were not used synthetically until recently [22]. Recent work in our laboratories demonstrated that these reactions also take place with **anti** stereochemistry [23].

3. Thermal Four-Center Eliminations. The gas-phase thermal elimination reactions of β-fluorosilanes and other β-halosilanes have been the topic of extensive kinetic and mechanistic research, from which it has been concluded that these reactions involve a unimolecular four-center transition state [2,24]. The thermal elimination of a β-methoxysilane has been shown to take place with **syn** stereochemistry [14]. Of more synthetic interest are the possibly related thermal elimination reactions of amine oxides [20] and sulfoxides [25], which have also been shown to be **syn**.

4. Base-induced Eliminations of β-Hydroxysilanes. Simple β-hydroxysilanes undergo **syn** β-elimination when treated with base in aprotic solvents (e.g. KH in THF or NaH in DMF). The corresponding lithium alkoxides do not undergo elimination unless electron withdrawing groups are present on silicon (e.g. Ph_3Si) [11b,26] or on the carbon attached to silicon. With mild electron withdrawing groups on carbon (-Z = -NHAc, -OR, -SPh [9a,d], -SePh [27], alkenyl, and alkynyl), the elimination reaction is still stereospecifically **syn** [28]. β-Hydroxy-silanes with more powerful carbanion stabilizing groups on carbon have not been well studied; they are difficult to prepare, and their elimination reactions are complicated by protiodesilylation or 1,3-silicon migration [29]. [Under suitable experimental conditions simple β-hydroxysilanes can also undergo base-induced protiodesilylation [30].]

[Since alkoxides (RO^-) are isoelectronic with alkyl fluorides (RF), syn elimination reactions of β-oxidosilanes may be mechanistically related to the thermal four-center elimi-nation reactions of β-halosilanes and related compounds dis-cussed in the previous section.]

5. Peterson Olefination Reaction. We use this term in a restricted sense, to refer to the overall olefination process depicted in Scheme 1, partly because this process is not as easily described as the other reactions described above, and partly because we believe that the mechanism of this reaction, at least under some conditions, may be **different** from those of the four types of elimination reactions summarized above.

This reaction has its origins in 1962, when Gilman and Tomasi reported the reaction of benzophenone with a silyl-substituted Wittig reagent to give tetraphenylallene [31]. In the following year, Krüger, Rochow, and Wannagat reported the reaction of sodium bis(trimethylsilyl)amide with several nonenolizable carbonyl compounds to give N-(trimethylsilyl)-imines [32a]. It was not until Peterson's classic paper of 1968 [3], however, that the generality and synthetic potential of the reaction were recognized [4,33].

Our recent interest in the mechanism of the Peterson olefination was stimulated by an observation made by Withers in our research group in 1976 [39]. As part of a study of the chemistry of bis(trimethylsilyl)epoxides, he found that 1,1-bis(trimethylsilyl)ethylene oxide reacted with Grignard reagents (PrMgBr and HexMgBr) in THF to give **trans** vinyl-silanes. The following reaction pathway was postulated.

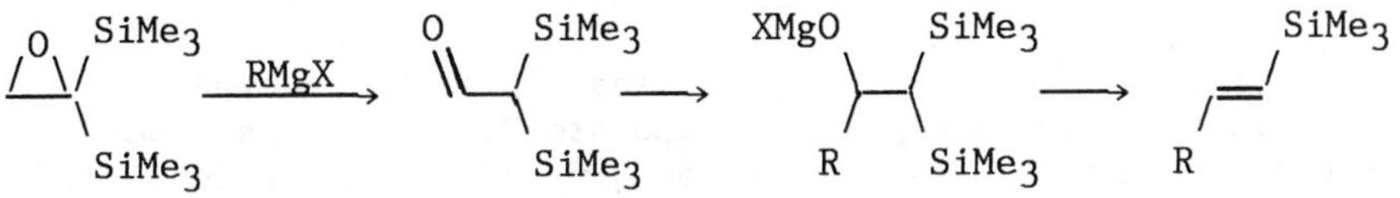

Reasoning that the β-elimination step would be slower in a less-coordinating solvent, we carried out the reactions (with HexMgBr and PhMgBr) in Et_2O, and obtained the β-hydroxysilanes. When these compounds were treated with a variety of bases, (e.g. KH, NaH) β-elimination occurred to give trans vinyl-silanes (reaction 1).

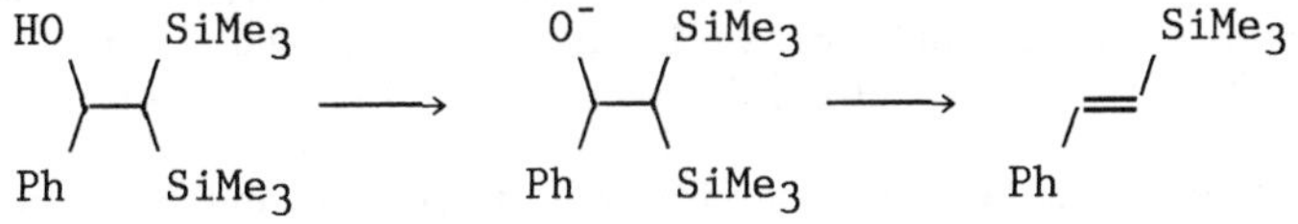

$$\tag{1}$$

We rationalized the formation of the trans products from these elimination reactions by considering possible conform-ations of the presumed intermediate β-oxidosilanes as shown below. Both conformations **4** and **5** have the favorable syn alignment of alkoxide and silicon for β-elimination; conform-ation **5**, which would lead to trans vinylsilane product is clearly preferable to **4** on steric grounds [40].

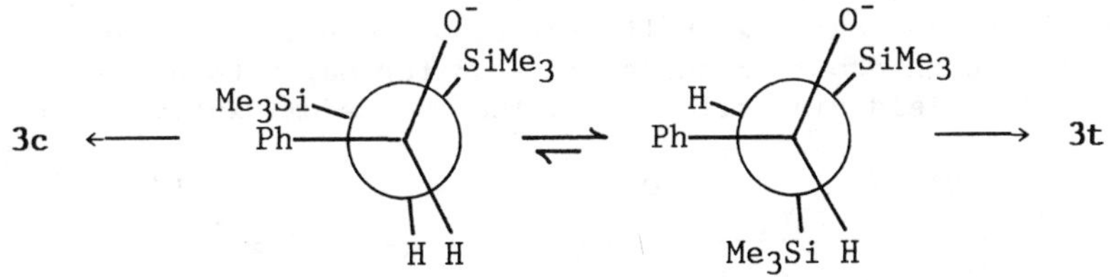

In 1974, Gröbel and Seebach reported the following exten-sion of the Peterson olefination [41]: the reaction of bis-(trimethylsilyl)methyllithium with benzaldehyde (and other aromatic aldehydes) to give a <u>mixture</u> of cis and trans vinyl-silanes **3c** and **3t** in a 1:1.4 ratio (reaction 2).

$$\underset{Ph}{/}\!\!=\!\!O \; + \; \text{Li}\!\!-\!\!\!\!\underset{SiMe_3}{\overset{SiMe_3}{\Big\langle}} \longrightarrow \underset{Ph \;\; SiMe_3}{/}\!\!=\!\!\backslash \; + \; \underset{Ph}{/}\!\!=\!\!\overset{SiMe_3}{/} \qquad (2)$$

$$\qquad\qquad\qquad\qquad\qquad\qquad \textbf{3c} \qquad\quad \textbf{3t}$$

Considering that both processes might be expected to involve the β-oxidosilane intermediate **2**, the contrast between the β-elimination reactions (reaction 1), giving predominantly **trans** product, and the Peterson olefination reaction (reaction 2), giving a cis-trans **mixture**, is striking. In order to duplicate the base-solvent system of the latter reaction, β-hydroxysilane **1** was treated with t-BuLi in THF/HMPA and with lithium diisopropylamide in THF/HMPA. In both cases, vinylsilane was formed as predominantly (about 95%) trans isomer **3t**. [Under some conditions, e.g. use of excess t-butyllithium, other products were also formed in significant amounts. We found that both cis and trans vinylsilanes **3c** and **3t** are consumed by t-butyllithium.]

We have repeated the experiment of Gröbel and Seebach several times, and confirmed that a mixture of cis and trans products is formed. To test for the possibility of cis-trans isomerization of the vinylsilanes, we carried out the Peterson olefination reaction of toluadehyde in the presence of trans vinylsilane **3t**; a mixture of cis and trans vinylsilanes derived from tolualdehyde was formed, while trans vinylsilane **3t** was unchanged.

We have recently investigated two additional approaches for the generation of the β-oxidosilane intermediate **2**: the reaction of bis(trimethylsilyl)acetaldehyde with phenyllithium (reaction 3) and the reaction of 1,1-bis(trimethylsilyl)-2-phenylethylene oxide with lithium aluminum hydride (reaction 4). Although other products are also formed, both of these reactions yield vinylsilane **3t** which is predominantly trans.

$$\underset{SiMe_3}{\overset{O\quad SiMe_3}{\big\backslash\!\!\big\langle}} \;\xrightarrow{\text{PhLi}}\; \underset{Ph\quad SiMe_3}{\overset{O^-\quad SiMe_3}{\big\rangle\!\!\big\langle}} \;\longrightarrow\; \underset{Ph}{/}\!\!=\!\!\overset{SiMe_3}{/} \qquad (3)$$

$$\qquad\qquad\qquad\qquad \textbf{2} \qquad\qquad\qquad\qquad \textbf{3t}$$

$$\underset{Ph\quad SiMe_3}{\overset{O\quad SiMe_3}{/\!\!\overset{\triangle}{\big\backslash}\!\!\big\langle}} \;\xrightarrow[\text{THF}]{\text{LiAlH}_4}\; \underset{Ph\quad SiMe_3}{\overset{O^-\quad SiMe_3}{\big\rangle\!\!\big\langle}} \;\longrightarrow\; \underset{Ph}{/}\!\!=\!\!\overset{SiMe_3}{/} \qquad (4)$$

$$\qquad\qquad\qquad\qquad \textbf{2} \qquad\qquad\qquad\qquad \textbf{3t}$$

It thus appears, from reactions 1, 3, and 4, that β-oxido-silane **2** undergoes elimination to give predominantly <u>trans</u> vinylsilane **3t**. Since the corresponding Peterson olefination (reaction 2) gives a <u>mixture</u> of cis and trans isomers, it must proceed, at least in part, by a pathway which does <u>not</u> involve the β-oxidosilane **2**.

Acknowledgments: We thank the National Science Foundation for financial support, and Dr. A. M. Hudrlik for assistance with the manuscript.

REFERENCES

1. L. H. Sommer, G. M. Goldberg, E. Dorfman, and F. C. Whit-more, J. Am. Chem. Soc., **68**, 1083-1085 (1946); F. C. Whit-more, L. H. Sommer, J. Gold, and R. E. Van Strien, J. Am. Chem. Soc., **69**, 1551 (1947); L. H. Sommer, D. L. Bailey, and F. C. Whitmore, J. Am. Chem. Soc., **70**, 2869-72 (1948).

2. A. W. P. Jarvie, Organometallic Chem. Rev. A, **6**, 153-207 (1970), and references cited therein.

3. D. J. Peterson, J. Org. Chem., **33**, 780-784 (1968).

4. For reviews, see (a) P. F. Hudrlik, "Organosilicon Com-pounds in Organic Synthesis", in "New Applications of Organometallic Reagents in Organic Synthesis", D. Sey-ferth, Ed., Elsevier, Amsterdam, 1976, pp 127-159; (b) T.-H. Chan, Accounts Chem. Res., **10**, 442-448 (1977); (c) P. D. Magnus, Aldrichimica Acta, **13**, 43-51 (1980); (d) E. W. Colvin, "Silicon in Organic Synthesis", Butterworths, London, 1981, pp 141-164; (e) P. D. Magnus, T. Sarkar, and S. Djuric, "Organosilicon Compounds in Organic Synthesis", in "Comprehensive Organometallic Chemistry", Volume 7, G. Wilkinson, F. G. A. Stone, and E. W. Abel, Eds, Pergamon Press, Oxford, 1982, pp 515-659; (f) W. P. Weber, "Silicon Reagents for Organic Synthesis", Springer-Verlag, Berlin, 1983, pp 58-78; (g) D. J. Ager, Synthesis, 384-398 (1984).

5. C. Trindle, J.-T. Hwang, and F. A. Carey, J. Org. Chem., **38**, 2664-2669 (1973); A. R. Bassindale, R. J. Ellis, and P. G. Taylor, Tetrahedron Lett., **25**, 2705-2708 (1984); A. R. Bassindale, R. J. Ellis, J. C.-Y. Lau, and P. G. Taylor, J. Chem. Soc., Perkin II, 593-597 (1986).

6. (a) P. F. Hudrlik and D. Peterson, Tetrahedron Lett., 1785-1787 (1972); (b) P. F. Hudrlik and D. Peterson, Tetrahedron Lett., 1133-1136 (1974); (c) P. F. Hudrlik and D. Peterson, J. Am. Chem. Soc., **97**, 1464-1468 (1975); (d) P. F. Hudrlik, D. Peterson, and R. J. Rona, J. Org. Chem., **40**, 2263-2264 (1975).

7. R. A. Ruden and B. L. Gaffney, Synth. Commun., **5**, 15-19
    (1975); K. Utimoto, M. Obayashi, and H. Nozaki, J. Org.
    Chem., **41**, 2940-2941 (1976).

8. P. F. Hudrlik, A. M. Hudrlik, R. N. Misra, D. Peterson, G.
    P. Withers, and A. K. Kulkarni, J. Org. Chem., **45**, 4444-
    4448 (1980); P. F. Hudrlik and A. K. Kulkarni, J. Am.
    Chem. Soc., **103**, 6251-6253 (1981).

9. (a) P. F. Hudrlik, A. M. Hudrlik, R. J. Rona, R. N. Misra,
    and G. P. Withers, J. Am. Chem. Soc., **99**, 1993-1996
    (1977); (b) A. P. Davis, G. J. Hughes, P. R. Lowndes, C.
    M. Robbins, E. J. Thomas, and G. H. Whitham, J. Chem.
    Soc., Perkin I, 1934-1941 (1981); (c) P. F. Hudrlik, A. M.
    Hudrlik, and A. K. Kulkarni, Tetrahedron Lett., **26**, 139-
    142 (1985); (d) P. F. Hudrlik, A. M. Hudrlik, A. K.
    Kulkarni, R. J. Rona, and S. Jain, unpublished work.

10. (a) P. F. Hudrlik, A. M. Hudrlik, G. Nagendrappa, T.
    Yimenu, E. T. Zellers, and E. Chin, J. Am. Chem. Soc.,
    102, 6894-6896 (1980); (b) W. Bernhard and I. Fleming, J.
    Organometal. Chem., **271**, 281-288 (1984).

11. For the use of these reactions for olefin inversion, see
    (a) P. B. Dervan and M. A. Shippey, J. Am. Chem. Soc., **98**,
    1265-1267 (1976); (b) M. T. Reetz and M. Plachky, Syn-
    thesis, 199-200 (1976).

12. C. Eaborn, F. Feichtmayr, M. Horn, and J. N. Murrell, J.
    Organometal. Chem., **77**, 39-43 (1974); S. G. Wierschke, J.
    Chandrasekhar, and W. L. Jorgenson, J. Am. Chem. Soc.,
    107, 1496-1500 (1985), and references cited therein.

13. A. W. P. Jarvie, A. Holt, and J. Thompson, J. Chem. Soc.
    B, 852-855 (1969); A. G. Brook, J. M. Duff, and W. F.
    Reynolds, J. Organometal. Chem., **121**, 293-306 (1976).

14. W. K. Musker and G. L. Larson, J. Am. Chem. Soc., **91**, 514
    (1969).

15. D. D. Davis and C. E. Gray, J. Org. Chem., **35**, 1303-1307
    (1970).

16. W. Kitching, Organometallic Reactions, **3**, 319-398 (1972).

17. β-Substituted organoboron compounds undergo three kinds of
    elimination reactions (acid-induced (anti), base-induced
    (anti), and thermal (syn)), which are superficially anal-
    ogous to the first three types of elimination reactions of
    organosilicon compounds discussed here: D. S. Matteson
    and J. D. Liedtke, J. Am. Chem. Soc., **87**, 1526-1531
    (1965); D. J. Pasto and R. Snyder, J. Org. Chem., **31**,
    2777-2784 (1966); D. J. Pasto and P. E. Timony,

J. Organometal. Chem., **60**, 19-29 (1973); G. L. Larson and A. Hernandez, J. Organometal. Chem., **102**, 123-127 (1975).

18. R. F. Cunico and E. M. Dexheimer, J. Am. Chem. Soc., **94**, 2868-2869 (1972).

19. R. B. Miller and G. McGarvey, J. Org. Chem., **43**, 4424-4431 (1978); F.-T. Luo and E. Negishi, J. Org. Chem., **48**, 5144-5146 (1983); I. Fleming and N. K. Terrett, J. Organometal. Chem., **264**, 99-118 (1984).

20. Thermal eliminations of β-silyl amine oxides (presumably syn) have been shown to take place with opposite stereochemistry to fluoride-induced (presumably anti) eliminations of the corresponding quaternary ammonium salts: N. V. Bac and Y. Langlois, J. Am. Chem. Soc., **104**, 7666-7667 (1982).

21. A. G. Brook, W. W. Limburg, D. M. MacRae, and S. A. Fieldhouse, J. Am. Chem. Soc., **89**, 704-706 (1967).

22. H. J. Reich, J. J. Rusek, and R. E. Olson, J. Am. Chem. Soc., **101**, 2225-2227 (1979).

23. P. F. Hudrlik, A. M. Hudrlik, and A. K. Kulkarni, J. Am. Chem. Soc., **107**, 4260-4264 (1985).

24. D. Graham, R. N. Haszeldine, and P. J. Robinson, J. Chem. Soc. B, 652-654 (1969), and references cited therein.

25. I. Fleming, J. Goldhill, and D. A. Perry, J. Chem. Soc., Perkin I, 1563-1569 (1982).

26. T. H. Chan, E. Chang, and E. Vinokur, Tetrahedron Lett., 1137-1140 (1970); T. H. Chan and E. Chang, J. Org. Chem., **39**, 3264-3268 (1974).

27. W. Dumont, D. Van Ende, and A. Krief, Tetrahedron Lett., 485-488 (1979).

28. If -Z is a good leaving group, epoxide formation may occur. For example, see [9a], and C. Burford, F. Cooke, E. Ehlinger, and P. Magnus, J. Am. Chem. Soc., **99**, 4536-4537 (1977).

29. K. Yamamoto, Y. Tomo, and S. Suzuki, Tetrahedron Lett., **21**, 2861-2864 (1980); M. Larchevêque and A. Debal, Chem. Commun., 877-878 (1981); K. Yamamoto and Y. Tomo, Chem. Lett., 531-534 (1983).

30. P. F. Hudrlik, A. M. Hudrlik, and A. K. Kulkarni, J. Am. Chem. Soc., **104**, 6809-6811 (1982).

31. H. Gilman and R. A. Tomasi, J. Org. Chem., 27, 3647-3650 (1962).

32. (a) C. Krüger, E. G. Rochow, and U. Wannagat, Chem. Ber., 96, 2132-2137 (1963); (b) G. Tuchtenhagen and K. Rühlmann, J. Liebig's Ann. Chem., 711, 174-183 (1968); (c) T. Morimoto, Y. Nezu, K. Achiwa, and M. Sekiya, Chem. Commun., 1584-1585 (1985).

33. Variations of the Peterson olefination which lead to C=N [32], C=S [34], and N=S [35] bonds are known. Analogous olefination reactions involving other elements also exist; for example, P (the Wittig reaction), B [36], Sn [37,38], Pb [38], Sb [38], and Te [38b].

34. M. van der Leij, P. A. T. W. Porskamp, B. H. M. Lammerink, and B. Zwanenburg, Tetrahedron Lett., 811-814 (1978); M. van der Leij and B. Zwanenburg, Tetrahedron Lett., 3383-3386 (1978).

35. P. A. T. W. Porskamp and B. Zwanenburg, Synthesis, 368-369 (1981).

36. A. Pelter, B. Singaram, and J. W. Wilson, Tetrahedron Lett., 24, 635-636 (1983).

37. T. Kauffmann, R. Kriegesmann, and A. Woltermann, Angew. Chem., Int. Ed. Engl., 16, 862-863 (1977); D. Seebach, I. Willert, A. K. Beck, and B.-T. Gröbel, Helv. Chim. Acta, 61, 2510-2523 (1978).

38. (a) T. Kauffmann, H. Ahlers, R. Joußen, R. Kriegesmann, A. Vahrenhorst, and A. Woltermann, Tetrahedron Lett., 4399-4402 (1978); (b) H.-J. Tilhard, H. Ahlers, and T. Kauffmann, Tetrahedron Lett., 21, 2803-2806 (1980).

39. G. P. Withers, Ph.D. Thesis, Rutgers University, New Brunswick, N. J., 1978.

40. For a similar argument, see H. Sakurai, K.-i. Nishiwaki, and M. Kira, Tetrahedron Lett., 4193-4196 (1973).

41. B.-T. Gröbel and D. Seebach, Angew. Chem., Int. Ed. Engl., 13, 83-84 (1974); Chem. Ber., 110, 852-866 (1977).

Chapter 10

Syntheses and properties
of phenylethynylsilanes
and polyphenylphenylsilanes

Chen Jianhua, Li Xinhua, Feng Shengyu, Chang Xiaoqing, Yin Shang and Du Zuodong – Department of Chemistry, Shandong University. People's Republic of China.

ABSTRACT

A series of new phenylethynylsilanes and polyphenylphenyl-silanes has been synthesized. Their compositions and structures are determined by ^{1}H NMR IR and elemental analysis. The factors which effect the Diels-Alder reaction are studied in detail. The properties of polyphenylphenylorganosilanes have been investigated.

INTRODUCTION

In recent years, some research work on the syntheses of poly-phenylphenylsilanes and bis(polyphenylphenyl)silanes has been done by our research group. These kinds of compounds have good heat-resistant qualities. They can be used as the stationary phase of high temperature gas chromatograph and additives of heat-vulcanized silicon rubber[1]. Some of them can be used to prepare heat resistance paints. All of them exhibit their specific properties. The polyphenylphenylsilanes are obtained mainly by the Diels-Alder reaction of dienophiles such as vinyl or phenylethynylsilanes with tetraphenylcyclopentadi-enone. Some effects of steric hindrance and electronegativity of different substituents attached to the dienophiles on the Diels-Alder reaction have been studied comprehensively. A series of new phenylethynylsilanes with different substituents has been synthesized.

SYNTHESES AND PROPERTIES OF PHENYLETHYNLORGANOSILANES

Phenylethynylorganosilanes containing different substituents are the raw materials of polyphenylphenylsilanes. In order to research into the effects of phenylethynylsilanes containing different groups on Diels-Alder reaction, we have synthesized a series of phenylethynylsilanes. To our knowledge there have been no reports about them until now. [2]-[4]

1. Methods of syntheses

We select mainly three methods to synthesize different phenylethynylsilanes. Here, the synthesis of $PhC{\equiv}CSiMe(OEt)_2$ is used as an example.

(1) By the Grignard reaction

$$PhC{\equiv}CH + EtMgBr \longrightarrow PhC{\equiv}CMgBr + EtH$$

$$PhC{\equiv}CMgBr + EtOSiMe(OEt)_2 \longrightarrow PhC{\equiv}CSiMe(OEt)_2 + BrMgOEt$$

(2) By the reaction of lithium phenylacetylide

$$PhC{\equiv}CH + BuLi \longrightarrow PhC{\equiv}CLi$$

$$PhC{\equiv}CLi + ClSiMe(OEt)_2 \longrightarrow PhC{\equiv}CSiMe(OEt)_2 + LiCl$$

(3) By the reaction of sodium phenylacetylide

$$PhC{\equiv}CH + Na \longrightarrow PhC{\equiv}CNa$$

$$PhC{\equiv}CNa + ClSiMe(OEt)_2 \longrightarrow PhC{\equiv}CSiMe(OEt)_2 + NaCl$$

The product can be obtained conveniently by the third method. In order to synthesize various phenylethynylchlorosilanes, the third method is modified. The sodium acetylide is dispersed in ether and then added slowly into the mixture of chlorosilane and ether. If monophenylethynylchlorosilanes need to be synthesized, the amount of chlorosilane should be much more than that of the sodium phenylacetylide in the solution.

The phenylethynylorganosilanols and phenylethynylorganodisiloxanes are obtained through the hydrolysis of monochlorophenylethynylorganosilanes. Monochlorosilanes are added to the alkaline mixture of water and ether at 0°C in an ice-salt bath. The Si-Cl bond is hydrolyzed and phenylethynylorganosilanols are obtained. When the hydrolysis is carried out in water at 70-80°C, the disiloxanes can be synthesized.

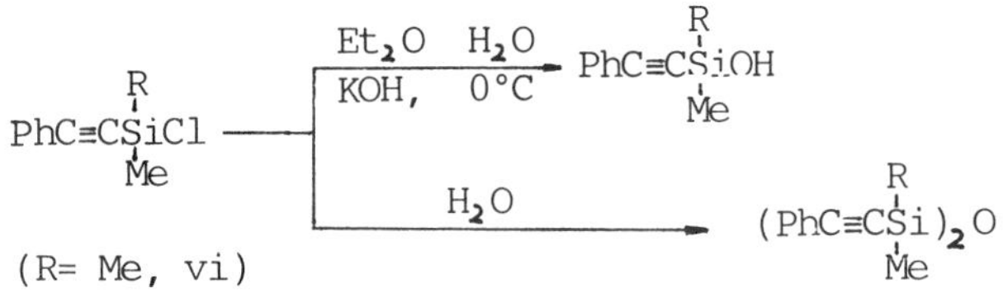

(R= Me, vi)

2. Physical Constants of Phenylethynylorganosilanes.
A series of phenylethynylsilanes is synthesized as shown in Table 1. The structures of these new compounds have been determined by IR, ^{1}H NMR and elemental analysis.

Table 1. Physical constants of phenylethynylsilanes.

Compounds	b.p. °C(mmHg)	n_D^{20}	d_4^{20}	^{1}H NMR δ ppm
$PhC\equiv CSiMe_2Cl$	68-69 (0.25)	1.5413	1.0245	0.61(CH_3) 7.25-7.52 (PhH)
$PhC\equiv CSiHMeCl$	68-69 (0.8)	1.5505	1.0384	0.65-0.72 (Me) 5.00-5.13(SiH) 7.22-7.55(PhH)
$PhC\equiv CSiCl\underset{Me}{\overset{CH=CH_2}{\mid}}$	81 (0.2)	1.5519	1.0272	0.7(Me) 6.2($CH=CH_2$) 7.3-7.58 (PhH)
$PhC\equiv CSiMe_2\overset{OH}{\mid}$	89-90 (0.15)	1.5481		0.43(Me) 2.48(OH) 7.3-7.55 (PhH)
$PhC\equiv CSiCl_2\overset{Me}{\mid}$	89-90 (1.0)	1.5496	1.1397	1.01(Me) 7.3-7.6(PhH)
$PhC\equiv CSiCl_2\overset{Ph}{\mid}$	126 (0.2)	1.5438	1.1926	7.3-7.9 (PhH)
$PhC\equiv CSiCl_2\overset{CH=CH_2}{\mid}$	119-120 (3.0)	1.5596	1.1522	6.4($CH=CH_2$) 7.5(PhH)
$PhC\equiv CSi(OEt)_2\overset{CH=CH_2}{\mid}$	162-163 (12.0)	1.5099	0.9845	1.3(Me) 3.9(CH_2) 6.2($CH=CH_2$) 7.4(PhH)
$(PhC\equiv C)_2SiCl\overset{Me}{\mid}$	179-180 (1.0)	1.6119		0.88(Me) 7.38-7.62 (PhH)

Table 1. continued. Physical constants of phenylethynylsilanes.

Compounds	b.p. °C(mmHg)	n_D^{20}	d_4^{20}	^{1}H NMR δ, ppm
$(PhC \equiv C)_2 SiCl_2$	214–215 (2.0)	1.5949	1.1845	7.2(PhH)
$(PhC \equiv CSiMe_2)_2O$	174–175 (1.2)	1.5666		0.41(Me) 7.3–7.55(PhH)
$PhC \equiv C\underset{\overset{\mid}{OEt}}{\overset{\overset{\mid}{OEt}}{Si}}PhPh_4$	43.5–44.5 (M,P)			o.82–1.22(Me) 3.8(CH$_2$) 6.8–7.6(PhH)

3. The Syntheses of Diphenylethynyldichlorosilane.

It is difficult to obtain diphenylethynyldichlorosilane directly by the sodium acetylide method by control of the molar ratio of reactants.

The molar ratio of SiCl$_4$ to PhC≡CH is 1:2, but no diphenyl-ethynyldichlorosilane can be obtained. The product is mainly the tetraphenylethynylsilane. The two-step method which must be used to synthesize the diphenylethynyldichlorosilane is as follows:

$$SiCl_4 + PhC \equiv CNa \longrightarrow PhC \equiv CSiCl_3$$

$$PhC \equiv CSiCl_3 + PhC \equiv CNa \longrightarrow (PhC \equiv C)_2 SiCl_2$$

4. ^{1}H NMR of Phenylethynylsianes

Table 2 shows some δ values which illustrate the chemical shift of methyl protons contained in phenylethynylsilanes and there exists a definite trend.

Table 2. Chemical shift of methyl protons.

Compounds	^{1}H NMR δ, ppm	Compounds	^{1}H NMR δ, ppm
$PhC \equiv CSiMeCl_2$	1.01	$PhC \equiv CSiMe_2OH$	0.43
$PhC \equiv CSiMe_2Cl$	0.61	$(PhC \equiv CSiMe_2)_2O$	0.41
$PhC \equiv CSiHMeCl$	0.65–0.72	$(PhC \equiv C)_2 SiMe_2$	0.2
$(PhC \equiv C)_2 SiMeCl$	0.88		
$PhC \equiv CSiViMeCl$	0.7		

Table 2 also shows that the electronegative groups attached to phenylethynylsilanes cause the chemical shift of methyl protons to low field. The stronger the electronegativity of groups, the higher the δ values. The order of δ values falls in the following sequence:

$$\underset{\underset{Me}{|}}{\overset{\overset{R}{|}}{PhC\equiv CSiCl}} > \underset{\underset{Me}{|}}{\overset{\overset{Me}{|}}{PhC\equiv CSiOH}} , \underset{\underset{Me}{|}}{\overset{\overset{Me}{|}}{PhC\equiv CSi-O-}} > (PhC\equiv C)_2SiMe_2$$

where R=Me, Cl, H, Vi or RC≡C-.

SYNTHESES AND PROPERTIES OF POLYPHENYLPHENYLORGANOSILANES

Reactions of syntheses

A series of polyphenylphenylorganosilanes has been synthesized by Diels-Alder reactions of vinyl or phenylethynylsilanes with cyclopentadienone[5-6].

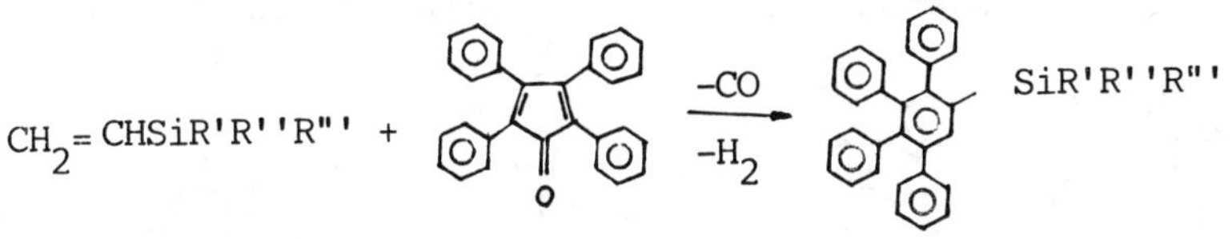

$$CH_2=CHSiR'R''R''' +$$ ... $$\xrightarrow[-H_2]{-CO}$$ ... $$SiR'R''R'''$$

abbr.: Ph$_4$PhSiR'R"R"'

$$PhC\equiv CSiR'R''R''' +$$ ... $$\xrightarrow{-CO}$$ ... $$SiR'R''R'''$$

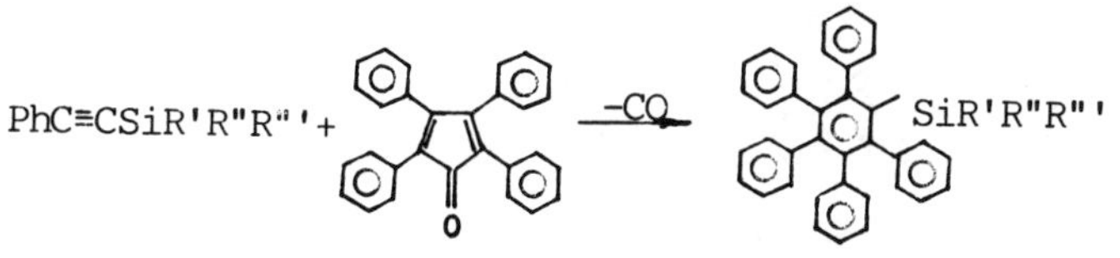

abbr.: Ph$_5$PhSiR'R"R"'

When vinyl or phenylethynylsilanes are used as dienophiles, the R', R" and R"' groups of dienophiles are:-Cl, -Ph, -Me, -OEt, -C≡CPh, -OSiMe$_2$CH=CH$_2$ and -OSiMe$_2$PhPh$_4$. The structures of dienes are: [7]

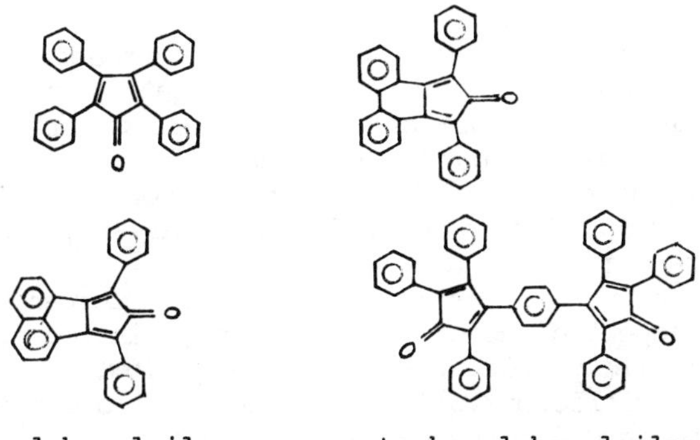

Tetraphenylphenylsilanes, pentaphenylphenylsilanes and bis (polyphenylphenyl)silanes of following structural types have been synthesized.

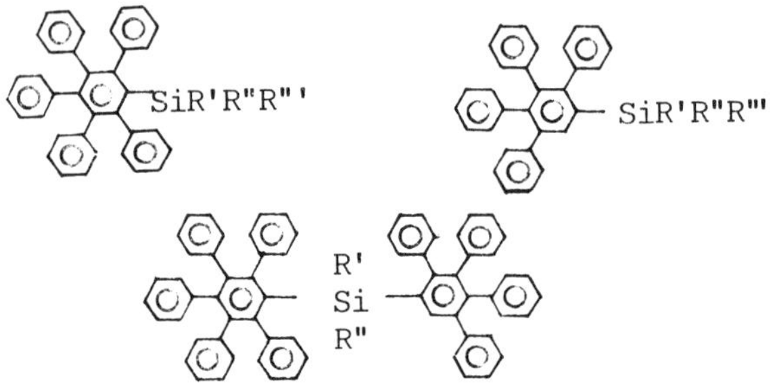

where R', R" and R"'=-H, -Cl, -Ph, -Me, -OH, -OEt, -Vi and -C≡CPh.

In addition, polynuclear hydrocarbonorganosilanes, polyphenyl-triphenyleneorganosiloxane and polyphenylphenylcyclosiloxanes have also been synthesized by our research group:[8-9]

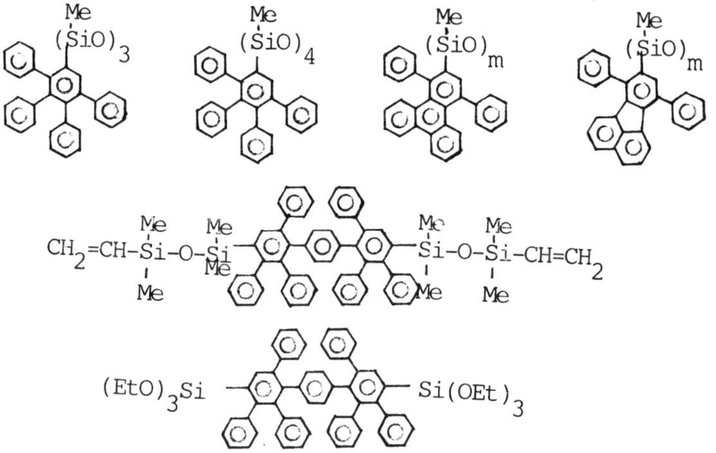

The structures of the new compounds are determined by elemental analysis, IR, ^{1}H NMR and mass spectrometry.

Some effects of electronegativity of substituents attached to dienophiles on reaction rate.

The Diels-Alder condensation is a cooperative reaction between dienes and dienophiles. Different electronegative substituents attached to dienophiles can affect the reaction rate. It is observed that the reactions are easy and fast if the electronegative substituents are linked to the dienophiles. The more the negative groups linked to silicon atoms are, the faster the reaction rate is.

For examples, the condensation of vinyltrichlorosilane with tetraphenylcyclopentadienone (tetracyclone) is completed at 225°C in 12 hr. the purple solution turns yellow. But the condensation of vinyltriethoxysilane with tetracyclone requires 24 hours at the same temperature.

The experimental results are summarized in Table 3.

Table 3. Effects of electronegativity on reaction rate.

Dienophiles	Products	Temp. °C	Time hr.
I. CH_2=CH-$SiCl_3$	$Ph_4PhSiCl_3$	225	12
II. CH_2=CHSi(OEt)$_3$	$Ph_4PhSi(OEt)_3$	225	24
III. CH_2=CHSiCl$_2$ 　　　C≡CPh	$Ph_4PhSiCl_2$ 　　C≡CPh	235	15
IV. CH_2=CHSiMeCl 　　　C≡CPh	$Ph_4PhSiMeCl$ 　　C≡CPh	235	20
V. CH_2=CHSi(OEt)$_2$ 　　　C≡CPh	$Ph_4PhSi(OEt)_2$ 　　C≡CPh	235	30

Table 3 shows the order of time required to complete the reaction

$$CH_2=CHSi(OEt)_3 > CH_2=CHSiCl_3$$

$$CH_2=\underset{\underset{C≡CPh}{|}}{C}HSi(OEt)_2 > CH_2=\underset{\underset{C≡CPh}{|}}{C}HSiMe\,Cl > CH_2=\underset{\underset{C≡CPh}{|}}{C}HSiCl_2$$

The reaction time is just contrary to the reaction rate.
Because the electronegativity of ethoxy group(OEt) is weaker than that of chlorine, the Diels-Alder reaction rate of dienophiles containing chlorine atom is faster than that of the ethoxy group.
The rate of reactions is concluded as below:

$$CH_2=\underset{|}{\overset{|}{C}}HSi-Cl > CH_2=\underset{|}{\overset{|}{C}}HSi-OEt$$

Some effects of steric hindrance of substituents attached to dienophiles on reaction rate.

The steric hindrance of substituents attached to dienophiles affects the reaction rate so greatly as to cease the reaction. The experimental results are listed in Table 4.

Table 4. Effects of steric hindrance.

Dienophiles + Tetracyclone	Products	Temp. °C	Time hr
I. PhC≡CSiHCl 　　　Me	Ph$_5$PhSiHCl 　　Me	230	20
II. PhC≡CSiCl$_2$ 　　　Me	Ph$_5$PhSiCl$_2$ 　　Me	230	89
III. PhC≡CSiCl$_2$ 　　　Ph	Ph$_5$PhSiCl$_2$ 　　Ph	230	100

Table 4. continued. Effects of steric hindrance.

Dienophiles + Tetracyclone	Products	Temp. °C	Time hr.
IV. $PhC \equiv CSiMe_3$	$Ph_5PhSiMe_3$	200	48
		225	48
V. $PhC \equiv CSiPh_3$	-------	310	72

The larger steric hindrance of phenyl group makes dienophile approach the diene with difficulty and, thus, the condensation proceeds slowly. Because -H, -Me and -Cl are small substituents attached to the silicon atom in compound I, the condensation is completed only in 20 hours. The reaction rate is shown as below:

$$PhC \equiv CSiHCl > PhC \equiv CSiCl_2 > PhC \equiv CSiCl_2$$
$$\quad\quad\quad\;\; Me \quad\quad\quad\quad Me \quad\quad\quad\quad Ph$$

The condensation of phenylethynyltrimethylsilane (IV) is carried out at 200 °C for 48 hours and, then, the temperature is increased to 225 °C and keeps so for 48 hours, the solution turns from purple to orange color and the reaction is complete. There are three phenyl groups in compound V, although the reactants are heated at 315 °C for 72 hours, no $Ph_5PhSiPh_3$ is formed. It is evident that the steric hindrance of phenyl group is much larger than that of methyl group. The order of reaction rate is as follows:

$$PhC \equiv CSiMe_3 > PhC \equiv CSiPh_3$$

A comparison of reaction rate between vinylorganosilanes and phenylethynylorganosilanes.
The condensation rate of vinyl or phenylethynylorganosilanes with tetraphenylcyclopentadienone is rather different. The reaction time is listed in Table 5.

Table 5. The comparison of reaction rate.

Dienophile +tetracyclone	Product	Temp. °C	Time hr.
I. $CH_2 = CHSiCl_3$	$Ph_4PhSiCl_3$	230	10
II. $PhC \equiv CSiCl_3$	$Ph_5PhSiCl_3$	230	54
III. $CH_2 = CHSiCl_2$ $\quad\quad\quad C \equiv CPh$	$Ph_4PhSiCl_2$ $\quad\quad C \equiv CPh$	235	15
IV. $CH_2 = CHSiMeCl$ $\quad\quad\quad C \equiv CPh$	$Ph_4PhSiMeCl$ $\quad\quad C \equiv CPh$	235	20

It is observed that the reaction rate of vinyl group is faster than that of phenylethynyl group either in the same compound or in different compounds. For example, vinyltrichlorosilane with tetracyclone can react completely at 230°C in 10 hours. The condensation of phenylethynyltrichlorosilane with tetracyclone is completed at the same temperature in 54 hours. When the vinyl group and phenylethynyl group are in the same compound, the molar ratio of dienophile to diene is 1:1 at 235°C in 15-20 hours, the products are determined by IR and ^{1}H NMR. The ^{1}H NMR proton signal of vinyl group (δ, 6.4 ppm) disappears, but the stronger peak of IR 2163cm^{-1} still occurs. It is obvious that the tetracyclone prefers to react with vinyl group rather than with phenylethynyl group.

5. Characteristics of Polyphenylphenylorganosilanes by ^{1}H NMR.

The different substituents on silicon can influence the chemical shift of protons in methyl groups attached to silicon. The bulky substituents have a great shielding effect on the resonance signals of methyl protons. The chemical shift of Si-CH$_3$ contained in different compounds is listed in table 6.

Table 6. ^{1}H NMR spectra of methyl protons in polyphenylphenylorganosilanes.

Compounds	^{1}H NMR(SiCH$_3$) δ, PPm	Substituents
I. Ph$_4$PhSiC≡CPh, Me, Cl	0.5	Ph$_4$Ph–, –C≡CPh, –Cl,
II. Ph$_5$PhSiC≡CPh, Me, Cl	−0.3	Ph$_5$Ph–, –C≡CPh, –Cl,
III. Ph$_5$PhSi–H, Me, Cl	0.6	Ph$_5$Ph–, –Cl, –H
IV. Ph$_5$PhSiCl$_2$, Me	−0.16	Ph$_5$Ph–, –Cl, –Cl
V. Ph$_5$PhSi(OH)$_2$, Me	−0.35	Ph$_5$Ph–, –OH, –OH
VI. Ph$_4$PhSi(OH)$_2$	−0.038	Ph$_4$Ph–, –OH, –OH

The signals of methyl protons appear in high field if the shielding effect is strong. Owing to the different structures of tetraphenylphenyl group and pentaphenylphenyl group, the deshielding effect of tetraphenylphenyl group is stronger than that of pentaphenylphenyl group. Thus, the chemical shift (δ value) of methyl protons in compound I is higher than that in compound II. Protons of SiCH$_3$ in compound I and II exhibit in low and high field respectively. The great steric hindrance of pentaphenylphenyl group is probably the main reason for its strong shielding effect. The difference of chemical shift between compound V and VI exists in the same way. The signal of compound VI is in the low field owing to the weak steric hindrance of tetraphenylphenyl group. The structural difference of compound II and III is $-C \equiv CPh$ and $-H$, compound II shows its signal in high field. This result can be explained in two ways: the first is the small steric hindrance of hydrogen atom and the second is the shielding effect of phenylethynyl group. The electronegativity of substituents attached to silicon atom can affect the chemical shift of methyl protons. If the electronegativity of substituents is strong, the density of electron surrounding the methyl proton decreases and the resonance signals appear in lower field. The structures of compound IV and V are different only in the chlorine atom and hydroxy group. Because electronegativity of chlorine atom is stronger than that of hydroxy group, the resonance signal of methyl protons in compound IV appear in the lower field [10-11].

REFERENCES

1. Du Zuodong, Chen Jianhua, Shi Baochuan and Wang Hao, Polymer Communications, 3 (1981) 174.
2. Wenzel E. Davidsohn and Malcolm C. Henry, Chemical Reviews, 67 (1967) 73.
3. Petrov, A. D., and Shchukovskaya, L. I., Zh. Obshch. Khim., 25 (1955) 1128 (Engl. Trans., 1083).
4. Kurt C. Frisch and Robert B. Young, J. Amer. Chem. Soc., 74 (1952) 4853.
5. Du Zuodong, Chen Jianhua and Shi Baochuan, Polymer Communications, 2 (1983) 110.
6. Freeburger M. E. and Leonard Spialter, J. Org. Chem. 35 (1970) 652.
7. Michael A. Ugliaruso, Michael G. Romanell and Ernest I. Becker, Chemical Reviews, 65 (1965) 261.
8. Chen Jianhua, Feng Shengyu and Du Zuodong, Chemical Journal of Chinese Universities, 12 (1986) 1150.
9. Chen Jianhua, Shi Baochuan and Du Zuodong, Chemical Journal of Chinese Universities, 9 (1986) 808.
10. Chen Jianhua, Li Xinhua. Zhang Xiaoqing, Chemical Journal of Chinese Universities, in press.
11. Chen Jianhua, Li Xinhua, Zhang Xiaoqing, Journal of Polymer, in press.

Chapter 11

Chemistry to synthesize
silicone intermediates from silica

John L. Speier, Senior Scientist – Corporate Research Scientist, Corporate Research Department, Dow Corning Corporation, Midland, Michigan 48640.

INTRODUCTION

Silicon is the second-most abundant element in the lithosphere of the earth but it is found only combined with the most abundant element, oxygen, as silica, SiO_2, or some silicate mineral such as kaolin clay, $H_2Al_2Si_2O_8 \cdot XH_2O$. The chemical conversion of one silicon-oxygen bond in a silicate to a silicon-carbon bond expends a minimum of 65 kilocalories per mole to form an organosilicon derivative that is meta-stable in the environment of the earth. Organosilicon compounds are unnatural substances existing only as products of human endeavors.

The history of these endeavors will be outlined as they led to a large world-wide "Silicone" industry and at least one example will be shown for every reaction that forms Si-C bonds from Si-inorganic reagents.

HISTORY

Scheele, a Swedish apothecary, discovered hydrofluoric acid, HF, a gas, in 1771 and found that it reacted with his glass apparatus to make the first volatile silicon compound ever seen, SiF_4.

J. J. Berzelius (1), fifty-three years later isolated the element, silicon and discovered the second volatile silicon compound, $SiCl_4$.

$$SiF_4 + 2\ KF \longrightarrow K_2SiF_6 \quad \text{a Hexafluorosilicate}$$

$$K_2SiF_6 + 4\ K \xrightarrow{\Delta} 6\ KF + Si^o \quad \text{the Element}$$

$$Si^o + 2\ Cl_2 \xrightarrow{\text{Flame}} SiCl_4 \quad \text{Tetrachlorosilane}$$

J. J. Ebelman (2) in 1839 made the first organic silicate.

$$SiCl_4 + 4\ EtOH \longrightarrow 4\ HCl + (EtO)_4Si \quad \text{Tetraethoxysilane.}$$

F. Wohler and H. Buff (3) carried out the first carbo-thermic reduction of silica (1857).

$$SiO_2 + 2\ C + 3\ HCl \xrightarrow{\text{hot}} 2\ CO + H_2 + Cl_3SiH \quad \text{Trichlorosilane.}$$

The truly great team of Friedel and Crafts (4) were the first to make organosilicon compounds (1863-66).

$$SiCl_4 + 2\ ZnEt_2 \xrightarrow{160°} 2\ ZnCl_2 + Et_4Si \quad \text{Tetraethylsilane}$$

They were most surprised to find that the Si–C bonds were not cleaved by chlorine and that tetraethylsilane chlorinated very like a typical hydrocarbon.

$$2\ Et_4Si + 3\ Cl_2 \xrightarrow{h\nu} Et_3SiC_2H_4Cl + Et_3SiC_2H_3Cl_2$$

$$Et_3SiC_2H_4Cl + KOAc \longrightarrow KCl + Et_3SiC_2H_4OAc \quad \text{an Ester}$$

$$Et_3SiC_2H_3Cl_2 + KOAc \longrightarrow (?) + (Et_3Si)_2O \quad \text{a Siloxane}$$

Friedel and Ladenburg (5) learned how to make alkyl-alkoxysilanes.

$$(EtO)_4Si + ZnEt_2 \xrightarrow{\Delta} Zn(OEt)_2 + Et_nSi(OEt)_{4-n} \quad n = 1,\ 2,\ 3$$

In 1872 Ladenburg (6) separated and identified all the possible ethyl-ethoxy-chlorosilanes shown in Table 1.

Table 1. Ethyl-ethoxy-chlorosilanes.

$SiCl_4$	$Si(OEt)Cl_3$	$Si(OEt)_2Cl_2$	$Si(OEt)_3Cl$
$EtSiCl_3$	$EtSi(OEt)Cl_2$	$EtSi(OEt)_2Cl$	$EtSi(OEt)_3$
Et_2SiCl_2	$Et_2Si(OEt)Cl$	$Et_2Si(OEt)_2$	$EtSi(OEt)_3$
Et_4Si	$Et_3Si(OEt)$		

And he made the first silicone polymers, e.g.,

$$Et_2Si(OEt)_2 + H_2O \longrightarrow 2\ EtOH + (Et_2SiO)_{poly}$$

and the first arylsilanes

$$SiCl_4 + Ph_2Hg \longrightarrow PhHgCl + PhSiCl_3$$

Polis (7) found a much better way to arylsilanes (1885).

$$SiCl_4 + 4\ PhCl + 8\ Na \xrightarrow{\text{Ether}} 8\ NaCl + Ph_4Si$$

Gatterman was the first to discover the catalytic effect of copper on a reaction of silicon (7) in 1889.

$$Si^0 + 3\ HCl \xrightarrow[\sim 300^\circ]{\text{Cu}} H_2 + HSiCl_3$$

F. S. Kipping (1899-1944) with a series of 54 papers laid the foundation for modern silicones made from a host of intermediates available by reactions of Grignard reagents with chloro or alkoxysilanes

$$RX + Mg \xrightarrow{\text{ether}} RMgX \quad X = Cl,\ Br,\ I. \quad R = alkyl\ or\ aryl$$

$$RMgX + ABCD\ Si \longrightarrow R\text{-}Si\ derivatives + MgX\ A,\ B,\ C,\ or\ D.$$

A, B, C, D = many combinations of halogen, alkoxy, alkyl, aryl or H substituents.

Dow Corning Corporation was founded in 1943 to commercialize silicones made by an etherless Grignard process.

$$(EtO)_4Si + n\ RX + n\ Mg \longrightarrow n\ MgXOEt + R_nSi(OEt)_{4-n}$$

MISCELLANEOUS ORGANOMETALLIC SYNTHESES   1945

$$(EtO)_4Si + n\ PhCl + 2\ n\ Na \longrightarrow n\text{-}NaCl + n\ NaOEt + Ph_nSi(OEt)_{4-n}$$

$$MeSiCl_3 + MeCl + Zn \xrightarrow[\text{vapor}]{375^\circ} ZnCl_2 + Me_2SiCl_2$$

1948

$$SiCl_4 + 2\ EtBr + Zn/Cu \xrightarrow{\text{ether}} ZnClBr + Et_2SiCl_2$$

$$SiCl_4 + 2\ t\text{-}BuLi \longrightarrow 2\ LiCl + t\text{-}Bu_2SiCl_2$$

1953

$$Si_2Cl_6 + 6\ p\ PhC_6H_4Li \longrightarrow 6\ LiCl + (p\text{-}PhC_6H_4)_6Si_2$$

$$SiCl_4 + 2\ Li(CH_2)_5Li \longrightarrow 4\ LiCl +$$

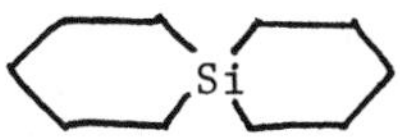

1957

$$3\ Me_2SiCl_2 + 3\ MeCl + 2\ Al \xrightarrow{345^\circ} 2\ AlCl_3 + 3\ Me_3SiCl$$

1960

$$6 \ SiO_2 + 4 \ Me_3Al_2Cl_3 \ \longrightarrow \ 4 \ Al_2O_3 + 3 \ SiCl_4 + 3 \ Me_4Si$$

1961

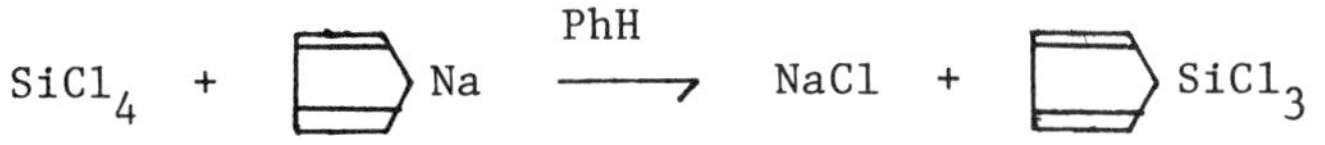

$$SiCl_4 + \ \boxed{\ } Na \ \xrightarrow{PhH} \ NaCl + \ \boxed{\ } SiCl_3$$

1963

$$3 \ MeSiCl_3 + Me_3Al_2Cl_3 \ \xrightarrow{200°} \ 2 \ AlCl_3 + 3 \ Me_2SiCl_2$$

THE DIRECT PROCESS

E. G. Rochow (9) got a patent in 1941 that revolutionized the industrial production of organochlorosilanes. The process is at its best with methyl chloride.

$$Si^0_{(s)} + MeCl_{(gas)} \ \xrightarrow[300°]{Cu}$$

	B.P.°C
Me_4Si	27
$HSiCl_3$	32
Me_2HSiCl	36
$MeHSiCl_2$	41
Me_3SiCl	57
$SiCl_4$	57.6
$MeSiCl_3$	66
Me_2SiCl_2	70

$$Si^0_{(s)} + MeCl + HCl \ \xrightarrow[HCl]{Cu} \ \text{increased amounts of } MeHSiCl_2, \ Me_2HSiCl, \ HSiCl_3$$

By far the best, most detailed study and review of this process is in the book by Voorhoeve (10). Two excellent papers were recently published on Catalysis of the Rochow Direct Process (11) and Active Site Formation in the Direct Process with Methyl Chloride (12).

Such a process has been extended to reactions of dialkylethers (13).

$$Si^0_{(s)} + 2 \ ROR_{(liq)} \ \xrightarrow[\substack{Pressure \\ \sim 250°C}]{Cat.} \ R_2Si(OR)_{2(liq)} \quad R = alkyl$$

HYDROSILATION BY FREE-RADICAL PROCESSES

In 1947 Barry, et al (13) showed that Cl_3SiH would add to olefins under pressure above about 350°C. About the same time Sommers (14) showed that the addition was accelerated by free-radical initiators and Burkhard (15) published many examples of free-radical additions.

$$Cl_3SiH + H_2C=CHC_6H_{13} \xrightarrow[60°C]{ACOOAc} 99\% \ Cl_3Sin-C_8H_{17}$$

Such free-radical addition processes were studied in some detail by myself and my friends (16) who found that the efficiency of such reactions was highly dependent upon the structures of the reagents. In descending orders of utility the systems can be qualitatively listed as:

Silanes:

Cl_3SiH, Br_3SiH, $PhSiH_3$ >> CH_3Cl_2SiH > $O[SiMe_2H]_2$ > R_3SiH

Olefins: $CH_2=CHAlk.$, ⬡ , ⬠ >> $CH_2=CHPh$

Initiators: Peroxides, Azo Cpds. > photolysis

Very little that is new has been published about such addition reactions in about 30 years.

HYDROSILATION CATALYZED BY TRANSITION METALS

The hydrogenation of ethylene is exothermic by −32.82 kcal/mole. The calculated heat of hydrosilation is about −38 kcal/mole. Thus hydrosilation should proceed more readily than hydrogenation if a mechanism exists which permits hydrosilation.

In 1953 a patent (17) was issued to Wagner which claimed that platinum on carbon was an effective catalyst above 130° to add Cl_3SiH to ethylene, acetylene, butadiene, vinyl chloride or vinyldiene fluoride.

In 1957, my friends and I published a paper (18) which showed that soluble Pt, Ru, and Ir compounds form superb homogeneous catalysts that are active at concentrations as low as 1 pt./million below room temperature.

A review of homogeneous catalysis of hydrosilation was written by me (19) and this process has become widely used everywhere to make alkyl and substituted alkylsilanes of the widest varieties, many available by no other means.

Our Technical Information Services at Dow Corning searched the literature for hydrosilations in December, 1986 and found patents and papers:

```
 325 by free-radicals        89 Reviews
2031 with platinum           57 with palladium
  92 with cobalt             38 with iridium
  85 with nickel             33 with copper
  72 with rhodium            11 with ruthenium
  64 with iron
```

CONDENSATION REACTIONS

(20) 1959

$$Cl_3SiH + PhH \xrightarrow[350°]{BCl_3} PhSiCl_3 + PhHSiCl_2 + H_2 + HCl$$

REDUCTIVE SILYLATION   (21) 1969

$$PhCH_2Cl + HSiCl_3 + R_3N \longrightarrow PhCH_2SiCl_3 + R_3NHCl$$

$$CHCl_3 + 3 HSiCl_3 + 2 R_3N \longrightarrow CH_2(SiCl_3)_2 + SiCl_4$$

$$PhCH=O + 2 HSiCl_3 + 2 R_3N \longrightarrow PhCH_2SiCl_3 + (Cl_2SiO)_x$$

$$PhCOOH + 3 HSiCl_3 + 2 R_3N \longrightarrow PhCH_2SiCl_3 + (Cl_2SiO)_x$$

MAGNESIUM REDUCTIVE SILYLATION   (22) 1969

$$\overset{\backslash}{\underset{/}{C}}=O + 2 Mg + 4 Me_3SiCl \longrightarrow \overset{\backslash}{\underset{/}{C}}\overset{SiMe_3}{\underset{SiMe_3}{}} + (Me_3Si)_2O + MgCl_2$$

$$PhCOOMe + Mg + 2Me_3SiCl \longrightarrow PhC\overset{OSiMe_3}{\underset{OMe}{-SiMe_3}} + MgCl_2$$

$$C_{10}H_8 + 3 Mg + 6 Me_3SiCl \longrightarrow$$

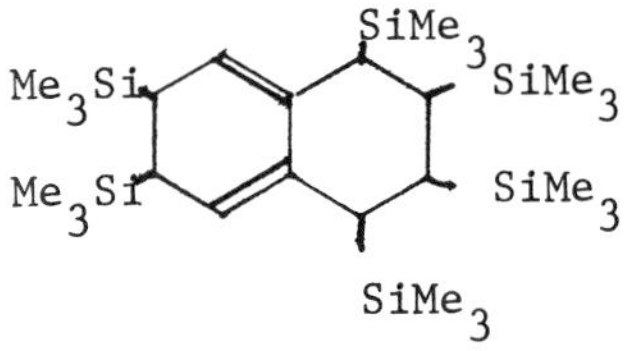

DIVALENT CARBON INSERTIONS   (23) 1970

$$R_3SiY + XCZ \longrightarrow R_3SiCXYZ$$

$$Cl_3SiH + N_2CH_2 \longrightarrow Cl_2HSiCH_2Cl + N_2$$

$$Et_3SiH + ICH_2HgI \xrightarrow{80°C} Et_3SiCH_3 + HgI_2$$

$$(EtO)_nEt_{3-n}SiH + N_2CH_2COOEt \xrightarrow{Cu} (EtO)_nEt_{3-n}SiCH_2COOEt$$

REACTIONS OF DISILANES   (24) 1973

$$Me_{6-n}Si_2Cl_n + RCl \xrightarrow[\Delta]{Pd} Me_{\frac{6-n}{2}}RSiCl_{\frac{n}{2}} + Me_{\frac{6-n}{2}}SiCl_{\frac{n-1}{2}}$$

R = ALLYL, BUTYL, PHENYL OR H   (25) 1975

$$Me_{6-n}Si_2Cl_n + YC_6H_4X \xrightarrow{Pd} 100\%\ YC_6H_4SiMe_{\frac{6-n}{2}}Cl_{\frac{n}{2}}$$

n = 0, 2, 4   X = Cl, Br, I   Y = NO$_2$, H

ELECTROCHEMICAL   (25) 1986

$$RX + R_3^1SiCl \xrightarrow{2e^-} RSiR_3^1 + ClX$$

RX = allyl chloride, carboxylate or sulfone

= aryl iodide, N⟨⟩Br, alkylCH=CHI

REFERENCES

1. Berzelius, J. J., Pogg. Ann. 1, 169 (1824).
2. Ebelman, J. J., Compt. rend., 19 398, (1844).
3. Wohler, F., Buff. H., Ann., 103, 218, (1857).
4. Friedel, C., Crafts, J. M., Compt. rend., 56, 592,
    (1863); Ann., 136, 203, (1865); ibid., 138, 19,
    (1866).
5. Friedel, C., Ladenburg, A., ibid., 143, 118, (1867);
    Compt. rend., 68, 923, 1869; Ann., 159, 259 (1871).
6. Laddenburg, A., ibid., 164, 300, (1872); 173, 143,
    (1874).
7. Polis, A., Ber. dtsch. chem. Ges. 18, 1540 (1885).
8. Gatterman, L., ibid., 22, 186 (1889).
9. Rochow, E. G., U. S. Pat. 2,380,995 (1941).
10. Voorhoeve, R. J. H., Organosilanes, Precursors to
    Silicones, Elsevier Publishing Co. (1967) 423 pp.
11. Ward, W. J., Ritzer, A., Carroll, K. M., Flock, J. W., J.
    of Catalysis 100, 240-249 (1986).
12. Banholzer, W. F., Lewis, N., Ward, W. J. ibid., 101,
    405-415 (1986).
13. Barry, A. J., DePree, L., Gilkey, J. W., Hook, D. E., J.
    Am. Chem. Soc., 69, 2916 (1947).
14. Sommers, L. H., Pietrusza, E. W., Whitmore, F. C., ibid.,
    69, 188 (1947).
15. Burkhard, C. A., Krieble, R. H., ibid., 69, 2687 (1947).
16. Speier, J. L., Zimmerman, R., Webster, J., ibid., 78,
    2278 (1956).
17. Wagner, G. H., U. S. Pat. 2,637,738 (1953).

18. Speier, J. L., Webster, J. A., Barnes, G. H., J. Am. Chem. Soc., 79, 974 (1957).

19. Speier, J. L., "Homogeneous Catalysis of Hydrosilation by Transition Metals, Advances in Organometallic Chemistry, Vol. 17, Academic Press, New York, 1979.

20. Barry, A. J., Gilkey, J. W., Hook, D. E., Ind. Eng. Chem., 51, 91 (1959).

21. Benkeser, R. A., Gaul, J. M., J. Am. Chem. Soc., 92(3), 729 (1970).

22. Dunogues, J., Calas, R., Biran, C., Duffaut, N., J. Organomet. Chem. 23, 50-52 (1970).

23. Seyferth, D., Pure Appl. Chem. 23(4) 391-412 (1970).

24. Atwell, W. H., Bokerman, G. N., U. S. Pat. 3,772,347 (1973).

25. Kawabata, N., et. al., J. Org. Chem., 51(21), 3996-4000 (1986).

Chapter 12

New approaches to organosilicon compounds

B. Kanner, W.B. Herdle and J.M. Quirk* — Union Carbide Corporation, Tarrytown, N.Y. *W.R. Grace, Columbia, Md.

In the absence of a commercially viable process
for the formation of carbon-silicon bonds, silicones
were little more than laboratory curiosities until
about 50 years ago. The discovery of the Direct
Process by Rochow provided an economic route to
methylchlorosilanes and thus greatly facilitated the
startup of the silicones industry. Subsequently,
this discovery spurred extensive research on the
nature of this complex process and, indirectly,
stimulated investigations in many other directions as
well. The effects are still with us today as witness
the enormous growth of organosilicon chemistry over
the past decade, especially in the service of organic
monomer and polymer synthesis.

Our continuing interest in the reactions of
simple organic molecules with silicon metal resulted
in the discovery of a new Direct Process involving
dimethylamine [1,2]:

$$3\ Me_2NH\ +\ Si\ \xrightarrow{250\ C}\ (Me_2N)_3SiH\ +\ 3/2\ H_2$$

The reaction conditions were generally similar to the reaction of methyl chloride but at a somewhat lower temperature and significantly faster rate. The reaction is inhibited by low levels of primary amines, oxygen and water. The mixed reaction of dimethylamine and ethanol is also observed under similarly mild conditions and even faster reaction rates [3]:

$$Me_2 NH + C_2 H_5 OH + Si \longrightarrow$$

$$(Me_2 NH)_x (C_2 H_5 O)_y SiH_z + H_2$$

where $x = 0-4$; $y = 0-4$
$z = 0-1$ and $x+y+z = 4$

The product mixture on reaction with excess ethanol is converted to ethyl silicate in excellent yield and to dimethylamine which can be recycled [3]. The vapor phase reaction of ethanol with silicon is initially also rapid but is gradually deactivated by the surface coating of silicon by by-products.

The product of the Direct Reaction of dimethylamine with silicon is tris(dimethylamino)-silane ("Tris"). The chemistry of this silane, which is otherwise tediously obtained from the reaction of trichlorosilane with excess dimethylamine, was largely unknown. The presence of reactive dimethylamino and hydrido groups suggested that a variety of reactions should be readily at hand.

SILYLAMINE REACTIONS

The alcoholysis of "Tris" was examined to determine the feasibility of selectively displacing the dimethylamino groups:

$$3 ROH + HSi(NMe_2)_3 \longrightarrow HSi(OR)_3 + 3 HNMe_2$$

The initial experiments using ethanol and t-butanol as models were disappointing:

$$3 C_2 H_5 OH + HSi(NMe_2)_3 \longrightarrow (C_2 H_5 O)_3 SiH$$

4%

$$+ Si(OC_2 H_5)_4 + HSi(NMe_2)_3$$

72%   22%

$$3 \ t\text{-}C_4H_9OH \quad + \quad HSi(NMe_2)_3 \quad \longrightarrow \quad NR$$

The reaction of ethanol with "Tris" was very rapid at room temperature but yielded primarily ethyl silicate with very little of the desired triethoxysilane. In an effort to obtain greater selectivity, the sterically hindered alcohol, t-butanol, was used. However, in this case, no reaction was observed. Evidently in the reaction with ethanol, the replacement of dimethylamino and hydrido groups proceeds rapidly and with little or no discrimination. However, as the "Tris" molecule is quite sterically hindered, replacement of the bulky dimethylamino groups by the still bulkier t-butoxy groups is unfavorable even at reflux.

We had previously observed that the reaction of siloxamines with polyols was effectively catalyzed by carbon dioxide [4]:

$$-Si(Me_2)_2NMe_2 \quad + \ ROH \quad \xrightarrow{\quad CO_2 \quad} \ -Si(Me_2)_2OR$$

When the reaction was repeated in the presence of a trace of carbon dioxide, an exothermic reaction of t-butanol with "Tris" was observed [5]:

$$3 \ t\text{-}C_4H_9OH \ + \quad HSi(NMe_2)_3 \quad \xrightarrow{\quad CO_2 \quad}$$

$$HSi(OC_4H_9)_3$$

$$> 95\%$$

The reaction proceeded rapidly to give tris(t-butoxy)silane in essentially quantitative yield. On prolonged reflux with excess t-butanol no tetrakis(t-butoxy)silane was observed. Carbon dioxide proved to be an effective and general catalyst for the selective formation of primary, secondary and tertiary trialkoxysilanes:

$$3 \ ROH \qquad + \qquad HSi(NMe_2)_3 \quad \longrightarrow \quad HSi(OR)_3$$

$$R = 1,2,3 \ Alkyl \qquad\qquad\qquad 88 - 98\%$$

A detailed mechanism for this process remains to be established but a key step is believed to involve the insertion of carbon dioxide to form the silyl carbamate followed by rapid displacement by alcohol:

$$\geq SiNMe_2 \quad + \quad CO_2 \quad \longrightarrow \quad \geq SiOOCNMe_2$$

$$\geq SiOOCNMe_2 \quad + \quad ROH \quad \longrightarrow \quad \geq SiOR \quad + \quad CO_2 \quad +$$
$$HNMe_2$$

Similarly, the silyl carbamate intermediate can be formed by reaction of the silylamine with dimethylammonium dimethylcarbamate which is likely formed in situ or which can be added as such:

$$CO_2 \quad + \quad 2 HNMe_2 \quad \longrightarrow \quad Me_2NCOO^- \quad ^+H_2NMe_2$$

$$"A"$$

$$\geq SiNMe_2 \quad + \quad "A" \quad \longrightarrow \quad \geq SiOOCNMe_2 \quad + \quad 2 HNMe_2$$

As long as the ROH:Tris stoichiometry is carefully controlled at 3:1, the yield of trialkoxy-silanes remains very high. When the stoichiometry falls below 3:1 the concentration of unreacted silylamine increases and as· the ratio increases above 3:1 there is a corresponding loss of silyl hydride.

Although carbon dioxide or its carbamate salt proved to be most effective catalysts for this process, a variety of other organic or inorganic acids were also useful catalysts [5].

A variety of other displacement reactions were then examined to establish the generality of the selectivity and steric effects observed in the alcoholysis process.

Oximes and hydroxylamines showed a similar pattern of catalyzed reactions with "Tris" [6]:

$$3 Me_2C=NOH \quad + \quad HSi(NMe_2)_3 \quad \xrightarrow{CO_2}$$

$$HSi(ON=CMe_2)_3$$
$$85 - 90\%$$

$$3(C_2H_5)_2NOH \quad + \quad HSi(NMe_2)_3 \xrightarrow{CO_2} HSi(ONEt_2)_3$$

$$90\%$$

The displacement reactions of acetic acid and acetic anhydride with "Tris" were distinctively different. Reaction with acetic acid yielded a complex mixture of monomeric and polymeric products while acetic anhydride reacted cleanly with "Tris" to give the triacetoxy derivative [7]:

$$HSi(NMe_2)_3 + 3 Ac_2O \xrightarrow{-20} HSi(OAc)_3 + 3 AcNMe_2$$

70%

The yield of triacetoxysilane was better than 70% when the reaction was run at -20 C but the product disproportionates rapidly when warmed to room temperature or above. The order of addition is critical. If the acetic anhydride is added to "Tris", polymeric products are obtained. Silylamines should be added to the acetic anhydride at -20 to 0 to obtain good yields of the acetoxy derivatives.

The transamination of "Tris" with secondary amines again evidenced the steric crowding already present in this molecule and its effect on the rate and completeness of this process [8]:

$$HSi(NMe_2)_3 + R_2NH \rightleftharpoons HSi(NMe_2)_x(NR_2)_{3-x}$$

In practice the reactions proved to be sluggish and gave the expected mixture of products. Transamination of "Tris" with diethylamine gave typical results:

$$HSi(NMe_2)_3 + 3 HNEt_2 \longrightarrow \quad HSi(NMe_2)_3 \qquad 22\%$$

$$HSi(NMe_2)_2(NEt) \qquad 51\%$$

$$HSi(NMe_2)(NEt_2)_2 \qquad 24\%$$

$$HSi(NEt_2)_3 \qquad 2\%$$

The transamination of "Tris" with ammonia and primary amines proceeds much more rapidly and can be driven more readily to completion because of the reduction of steric crowding around silicon. These reactions will be discussed in a later section in connection with the formation of silicon nitride.

The diethylamino- and piperidino- analogs of "Tris" behaved very much like "Tris" itself including the ready insertion of carbon dioxide at ambient temperatures to form the tris(carbamato)silane derivatives [9]:

$$HSi(NR_2)_3 \xrightarrow{\quad CO_2 \quad} HSi(OOCNR_2)_3$$

These compounds were stable at room temperature but decomposed gradually upon moderate warming.

SILYLHYDRIDE REACTIONS

The silyl hydride group in "Tris" could be expected to undergo a number of reactions of which hydrosilation is perhaps the most important. The initial results were disappointing. Attempts to react "Tris" with several typical olefins using standard hydrosilation reaction conditions gave essentially no yield of the expected products. The results with acetylenes were somewhat better but the yields were modest and with long reaction times:

$$HSi(NMe_2)_3 + HC{\equiv}CH \xrightarrow[\text{5 Hrs.}]{\quad Pt \quad} H_2C{=}CHSi(NMe_2)_3$$

$$12\%$$

$$HSi(NMe_2)_3 \;+\; C_4H_9C{\equiv}CH \xrightarrow[\text{13 Hrs.}]{\quad Pt \quad}$$

$$C_4H_9CH{=}CHSi(NMe_2)_3$$

$$23\%$$

The low reactivity of "Tris" in hydrosilation processes could be rationalized as resulting from a strong interaction of the silylamine with the platinum catalyst [10] thereby interfering with key steps in the hydrosilation catalytic cycle. Another, and perhaps more compelling reason for the sluggishness of these hydrosilation reactions is the steric crowding around silicon in the "Tris" molecule which effectively blocks access to Si-H. This could significantly inhibit the oxidative addition of the Si-H to the platinum catalyst, thus interrupting the catalytic cycle:

$$HSi(NMe_2)_3 + Pt \longrightarrow H-Pt-Si(NMe_2)_3$$

In attempting to offset these factors it was believed that carrying out the reaction at substantially higher temperatures might be helpful, providing that the catalyst was not deactivated under these conditions. After a few scouting experiments, excellent results were obtained [11]:

$$HSi(NMe_2)_3 + RC{\equiv}CH \xrightarrow[200°C]{Pt} RCH=CHSi(NMe_2)_3$$
$$97\%$$

$$HSi(NMe_2)_3 + HC{\equiv}CH \xrightarrow[225°C]{Pt} HC=CHSi(NMe_2)_3$$
$$95\%$$

Based on previous work, we had believed that hydrosilation reactions run at these high temperatures could result in rapid deactivation of the platinum catalyst. However, no evidence for this was found. Unexpectedly, excellent results were obtained using quite low levels of platinum catalyst. The yields obtained were quite sensitive to the reaction temperature employed. Thus, at 150°C, the conversion of "Tris" to its vinyl derivatives was only 8% after two hours.

An unexpected feature of the above reactions was the complete absence of a bis-adduct which is characteristic of acetylene hydrosilations:

$$HSi= + HC{\equiv}CH \longrightarrow =SiCH=CH + =SiCH_2CH_2Si=$$

Although platinum   was an excellent catalyst for acetylene hydrosilations it was not an effective catalyst for the hydrosilation of "Tris" with olefins. However much more promising results were obtained with rhodium catalysis [12]:

$$HSi(NMe_2)_3 + H_2C=CH_2 \xrightarrow{Rh_2Cl_2(CO)_4} C_2H_5Si(NMe_2)_3$$
$$70\%$$

$$+ \qquad\qquad H_2C{=}CHSi(NMe_2)_3$$

$$20\%$$

The apparent hydrosilation product, ethyl tris(dimethylamino)silane was obtained along with a significant quantity of vinyl tris(dimethylamino)silane. The presence of the latter product strongly suggested that the process was more complex and that one or both products might have arisen from a condensation reaction between "Tris" and ethylene:

$$HSi(NMe_2)_3 + H_2C{=}CH_2 \longrightarrow H_2C{=}CHSi(NMe_2)_3 + H_2$$

The ethyl derivative could have then arisen from the hydrogenation of the initially formed vinyl "Tris", as the rhodium complex is also an effective hydrogenation catalyst

To determine the relative importance of the potentially competing hydrosilation and condensation reactions between "Tris" and ethylene, the effects of excess ethylene on product distribution were determined. When reacted with excess ethylene, "Tris" was converted to vinyl "Tris" in excellent yield. The dramatic effects of ethylene concentration on its reactions with "Tris" and tri-t-butoxysilane which are both sterically hindered silanes is summarized in the following equations:

$$HSiR_3 \quad + \quad H_2C{=}CH_2 \quad \longrightarrow \quad H_2C{=}CHSiR_3 \quad +$$
$$\text{excess}$$
$$95\%$$

$$R = O{-}tBu; \; NMe_2 \qquad C_2H_5SiR_3$$

$$5\%$$

Clearly when an excess of ethylene was used, the dominant process was condensation resulting in vinylsilane formation [12].

Very similar results have been observed for propylene, styrene and isobutylene. The condensation process also yields excellent results with conjugated olefins such as butadiene:

$$HSi(NMe_2)_3 \quad + \quad H_2C{=}CH{-}CH{=}CH_2 \quad \xrightarrow{\text{Rh}}$$

$$H_2C{=}CH{-}CH{=}CHSi(NMe_2)_3$$

The hydrosilation of methyl methacrylate by "Tris" and trialkoxysilanes can give 1,2- or 1,4-hydrosilation. The results which are summarized below again demonstrated the dominant influence of steric crowding in determining the reaction path [12]

$$HSiR_3 \quad + \quad CH_2 = CMeCOOMe \quad \xrightarrow{\quad Rh \quad}$$

$$R_3 SiCH_2 CHMeCOOMe$$

1,2 - Hydrosilation

+

$$Me_2 C = C (OMe) OSiR_3$$

1,4 - Hydrosilation

	% 1,2-	% 1,4-
R = OMe	50	50
= OEt	50	50
= OiPr	5	95
= OtBU	0	100
= NMe$_2$	0	100

With increasing steric crowding around silicon, 1,4-hydrosilation becomes the dominant process [12].

OTHER CHEMICAL PROPERTIES

The reactivity of "Tris" in a variety of processes which have already been described suggests that it should have practical utility in a number of ways. Some initial results follow:

The ease of reaction of "Tris" with alcohols led to the expectation that it would react readily with paper, cotton, other textiles and inorganics such as hydrophilic silicas. This was confirmed with the observation of rapid reactions at ambient temperatures. Somewhat unexpectedly, it was observed that the treatment of these materials with "Tris" rendered them completely hydrophobic. Furthermore the treated materials retained their hydrophobic properties for extended periods with surprising resistance to hydrolysis. The behavior of these materials is still under investigation.

"Tris" should be a preferred starting material for the preparation of silicon nitride. The transamination of "Tris", as mentioned earlier, with primary amines or ammonia proved to be a facile process for the preparation of prepolymer intermediates [8]:

$$HSi(NMe_2)_3 + RNH_2 \longrightarrow [-HSiNR-]_x + 3\ Me_2NH$$

Upon pyrolysis the prepolymers were converted to silicon nitride of high purity and in good conversion.

REFERENCES

1. B. Kanner, XV Organosilicon Symposium, Duke University, March 27, 1981.
2. B. Kanner and W.B. Herdle, U.S. Patent 4,255,348 March 10,1981.
3. B. Kanner, W.B. Herdle and D.L. Bailey, U.S. Patent 4,289,889 September 15, 1981.
4. B. Kanner and B. Prokai, U.S. Patent 3,957,842 May 18, 1976.
5. B. Kanner and S.P. Hopper, U.S. Patent 4,395,564 July 13, 1982.
6. B. Kanner and S.P. Hopper, U.S. Patent 4,384,131 July 13, 1982.
7. J.M. Quirk, A.P. DeMonte, B. Kanner and S.P. Hopper, U.S. Patent 4,556,725 December 3, 1985.
8. B. Kanner, S. P. Hopper, R.E. King and C.L. Schilling, U.S. Patent pending.
9. B. Kanner, S.P. Hopper and D.J. Sepalak, U.S. Patent, 4,400,526 September 30,1982.
10. W.E. Dennis and J.L. Speier, J. Org. Chem. 35, 3879 (1970).
11. B. Kanner, J.M. Quirk, A.P. DeMonte and K.R. Mehta, U,S. Patent 4,558,146 December 10, 1985.
12. J.M. Quirk, XX Organosilicon Symposium, Union Carbide, April 19,1986.

PART III

SILICON IN LIVING SYSTEMS

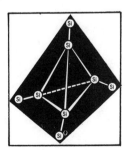

Chapter 13

The value and new directions of silicon chemistry for obtaining bioactive compounds

Sandor Barcza — Sandoz Research Institute, East Hanover, NJ 07936 USA.

This is a review of potential and realized application of organosilicon chemistry to biological goals. Biologically relevant Si chemistry centers around bioactive compounds, medical polymers and the occurrence and role of Si in living nature. Bioactivity can be applied in <u>medicine</u>, <u>agriculture</u>, (herbicides, fungicides, insecticides etc.) and <u>flavors/fragrances</u>. I plan to sample significant and instructive prototype applications from the past, and give pointers to the future. More complete earlier reviews of work done exist [1]. Some practical <u>industrial aspects</u>, from conception to market, will be included.

SYNTHESIS

The widespread application of Si chemistry to organic synthesis [2] is no less important to medicinals. There is hardly a multistep drug synthesis or analysis without the involvement of silicon. Consider the ubiquitous $[(iPr)_2Si]_2O$ cyclo protecting group for nucleosides and carbohydrates and the large scale use of Me_2Si on penicillins.

Going beyond protecting and other reagents is to <u>incorporate Si in the products</u>. The variety of Diels-Alder and 1,3 dipolar cycloadditions are as yet impractical for Si cycles, since the mostly high energy chemistry of double-

bonded Si species is difficult to scale up. But the rich
substitution chemistry of Si compensates: 1) substitutions on
Si, now also stereochemically understood [3], 2) clean SN_2 on
C alpha to Si, uncomplicated by Wagner-Meerwein rearrangement
and elimination and 3) reasonably understood polarity of
vinylsilanes and beta hetero substituted silanes.

We synthesized a variety of heterosilacycles by combining 1) and 2) above into "bifunctional cyclizations" [4] of
organic dinucleophiles with Si dielectrophiles in an ordered,
regioselective manner, e.g.:

Sleep inducer

About 10^5 organic, but only about 10^3 organosilicon
chemicals are commercially available, and the latter are
probably more expensive. This is rarely an obstacle for
drugs required in small amounts. Our own two clinical substances were prepared in ~10-100 kg amounts for clinical,
toxicity, stability, metabolism and formulation studies.
Even a Si-agrochemical fungicide reached the market ((p-F-
Ph)$_2$Si(CH$_3$)CH$_2$-triazole, DPX-H6573)[5].

Substructural analysis [6] of commercial Si compounds
revealed good diversity, but dominance of being monomer-by-
product derived. A complement to the "direct synthesis" of
methylchlorosilanes is highly desirable.

Finding novel and suitable bioactive compounds is so
difficult and so important, that it is a must to use the
complementary diversity that Si structures offer.

BIOACTIVE SILICON COMPOUNDS

The idea of silicon drugs conjured up visions of kidney
stones and silicosis for early pharmacologists and clini-
cians, who feared such "unnatural" substances.

Rochow prophetically remarked in 1947 that there
appeared to be no toxicity associated with the element Si or

Si-C bonds in organosilicon compounds [7]. Since then, many examples have accumulated to support this. Some "natural" compounds are highly toxic (e.g., botulism toxins, composed of amino acids = food), and many "unnatural" (synthetic) substances have been found safe. Even within the class of silatranes the LD_{50} varies from 0.1 to ~5000 mg/kg [8].

It is clear that each organosilicon developmental candidate must be studied for its own toxicity, side effects; regulatory authorities seem to be in complete agreement.

Fear of Si is additionally reduced by observing examples showing that the body - having the enzymes - metabolizes organosilicon structures via the organic groups. Microbes have been found that even do this stereospecifically [9]. Even if a typical drug dose would be completely degraded to silicic acid, the Si released would be of the same order or magnitude as intake from food and water (e.g., at Sandoz:22 mg/l). Steers on high silicon diet cleared much Si in the urine, and did not develop urinary calculi. Even quartz implants were measured to be thinned by the body. [However, silicates inhaled in larger amounts do cause silicosis, and asbestos causes lung cancer].

Converse fears in agrochemistry, that siloxanes may persist forever, were dispelled by C.L. Frye who showed that the organic groups are photolyzed off, while clay surfaces help.

Pioneering studies of Fessenden et al [10] generated early examples of Si bioisosteres of mostly central nervous system drugs, found similar activities, but also some differences (e.g., sila-meprobamate).

They also showed the first examples of metabolism: hydroxylation on CH_3Si, PhSi, oxidation of HSi and alkyl Si degradation.

Other Si analogs of active organic structures followed: our total synthesis of the first silasteroids [11] provided estrogens, and was extended by Pitt et al [12] to ring-A non-aromatic structures, intended to be progestins.

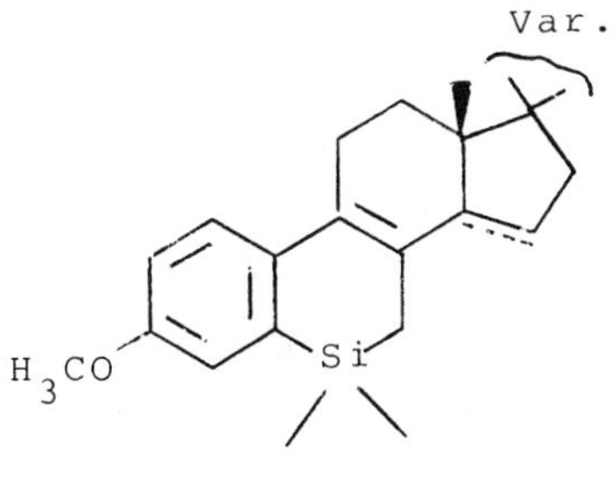

Silaestrogens 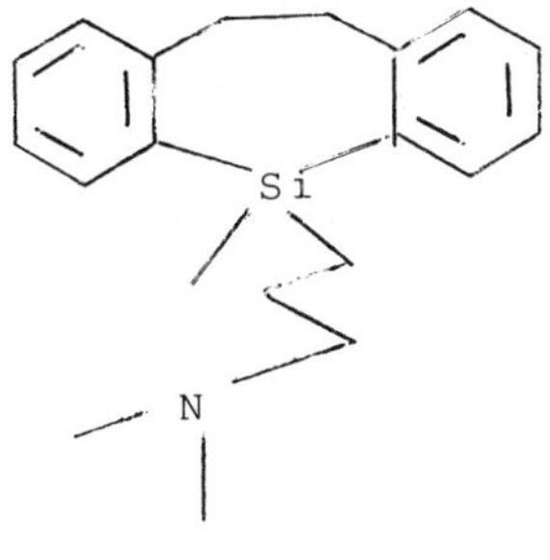   Silipramine

Very significant entry by the Coreys et al [13] prepared
and studied the structures of analogs of tricyclic
anti-depressants, e.g. silipramine.

Systematic investigation of sila-carba pairs by
Wannagat and Tacke and more recently by Tacke et al [14]
yielded some more highly active sila analogs, differing
activity profiles, distinction of receptor subtypes via the
sila compound, optically active sila drugs, etc.

Meanwhile, Wannagat prepared sila fragrances [15].

Even greater contributions are made by silicon compounds
whose carbon analogs are prepared with difficulty or do not
exist at all: silatranes, which elicit a wide variety of biological responses [1e] and cisobitan, the estrogenic 2,6-cis-
diphenylhexamethyl-cyclotetrasiloxane [16], and the hypnotic
diphenylsilanediol [17].

Si AS A TOOL TOWARDS BIOLOGICAL GOALS

The main message of this review is that Si, having
proven itself as a nontoxic element (unlike B, Be, Sn, Hg,
Cd, Pb etc), is a tool complementary to organic chemistry to
realize biological action.

Parameters which influence biological activity at the
molecular level are: molecular shape, flexibility, bulk,
surface, the properties of the surface, e.g., lipophilicity/
hydrophilicity, polarizability, charges, electrostatic potential, hydrogen bonding donor and acceptor properties, lone
pairs, pi electron density and delocalizability, pKa
(acidity, basicity), dipole moments, etc.

Some of these are related, are location-specific,
vectorial (directional) and dynamic.

The chemical and physical consequences of the structure
of Si compounds also make them useful at the macroscopic,
physical and physicochemical level.

For exploiting silicon structures, it must be considered
which ones exist under the range of applications. Surviving
isolation, purification, formulation and stability during
storage are important. Most Si-C bonds and many Si-O bonds
are stable enough for the purpose, in the order Si-O-Si >
Si-OH ~ Si-OR > Si-O-aryl > Si-O-acyl, etc. Structures with
Si-N and better leaving groups are hydrolized too easily.
Si-H bonds are likely to be oxidized chemically and metaboli-
cally. This still leaves ample opportunities in the largely
unexplored field.

Importantly, modern approaches to biological agents include cases where not stability, but specified action, transformation is purposely included, such as in pro-drugs and suicide substrates.

Some exemplification of the above principles follow.

The _physical_ properties and chemical-biological inertness are exploited in using polyorganosiloxanes as prosthetic devices, e.g. bone replacement, heart valves.

Physicochemical effects are applied in silicone membrane sustained release drug depots.

Chemical (covalent)-_physicochemical_ effect is put to use by trialkylsilylating hydroxyl groups of active molecules. The drastic increase in _lipophilicity_ causes long residence in lipid tissues and sustained release and activity. Testosteroxy-silane derivatives were not only longer acting, but had higher time-activity integral. A single dose of our 3-trimethylsilyl-methyl-estrogen had at least three months of activity.

Bonds to Si are longer and more flexible than bonds to C, with corresponding geometric consequences. Since the Si is often alkyl (methyl) substituted, the effective lipid bulk and resultant lipophilicity of sila analogs is higher. An increase of 0.6 in the log of the lipid/aqueous partition coefficient can be expected upon $(alkyl)_n$-C $\rightarrow$ $(alkyl)_n$-Si substitution.

Our own disila-morpholine derivative [18]

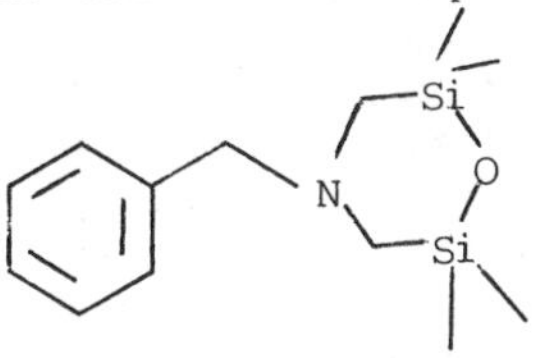

X=Si   vs   X=C

showed high muscle relaxant activity in central nervous system related animal testing, whereas there was no measurable activity at equal doses of the X=C analog. QSAR (quantitative structure-activity relation) analysis on the series showed that this could be accounted for in terms of lipophilicity.

A special benefit of Si versus C chemistry is the greater ease with which many types of _quaternary, including spiro centers_ can be constructed. The cholesterol absorption inhibiting agent that I synthesized

SAN-58-035

had 8.7 times the activity of the C-analog, and was considerably easier to construct. Corresponding straight (unbranched) saturated carbon chain (fatty acyl) analogs were
stated to be inactive.

Some other quaternary structures constructed in our
laboratory are

Vasodilator [19]   Antifungal [20]   Growth Hormone
                                                 Inhibiting [21]

From quaternary Si centers potentially C-centered drug
leads may be generated, based on bioisosterism.

The high energy of double bonds to Si prevents their
existence in vivo. Thus, Si could act as its own "aromatase
inhibitor" or preventor of dehydrogenation. In some
molecules this may preclude potentially carcinogenic polyaro-
matic systems from forming. This was one of the goals of our
6-silasteroid total synthesis [11]. Side chain Si-substitu-
ted steroids have also been prepared as aromatase inhibitors
[22].

MODERN ASPECTS AND FUTURE POINTERS OF BIOSILICON CHEMISTRY

As drug research begins to apply more modern tools, so
does (should) organosilicon chemistry. This not only brings
the obvious benefit of greater efficiency and insight, but
has nonobvious synergistic benefits and feedback, including:
1) the designed expansion of scope of structural and other
property variation of molecules through unique properties of
silicon, and 2) enrichment and fine tuning of insight by Si/C
comparisons, since the [physico] chemical differences of Si/C
substitution are already better understood.

Of prime importance is <u>mechanistic insight</u> and rational design based on mechanistic thinking. Such was the case for the <u>suicide inhibitor</u> of a P450 steroid side chain oxidizing enzyme [23]. The unique properties of Si allow the 17-C $(CH_3)(OH)-CH_2-CH_2-SiMe_3$ substituted steroid to become a suicide substrate.

Other modern aspects of biosilicon applications involve structural details, quantitative evaluations and computer assistance.

The hand-in-hand application of synthesis and <u>X-ray</u> crystal and molecular <u>structure determination</u> was most widely/earliest applied by the Coreys [24]. Results were discussed in the context of benzo wing angles of tricyclic antidepressants.

For [anticipated] structure-activity correlation it is important to compute molecular structures before even a substance is synthesized. The most practical tool for that is <u>molecular mechanics</u> [25], if well parametrized for the Si structure.

Molecular geometry as well as electrostatic potential, charge distribution, dipole moment, etc. can be calculated by <u>semiempirical molecular orbital</u> methods. These were done by Höltje et al [26] for Si-anticholinergics. The results were also displayed by sophisticated molecular graphics. This work, to my knowledge, represents the most sophisticated study so far of computer-assisted <u>molecular modeling</u> and property-activity correlation applied to Si compounds. Similar modeling was done on cisobitan, comparing it with other estrogen geometries, e.g. that of diethylstilbestrol [27].

More rigorous, but computationally much more intensive <u>ab initio</u> calculations have been done on smaller Si structures [28]. The main value of these is to provide rigorous foundation and parameters for more empirical computations on more complex systems.

For series of bioactive compounds <u>statistical correlations</u> are useful, in order to establish Quantitative Structure-Activity Relationships. For these (e.g. "Hansch analysis") too, more physicochemical parameters have to be determined. <u>Pattern recognition</u> is also of interest, and a certain type of it was applied to antimicrobial aminoalkyl-silicon compounds.

A new dimension in bioorganosilicon chemistry is <u>interfacial systems</u>, such as

1) antimicrobial bonded aminoalkylsilanes on glass, etc.

2) non-thrombogenic surfaces

3) bonded silica and glass surfaces for reverse phase separation of [bio]molecules, enantiomer (chiral) separations and affinity columns

4) biosensors, electroactive surfaces, enzyme electrodes

Silicon in living nature provides a tantalizing area. Plants require silicon, E. Carlisle has shown [29] Si to be essential for bone development in chicks, B. Volcani [30] furthered the understanding of the diatom's Si metabolism. Atherosclerosis has an inverse correlation with Si-content of blood vessels.

It is hoped that a bridge will be built by future researchers between Si in living nature and organosilicon chemistry.

Acknowledgment: Although the main purpose of this paper was to make general points, review trends and sample other's work, it is my pleasant obligation to thank my co-workers in silicon chemistry:C.W. Hoffman, M. Hidalgo Myers, B. Lewis, R. Grotenstein, M.A. Thiede, R. Barth, Dr. R. Ruprecht, and thanks are due also to the late Dr. A.J. Frey for enlightened support of my proposals, and to Mr. F.H. Weinfeldt et al. for formulating patents.

REFERENCES

[1]a) Biochemistry of Silicon and Related Problems; Ed:Bendz, G., Lindqvist, I., Runnström-Reio, V., Plenum, New York, 1978. b) Silizium und Leben, Voronkov, M.G., Zelchan, G.I., Lukevitz, E.; Akademie-Verlag, Berlin, 1975. Second Ed: Kremnii i Zhizn, Ziatne, Riga, 1978. c) Fessenden, R.J., Fessenden, J.S., Adv. Organometallic Chem. 1980, 18, 275. d) Topics in Curr. Chem. 1979, 84, p1: Tacke, R., Wannagat, U.; e) ibid., p 77, Voronkov; f) Barcza, S, Actual. chim., 1986, 83.

[2]a) Silicon Reagents for Organic Synthesis, Weber, W.P., Springer-Verlag, New York, 1983. Also: Am. Chem. Soc. Short Course. b) Silicon in Organic Synthesis, Colvin, E., Butterworths, London, 1980.

[3]a) Corriu, R.J.P., Guerin, C., Moreau, J.J.E., Topics in Stereochem. 1984, 15, 43; b) Stereochemistry, Mechanism and Silicon, Sommer, L.H., McGraw-Hill, New York, 1965.

[4]   Barcza, S., Thiede, M.A., 5th Int. Symp. Organosilicon
        Chem. Karlsruhe, 1978, Abstracts, p 148.   Barcza, S.,
        US Pat. 4,132,725, Jan. 2, 1979; 4,175,091, Nov. 20,
        1979.

[5]   Moberg, W.K., Basarab, G.S., Cuomo, J., Liang, P.H.,
        Am. Chem. Soc. Natl. Mtg., Chicago, Sept. 1985,
        Abstracts, Agro 30; Chem. Eng. News, Sept. 23, 1985,
        p. 43.

[6]   Barcza, S., XIX Organosilicon Symp., Baton Rouge, 1985,
        Abstracts, p. 56.

[7]   Burkhard, C.A., Rochow, E.G., Booth, H.S., Hartt, J.,
        Chem. Rev. 1947, 41, 97.

[8]   Voronkov, M.G., p. 395 in Ref 1a.

[9]   Tacke, R., in Organosilicon and Bioorganosilicon Chem.,
        Ed., Sakurai, H., p. 251.   Ellis Horwood Ltd.
        Chichester, 1985.

[10]   Ref. 1c; Fessenden, R.J., Coon, M.D., J. Med. Chem.
        1965, 8, 604.

[11]   Barcza, S., Hoffman, C.W., 2nd Int. Symp. Organosilicon
        Chem. Bordeaux, 1968, Abstracts, p. 12; Tetrahedron
        1975, 31, 2363; Barcza, S., US Pat. 3,529,005, Sept.
        15, 1970; US Pat. 3,637,782, Jan. 25, 1972.

[12]   Pitt, C.G., Friedman, A.E., Rector, D., Wani, M.C.,
        Tetrahedron, 1975, 31, 2369.

[13]   Corey, J.Y., Farrell, R.L., J. Organomet. Chem., 1978,
        153, 15. Also Ref. 24.

[14]   Ref 1d; Tacke, R., Zilch, H., Endeavour, 1986, 10, 191;
        also: Actual. Chim., 1986, 75.

[15]   Wannagat, U., Nova Acta Leopold., 1985, 59, 353.

[16]   Bennett, D.R., Aberg, B., Eds: Acta Pharm. Tox. 1975,
        36, Suppl. III, p 1-147 (12 chapters); LeVier, R.R.,
        Chandler, M.L., Wendel, S.R., p 473-514 in Ref 1a;
        Strindberg, B., ibid 515.

[17]   LeVier, R.R., Actual. chim. 1986, 89.

[18]   Barcza, S., Thiede, M.A., 6th Int. Symp. Organosilicon
        Chem., Budapest, 1981, Abstracts, p 203; Barcza, S., US
        Pat. 4,224,317, Sept. 23, 1980; US Pat. 4,208,408, June
        17, 1980.

[19]   Barcza, S., Hidalgo, M.G., 3rd Int. Symp. Organosilicon
       Chem., Madison, 1972, Abstracts, p 20; Barcza, S., US
       Pat. 3,585,228, June 15, 1971, also US Pat. 3,592,831,
       July 13, 1971.

[20]   Barcza, S., Lewis, B., Thiede, M.A., 4th Int. Symp.
       Organosilicon Chem., Moscow, 1975, Abstracts, Vol I,
       Part 2, p 24; Barcza, S., US Pat. 3,692,798, Sept. 19,
       1972.

[21]   Barcza, S., Thiede, M.A., 5th Int. Symp. Organosilicon
       Chem., Karlsruhe, 1978, Abstracts, p 210.

[22]   Burkhart, J.P., Weintraub, P.M., Wright, C.L.,
       Johnston, J.O., Steroids, 1985, 45, 357.

[23]   Nagahisa, A., Orme-Johnson, W.O., Wilson, S.R.,
       J. Am. Chem. Soc., 1984, 106, 1166.

[24]   Corey, E.R., Corey, J.Y., Glick, M.D., Acta
       Cryst., 1976, B32, 2025; Corey, E.R., Paton, W.F.,
       Corey, J.Y., J. Organomet. Chem., 1979, 179, 241.

[25]   Cartledge, F.K., J. Organomet. Chem., 1982, 225,
       131; Frierson III, M.R., Diss. Abstr. Int. (Chem.),
       1984, 45, 558.

[26]   Höltje, H.-D., Busch, T., Deutsche Apoth. Ztg., 1986,
       126, 2007.

[27]   Grigoras, S., LeVier, R., Lane, T., XX Organosilicon
       Symp., Tarrytown, 1986, Abstracts, A9.

[28]   Luke, B.T., J.A. Pople, M-B. Krogh-Jespersen, Y.
       Apeloig, J. Chandrasekhar and V. Rague Schleyer, P.,
       J. Am. Chem. Soc., 1986, 108, 260-9, also see p 270-84.

29]   Carlisle, E.M., p 231 in Ref 1a; also Ch4, p 69 in Ref
       30.

[30]   Silicon and Siliceous Structures in Biological Systems,
       Eds.: Simpson, T.L., Volcani, B.E., Springer-Verlag,
       Berlin, 1981.

Chapter 14

Silicon in living systems

M.G. Voronkov – Institute of Organic Chemistry, Siberian Division, Academy of Sciences, 664033 Irkutsk, USSR.

Silicon is the basis of inanimated nature. Oxygen compounds of silicon, silica and silicates, constitute more than half of the earth's crust which contains 27.8 % of silicon, as much as other elements taken together.

Silicon is an ubiquitous element. It is present in minor or more or less significant quantities in nearly all living organisms and materials of natural or artificial origin. Silicon compounds are daily taken up by the organisms of human with water (where they are always in the soluble form), with inhaled air (as fine particles of silicon-containing dust) and, finally, with food.

The daily silicon uptake by man with food (especially vegetable) is about 0.5 g Si. Considerably greater quantities of this element are digested by herbivorous animals. For all that, silicon does not seem to be accumulated in the organism of mammals (the silicon content of the human body is several grams) and usually does not produce either adverse or beneficial effect. The one exception is silicosis, lung disease caused by chronic inhalation of comparative coarse particles of silica.

Almost three decades ago all this led scientists to a strong conviction that silicon compounds are biologically inert, useless and even harmful. The presence of silicon in the organism of mammals was considered occasional and unessential. The ideas, widely accepted at that time, that silicon is of no importance in living systems has contributed much to the introduction of organosilicon polymers, polyorganosiloxanes (silicones) into medicine and cosmetology.

At present, polyorganosiloxanes are extensively employed in plastic and restorative surgery. This is due, first of all, to the fact that they are soft implant materials which do not produce any hazardous effect nor undergo decomposition in the organism. The stability of these compounds to aging and sterilization as well as high adhesive properties, minimal tissue response to silicone-based implants make them rather promising for medical use (prosthetic devices, heart valves, artificial arteries, bone substitute, etc.). Silicones have also found application as a "depot" of therapeutic preparations (drugs of prolonged action).

The discovery of bioactivity of organosilicon compounds, silatranes, $RSi(OCH_2CH_2)_3N$, made by us in 1963 changed the prevailed concept on the biological inertness of this element. Some of these compounds, 1-arylsilatranes proved to be highly toxic for warm-blooded animals and exerted a stimulating effect on the central nervous system.

The clearly-displayed physiological effect of silatranes has convinced us that many other silicon derivatives have to possess a specific bioactivity and that a number of derivatives of this element should play a significant role in certain vital processes.

Natural silicon compounds played an essential part in the origin of life on earth. Possessing a specific structure, adsorbtion and catalytic properties, they served as a matrix on which simple organic molecules from the world ocean were transformed to complex protein bodies, nucleotides and nucleic acids. Besides, optically active forms of quartz ensured the appearance of existing at present asymmetry of natural organic compounds. In the course of biogenesis on the surface of silica, silicates and alumosilicates (clays, first of all) silicon could not but insert in the organic structures formed, and in that remote epoch it was one of the principle elements in living nature and its protoorganisms. This is confirmed by the fact that nowadays silicon is of great importance for numerous organisms left at the lower evolutional stage and occurs there in large quantities.

Numerous representatives of lower life-forms in-
habiting oceans, seas and lakes and composing the
bulk of phyto- and zooplankton take up silica from
the external aqueous environment and use it for the
formation of their shells, frustules and frameworks.
These are relic simple organisms such as diatoms,
foraminifera, radiolaria, siliceous sponges, etc.
    Much silicon is present in existing nowadays
"prehistoric" plants (horsetail, moss, ferns). There
are widely met "silicate bacteria" which break up
rocks by liberating from minerals potassium, phosphorus, silicon and some trace elements assimilated
by plants. Therefore, they can be used as valuable
fertilizers. Moreover, the biomass of these bacteria
proved to be an effective fodder additive.
    We have established silicon to be an inherent
component of nucleic acids where its content amounts
to 0.15-0.30 %. The silicon:phosphorus atomic ratio
is 1 per 20-30 in desoxyribonucleic acid and 1 per
25-46 in ribonucleic acid (RNA). Silicon in nucleic
acids is assumed to be isomorphous to phosphorus and
able to substitute some phosphorus atoms. It is
quite possible, however, that orthosilicis acid is
bound with macromolecules of nucleic acid via hydrogen bonding.
    Recently Soviet scientists N.E. and E.P. Aleshins
and E.R. Avakyan have given evidence for the essential role of silicon in functioning of nucleic
acids. They have found than at even lowest Si levels
in the nutrient medium the Si:P atomic ratio in RNA
of rice is 1:7. The administration of digestible
silicon into the medium changes this ratio to 1:3
and the elimination of phosphorus from the nutrient
medium leads to a predominance of silicon (3:1). So,
rice possesses a kind of fermentative system utilizing for the formation of nucleic acids not only
phosphorus but silicon as well.
    Silicon compounds also play an important part in
the life activity of organisms at a higher stage of
evolutionary development. These include, for example,
higher cultivated plants (wheat, oats, barley,
millet, rice, sugar-beet, etc.), bamboo and some
trees (fir, larch, palm). These and other plants
with comparatively high silicon contents intensively
assimilate silicon from soil.
    The role of silicon in the life of plants is not
completely understood yet. It is well known, however,
that silicon is present in plant tissue partly as
organosilicon compounds (silicic esters of hyrdo-
carbons and proteins) and partly as mineral silicon
compounds (silica, silicic acids and silicates).
There have been found in plants certain enzymes

"silicases" which facilitate the conversion of inorganic compounds of silicon to organosilicon derivatives.

The importance of silicon in the growth and
development of birds and mammals has been confirmed
by E.M. Carlisle and K. Schwartz. They observed
drastic changes in the skeleton, feather, skin and
weight of chicks and rats fed on silicon-deficient
diet and kept in plastic chambers completely pre-
venting the penetration of environmental silicon.

The addition of some organic and readily assimilated inorganic compounds of silicon to fodder leads
to weight gains and increased vitality of farm
animals.

In the organism of higher animals and human silicon occurs in trace amounts (0.001 % of body weight),
although it is found in all tissues and organs. Especially high silicon contents are observed in connective tissue, skin, lungs, glands, bones, dental
enamel, teeth, hair.

In the organism of mammals and man silicon compounds are present in three principal forms:

1. Water-soluble inorganic compounds capable of
passing through cell walls which can be readily
eliminated from the organism (organosilicic acids,
ortho- and oligo-silicic acid anions).

2. Organosilicon compounds and complexes soluble
and insoluble in organic solvents and containing
Si-O-C groups (ortho- and oligo-silicic esters of
carbohydrates, proteins, steroids, lipids, etc.).
The formation of organic derivatives of silicic acid
in living organisms involves hydroxy groups of not
only carbohydrates but of hydroxyamino acids and
other hydroxyl-containing compounds. One of the
possible forms of organosilicon compounds in the
organism of mammals are silicic esters of bi- and
polyatomic phenols and their derivatives occurring
in living cells and concentrating in nuclei and
mitochondria. The silicon atom coordination to the
nitrogens of aminoacids is also quite possible.

3. Usually foreign insoluble specific polymers
(polysilicic acids, silica, silicates) whose surface
is always covered with a chemically bonded (chemisorbed) layer of organic substances.

The absorption and localization of silicon in the
organism of mammals have been studied at the cellular
level. Silicon was found in nuclei, mitochondria,
vesicles and microsomes of liver, spleen and kidney
cells of the rat. In mitochondria silicon-containing
granules were revealed.

It is established at the molecular level that
silicon participates in the synthesis of mucopoly-

saccharides, in the formation of cartilage and connective tissue. Indeed, silicon is an inherent component of collagen contained in connective tissue and mucoplysaccharide-protein complex, which performs an essential role in bone mineralization.

Silicon is essential for normal functioning of epithelial and connective tissues to which it imparts, in particular, strength, elasticity and impermeability by serving as an agent which holds the protein and keratin molecules together. In blood vessels silicon, concentrated mainly in elastin and, to a lesser degree, in collagen, prevents the deposition of lipids, normalizes the permeability of walls and increases their elasticity.

The breakdown of silicon metabolism is associated with diseases such as atherosclerosis, cancer, leprosy, tuberculosis, diabetis, hepatitis, encephalitis, goitre, certain types of dermatites, etc., as well as with some processes of ageing. For example, in the case of atherosclerosis the silicon content in the connective tissue of blood vessels decreases sharply. This results in a drop in elasticity of arterial walls due to the disappearance of elastin responsible for their resilience and in an increase of their permeability to lipids. The silicon content of skin, arterial vessels and bony tissue decreases with age and is closely connected with the processes of senescence. At the same time, the concentration of silicon compounds in the blood of pregnant women, nursing mothers and new-born children is extremely high.

Silicon takes an active part in the metabolism of calcium and some other elements (P, Cl, F, Na, S, Al, Mo, Co) and lipids.

Our many year research has led to the discovery of numerous classes of organosilicon compounds displaying various types of bioactivity. Among them non-toxic silatranes, $RSi(OCH_2CH_2)_3N$ distinguished for a broad spectrum of action on many living organisms are of special interest. The character of the biological activity of silatranes is determined, first of all, by the substituent at the silicon atom. These compounds easily penetrate cell membranes which seems to be due to the high dipole moment, spheric form and electronic structure of their molecule. Some silatranes intensify the biosynthesis of protein and nucleic acids and, as a consequence, the growth of connective tissue and hair of man and animals.

Laboratory and clinical studies have shown some 1-alkoxysilatranes and 1-organylsilatranes (including 1-ethoxysilatrane and 1-chloromethylsilatrane)

to accelerate considerably (1.5-fold) the healing of
non-infected flesh wounds and burns and to display
an antiulcerous effect. The addition of 1-chloro-
methylsilatrane to the diet of hens at the pre-lay-
ing time results in weight gains, improves haemo-
poietic processes and increases the metabolic level
of proteins, fats, carbohydrates and mineral substances. The metabolic level improvement produces a
beneficient effect on the quality of eggs and the
chemical composition of fowl-flesh. The hen produc-
tivity (egg production) was increased on average by
12-13 % and the egg mass by 1.5-2 g.

1-Ethoxysilatrane and 1-chloromethylsilatrane increase by 17-21 % the silk-production of the silkworm
and improve the silk thread quality. At the same
time, unlike other known stimulants, these preparations do not reduce the productive activity of the
silkworm females.

These silatranes are effective growth stimulants
enhancing the yield of cotton, potato, tomato,
cereals and the green mass of maize, etc. The wide
spectrum of biological action of silatranes as bio-
stimulants allows them to be successfully used in
agriculture.

Some other organosilicon regulators of plant
growth, herbicides and algicides, have been found.
The former include, first of all, compounds of the
general formula $XCH_2CH_2Si(OR)_2Y$, in particular,
practically non-toxic Etephon of prolonged action.
This chemical increases by 70 % the cropping power
of tomatoes and by 160-170 % the yield of latex
from rubber-bearing plants.

The herbicidal effect is exhibited, for example,
by organosilicon compounds of the type $(BrCH_2)_nSiR_{4-n}$
(n = 2-3) and $Cl_3C(CH_3)Si(OCH_3)_2$.

A strong fungicidal effect of organosilicon
amines of the type $XYZSi(CH_2)_3NRR'$ which kill patho-
genic fungi in concentration 4-17 mkg/ml has been
revealed.

Complexes of 3-aminopropyltrimethoxysilane and
its N-substituted derivatives with pentahalophenols
rank among interesting antimicrobial preparations
possessing insecticidal, fungicidal and algicidal
activity.

Recently organosilicon compounds displaying an
antiviral effect have been obtained.

An extensive search for biologically active
organosilicon compounds has led to the discovery of
silicon derivatives which show antiblastic, anti-
sclerotic, membranostabilizing, anticoagulative,
analeptic, adaptogenic, narcotic, psychotropic,
antipyretic, ganglion-blocking, pilotropic (hair

growth stimulating), bactericidal, fungicidal, zoo-
cidal, insecticidal, chemosterilizing and insect
repelling effects.

During the latest years organosilicon compounds
with hypotensive, spasmolytic, anticonvulsive, anti-
arrhythmic, sedative, antitumorous, antigonadotropic,
antiholinergic and other types of biological effect
have been obtained.

The data on the bioactivity of organosilicon compounds accumulated until present allow the existing
or potential drugs therefore to be divided into four
types:

1. Donors of vitally essential silicon (1-alkoxy-
silatranes, orthosilicic acid esters, etc.). To the
compounds of this type one may attribute tetrakis-
(N,N-dialkylaminoethoxy)silane hydrochloride hydrates
$Si(OCH_2CH_2NR_2)_4 \cdot 4 HCl \cdot 4 H_2O$ (synthesized by P. Rosci-
szewski in Poland))stable in neutral and acid media,
and some other their analogs. These compounds show
an antiulcerous effect, stimulate the growth of
microorganisms and, at the same time, increase the
activity of antibiotics.

2. Compounds whose physiological effect is condi-
tioned by specificity of their molecular and electronic structure due to the presence of one or more
silicon atoms in their molecules (1-arylsilatranes,
cis-2,6-diphenylhexamethylcyclotetrasiloxane, etc.).

3. Iso-structural sila-derivatives of bioactive
organic compounds including "sila-pharmaca".

4. O-, N-, S- and C-silylated bioactive organic
compounds.

The above calssification outlines the trend in
further search for new types of bioactive organosilicon compounds in general and drugs in particular.

The extensive research of silicon-containing
analogs of numerous known drugs carried out by
U. Wannagat and R. Tacke has shown organosilicon
compounds in most cases to have a physiological effect similar to and simetimes stronger than that
of iso-structural organic substances. Unlike the
latter, however, these compounds, having performed
their therapeutic functions, are decomposed in the
organism inducing no side-effects. Besides, they
are, as a rule, less toxic than the corresponding
organic analogs.

The similarity in biological effect between iso-
structural compounds of silicon and carbon is of
great interest in designing new drugs, pesticides
and other types of bioactive substances since the
corresponding silicon compounds are often more accessible and cheaper than their carbon analogs.

The latter of the above mentioned types of bioactive organosilicon compounds is represented by known drugs modified by silylation of functional groups. These triorganylsilyl moieties substitute for the hydrogen atom attached to oxygen, nitrogen, sulfur or carbon. This substitution performs a dual function. First, it drastically changes the character of most reactive functional OH, NH_2 and SH groups. This increases lipophilicity (trimethyl-silylation enhances 2-5 fold the solubility in lipids) and affects the acid-basic properties of molecules and, consequently, the interaction with receptors and the metabolism of silylated compounds. Since the Si-O-C, Si-N-C and Si-S-C groups in these molecules are readily subjected to hydrolytic cleavage they may be regarded as precursors or donors of the corresponding pharmaceuticals. The rate of their liberation in the organism may be controlled by varying hydrocarbon substituents at the silicon atom.

Second, these bulky substituents considerably change the molecular parameters which, naturally, influences the orientation of the molecule on biological receptors and its ability to penetrate the cell membranes.

C-Trialkylsilyl derivatives of the known drugs usually display an analogous physiological effect. However, the degree of this effect can adequately be controlled. In this way it was possible to increase the activity of antibiotics, steroid hormones and insect-repellents.

There is no doubt at present that in the near future a series of bioactive organosilicon compounds will assume their proper place among the existing therapeuticals and agrochemicals.

The successful development of bioorganosilicon chemistry will also widen our ides on the ways of origin of life, on the mechanism of physiological processes and the structure of living matter.

Chapter 15

Permethylated siloxane insect toxicants

Robert R. LeVier – Dow Corning Corporation, Midland, Michigan 48686.

ABSTRACT

Linear and cyclic dimethylsiloxanes over a broad range of molecular weights are known for their low order of toxicity to man and animals. In contrast to this relative absence of activity in mammals, nonfunctional siloxanes have been found to be toxic to an array of insects including crickets, ants, cockroaches, alfalfa weevils, fleas, lice and others. These materials are effective as contact poisons and the onset of action ranges from seconds to several minutes depending on concentration and the anatomic site of exposure. Activity is relatively independent of molecular weight. While the mechanism is unknown, it does appear that activity depends upon some physical property rather than specific structural features.

INTRODUCTION

Linear and cyclic polydimethylsiloxanes (PDMS) ranging in viscosity from 0.65 centistokes (cs) to greater than 60,000 cs have found broad applications because of the stable physicochemical properties of these polymers and a uniformly low order of mammalian toxicity. In addition to many industrial applications, PDMS materials have well-established uses in food, cosmetic, pharmaceutical and medical device areas.

In contrast to PDMS, a number of low molecular weight functional silanes and siloxanes have been shown to possess biological activity [1]. In the majority of cases active structures bear organic functionalities typically found in medicinal chemistry. Pharmacologically active organosilicon compounds have potential for associated side-effects of greater or lesser significance. For example, an organosilicon estrogen caused gynecomastia in some prostatic cancer patients [2] and a silane anticonvulsant has been shown to have limiting hepatotoxicity in epileptic dogs [3].

The low mammalian toxicity of PDMS spans a viscosity range of about one to greater than 60,000 cs [4]. The acute oral LD_{50} exceeds 30 gm/kg body weight and one to two percent polymer in the diet for two years results in no detectable adverse effects in several species. PDMS administered orally has no effect on reproduction nor is it teratogenic, mutagenic or carcinogenic. PDMS is not a skin sensitizer and eye exposure elicits only transient conjunctival irritation. Both linear and cyclic permethylated siloxanes induce little toxicity via inhalation exposure.

It is of interest, then, to find that dimethylsiloxanes and other relatively unreactive siloxanes are toxic to a variety of insects. In 1979 Lover, et al. [5] reported that aqueous emulsions of dimethyl and phenylmethyl linear siloxanes in the viscosity range 100 to 12,000 cs are pediculicidal. More recently, Ito and Nishimura [6] demonstrated that a variety of cyclosiloxanes are insecticidal. An association of broad spectrum insecticidal activity with a body of polymers that are essentially nontoxic is unusual. The objective of the present study was to further characterize the insecticidal properties of linear and cyclic siloxanes.

MATERIALS AND METHODS

Four species of insects were used. Adult male and female brown crickets, Acheta domesticus, were obtained from a local supplier. Adults of both sexes, larvae and eggs, as appropriate, of alfalfa weevils (Hypera postica), cat fleas (Ctenocephalides felis), and bovine lice (Damalinia bovis) were obtained from their natural hosts.

Compounds not soluble in water were applied without dilution to the ventral thorax of adult forms with a microliter syringe or exposure was achieved by immersion. Eggs and larvae were either immersed or exposed to compound-treated filter paper. Water soluble compounds were diluted in water (V/V) without surfactants. Excess compound was removed after a specified exposure period by water rinsing and/or blotting with filter paper.

Adult crickets were also exposed to compounds in an aquarium test. For this purpose two groups of 45 crickets were placed in an aquarium measuring 75 cm X 30 cm X 28 cm along with several paper towels and a plastic disk containing a mixture of bait and test compound. The bait consisted of a 2:1 mixture of flour and sugar to which test compound was added to make a thick slurry. Mortality was recorded over a period of up to 48 hours.

In the case of adults mortality was defined as loss of the righting reflex and all movement with no recovery within 48 hours. For larvae death was defined as loss of all movement for at least 24 hours. Toxicant activity against eggs constituted failure to hatch or larval death shortly after hatching. Regression analysis was used to estimate the lethal dose for 50 percent of the insects tested (LD_{50}).

RESULTS AND DISCUSSION

CRICKETS-DIRECT APPLICATION

PDMS, 10 cs, was evaluated for toxicant activity in crickets by direct application to the ventral thorax. Application to sites caudad or cephalad to the thoracic segment was effective but the time to effect increased with increasing distance from this segment. Under these conditions the calculated LD_{50} was 0.87 ul (Figure 1). The righting reflex was lost between 5 and 15 sec. after PDMS application. Dependence on PDMS viscosity and, hence, molecular weight was estimated by applying 5 ul of polymer over a viscosity range of 5 to 1000 cs. As shown in Figure 2, there was only a two-fold decrease in mortality over a 200-fold increase in viscosity. The time to loss of the righting reflex did increase as a function of polymer viscosity.

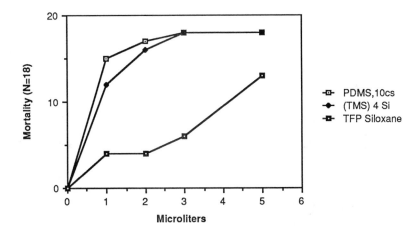

Fig. 1. Direct application of selected siloxanes to the ventral thorax of the brown cricket.

The results for some selected materials tested against crickets are summarized in Table 1 and Figure 3. All of the materials were active and of approximately equal potency. The water soluble silicone polymers are effective toxicants by virtue of their dimethylsiloxane content and the glycol fraction functions to impart water solubility. The glycol polymer fraction is equally functional in either pendant or intercalary configurations. The toxicant activity of dimethylsiloxane-ethylene oxide/propylene oxide polymers in water dilution is lost over a period of 12 to 24 hours but activity is retained when propylene glycol is used as the solvent. To illustrate, the LD_{50} in water was 3.4 percent for a freshly prepared solution with no activity retained after about 12 hours of storage at room temperature. The LD_{50} was 0.03 percent using propylene glycol as solvent and activity was stable over time.

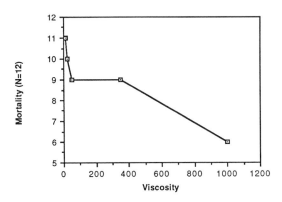

Fig. 2. Direct application of PDMS to the ventral thorax of the brown cricket. Effect of viscosity.

Table 1. Toxicant activity against the adult brown cricket.

Compound *	LD_{50}
PDMS, 10 cs	0.87 ul
$[(CH_3)_2SiO]_7$	1.07 ul
$[(CH_3)_3SiO]_4Si$	1.07 ul
$[CH_3,Ethyl\ SiO]_5$	1.56 ul
$[(CH_3,Hexyl)_2SiO]_3$	1.08 ul
$[(CH_3,Trifluoropropyl)_2SiO]_{3-4}$	0.70 ul
Poly(CH_3,Trifluoropropyl) Siloxane	3.70 ul
$[CH_3,(HOCH_2CH_2)_2Si]_2O$	2.71 ul
$CH_3,(CH_3O)_2$, Phenyl Si	2.09 ul
Dimethylsiloxane-EO/PO polymers	0.25 %
Dimethylsiloxane-Ethylene Oxide Copolymer	0.004 %

* Direct application of test samples to the ventral thorax except
  for immersion exposure to the two water soluble polymers. In all
  cases distilled water served as the control treatment.

CRICKETS – INDIRECT APPLICATION

The toxicant activity of cyclic dimethylheptamer (D_7) and a dimethylsiloxane-EO/PO polymers in the aquarium test paralleled their activity by direct application (Figure 4). That is, the toxicant activity of the silicone glycol emerged slowly compared to the time course for D_7 when the materials were presented in a bait. Neither material appeared to have attractant or feeding inhibition activity.

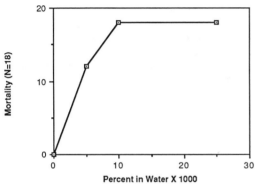

Fig. 3. Direct application of dimethylsiloxane-ethylene oxide copolymer to the ventral thorax of the brown cricket.

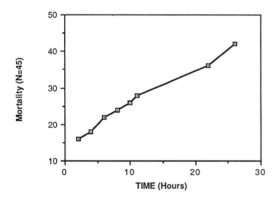

Fig. 4. Toxic effect of cyclic dimethylheptamer to the brown cricket. The silicone was incorporated in a bait.

ALFALFA WEEVILS – DIRECT APPLICATION

The results of exposing adult alfalfa weevils and weevil larvae and eggs to the dimethylsiloxane-glycol copolymers in water dilution are summarized in Table 2. Both of the polymers tested contained about 75 percent non-siloxane. A lower non-siloxane content diminishes water solubility but did not decrease insecticidal activity.

The greater insecticidal activity of a dimethylsiloxane-glycol block copolymer (dimethylsiloxane-ethylene oxide copolymer) compared to a branched copolymer variant (dimethicone copolyol) is probably dependent upon the configuration of the polymer in solution. That is, these polymers may orient to maximize the glycol/water interface by internalizing the siloxane phase. More dimethylsiloxane character is presented to the insect integument in the case of the siloxane-ethylene oxide copolymer. Additionally, dimethicone copolyol is probably hydrolytically unstable in water solution over a period of 12 to 24 hours.

Aqueous solutions of dimethicone copolyol pass through a viscosity increase cycle as the polymer concentration increases. It was observed that the toxicity to weevil eggs declined sharply above a polymer concentration of 70 percent in water and that this decline coincided with the peak in the viscosity cycle. Adherence of the silicone to the weevil eggs appeared to be greatest at the 70 percent polymer concentration.

Table 2. Toxicant activity against the alfalfa weevil.

Compound	Form Treated	LD_{50} *
Dimethylsiloxane-EO/PO polymers	Adults	20.4
	Larvae	8.4
	Eggs	59.6
Dimethylsiloxane-Ethylene Oxide Copolymer	Adults	3.2
	Larvae	0.9
	Eggs	29.2

* Percent polymer in water for a 30 sec. immersion. Distilled water served as the control treatment.

FELINE FLEAS – DIRECT APPLICATION

A one minute exposure of feline adult fleas and eggs to the water soluble siloxane copolymers was sufficient to express insecticidal activity (Table 3). It is evident that this species does not differ qualitatively from the other species tested.

Table 3. Toxicant activity against the feline flea.

Compound	Form Treated	LD_{50} *
Dimethylsiloxane-EO/PO polymers	Adults	25.0
Dimethylsiloxane-Ethylene Oxide Copolymer	Adults	0.17
	Eggs	7.8

* Percent polymer in water for a one minute immersion. Distilled water served as the control treatment.

BOVINE LICE – DIRECT APPLICATION

The activity of permethylated cyclosiloxanes was tested using two siloxane preparations and two methods of application. Material A was a 1:1 mixture of cyclic tetramer (D_4) and pentamer (D_5). Material B was a similar mixture of cyclics presented as an approximately 50 percent microemulsion in water. In the first experiment the test samples were applied as four drops delivered to groups of five lice on a filter paper pad followed with a water rinse after 30 sec. In the second design test samples were flooded on groups of five lice on a copper screen. The lice were rinsed twice with water after a ten second exposure period.

The results are summarized in Tables 4 and 5. Both the undiluted mixture of dimethyl cyclics and an aqueous emulsion of these cyclics were effective against bovine lice. In the form of a water emulsion the low molecular weight cyclics provided 67 percent kill (14 hours) at a one percent concentration.

Table 4. Toxicant activity against bovine lice. Experiment 1*.

Time (Min.) After Treatment	A a b c	B	C	D	E	W
0.5	0 0 0	0 0 0	0 0 0	0 0 0	0 0 0	0 0 0
1	5 Dead	5 Dead	0 0 0	0 0 0	1 0 0	5 Normal
2			0 0 0	1 0 0	1 0 0	
5			1 0 0	0 1 0	1 0 1	
10			0 0 2	0 1 1	0 1 1	
20			0 0 3	1 1 1	0 0 3	
30			0 0 3	1 1 0	0 0 4	
40	5 Dead	5 Dead	2 Dead	4 Dead	1 Dead	0 Dead

* Groups of 5 lice on a filter pad treated with 4 drops. A = 1:1 D_4 and D_5 applied without dilution. B = Mixed cyclics emulsion, 50 percent siloxane in water. E = Emulsion diluted 1:50 in water. W = Water. a = Movement of antennae only. b = Movement of antennae and legs. c = Crawling.

Table 5. Toxicant activity against bovine lice. Experiment 2*

Time (Hour) After Treatment	A	B	C	D	E	W
			Percent Mortality			
1	40	100	67	53	7	0
14	90	--	80	73	67	0

* Groups of 2-3 groups of 5 lice were flooded with treatment samples for 10 sec. See Table 4 for explanation of the samples.

The results of these experiments demonstrate that a range of silicon-based compounds and materials are lethal to four species of insects and that these materials are contact poisons rather than stomach poisons. Other experiments have demonstrated a toxic action against several species of ants, cockroaches, spiders, moths, beetles, centipedes, mealworm adults and larvae and feline ear mites. The mechanism of action of these relatively unreactive silicones remains unknown but the uniform activity and potency of the tested materials suggests that activity may well be dependent upon some thermodynamic property rather than a specific structural feature.

None of the silicones tested possess notable potency compared to currently available organic insecticides. However, the high order of safety to man and animals may make silicone insecticides attractive for use in certain applications. Incorporation of an insecticidal silicone in bait devices could provide activity over long periods of time because of the stability of silicones and such preparations would be virtually nontoxic to man and animals. These features of silicones along with their favorable cosmetic properties and low heat of vaporization, particularly for low molecular weight dimethyl cyclics, could offer a safe, effective and easy to use alternative to available human-use pediculicides and treatments for animal ectoparasites.

ACKNOWLEDGEMENTS

The author gratefully thanks Dr. T. W. Schillhorn van veen (bovine lice) and Dr. R. Bland (alfalfa weevil) for their experimental assistance and V. A. LeVier for technical assistance.

REFERENCES

1. LeVier, R. R., Chandler, M. L., and Wendel, S. R., 1978. Biochemistry of Silicon and Related Problems, Ed. by Bendz, G., and Lindquist, I.,Plenum Publishing Corp., page 473.

2. Alfthan, O., Anderson, L., Esposti, P. L., Fossa, S. D., Gammelgaard, P. A., Gjores, J. L., Issacson, S., Rasmussen, F., Ruutu, M., van Schreeh, T., Setterberg, G., Strandell, P., and Strindberg, B., 1983. Scand. J. Urol. Nephrol. 17:37.

3. Cunningham, J. G., Ford, R. B., Gifford, J. A., Hulce, V. D., Chandler, M. L., and LeVier, R. R., 1981. Amer. J. Vet. Res. 42:2178.

4. Calandra, J. C., Keplinger, M. L., Hobbs, E. J., and Tyler, L. J., 1976. Amer. Chem. Soc., Polymer Preprints 17:1.

5. Lover, M. J., 1979. U. S. Patent No. 4,146,619.

6. Ito, K., and Nishimura, Y., 1984. Japanese Patent No. 59-39809.

PART IV

SILICON REACTIVE INTERMEDIATES

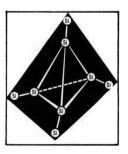

Chapter 16

Mechanism of photochemical reactions of aryldisilanes. The $^1(\sigma\pi)$ orthogonal intramolecular charge-transfer (OICT) states

Hideki Sakurai — Department of Chemistry, Faculty of Science, Tohoku University, Sendai 980, Japan.

ABSTRACT

Several arylpentamethyldisilanes and related compounds reveal dual fluorescence corresponding to two singlet excited states. The high polarity of the lowest excited singlet states with charge-transfer nature is evidenced by the solvent shifts. Structural effects clearly demonstrate that Si-Si bonds and aryl groups function as electron-donors and acceptors, respectively. The orthogonality of σ and π orbitals in the excited states is also proved by the stereoelectronic effects on the fluorescence. Product studies of photolysis of several aryldisilanes, carried out in detail under a variety of conditions, are discussed. Major products in nonpolar solvents arised mainly from the silyl radical pair derived by homolysis of the Si-Si bond, whereas in polar solvents containing methanol, products of methanolysis of the charge-separated $(\sigma\pi^*)$ excited states were obtained as well.

INTRODUCTION

Since the first discovery of of unusual electronic spectra of aryldisilanes [1], photochemistry of aryldisilanes has been an intriguing problem. Photolysis of an aryldisilane was reported for the first time by Sommer et al. on pentaphenylmethyldisilane in methanol [2]. Ishikawa, Kumada and their coworkers have studied later the photochemical behavior of phenylpentamethyldisilane and 4-methylphenylpentamethyldisilane rather extensively [3]. They have observed the formation of main products derived by an intermediate formed by 1,3-trimethyl-silyl shift. Weber et al. [4] have investigated the photo-reactions of arylpentamethyldisilanes with dimethyl sulfoxide in dioxane.

The primary photoreactions of aryldisilanes known until now can be classified into four types as shown in Scheme I; the formation of a silaethene via elimination of a hydrosilane (type 1), the formation of a C=Si species (type 2), elimina-tion of a silylene (type 3), and the attack of nucleophiles to an SiSi bond of the photoexcited phenyldisilane (type 4).

Scheme I

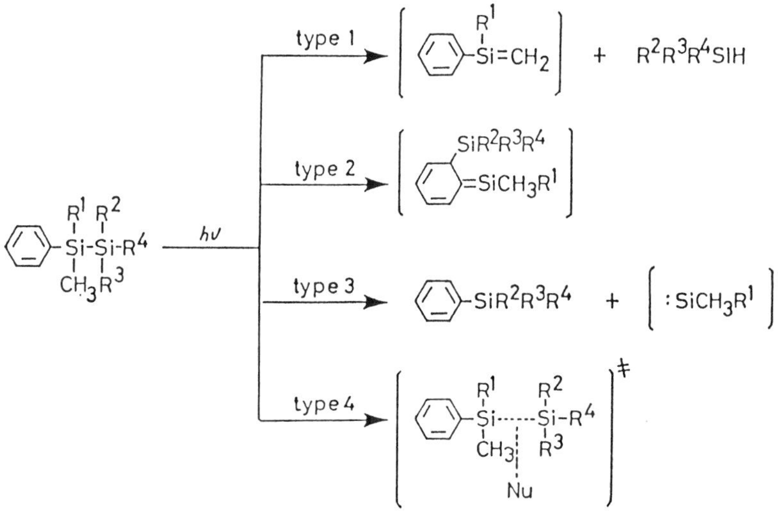

We have previously demonstrated that the silyl radical pair derived from homolysis of the Si-Si bond should play an important role in the photolysis under certain conditions [5], and also have revealed that the lowest excited singlet states of arylpolysilanes are assigned to the ($\sigma\pi^*$) orthogonal intramolecular charge-transfer (OICT) states [6]. In this paper, the author will describe (i) the nature of the excited states of aryldisilanes, (ii) the effect of solvent polarity on the photoreactions of 4-methylphenylpentamethyl- and 4-isopropenylphenylpentamethyldisilane, (iii) photochemistry of 2-pentamethyldisilanylnaphthalene and will demonstrate that a great diversity of the mode of photoreactions, caused by the solvent and substituent effects, can be understood by taking the ($\sigma\pi^*$) OICT states into consideration.

DUAL FLUORESCENCE OF ARYLDISILANES
AND RELATED COMPOUNDS

Fluorescence spectra were measured in several solvents at room temperature for compounds 1-14 listed below. Some representa-tive data are listed in Table 1 [6].

⟨⟩-SiMe₃   ⟨⟩-SiMe₂CMe₃   MeO-⟨⟩-Si₂Me₅   Me₂N-⟨⟩-Si₂Me₅

1   **2**   **9**   **10**

⟨⟩-SiMe₂SiMe₃   ⟨⟩-SiMe₂GeMe₃

3   **4**

⟨⟩-GeMe₂SiMe₃   ⟩-⟨⟩-Si₂Me₅

5   **6**

Me-⟨⟩-Si₂Me₅   Me-⟨⟩-Si₂Me₅

7   **8**

(structure 11) Me, Si₂Me₅   **11**

(structure 12) SiMe₂, SiMe₂   **12**

(structure 13) Si Me₂, Me₂   **13**

(structure 14) Si, Me, SiMe₃   **14**

Compounds except for 1, 2, 9, and 10 exhibit dual fluorescence especially in moderately polar solvents. Figure 1 shows dual fluorescence of 3 as a representative example. The emission at longer wave length does not originate from an excimer because the change in the concentration, for example of 13 in THF, did not affect the relative intensities of the two fluorescence bands. There is almost no solvent effect on the normal fluorescence 'b' bands observed at ca. 34,000 cm^{-1}. In contrast, the red shift was observed for the broad fluorescence 'a' bands at longer wavelength in polar solvents, indicating that the emitting states are polar.

Shizuka et al. have observed the dual fluorescence for some arylpolysilanes for the first time and have assigned the polar emitting state to be derived from the $2p\pi \rightarrow 3d\pi$ CT state [7]. However, we will demonstrate that the polar emitting state should be assigned in fact to the $\sigma(Si-Si)-\pi^*$ state.

For fluorescence, Si-Si or Si-Ge catenation is necessary in order to obtain the capacity as the electron donor with low ionization potentials. First, fluorescence spectra of phenyl-substituted group IVB catenates; phenylpentamethyldisilane (3), trimethylgermylphenyldimethylsilane (4), and trimethylsilylphenyldimethylgermane (5), measured in acetonitrile, were compared as shown in Figure 2. The CT fluorescence of 4 exhibits red shift in comparison with those of 3 and 5. Since the value of the ionization potential of Me₃SiGeMe₃ (8.65 eV) is similar to but smaller than that of Me₃SiSiMe₃ (8.69 eV), the $\sigma(Si-Ge)-\pi^*$ intramolecular charge-transfer excitation should be even more favorable than the $\sigma(Si-Si)-\pi^*$. This fact clearly indicates that the requirement of more than two silicon atoms necessary for CT fluorescence is not owing to lowering the level of the Si-Si acceptor orbitals due to $3d\pi-3d\pi$ conjugation but to raising the highest occupied orbital levels of the Si-Si donor group.

Next, the roles of acceptor orbitals are examined on the basis

Table 1. Fluorecence spectral data of arylpentamethyldisilanes and related compounds.

Compound	Fluorescence $(\nu \cdot 10^{-4} \ cm^{-1})$ [a]			
	i-Oct.	THF	EtOH	CH_3CN
3	33.8	33.8	33.8	----
	29.0	27.6	27.5	26.5
4	34.1	33.7	33.8	33.5
	28.5	27.2	26.7	26.2
5		33.5	33.9	33.3
		27.3	27.9	26.6
6	32.9	32.0	32.0	32.5
	----	26.3	26.4	24.7
7	33.7	33.3	33.3	----
	----	28.3	28.0	27.0
11	34.0	33.7	33.9	33.5
	27.5	27.1	27.0	26.5
12	33.4	33.2	33.5	33.4
	27.5	26.5	27.0	26.0
13		34.1	34.0	34.0
	29.0	26.8	27.0	26.0
14	34.5	33.9	34.3	33.8
	----	28.3	28.0	27.4

[a]
Upper and lower columns are for fluorescence 'b' and 'a', respectively.

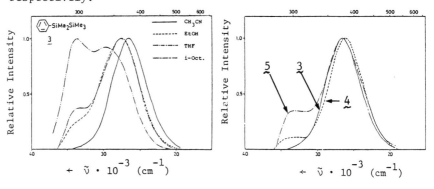

Fig. 1. Fluorescence spectra of **3** in various solvents.

Fig. 2. Fluorescence spectra of **3**, **4**, and **5** in acetonitrile.

of the p-substituent effects on the fluorescence spectra of phenylpentamethyldisilane (3). Figure 3 shows the effects of various p-substituents on the fluorescence of 3: isopropenyl- (6), methyl- (7), methoxy- (9), and N,N-dimethylamino-derivatives (10). 6 has a CT fluorescence band at the longest wavelength. The blue shift in CT emission bands was observed from 6 to 3 to 7. From these facts it was concluded that effects of p-substituents on the fluorescence of phenylpentamethyldisilane (3) should be taken as decisive evidence that the CT emission originates from the $^1(\sigma\pi^*)$ ICT.

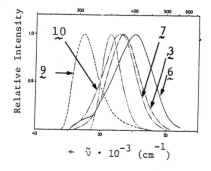

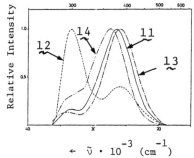

Fig. 3. Fluorescence spectra of 3, 6, 7, 9, and 10.

Fig. 4. Fluorescence spectra of 11, 12, 13, and 14.

Geometry of $^1(\sigma\pi^*)$ intramolecular charge-transfer state was then considered. Figure 4 shows fluorescence spectra of 11, 12, 13, and 14 in acetonitrile at room temperature. Remarkable stereoelectronic effects are found; thus, 12 exhibits CT fluorescence but with relatively weak intensity as compared to the normal fluorescence, while 13 emits preferentially from the CT state. The positions of CT fluorescence bands of 12 and 13 are nearly the identical and lie in the longest wavelength region of the four. 14 also exhibits CT fluorescence, however, the band shifts to blue region. The CT fluorescence band of 11 lies in an intermediate region between those of 14 and 12 (and 13). This structural evidence indicate that the most favorable geometry of (σπ*) ICT states is the form in which the σ(Si-Si) orbital is orthogonal to the π orbital (and therefore in the benzene nodal plane) as shown below.

In conclusion, dual fluorescence of a number of aryldisilanes and related compounds are characterized by the existence of the polar low-energy (σπ*) ICT states in which donor σ and acceptor π orbitals are favorably orthogonal (OICT). The $^1(\sigma\pi^*)$ OICT states resemble, in concept, both $^1(n\pi^*)$ TICT states in p-dimethylaminobenzonitrile [8] and sudden polarization in the $^1(\pi\pi^*)$ states of olefins [9], in which two interacting orbitals are orthogonal in the twisted molecular conformation. These phenomena are in good harmony and we propose OICT as a general term.

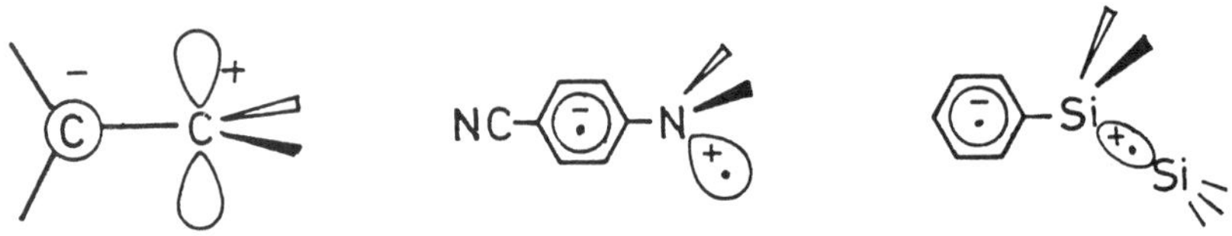

"sudden-polarized" $^1(\pi\pi)$ $^1(n\pi)$TICT $^1(\sigma\pi)$OICT

EFFECTS OF SOLVENT POLARITY ON THE PHOTOREACTION OF 4-METHYL-PHENYLPENTAMETHYLDISILANE

Methanol is often added for trapping the C=Si double bonded species, but it should be taken to note that the addition of methanol changes the polarity of the reaction medium at the same time.

When 4-methylphenylpentamethyldisilane (7) was irradiated in a mixed solvent of benzene and methanol at room temperature methanol adducts (16 and 17), 4-methylphenyldimethylmethoxysilane (19), and 4-methylphenyltrimethylsilane (20) were obtained [3] In addition to these compounds, 4-methylphenyldimethylsilane (21), trimethylsilane (22), and trimethylmethoxysilane (23) were newly detected as products [6].

Scheme II

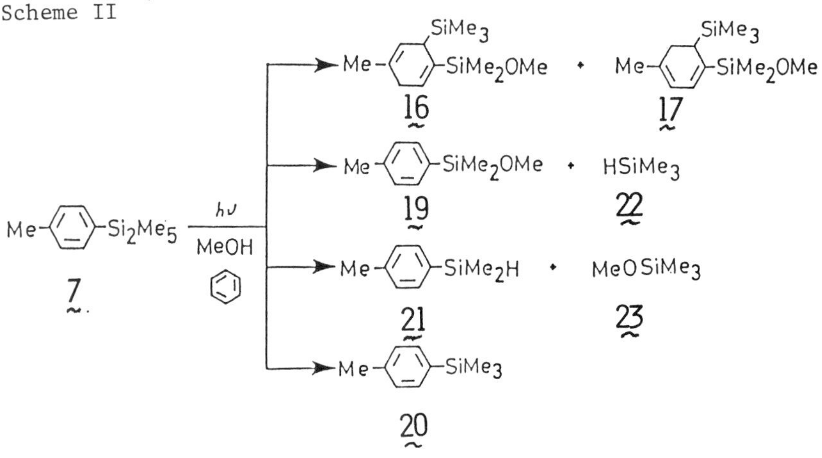

The reaction is likely to occur from the lowest excited singlet state, because the reaction could not be quenched by piperylene. Whereas the origins of 16, 17, and 20 are easily explained as produced via the type 2 and 3 reactions, respectively, it is not unequivocal, at the first glance, to account for the formation of other products 19, 21, 22, and 23 by whether they are formed via the type 1 pathway or via the solvolytic cleavage of the Si-Si bond similar to the type 4 reaction.

Distribution of deuterium as well as the solvent effects on the D-contents in the products were investigated in the reaction with methanol-d_1. Methanol adducts 16 and 17 were found to contain deuterium irrespective of the concentration of methanol-d_1, while the sum of the yields of 16 and 17 decreased as the ratio of methanol to benzene increased. Therefore, the formation of 16 and 17 may be explained by the type 2 pathway. Namely, addition of methanol to the intermediate 15 (or equivalent) explains the formation of 16 and 17. On the other hand, D-contents of the products 19–23, except for 20, change depending on the concentration of methanol-d_1.

Scheme III

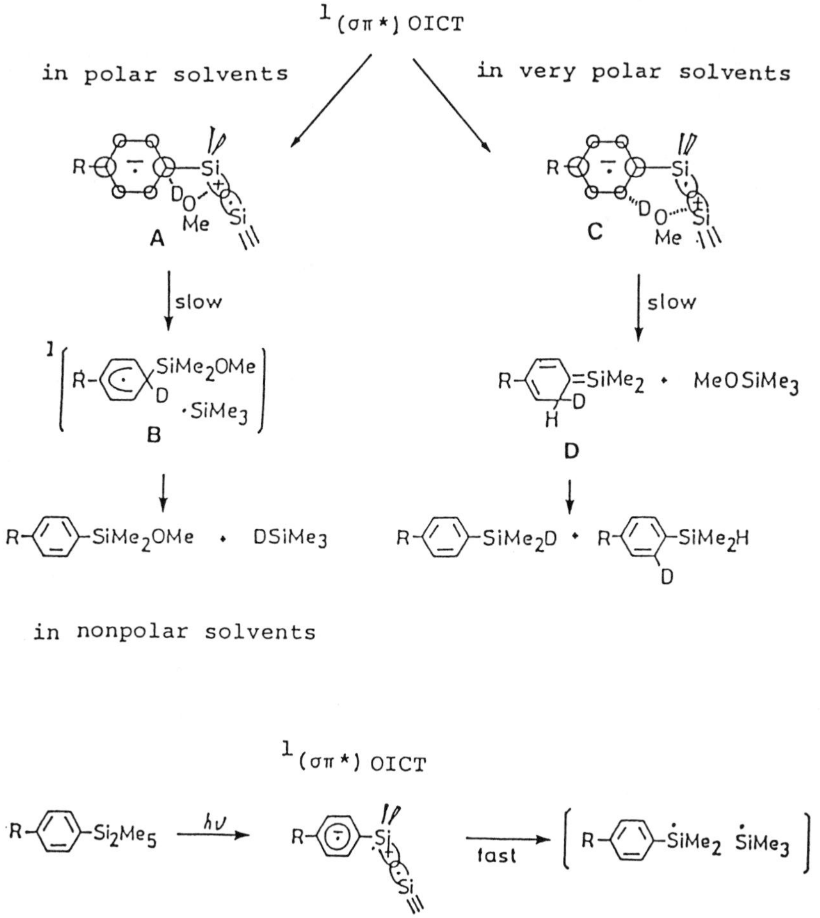

1(σπ*) OICT
in polar solvents
in very polar solvents
A
C
slow
slow
B
D
in nonpolar solvents
hν
fast

(Hereafter, deuterated and undeuterated products are distinguished by adding symbols -H and -D to the product numbers, respectively.)

The formation of methoxysilanes, 19 and 23, can be explained by addition of methanol to Si=C double-bonded species, which should be formed by disproportionation of two silyl radicals (type 1 reactions) [5].

$$RMe_2Si\cdot \ + \ R'Me_2Si\cdot \ \longrightarrow \ RMeSi=CH_2 \ + \ R'Me_2SiH$$

$$RMeSi=CH_2 \ + \ MeOX \ \longrightarrow \ RMeSiCH_2X(OMe) \quad (X = H \ or \ D)$$

Thus the formation of 19-D and 23-D in MeOD may be explained straightforwardly by the mechanism. However, the formation of 19-H and 23-H in MeOD is not compatible with this mechanism. The D-contents of both 19-D and 23-D decrease even at higher concentration of MeOD! The formation of deuterated hydrosilanes also cannot be explained simply by the type 1 reaction only.

However, a mechanism involving methanolysis of the charge-separated OICT state can explain these facts quite satisfactorily. Thus, in polar solvents, the attack of methanol on the OICT intermediate occurs at silicon bound to benzene. Simultaneous proton transfer at carbon of benzene bound to silicon leads to the formation of an intermediate (A) which collapses to 19-H and 22-D in MeOD through an intermediate B. In very polar solvents, charges are diffused more in OICT intermediates and hence the attack of MeOD and proton transfer can also occur at the terminal silicon and the ortho carbons, respectively, to lead to an intermediate (C) which further gives 23-H and 21-D.

In nonpolar solvents, the OICT state will not be stabilized, so that homolysis of the Si-Si bond occurs readily to lead to the silyl radical pairs which gives products of types 1 and 2 [5]. Experimental details can be accounted for by the mechanism quite nicely as shown in Scheme III.

PHOTOREACTION OF 4-ISOPROPENYLPHENYLPENTAMETHYLDISILANE

Scheme IV

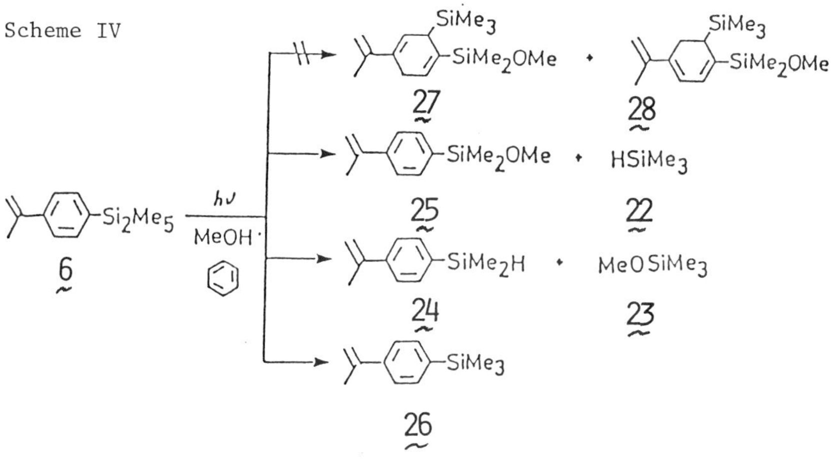

4-Isopropenylphenylpentamethyldisilane (6) has the most stable ($\sigma\pi^*$) OICT state among the compounds investigated on fluorescence spectra. The stabilization of ($\sigma\pi^*$) OICT states by an electron-withdrawing isopropenyl group may be expected to exert an effect similar to those of polar solvents in the photolysis, since both effects function as stabilizing polar charge-separated structure at the excited state. Hence, it is of interest to investigate the photolysis of 6 in connection with the photolysis of 7 in polar media.

When 6 was irradiated in a mixed solution of methanol and benzene under similar conditions to the photolysis of 7, 4-isopropenylphenyldimethylsilane (24), 4-isopropenylphenyl-dimethylmethoxysilane (25), and 4-isopropenylphenyltrimethyl-silane (26) were produced in addition to trimethylsilane (22) and trimethylmethoxysilane (23), as shown in Scheme IV. Significantly, the compounds 27 and 28, expected to arise from the methanol addition to a silaethene intermediate similar to 15, are not found at all in the experiments under a variety of conditions. Moreover, the D-contents in the products indicate that the formation of (24-D and 22-H) and (25-H and 23-D) is negligible although a small amount of (24-D and 22-H) was observed in a relatively nonpolar medium. Thus, in the photolysis of 6, only two types of methanolysis occurred on the photoexcited 6.

$SiR^2R^3R^4$

$\langle\text{ring}\rangle=SiCH_3R^1$

15

Thus the mechanism of the photoreaction of 6 is satisfactorily described in a similar way as that of photolysis in polar media shown in Scheme III. Methanolysis of the (σπ*) OICT states is a key step in the photolysis.

In conclusion. the diversity of the photoreactions of aryldi-silanes can be satisfactorily explained by the concept of (σπ*) OICT. In nonpolar solvents, major products are derived by silyl radical pairs after homolysis of the Si-Si bond, whereas in polar solvents, especially for aryldisilanes with an electron-withdrawing substituent, solvolytic reactions on the (σπ*) OICT states determine the products.

PHOTOREACTION OF 2-PENTAMETHYLDISILANYLNAPTHALENE

Photolysis of 2-pentamethyldisilanylnaphthalene (29) was reported to afford a 2,3-disilylnaphthalene derivative as a sole volatile product after trapping with isobutene [10], contrary to the expected 1,2-disilyl derivative (31) by the free-radical mechanism indicated above. Based on CNDO/2 calculations, the results have been explained as a consequence of the orbital symmetry control. HOMO(π)-LUMO(Si-Siσ*) interaction is favorable for silyl migration only to the C(3) atom of naphthalene. The phase relationship between orbitals on C(1) and the migrating Si is antibonding and hence migration to the C(1) atom is said not allowed.

Unfortunately, however, our own examination of the photoreaction of 29 shows that this explanation was led by an erroneous assignment of the product and the free-radical reaction is indeed the case. The previous workers have assigned the structure of the photoproduct to be 2a indirectly by protode-silylation. However, the previous workers failed to detect that protodesilylation of 1,2-disilylnaphthalene actually occurred with migration of the 1-silyl group. Direct NMR examination of the photoproduct also clearly indicate that the product was indeed 1,2-disilylnaphthalene [11].

Thus photolysis of 2-pentamethyldisilanylnaphthalene proceeds not by a concerted but by a free-radical mechanism.

REFERENCES

(1) (a) H. Sakurai, M. Kumada, Bull. Chem. Soc. Jpn., 37, 1884 (1964). (b) H. Gilman, W. H. Atwell, G. L. Schwebke, J. Organomet. Chem., 2, 369 (1964). (c) D. N. Hague, R. H. Prince, Chem. Ind. (London), 1492 (1964).

(2) P. Boudjouk, J. R. Roberts, C. M. Golino, and L. H. Sommer, J. Am. Chem. Soc., 94, 7926 (1972).

(3) (a) M. Ishikawa and M. Kumada, Adv. Organomet. Chem., 19, 51 (1981). (b) M. Ishikawa, Pure Appl. Chem., 50, 11 (1978), and references cited therein.

(4) (a) H. Okinoshima and W. P. Weber, J. Organomet. Chem., 149, 279 (1978). (b) H. S. D. Soysa and W. P. Weber, J. Organomet. Chem., 173, 269 (1979).

(5) H. Sakurai, Y. Nakadaira, M. Kira, H. Sugiyama, K. Yoshida, and T. Takiguchi, J. Organomet. Chem., 184, C36 (1980).

(6) H. Sakurai, M. Kira, H. Sugiyama, submitted for publication.

(7) (a) H. Shizuka, H. Obuchi, M. Ishikawa, and M. Kumada, J. Chem. Soc. Chem. Commun., 1981, 405. (b) H. Shizuka, Y. Sato, M. Ishikawa, and M. Kumada, ibid., 1982, 439. (c) H. Shizuka, Y. Sato, Y. Ueki, M. Ishikawa, and M. Kumada, J. Chem. Soc. Faraday Trans. 1, 80, 341 (1984). (d) H. Shizuka, H. Obuchi, M. Ishikawa, and M. Kumada, ibid., 80, 383 (1984).

(8) Z. R. Grabowski, K. Rotkiewicz, A. Siemiarczuk, D. J. Cowley, and W. Baumann, Nouv. J. Chem., 3, 443 (1979).

(9) (a) W. G. Dauben and J. S. Ritsher, J. Am. Chem. Soc., 92, 2925 (1970). (b) V. Bonacic-Koutecky, P. Bruckmann, P. Hiberty, J. Koutecky, C. Leoforestier, and L. Salem, Angew. Chem. Internat. Engl. Ed., 14, 575 (1975).

(10) M. Ishikawa, M. Oda, N. Miyoshi, L. Fabry, M. Kumada, T. Yamabe, K. Akagi, and K. Fukui, J. Am. Chem. Soc., 101, 4612 (1979).

(11) H. Sakurai, M. Yamaguchi, and M. Kira, to be published.

Chapter 17

Quantitative aspects of silylene reactions

Iain M.T. Davidson — Department of Chemistry, The University, Leicester, LE1 7RH, Great Britain.

The progress of silylene chemistry has been well chronicled by excellent and wide-ranging reviews written by leading practitioners [1,2]. This contribution will concentrate on some recent developments in the kinetics of silylene reactions; other aspects of this topical subject feature in the contributions by M. A. Ring and R. Walsh.

Silylenes are cleanly produced by pyrolysis of disilanes with at least one hydrogen, halogen, or alkoxy group attached to silicon [1], for reasons that are well understood [3]. Recent kinetic results for silylene-forming pyrolyses of some disilanes [4] are given in Table I, together with some earlier results. The latter were obtained before it was appreciated that silylenes could undergo a number of secondary reactions besides insertion, as illustrated for dimethylsilylene in Scheme I.

Scheme 1: Typical reactions of dimethylsilylene.

$Me_2Si: \rightleftharpoons HMeSi=CH_2$ Ref. 5

$2\ Me_2Si: \rightleftharpoons Me_2Si=SiMe_2$ Ref. 6

$Me_2Si: + HMeSi=CH_2 \rightarrow Me_2Si \overset{CH_2}{\underset{SiMeH}{\triangle}} \rightarrow HMe_2SiCH_2\ddot{S}iMe$ Ref. 7

These complications were reduced in the latest work [4] by carrying out pyrolyses in the presence of butadiene, a widely used silylene trap, and by measuring rate constants and

Arrhenius parameters for decomposition of the original
disilane, formation of the concomitant monosilane, and of the
silacyclopentenes resulting from silylene addition to butadiene; these are known to be thermally stable under the
conditions used [8].

Table 1. Arrhenius parameters for formation of silylenes from disilanes.

Reaction	log A	E/kcal	$k_{350°C}/s^{-1}$	Ref.
$Me_3SiSiMe_2Cl \rightarrow :SiMe_2 + Me_3SiCl$	11.7±0.3	50.1±1.1	1.3×10^{-6}	9
$Me_3SiSiMe_2H \rightarrow :SiMe_2 + Me_3SiH$	12.9±0.3	47.4±1.0	1.9×10^{-4}	10
	13.1±0.1	48.5±0.3	2.1×10^{-4}	4
$HMe_2SiSiMeH_2 \rightarrow :SiMe_2 + MeSiH_3$	12.6±2	46±5	2.9×10^{-4}	10
$Me_3SiSiMe_2OMe \rightarrow :SiMe_2 + Me_3SiOMe$	12.7±0.2	44.5±0.5	1.2×10^{-3}	4
$MeOMe_2SiSiMe_2OMe \rightarrow :SiMe_2 + Me_2Si(OMe)_2$	12.7±0.2	39±1.5	1.0×10^{-1}	4
$Me_3SiSiMeOMe^nPr \rightarrow :SiMe^nPr + Me_3SiOMe$	12.7±1.0	43.6±0.3	2.5×10^{-3}	4
$H_2MeSiSiMeH_2 \rightarrow :SiMeH + MeSiH_3$	14.3±0.3	47.9±0.6	3.1×10^{-3}	4
$H_2MeSiSiH_3 \rightarrow :SiH_2 + MeSiH_3$	15.3±0.2	50.7±0.4	3.1×10^{-3}	11
$H_2MeSiSiH_3 \rightarrow :SiMeH + SiH_4$	14.1±0.2	49.9±0.4	4.3×10^{-4}	11
$HMe_2SiSiMeH_2 \rightarrow :SiMeH + Me_2SiH_2$	13.7±0.6	46.2±1.4	2.8×10^{-3}	10

It should be noted that rate constants in Table I for
pyrolyses forming $:SiMe_2$ at 350°C cover a range of 10^5, with
methoxydisilanes as the most thermally labile [1]. There
appears to be a downward trend in the A factors with
increasing methylation of the eliminated silylene; the A
factors for elimination of $:SiMe_2$ are unusually low [12], but
there is good agreement in Table I between compounds, as there
was between the different types of kinetic measurement for
each compound in the presence of butadiene [4].

In co-operation with H. E. O'Neal we have improved and
extended our kinetic modeling of the complex and interesting
intramolecular reactions of the series of silylenes
$R(Me_3Si)Si:$ ($R = H$, Me, $SiMe_3$), revising earlier work on
$Me_3SiMeSi:$ [8] (Scheme II) and $(Me_3Si)_2Si:$ [13] (Scheme III),
and modeling for the first time the related reactions of
$Me_3SiHSi:$ [14] (Scheme IV), with particular emphasis on
internal consistency.

Scheme 1 - Reactions involving Me₃SiSiMe

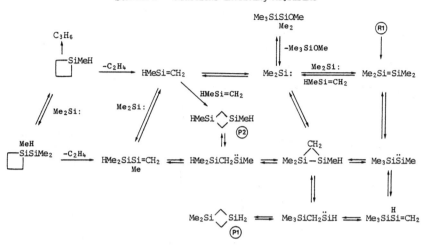

Scheme II provides for the modeling of experiments starting from methoxypentamethyldisilane (precursor to :SiMe₂), 1-methyl-siletane (precursor to HMeSi=CH₂) and the precursor to Me₂Si=SiMe₂ used by Roark and Peddle [15]. Inclusion of insertion reactions of :SiMe₂ into the first two precursors makes Scheme II more realistic than the original [8]. Scheme III is as previously suggested [13], while Scheme IV closely follows that suggested by Boo and Gaspar [14], including the transposition reactions converting Me₃SiHSi: to HMe₂SiMeSi: via HMeSi=SiMe₂. Although Me-shifts require higher activation energies than H-shifts [8], they have a part to play in these isomerizations, especially in Scheme IV. Enthalpies and entropies of formation of all molecules and intermediates in Schemes II-IV were first estimated from a new version [16] of Ring and O'Neal's additivity scheme [17], revised and extended in keeping with the following recent developments in silicon kinetics and thermochemistry. The thermochemistry of the series of monosilanes from SiH₄ to SiMe₄ has been revised [18], as has the enthalpy of formation of :SiH₂ [19]. Energy barriers have been calculated for silylene ⇌ silene [20] and silylsilylene ⇌ disilene [21-23] isomerizations, and for :SiH₂ insertion into H₂ [24]. New rate constants have been measured for :SiMe₂ insertions [25] and for silylene-forming pyrolyses of monosilanes [26] and disilanes [27]. Not only did these developments enable the original thermochemical scheme [17] to be improved and extended to cover silylenes, silenes, and disilenes, but it enabled kinetic "rules" for reaction types to be developed. They are no more than guide-lines because of uncertainties and inconsistencies in the available data, and may well need to be revised in the light of future developments, but their consistent application to the reactions in Schemes II-IV gave the encouraging results shown in Table II. Full details of the thermochemical and kinetic estimates will be pubished elsewhere [16].

Scheme 3: Isomerization of Me₃SiSiSiMe₃.

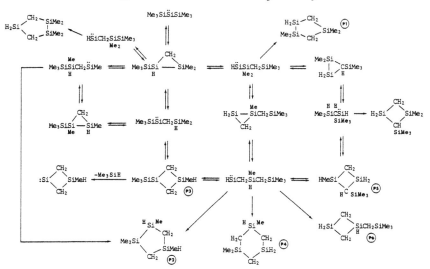

Scheme 4: Isomerization of Me₃SiSiH.

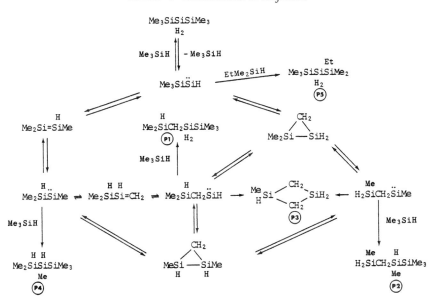

Table 2. Comparison of experimental and calculated product yields.

Scheme 2

Initial Intermediate	T/K	[P1]/[P2]	
		Experiment	Calculated
$Me_2Si:$	760	1.5	1.49
	850	>0.7	0.94
$HMeSi=CH_2$	760	0.4	0.41
	850	~1.0	0.98
$Me_2Si=SiMe_2$	633	3	2.96

Scheme III at 773 K

Products	P1	P2	P3	P4	P5	P6
Experimental %	60	small	small	<5	<5	~5
Calculated %	60	2.7	2.3	3.1	4.9	5.7

Scheme IV at 783 K with R_3Si-H trapping

Products	P1	P2	P3	P4	P5
Experimental %	11	4	4	7	9
Calculated %	11	4.0	4.5	9.2	8.6

The mechanism of addition of silylenes to dienes is of long-standing interest [1]. Gaspar and co-workers have recently [28] extended earlier studies [29] on addition of :SiMe₂ to hexa-2,4-dienes to provide good evidence that silylene addition to dienes proceeds stepwise by initial 1,2-addition to form a vinylsilacyclopropane followed by ring-opening homolysis to give biradicals which rearrange to silacyclopentenes; in suitable circumstances another important pathway is a concerted 1,5-H-shift converting the vinylsilacyclopropane to a propenylvinylsilane. These reactions are illustrated in Scheme V. The reverse reactions are also

Scheme 5

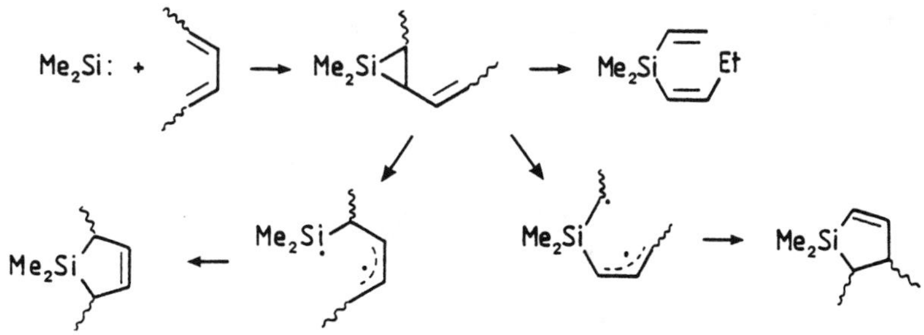

attracting attention. Dimethyl-cis-1-propenylvinylsilane has been shown to extrude :SiMe₂ and to rearrange to silacyclopentenes [30]. The report that dimethylsilacyclopent-3-ene extruded :SiMe₂ on pyrolysis [8] has been followed by more detailed mechanistic [31] and kinetic [32] studies. It emerged from both of the latter studies that formation of

:SiMe$_2$ and butadiene was accompanied by isomerization of
dimethylsilacyclopent-3-ene to dimethylsilacyclopent-2-ene,
the latter amounting to 30% of the observed butadiene under
FVP conditions at 750°C [31], and 41% in a stirred-flow system
with longer contact time at 570-650°C [32]; Arrhenius
parameters (log A, E/kcal. mol^{-1}) for formation of butadiene
and dimethylsilacyclopent-2-ene were (14.0±0.6, 64±3) and
(13.7±0.6, 64±3) respectively [32], consistent with both
products having a common precursor. Pyrolysis of 1,1,2-
trimethyl-1-silacyclo-pentene [31] gave a range of products
that mirrored those from addition of :SiMe$_2$ to a diene with
terminal Me-substitution [28], consistent with thermal
reactions from either starting point proceeding through a
vinylsilacyclopropane intermediate that decomposed mainly by
carbon-carbon bond rupture [31], as expected from the relative
strengths of carbon-carbon and silicon-carbon bonds.
Likewise, pyrolysis of dimethylsilacyclopent-3-ene would
proceed as shown in Scheme VI; but when we use Scheme VI to

Scheme 6

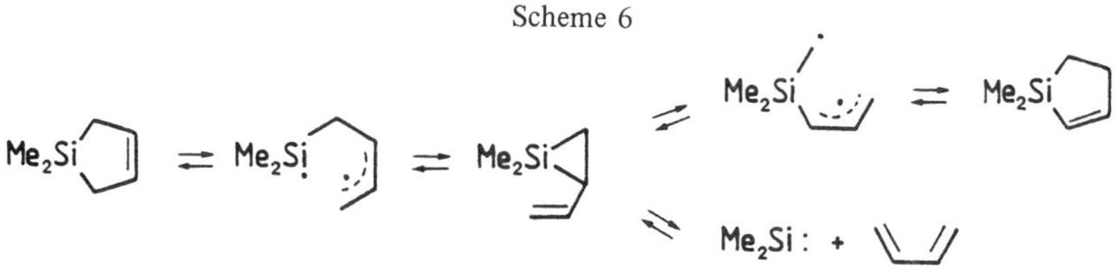

rationalise the reverse process, addition of :SiMe$_2$ to
butadiene, we would expect dimethylsilacyclopent-2-ene to be
the main product, resulting from the favoured carbon-carbon
bond rupture. In fact, addition of :SiMe$_2$ to butadiene around
400°C (a temperature region where silacyclopentenes are
thermally stable [32]) gives ≤10% dimethylsilacyclopent-2-ene,
but ≥90% dimethylsilacyclopent-3-ene [4]. Addition of :SiMeH
to butadiene under similar conditions gives methylsilacyclo-
pent-3-ene as the only 1:1 adduct [4], in contrast to the
products expected in Scheme VII, the counterpart to Scheme VI;

Scheme 7

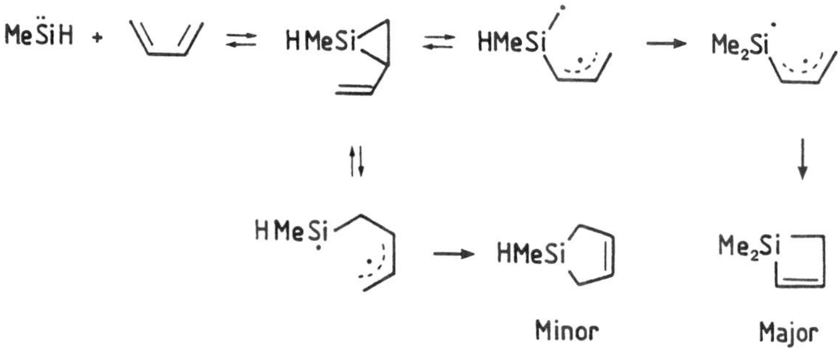

Minor   Major

dimethylsilacyclobutene would be stable [33], while the
radical rearrangement by a 1,2-H shift leading to it would be
rapid under the experimental conditions used [34].   An
analogous problem was pointed out some years ago by Barton,
who found that the only cyclization product from 4-butenyl-
methylsilylene was methylsilacyclopent-3-ene [35]; he
suggested that it might be formed from a vinylsilacyclopropane
intermediate by a formal 1,3-silyl shift, as shown in Scheme
VIII.   This type of mechanism could resolve the problems
raised in Schemes VI and VII.

Scheme 8

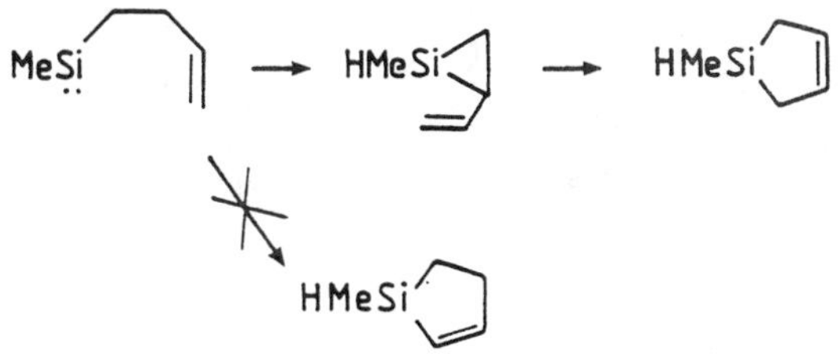

We have recently built an improved stirred-flow pyrolysis
apparatus attached to a HP 5995C gc/mass spectrometer, which
we have begun to use to investigate the addition of silylenes
to butadiene [36]; our initial results for addition of :SiMe$_2$
and :SiMeH are illustrated in Figures I and II respectively.
Figure I shows the formation of 1,1-dimethylsilacyclopent-3-
ene as the major trapping product from :SiMe$_2$, with 1,1-
dimethylsilacyclopent-2-ene as the minor; formation of a
trisilane by insertion of :SiMe$_2$ into its precursor amounted
to 0-2% of the cyclic products between 350°C and 450°C, thus
confirming the efficacy of butadiene as a silylene trap over
that temperature range (cf. Table I).   Figure II shows
interesting differences from Figure I; :SiMeH gives only the
1-methylsilacyclopent-3-ene (confirmed by proton NMR) with no
1-methylsilacyclopent-2-ene, but with two new products which
appear to be butadiene adducts of butenylmethylsilylene. This
product distribution may be explained by Scheme IX, strongly
reminiscent of Barton's Scheme VIII.   The silyl and hydrogen

Scheme 9

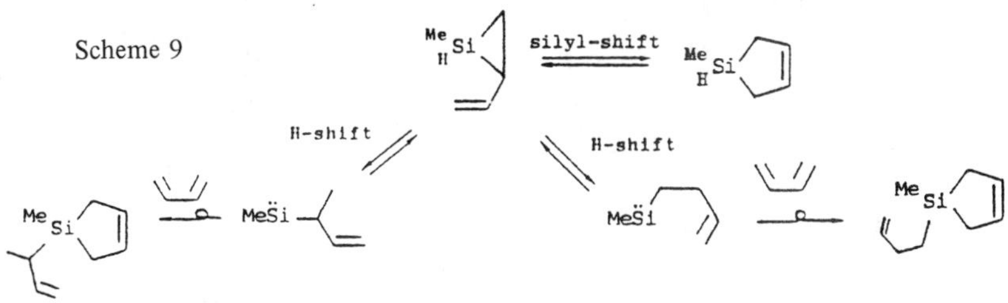

shifts in Scheme IX compete effectively with the homolysis
reactions in Scheme VII because the major product of the
latter, dimethylsilacyclobutene, requires a significant

Fig. 1.

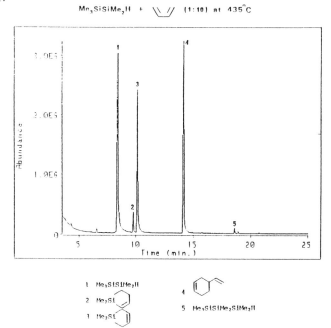

Fig. 2.

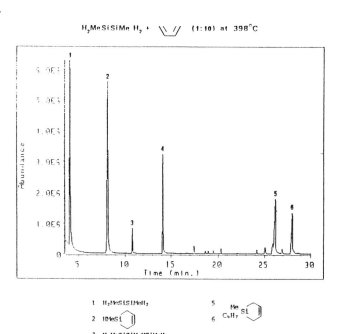

activation energy for its formation to overcome the ring
strain [16]. These studies are continuing.

ACKNOWLEDGEMENTS

It is a pleasure to acknowledge valuable discussions with
Peter Gaspar, Mark Gordon, Ed O'Neal, Morey Ring, and Robin
Walsh in the preparation of this contribution; they share none
of the author's responsibility for any misguided ideas
herein! Experimental results from our laboratory owe much to
the enthusiasm and skill of my co-workers, especially Michael
P. Clarke, and to financial support from the SERC and Dow
Corning (Europe).

REFERENCES

1. Atwell, W.H., Weyenberg, D., Angew. Chem., Int. Ed. Engl.,
   1969, **8**, 469.
2. Gaspar, P.P., React. Intermed. (Wiley), 1978, **1**, 229;
   1981, **2**, 335; 1985, **3**, 333.
3. Davidson, I.M.T., J. Organometallic Chem., 1970, **24**, 97.
4. Davidson, I.M.T., Hughes, K.J., Ijadi-Maghsoodi, S.,
   Organometallics, 1987, **6**, 639.
5. Conlin, R.T., Wood, D.L., J. Am. Chem. Soc., 1981, **103**,
   1843.
6. Conlin, R.T., Gaspar, P.P., J. Am. Chem. Soc., 1976, **98**,
   868.
7. Wulff, W.D., Goure, W.F., Barton, T.J., J. Am. Chem. Soc.,
   1978, **100**, 6236.
8. Davidson, I.M.T., Scampton, R.S., J. Organometallic Chem.,
   1984, **271**, 249.
9. Davidson, I.M.T., Delf, M.E., J. Chem. Soc., Faraday Trans.
   I, 1976, **72**, 1912.
10. Davidson, I.M.T., Matthews, J.E., J. Chem. Soc., Faraday
    Trans. I, 1976, **72**, 1403.
11. Vanderwielen, A.J., Ring, M.A., O'Neal, H.E., J. Am. Chem.
    Soc., 1975, **97**, 993.
12. Walsh, R., J. Phys. Chem., 1986, **90**, 389.
13. Davidson, I.M.T., Hughes, K.J., Scampton, R.S., J. Organo-
    metallic Chem., 1984, **272**, 11.
14. Boo, B.H., Gaspar, P.P., Organometallics, 1986, **5**, 698.
15. Roark, D.N., Peddle, G.J.D., J. Am. Chem. Soc.,1972, **94**,
    5837.
16. O'Neal, H.E., to be published.
17. O'Neal, H.E., Ring, M.A., J. Organometallic Chem., 1981,
    213, 419.
18. Doncaster, A.M., Walsh, R., J. Chem. Soc., Faraday Trans.
    2, 1986, **82**, 707.
19. Walsh, R., Pure Appl. Chem., 1987, **59**, 69, and references
    therein.

20. Nagase, S., Kudo, T., J. Chem. Soc., Chem. Commun., 1984, 141, 1392.
21. Nagase, S., Kudo, T., Organometallics, 1984, 3, 1320.
22. Gordon, M.S., Truong, T.N., Bonderson, E.K., J. Am. Chem. Soc., 1986, 108, 1421.
23. Olbrich, G., Chem. Phys. Lett., 1986, 130, 115.
24. Gordon, M.S., Gano, D.R., Binkley, J.S., Frisch, M.J., J. Am. Chem. Soc., 1986, 108, 2191.
25. Baggott, J.E., Blitz, M.A., Frey, H.M., Lightfoot, P.D., Walsh, R., Chem. Phys. Lett., 1987, 135, 39.
26. Ring, M.A., O'Neal, H.E., Rickborn, S.F., Sawrey, B.A., Organometallics, 1983, 2, 1891.
27. Davidson, I.M.T., Hughes, K.J., Ijadi-Maghsoodi, S., Organometallics, 1987, 6, 639, and references therein.
28. Lei, D., Gaspar, P.P., J. Chem. Soc., Chem. Commun., 1985, 1149.
29. Lei, D., Hwang, R.-J., Gaspar, P.P., J. Organometallic Chem., 1984, 271, 1.
30. Gaspar, P.P., Lei, D., Organometallics, 1986, 5, 1276.
31. Lei, D., Gaspar, P.P., Organometallics, 1985, 4, 1471.
32. Davidson, I.M.T., Hughes, K.J., to be published.
33. Auner, N., Davidson, I.M.T., Ijadi-Maghsoodi, S., Organometallics, 1985, 4, 2210.
34. Barton, T.J., Revis, A., Davidson, I.M.T., Ijadi-Maghsoodi, S., Hughes, K.J., Gordon, M.S., J. Am. Chem. Soc., 1986, 108, 4022.
35. Barton, T.J., Burns, G.T., Organometallics, 1983, 2, 1.
36. Clarke, M.P., Davidson, I.M.T., unpublished work.

Chapter 18

Silyl cations

Joseph B. Lambert, William J. Schulz, Jr., JoAnn A. McConnell and Wojciech Schilf — Department of Chemistry, Northwestern University, Evanston, IL 60205, USA.

ABSTRACT

Abstraction of hydride from silanes by trityl perchlorate produces silyl perchlorates. In dilute CD_2Cl_2 or sulfolane, these materials are best described as ionic when the silicon atom carries phenyl or alkylthio substituents (Ph_3SiClO_4 and $(R'S)_3SiClO_4$). These conclusions are based on conductance, cryoscopic, and NMR spectroscopic (^1H, ^{13}C, ^{15}N, ^{29}Si, ^{35}Cl) evidence. At higher concentrations, the ionic form is in equilibrium with an associated form.

INTRODUCTION

Although trivalent silyl cations have been known for over 20 years from gas phase studies [1-5], attempts to observe these species in solution have been fraught with difficulties [6]. These problems have derived from the high affinity of electrophilic silicon for oxygen and halogens. Observation of trivalent silyl cations in solution consequently depends on careful consideration of the solvent and of the accompanying anion. Moreover, the substituents on silicon must be chosen carefully.

The solvent must have extremely low nucleo-

philicity, particularly if oxygen or halogens are
present in the solvent molecule. In addition, the
solvent must have a high dielectric constant in order
to support the ionic species. We expect that all
hydroxylic solvents are excluded, but there is a wide
range of highly polar, nonhydroxylic solvents, including dichloromethane, sulfolane, and acetonitrile.
Dipolar, nonhydroxylic solvents with high nucleo-
philicity such as pyridine and hexamethylphosphor-
amide, however, are excluded. We have carried out
most of our experiments in dichloromethane and
sulfolane, which have very low nucleophilicities but
can stabilize ionic species.

Because of their high affinity for silicon,
chloride and bromide are probably excluded as anions,
but iodide is a possibility. Other favorable anions
include triflate ($^-OSO_2CF_3$) and perchlorate (ClO_4^-),
but not BF_4^-, which can make a fluoride ion available
to silicon. Although BPh_4^- appears to be an excellent choice, synthetic problems have prevented its
use up to the present. We have concentrated our
efforts on triflate and perchlorate.

Stabilization of positive charge through
appropriate choice of substitution is the time-
honored method in carbenium ion chemistry. Requirements are different for silylenium (trivalent silyl
cation) chemistry. Overlap between 2p orbitals,
which is very useful for carbenium ions (benzyl,
alkyl), has been replaced by the much weaker 3p-2p
overlap for analogous silylenium ions. Thus resonance interactions, although present, should not
provide as much stabilization as in carbon cases.
Stabilization of positive charge still may be obtained through alternative approaches. Polarization
interactions through the presence of induced dipoles
are known to be very important in the gas phase.
Thus polarizable atoms such as sulfur, phosphorus, or
selenium may be effective in stabilizing positive
charge on sulfur. We have thoroughly examined the
case of sulfur, i.e., $(R'S)_3Si^+$, which may provide
stabilization not only through induced dipoles but
also possibly from 3p-2p overlap [7]. Phenyl rings
can stabilize charge not only through traditional
resonance effects but also through π polarization
(creation of negative charge at one end of the phenyl
ring and positive charge at the other, but no net
increase or decrease of charge as occurs in
resonance). Consequently, we also have examined the
case of Ph_3Si^+ (sityl) [8], in which phenyl may
stabilize charge through π polarization or possibly π
donation. We report herein a brief discussion of

these results.

PROTON AND CARBON-13 SPECTRA

Silyl perchlorates were prepared by the reaction of silyl hydrides with trityl perchlorate. The stronger C-H bond replaces the weaker Si-H bond: $R_3SiH + Ph_3CClO_4 \longrightarrow R_3SiClO_4 + Ph_3CH$. This reaction was first used by Corey for this purpose [9]. Others pointed out [10] that the resulting perchlorate can be ionic, ion paired, or covalent. We have carried out this reaction for several substituents, including alkylthio (R = MeS, EtS, iPrS) and phenyl (R = Ph). The ^1H and ^{13}C spectra of the reacted solution are very clean, indicating production of a single species. Thus the ^1H spectrum is characterized by complete loss of the silyl hydride resonance and the concomitant appearance of the Ph_3CH resonances. For the alkylthio substrates, the alkyl resonances from the starting material are replaced by a new set of alkyl resonances from the silyl perchlorate. For the phenyl case, the aromatic resonances undergo a similar transformation. If excess trityl perchlorate is added, its resonances are clearly discerned at a distinct position. Thus the Corey procedure provides a very clean entry into silyl perchlorates. The only by-product is the hydrocarbon triphenylmethane, which should not cause problems.

CONDUCTANCE EXPERIMENTS

Although dichloromethane is polar and nonnucleophilic, its ionizing power is low. Nonetheless, trityl perchlorate (Ph_3CClO_4) is highly conducting (equivalent conductance $\Lambda = 43.5$ mho cm^2 eq^{-1} at about 1 mM). The stability of this ion permits its free existence under these conditions. Reaction of tris(2-propylthio)silane, $(iPrS)_3SiH$, with an equivalent of trityl perchlorate in CH_2Cl_2 gives a solution of tris(2-propylthio)silyl perchlorate that has nearly the same equivalent conductance ($\Lambda = 33.4$) as trityl perchlorate. This silyl perchlorate therefore is a very strongly conducting entity. This observation is clearly not the result of accidental hydrolysis, which would yield the silanol and perchloric acid. In CH_2Cl_2, perchloric acid is either covalent or highly ion paired and gives a conductance of about 0.01 mho cm^2 eq^{-1}. A negligible amount of the observed conductance from tris(2-propylthio)silyl perchlorate can derive from $HClO_4$.

Triphenylsilyl (sityl) perchlorate is much less

conducting in CH_2Cl_2, $\Lambda = 1.13$ mho cm^2 eq^{-1}. Although this value is some 3% that of trityl, it is still a hundred times that of $HClO_4$ under the same conditions. This result suggests that sityl may be ion paired under these conditions. A plot of log Λ vs. log M, in which M is concentration, provides information about such species. Free ions give a horizontal plot, since the increase in conduction is canceled out through division by concentration (Λ is conductance per concentration). Such plots are obtained for trityl and tris(2-propylthio)silyl in CH_2Cl_2. When ion pairs are present, the plot is linear with a negative slope, because a decrease in concentration breaks up ion pairs and the cancelation of conductance by concentration is no longer in effect. Such a plot is obtained for sityl perchlorate, so that this species exists as an equilibrium between ion pair and free ion in CH_2Cl_2. Conductance of all these ions in the highly ionizing sulfolane is high, so that this solvent permits free ions even for sityl at the concentration ranges of the conductance experiments.

MOLECULAR WEIGHT

Cryoscopic methods are useful in distinguishing various possible structures in solution that differ in the number of particles. The covalent form (or a tight ion pair) has one particle, a free ion two, and a bridged dimer three. Moreover, sulfolane, which qualifies as an excellent solvent for electrophilic organic ions, has a very large cryoscopic constant and hence is very suitable for freezing point experiments. Consequently, we have carried out freezing point depression measurements on solutions of silyl perchlorates in sulfolane. As controls, we ran triphenylmethane (1 particle found: observed molecular weight 232, calculated 244) and trityl perchlorate in the presence of an equivalent of triphenylmethane (2 particles found: obs. 367, calc. 343).

The sulfur-substituted cation, tris(2-propylthio)silyl perchlorate, gave the expected result for an ionic monomer (2 particles found: obs. 377, calc. 353). The error was much larger in the one particle (covalent) model (obs. 188, calc. 353) and the three particle dimer (obs. 565, calc. 706), as well as for a molecule in which trityl was attached to sulfur to form a two particle sulfonium ion (obs. 411, calc. 597).

Similar results were obtained for sityl

perchlorate. Calculations based on a two particle,
ionic monomer were within 6% of the theoretical
value, whereas the one particle covalent monomer was
47% off and the three particle ionic dimer was 20%
off. In contrast, sityl azide and sityl triflate
gave one particle molecular weights.

Thus the molecular weight measurements confirm an
ionic structure containing two particles for both the
tris(2-propylthio)silyl and the sityl perchlorates.
The covalent monomer and the ionic dimer are not
consistent with the observations.

NMR RESULTS FOR OTHER NUCLEI ⡀

Chlorine-35. As Olah has pointed out, ^{35}Cl should
be a useful nucleus for distinguishing ionic and
covalent perchlorates. In the ionic form, the
chlorine nucleus exists in a tetrahedral environment.
The resulting absence of an electric field gradient
should result in very narrow linewidths. In a
covalent form, the environment should have a considerable electric field gradient and hence very
large linewidths. We have carried out a detailed
study of the ^{35}Cl resonances for sityl perchlorate in
sulfolane. At dilute concentrations (0.0020 M),
corresponding to those of the conductance and
cryoscopic experiments, the resonance occurs at δ 4.5
(δ 0 is external dilute $HClO_4$ in H_2O) with a
linewidth at half height of 18 Hz. This is a very
narrow line for ^{35}Cl and clearly corresponds to an
ionic perchlorate. As the concentration is raised,
the single signal broadens and moves to higher field.
At 0.0042 M it occurs at δ 0.1 with a linewidth of
220 Hz, and at 0.0088 M it is at -17.5 with a
linewidth of 380 Hz. The change in chemical shift
and linewidth is evidence for a dynamic equilibrium, ·
presumably between the ionic form and either an ion-
paired or a covalent form. A covalent perchlorate
should have linewidths of kHz (thus the linewidth of
CCl_4 is 13,000 Hz). Consequently, the silyl
perchlorates we are studying bear no resemblance to
covalent materials, but the spectral changes indicate
that the ionic forms are in equilibrium with an
associated form.

Nitrogen-15. In order to study the interaction of
nitrogen nucleophiles with the silyl perchlorates, we
have used ^{15}N NMR spectroscopy. There can be an
equilibrium not only between ionic and covalent forms
but also between ionic and complexed forms (eq 1).

$$R_3Si^+ + :X \rightleftharpoons R_3Si-X^+ \tag{1}$$

Nitrogen-15 is a particularly apt probe because there
is a 100 ppm difference between the chemical shifts
of lone-pair bearing (uncomplexed) nitrogen and
quaternary (complexed) nitrogen. We have used
acetonitrile (a weak nucleophile) and pyridine
(a strong nucleophile) to study the complexation
equilibrium. Free CH_3CN in CD_2Cl_2 resonates at δ 244
(δ 0 is anhydrous NH_3). When 1 equivalent of CH_3CN
is added to sityl perchlorate in CD_2Cl_2, it resonates
at δ 245, essentially unchanged from free CH_3CN.
Protonated CH_3CN resonates at δ 141 and methylated
CH_3CN at δ 132. Clearly, acetonitrile does not
complex with sityl under these conditions. Free
pyridine in CD_2Cl_2 resonates at δ 314, whereas protonated pyridine is at 205 and methylated pyridine at
200. When 1 equivalent of pyridine is added to sityl
perchlorate, the ^{15}N shift moves to δ 217. Thus
pyridine does form a complex with sityl. Similar
conclusions could be derived from ^{13}C experiments,
but the complexation shifts are not nearly so large.
The combination of ^{13}C and ^{15}N NMR experiments provides reliable criteria for determining whether
specific nuclei form complexes with sityl or other
silyl perchlorates.

Silicon-29. The [29]Si nucleus has extremely
low receptivity because of low natural abundance, low
natural sensitivity, and extremely long relaxation
times in the absence of attached protons. Enhancement of sensitivity through polarization transfer
experiments is hampered by the negative gyromagnetic
ratio of [29]Si and by frequently unknown [29]Si-[1]H
coupling constants. Thus we have only been able to
carry out successful [29]Si experiments at relatively
high concentrations (> 0.1 M), at which association
probably occurs. We have observed no useful
resonances for tris(2-propylthio)silyl perchlorate.
Sityl perchlorate in CH_2Cl_2 gives a weak resonance at
δ 2.0 and in sulfolane at δ 3.1. These resonances
may be attributed to associated silyl perchlorates,
either tight ion pairs or the covalent form. As the
concentration is lowered, the resonance disappears
quickly.

ALTERNATIVES

A number of alternative structures for the silyl
perchlorates have been considered and eliminated.

Covalent form. The high conductivity of the
tris(2-propylthio)silyl perchlorate eliminates the
covalent alternative. Moreover, the molecular weight
clearly indicates two particles. The cryoscopic

results for triphenyl perchlorate also indicate two
particles, but the conductance is not so high. The
increase in equivalent conductance at lower concentrations suggests an equilibrium between free and
associated ions. At low concentrations the sharp
^{35}Cl resonance is characteristic of an ionic perchlorate. An increase in concentration results in a
broadening of the ^{35}Cl resonance and movement to
higher field. Thus both conductance and ^{35}Cl
resonance indicate some sort of association at higher
concentration for sityl perchlorate. It is this
associated form apparently that gives the only
observable ^{29}Si resonance, around δ 3 for sityl
perchlorate in relatively concentrated sulfolane
solution. It is not clear whether the associated
form is a tight ion pair or a covalent form.

Bridged dimer. Two silylenium ions might be in
equilibrium with a bridged dimer such as **1**. The

1          $(R'S)_2Si$ $Si(SR')_2$ $2ClO_4^-$

observation of only one isopropyl resonance would
seem to eliminate such a structure, unless the dimer
were rapidly interconverting with the monomers. The
cryoscopic measurements, however, are not in agreement with a three particle structure.

Onium ion. Reaction of trityl with the sulfur
lone pairs can give rise to a structure such as **2**.

2          $H-Si-S$ ClO_4^-

When R' = Me, there is some of this form present in
CD_2Cl_2, but for R' = Et or iPr there is very little.
The ^1H spectrum provides very clear evidence for the
loss of the SiH resonance as trityl perchlorate is
added.

Solvent complex. Complexation with solvent has
been demonstrated by Bassindale and co-workers [11]
for the case of trimethylsilyl perchlorate. Such

complexes are not known for dichloromethane or sulfolane solvent but are conceivable for aceto-nitrile. The addition of CH_3CN to solutions of sityl perchlorate in CD_2Cl_2, however, gives rise to no shift in the ^{15}N resonance from that of free CH_3CN, so that there is no complexation between sityl and CH_3CN.

Having disposed of all viable alternatives, we conclude that the silylenium ion is the best description for tris(2-propylthio)silyl perchlorate and triphenylsilyl perchlorate in dilute solutions in dichloromethane and sulfolane.

ACKNOWLEDGEMENTS

This work was supported by the National Science Foundation (Grant No. CHE86-09899).

REFERENCES

(1)   B. G. Hobrock and R. W. Kiser, *J. Phys. Chem.*, **66**, 155-158 (1962).

(2)   G. G. Hess, F. W. Lampe, and L. H. Sommer, *J. Am. Chem. Soc.*, **86**, 3174-3175.

(3)   W. P. Weber, R. A. Felix, and A. K. Willard, *Tetrahedron Lett.*, 907-910 (1970).

(4)   M. K. Murphy and J. L. Beauchamp, *J. Am. Chem. Soc.*, **99**, 2085-2089 (1977).

(5)   W. J. Pietro and W. J. Hehre, *J. Am. Chem. Soc.*, **104**, 4329-4332 (1982).

(6)   R. J. Corriu and M. J. Henner, *J. Organomet. Chem.*, **74**, 1-28 (1974).

(7)   J. B. Lambert, W. J. Schulz, Jr., *J. Am. Chem. Soc.*, **105**, 1671-1672 (1983).

(8)   J. B. Lambert, J. A. McConnell, and W. J. Schulz, Jr., *J. Am. Chem. Soc.*, **108**, 2482-2484 (1986).

(9)   J. Y. Corey, *J. Am. Chem. Soc.*, **97**, 3237-3238 (1975).

(10) J. B. Lambert and H.-n. Sun, *J. Am. Chem. Soc.*, **98**, 5611-5615 (1976).

(11) A. R. Bassindale and T. Stout, *J. Chem. Soc. Perkin II*, 221-225 (1986).

Chapter 19

The chemistry of unsaturated silicon compounds. Nickel-catalyzed reactions

Mitsuo Ishikawa — Department of Applied Chemistry, Faculty of Engineering, Hiroshima University, Higashi-Hiroshima, 724, Japan.

INTRODUCTION

In 1967, the first evidence for the existance of a silene as a transient intermediate was provided in the pioneering researches of Gusel'nikov and Flowers, who thermolyzed 1,1-dimethyl-1-silacyclobutane and observed that the 1,1-dimethylsilene reacted with trapping agents such as alcohol[1]. Since that time, numerous papers on the chemistry of the silenes have been published[2-4], and many types of the silene have been prepared by the methods which involve thermolysis[1], photolysis[3,5], and salt elimination[6]. In spite of these studies, no interest has been shown in the transition-metal-catalyzed formation of the silenes[7].

We report here the reactions of silacyclopropenes and (phenylethynyl)polysilanes with phenyl-(trimethylsilyl)acetylene in the presence of a catalytic amount of dichlorobis(triethylphosphine)-nickel(II). We also describe the stoichiometric reactions of the silacyclopropene with tetrakis(triethylphosphine)nickel(0). These reactions afford the products which can be best explained by assuming the transient formation of 1-silapropadiene-nickel complexes.

NICKEL-CATALYZED REACTIONS

When a mixture of 3-phenyl-1,1,2-tris(trimethylsilyl)-1-silacyclopropene (1) which is obtained
photochemically from 2-(phenylethynyl)-2-(trimetyl-
silyl)hexamethyltrisilane and phenyl(trimethylsilyl)-
acetylene is heated at 135°C in the presence of a
catalytic amount of dichlorobis(triethylphosphine)-
nickel(II) in a sealed tube, two crystalline products, 3-phenyl-4-[phenyl(trimethylsilyl)methylene]-
1,1,2-tris(trimethylsilyl)-1-silacyclobut-2-ene (2)
and 1,4,4-trimethyl-3,6-diphenyl-1,2,5-tris(trimethylsilyl)-1,4-disilacyclohexa-2,5-diene (3) are
obtained in 36 and 56% yields, respectively[11,12].
No silole derivative can be detected by spectroscopic
analysis.

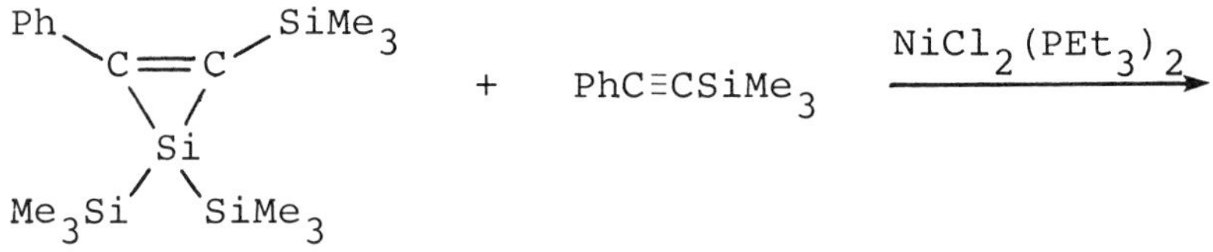

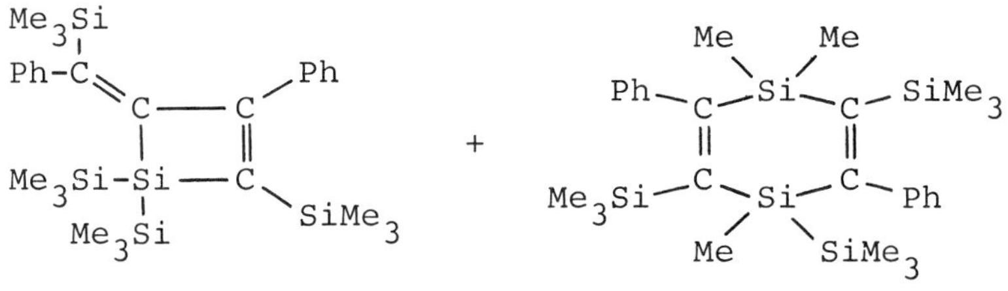

We have found that the nickel-catalyzed reactions of 1,1-dimethyl-3-phenyl-2-(trimethylsilyl)-,
1-methyl-1,3-diphenyl-2-(trimethylsilyl)-, and 1,1,3-
triphenyl-2-(trimethylsilyl)-1-silacyclopropene with
phenyl(trimethylsilyl)acetylene afford 3,4-diphenyl-
2,5-bis(trimethylsilyl)siloles in high yields[13].
We proposed a nickelasilacyclobutene as a key intermediate for the formation of the siloles. In the
nickel-catalyzed reaction of 1, however, two bulky
trimethylsilyl groups on the ring silicon atom in
the nickelasilacyclobutene arising from the reaction

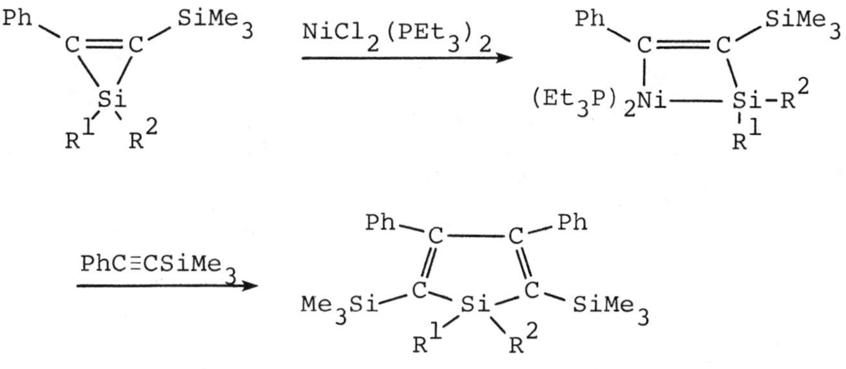

$$R^1=R^2=Me, \quad R^1=Me, R^2=Ph, \quad R^1=R^2=Ph$$

of 1 with the nickel catalyst presumably prevent the formation of the silole. Therefore, it seems likely that the nickelasilacyclobutene (4) would isomerize to give another type of intermediates such as sila-propadiene-nickel complex (6) and nickeladisilacyclo-pentene (7). In fact, the formation of 2 and 3 may be understood by the reaction of 6 and 7 with phenyl-(trimethylsilyl)acetylene.

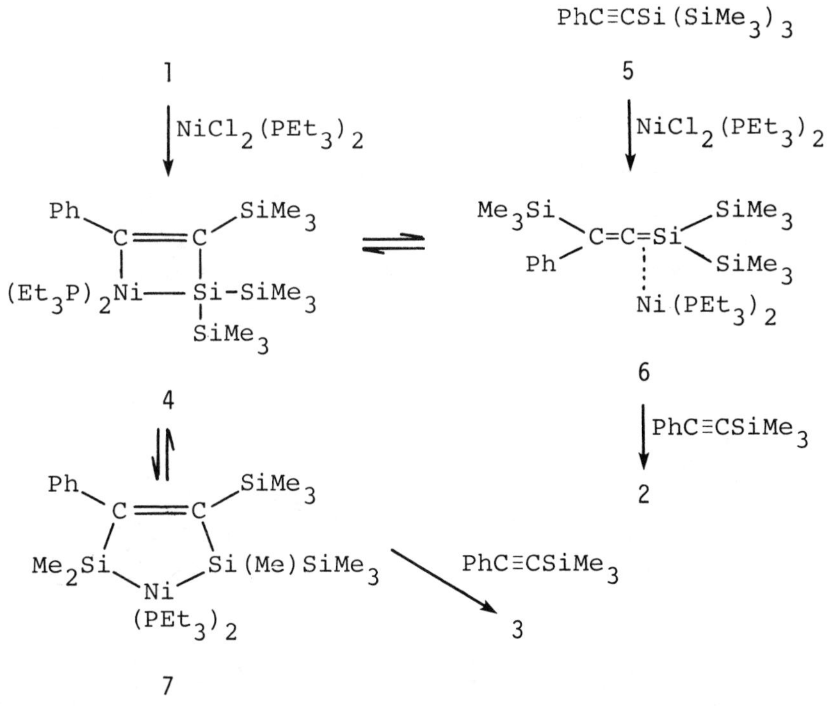

The reaction of 2-(phenylethynyl)-2-(trimethyl-silyl)hexamethyltrisilane (5) with phenyl(trimethyl-silyl)acetylene in the presence of the nickel cata-lyst at 200°C in a degassed sealed tube again affords 2 and 3 in 19 and 58% yields, respectively[11]. The fact that the reactions of 1 and 5 with phenyl(tri-methylsilyl)acetylene in the presence of the nickel catalyst produce the same products 2 and 3 indicates that both reactions involve   common intermediates. The formation of the silapropadiene-nickel complex 6 from 5 may be understood in terms of the nickel-catalyzed isomerization of 5 involving a 1,3-trimeth-ylsilyl shift.

Similar nickel-catalyzed reaction of 1-methyl-3-phenyl-1,2-bis(trimethylsilyl)-1-silacyclopropene (8) with phenyl(trimethylsilyl)acetylene gives four products, 1-methyl-3,4-diphenyl-1,2,5-tris(trimethyl-silyl)silole (9) (41%), two isomers of 1,1,4,4-tetra-methyl-3-phenyl-5-[phenyl(trimethylsilyl)methylene]-

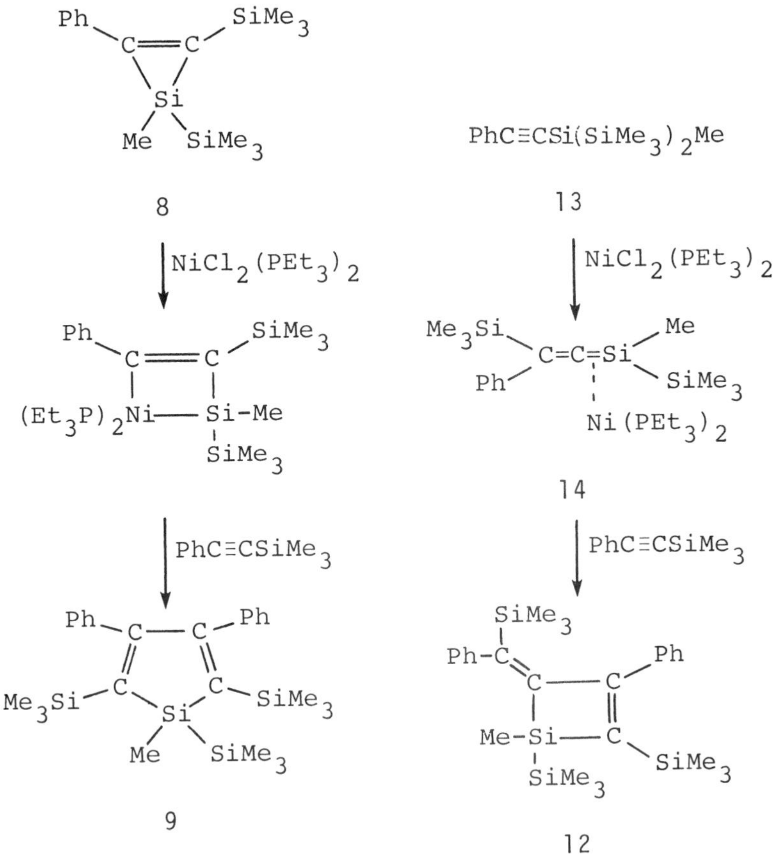

$$PhC{\equiv}CSi(SiMe_3)_2Me$$

13

$$NiCl_2(PEt_3)_2$$

14

$$PhC{\equiv}CSiMe_3$$

8

$$NiCl_2(PEt_3)_2$$

9

$$PhC{\equiv}CSiMe_3$$

12

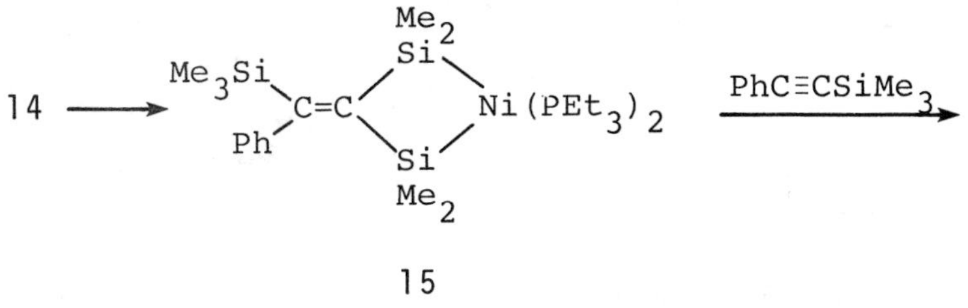

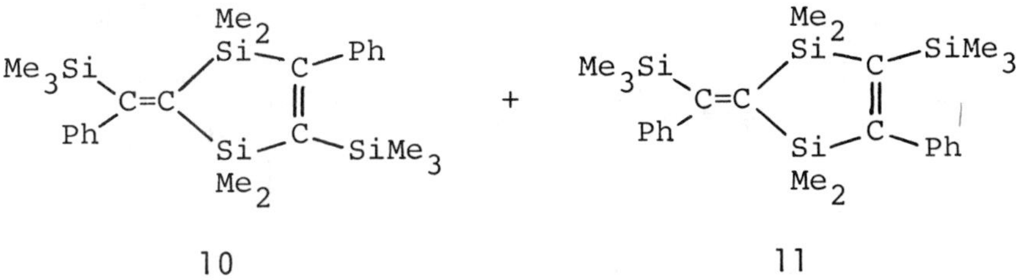

15

10   11

2-trimethylsilyl-1,4-disilacyclopent-2-ene (10 and
11) (36%), and 1-methyl-2-phenyl-4-[phenyl(trimethylsilyl)methylene]-1,2-bis(trimethylsilyl)-1-silacyclo-
but-2-ene (12) (3%). When a mixture of 2-(phenylethynyl)heptamethyltrisilane (13) and phenyl(trimethylsilyl)acetylene in the presence of the nickel
catalyst is heated at 200°C, products 10 - 12 are
obtained in 23, 23, and 33% yields, respectively[12].
The silole 9 can not be detected in this reaction.
The formation of 10 and 11 can best be explained by
the reaction of a nickeladisilacyclobutane intermediate (15) with phenyl(trimethylsilyl)acetylene,
while product 12 corresponds to the addition of silapropadiene-nickel complex (14) to phenyl(trimethylsilyl)acetylene. Here again, common intermediates
14 and 15 seem to be involved in both nickel-catalyzed reactions of 8 and 13.

The reaction of 1-methyl-1,2-diphenyl-3-(trimethylsilyl)-1-silacyclopropene (16) with phenyl(trimethylsilyl)acetylene in the presence of the nickel
catalyst affords 1-methyl-1,3,4-triphenyl-2,5-bis-
(trimethylsilyl)silole (17) in 73% yield[13]. The
nickel-catalyzed reaction of 1-phenyl-1-(phenylethynyl)tetramethyldisilane (18) with phenyl(trimethylsilyl)acetylene at 200°C gives the silole 17
and 1,1,4,4-tetramethyl-3-phenyl-5-(diphenylmeth-
lene)-2-(trimethylsilyl)-1,4-disilacyclopent-2-ene
(20) in 10 and 52% yields, respectively. Again, the
formation of 20 may be understood by the reaction of
nickeladisilacyclobutane (19) with phenyl(trimethylsilyl)acetylene.

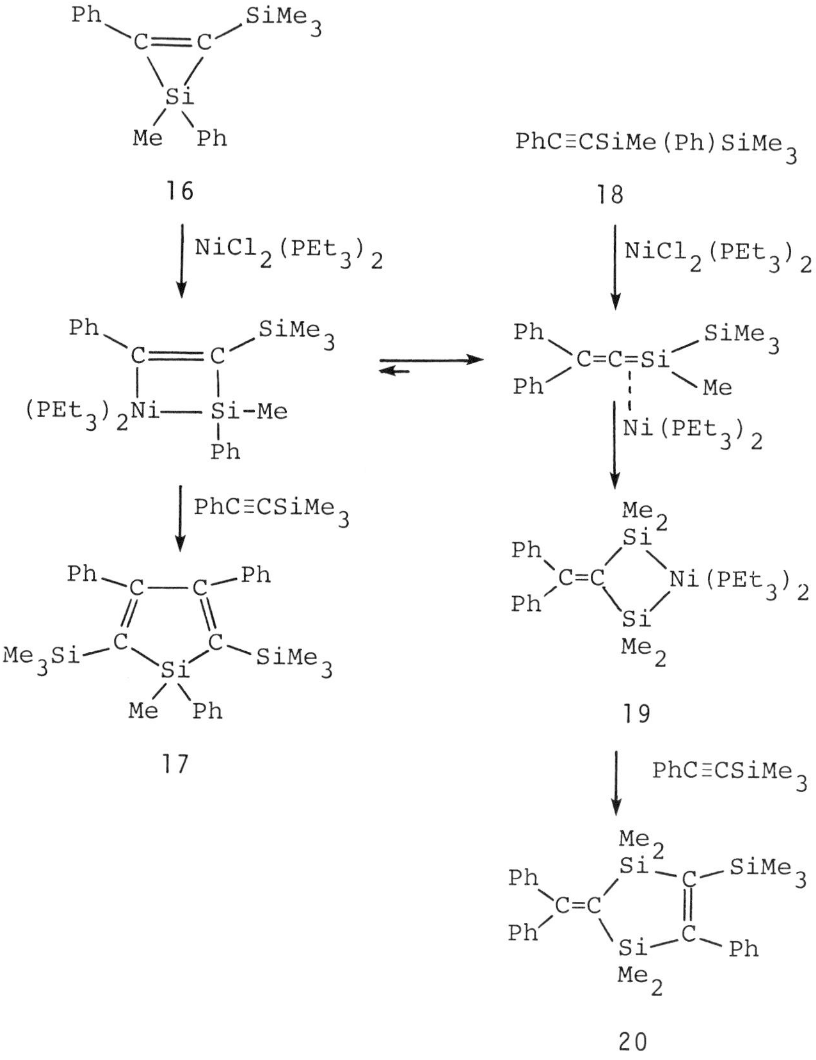

ISOMERIZATION OF A 1-SILAPROPADIENE

When 2-mesityl-2-(phenylethynyl)hexamethyltri-
silane (21) is heated with phenyl(trimethylsilyl)-
acetylene in the presence of a catalytic amount of
tetrakis(triethylphosphine)nickel(0) at 195°C, two
products, 1-mesityl-3-phenyl-2-[phenyl(trimethylsilyl)methylene]-1,4-bis(trimethylsilyl)-1-silacyclo-
but-3-ene (22) and 2,5-diphenyl-1-mesityl-4,4,6-tri-
methyl-1,3-bis(trimethylsilyl)-1,4-disilacyclohexa-
2,5-diene (23) are obtained in 77 and 11% yields,
respectively.

The formation of 22 clearly indicates that the

1-silapropadiene-nickel complex (24) must be formed
in this reaction. The product 23 would be produced
by the reaction of the nickeladisilacyclopentene
intermediate (25), with phenyl(trimethylsilyl)-
acetylene. It seems likely that the intermediate 25
would be formed from the silapropadiene-nickel
complex 24[14].

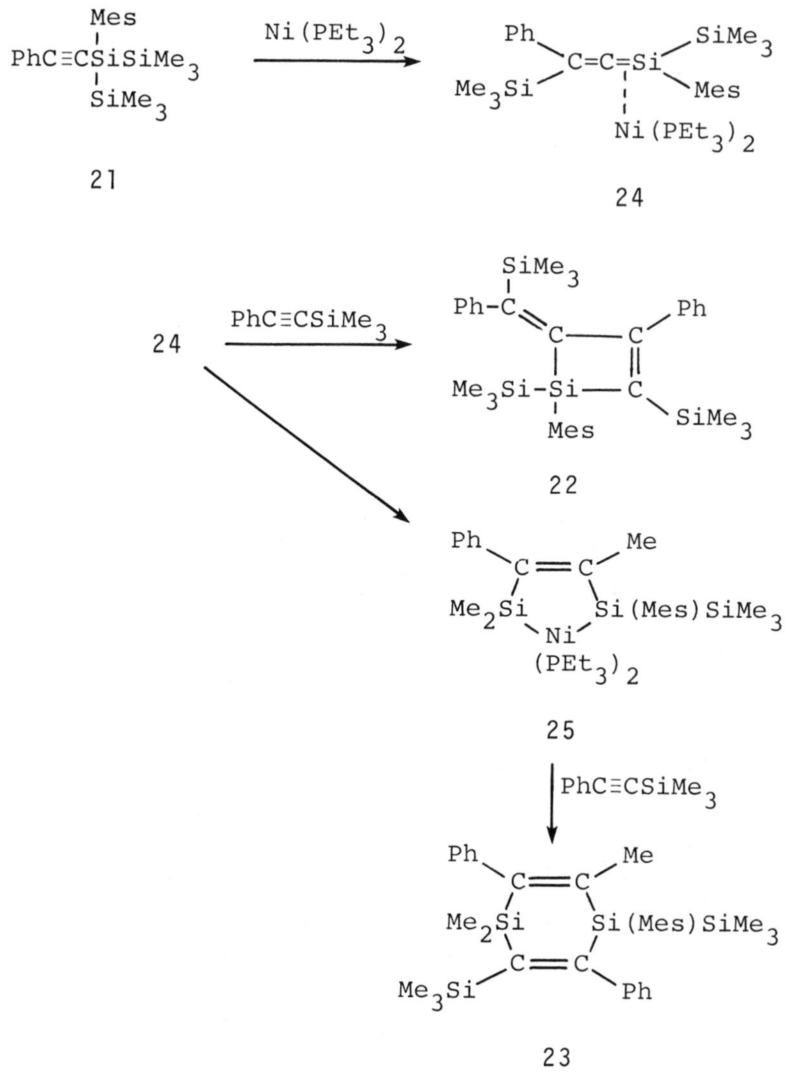

21

24

24

22

25

23

Interestingly, when compound 21 is heated in
the absence of phenyl(trimethylsilyl)acetylene under
the same conditions, 1-mesityl-1-phenyl-1-(trimethyl-

silylethynyl)trimethyldisilane (26) and two isomers
of 5,6-benzo-1,3-disilacyclohexane derivative (27
and 28) in 36, 36, and 22% yields. The production
of 26, 27, and 28 may be explained in terms of the
isomerization of the 1-silapropadiene-nickel complex
24.

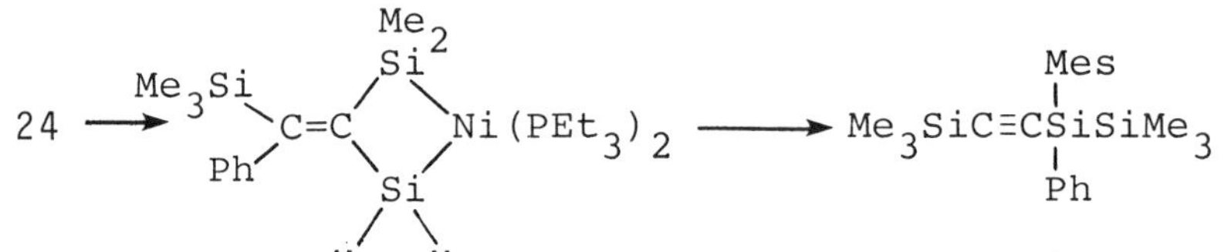

FORMATION OF A NICKELASILACYCLOBUTENE

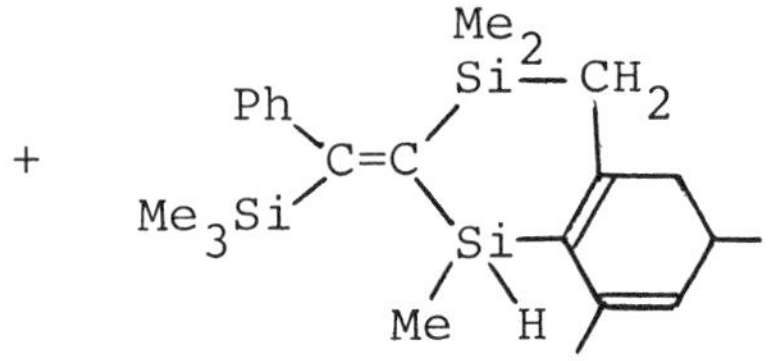

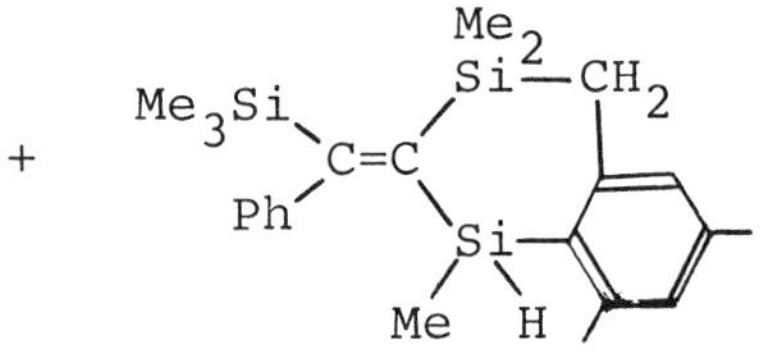

Next, we have investigated the stoichiometric
reaction of silacyclopropenes with tetrakis(triethylphosphine)nickel(0) in the hope of obtaining a
nickelasilacyclobutene[15]. Treatment of 1-mesityl-
3-phenyl-1,2-bis(trimethylsilyl)-1-silacyclopropene
(29) obtained photochemically from the phenylethynyl-
disilane 21 with 1 equiv of the nickel(0) catalyst
in benzene at room temperature, produces 2-mesityl-
4-phenyl-1,1-bis(triethylphosphine)-2,3-bis(trimethyl)-1-nickela-2-silacyclobut-3-ene (30). The
reaction is very clean and the silacyclopropene 29
is completely transformed into the 1-nickela-2-sila-
cyclobut-3-ene 30 within 1 n at room temperature.
The nickelasilacyclobutene 29 can also be prepared
in a xylene solution from the reaction of 29 with
the nickel compound at room temperature. Compound
30 is stable in solutions, although all attempts to
isolate this compound were unsuccessful.

When a xylene solution of 30 is heated to reflux
with phenyl(trimethylsilyl)acetylene, 1-mesityl-3,4-
diphenyl-1,2,5-tris(trimethylsilyl)silole in 32%
yield, in addition to 6% of 22.

To our surprise, heating the xylene solution of
30 at 110°C in the absence of phenyl(trimethylsilyl)-
acetylene gives 5,6-benzo-1,3-disilacyclohexenes 27

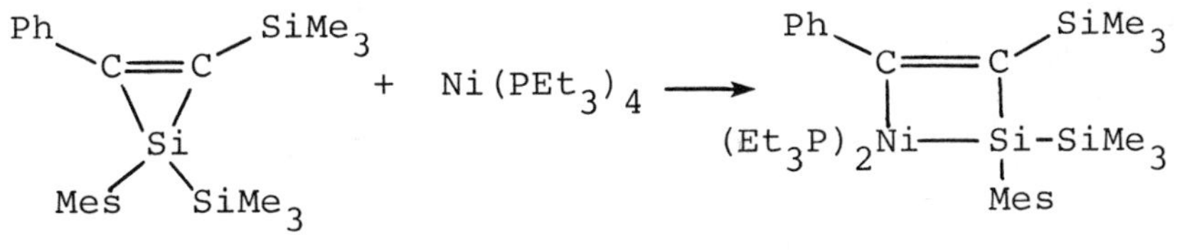

$$
\begin{array}{ccc}
\underset{\text{Mes}}{\overset{\text{Ph}}{\diagdown}}\text{C}=\text{C}\overset{\text{SiMe}_3}{\diagup} & + \ \text{Ni(PEt}_3)_4 \longrightarrow & \underset{\text{Mes}}{\overset{\text{Ph}}{\diagdown}}\text{C}=\text{C}\overset{\text{SiMe}_3}{\diagup}
\end{array}
$$

29   **30**

$$
\textbf{30} \longrightarrow \underset{\text{Me}_3\text{Si}}{\overset{\text{Ph}}{\diagdown}}\text{C}=\text{C}=\text{Si}\overset{\text{SiMe}_3}{\underset{|}{\diagup}}\underset{\text{Mes}}{\overset{|}{\text{Ni(PEt}_3)_2}} \longrightarrow \textbf{27} \ + \ \textbf{28}
$$

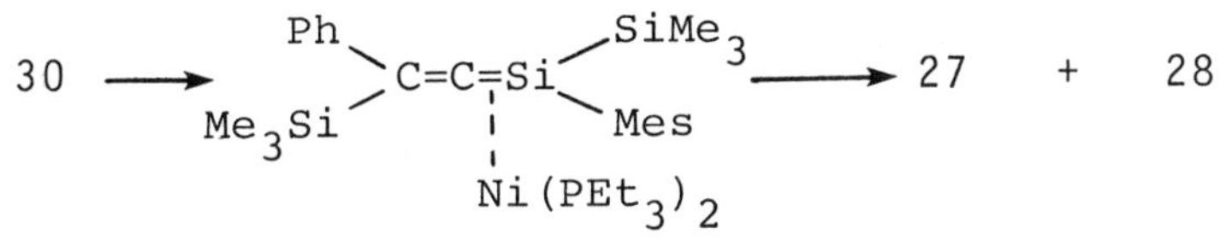

and 28 in 47 and 41% yields, respectively. The results clearly indicate that the formation of 27 and 28 must involve the isomerization of the nickela-silacyclobutene 30 to the silapropadiene-nickel complex 24.

ACKNOWLEDGEMENT

I wish to acknowledge the splendid work of my coworkers, and in particular Hiroshi Sugisawa, Shigeji Matsuzawa, and Joji Ohshita in carring out much of the work described here. Financial support from Nitto Electric Industrial Co. Ltd., and shin-etsu Chemical Co. Ltd. is also acknowledged.

REFERENCES

1. Gusel'nikov, L.E., Nametkin, N.S. Chem. Rev., 1979, 79, 529.
2. Barton, T.J. Pure Appl. Chem. 1980, 52, 615.
3. Ishikawa, M., Kumada, M. Adv. Organomet. Chem., 1981, 19, 51.
4. Raabe, G., Michl, J. Chem. Rev., 1985, 85, 419.
5. Brook, A.G. J. Organomet. Chem., 1986, 300, 21.
6. Wiberg, N. J. Organomet. Chem., 1984, 273, 141.
7. It has been reported that the reaction of tran-sition-metal halides with silylmethyl Grignard reagents gives products which may be explained in terms of a silene intermediate[8-10].
8. Tamao, K., Yoshida, J., Okazaki, S., Kumada, M. Isr. J. Chem., 1976/1977, 15, 265.
9. Pannell, K.H. J. Organomet. Chem., 1970, 21, 17.
10. Cundy, C.S., Lappert, M.F., Pearce, R. J. Organo-met. Chem., 1973, 59, 161.

11. Ishikawa, M., Matsuzawa, S., Hirotsu, K.,
    Kamitori, S., Higuchi, T. Organometallics, 1984,
    3, 1930.
12. Ishikawa, M., Matsuzawa, S., Higuchi, T.,
    Kamitori, S., Hirotsu, K. Organometallics, 1984,
    4, 2040.
13. Ishikawa, M., Sugisawa, H., Harata, O., Kumada,
    M. J. Organomet. Chem., 1981, 217, 43.
14. Ishikawa, M., Ohshita, J., Ito, Y.
    Organometallics, 1986, 5, 1518.
15. Ishikawa, M., Ohshita, J., Ito, Y., Iyoda, J.
    J. Amer. Chem. Soc., 1986, 108, 7417.

Chapter 20

Unexpected chemistry in the production of multiple bonds to silicon by thermal beta-eliminations†

T.J. Barton, S. Bain, S. Ijadi-Maghsoodi, G.R. Magrum, G. Paul, L.R. Robinson and M.H. Yeh –

Department of Chemistry, Iowa State University, Ames, Iowa, USA. 50011.

The generation of silenes by thermally-induced beta-elimination of Me_3SiOMe from silyl ethers of type 1, first reported by Gusel'nikov (ref. 1), has proved to be a convenient route to a variety of interesting silenes (ref. 2). For some time we were confused as to why this route worked so well while the isomeric systems 2, for which extrusion of Me_3SiOMe should be much more thermodynamically favorable (ca. 45 kcal), failed to produce silenes upon thermolysis. Thus we have now carried out an extensive kinetic study of the thermolyses of these and related systems which has revealed that 2 eschews beta-elimination in favor of alpha-elimination to produce a carbene. It appears that this surprising behavior is largely due to a more favorable Arrhenius preexponential factor for the 3-centered (alpha) elimination.

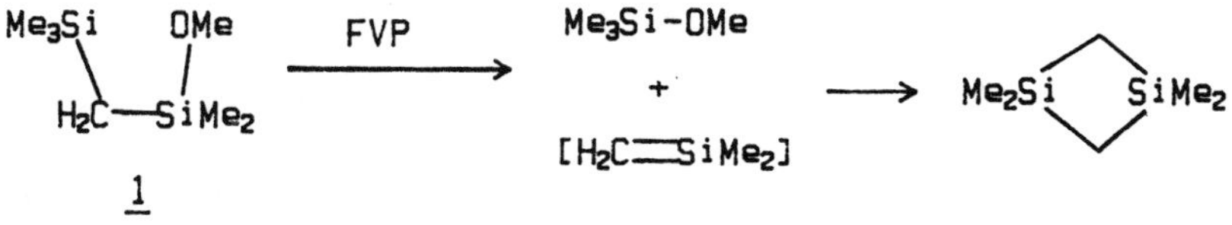

†The National Science Foundation is gratefully acknowledged for support of this work.

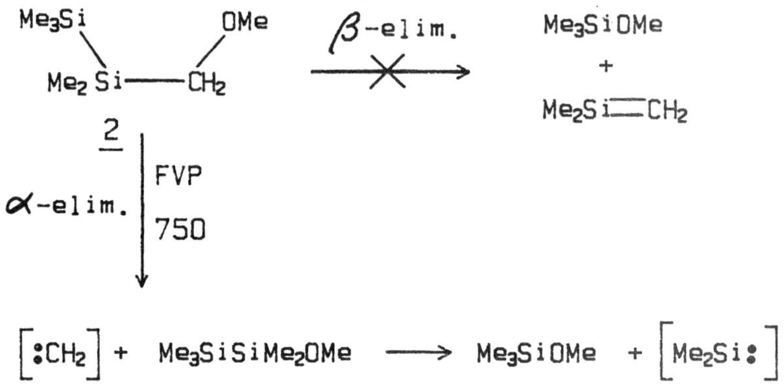

$$Me_3Si \diagdown \qquad \diagdown OMe \qquad \xcancel{\beta\text{-elim.}} \qquad Me_3SiOMe$$
$$Me_2Si\text{——}CH_2 \qquad\qquad\qquad +$$
$$\qquad\qquad\qquad\qquad\qquad Me_2Si\text{==}CH_2$$

$$\underline{2}$$

$$\alpha\text{-elim.} \Bigg| \begin{array}{c} FVP \\ 750 \end{array}$$

$$\Big[{:}CH_2 \Big] + Me_3SiSiMe_2OMe \longrightarrow Me_3SiOMe + \Big[Me_2Si{:} \Big]$$

Recently we attempted to extend this route to the generation of 1-silaallene $\underline{4}$ through the pyrolysis of vinyl silyl ether $\underline{3}$. Surprisingly, flash vacuum pyrolysis (FVP) of $\underline{3}$ instead produced isopropenylsilane $\underline{5}$ and disiloxane $\underline{6}$ along with lesser amounts of cyclosiloxanes D_3 and D_4.

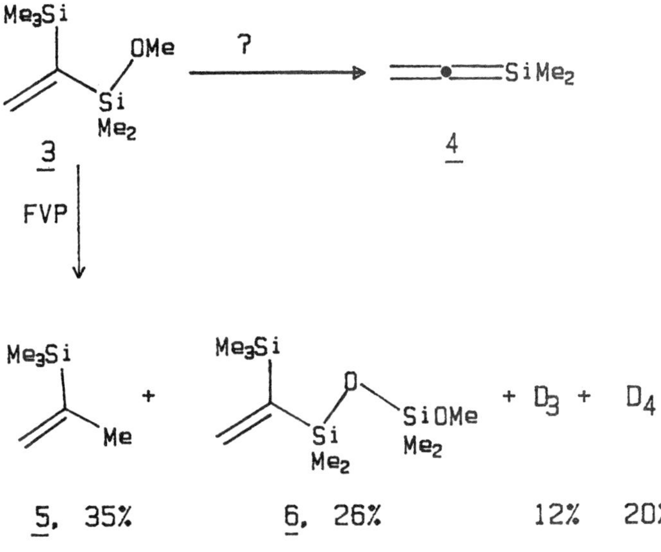

$$Me_3Si \diagdown \qquad \diagup OMe \qquad \xrightarrow{?} \qquad \text{==•==}SiMe_2$$
$$\diagup\diagdown_{\substack{Si \\ Me_2}}$$

$$\underline{3} \qquad\qquad\qquad\qquad \underline{4}$$

$$FVP \Bigg|$$

$$Me_3Si \diagdown \qquad\qquad Me_3Si \diagdown \qquad\quad O$$
$$\diagup\diagdown_{Me} \qquad + \qquad \diagup\diagdown_{\substack{Si \\ Me_2}}\diagdown_{\substack{SiOMe \\ Me_2}} \quad + \; D_3 \; + \; D_4$$

$$\underline{5}, \; 35\% \qquad\qquad \underline{6}, \; 26\% \qquad\qquad 12\% \quad 20\%$$

A wide variety of mechanistic possibilities for the apparent extrusion of dimethylsilanone ($Me_2Si=O$) from $\underline{3}$ were experimentally tested. However virtually all candidates were eliminated by the observation that FVP of the -OCD3 labeled ether, $\underline{3}$-d$_3$, afforded $\underline{5}$-d$_3$ as the exclusive isopropenylsilane product. We rationalize this by an initial 1,5-shift of deuterium to the olefinic terminus to afford 1,4-diradical $\underline{7}$, closure to silaoxetane $\underline{8}$, and finally decomposition to dimethylsilanone and $\underline{5}$-d$_3$. To our knowledge this represents the first example of the reaction of an olefin and an alkane to produce two radicals via a hydrogen atom transfer.

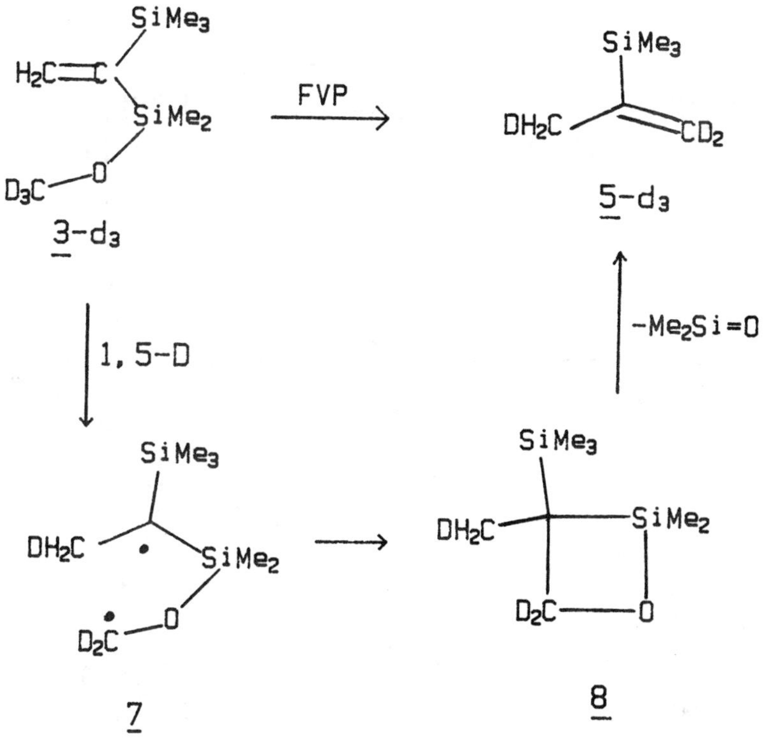

Another seemingly trivial extension of a well established reaction is the isomerization of 2-trimethylsilylpropenal ($\underline{9}$) to allenyl ether $\underline{10}$. Although the thermal isomerization of alpha-silylketones to silyl enol ethers is normally a very well behaved reaction, FVP of $\underline{9}$ at 750 °C rather

cleanly affords butadiene along with the cyclosiloxanes D_3 and D_4. We rationalize this
surprising extrusion of $Me_2Si=O$ as arising from
initial isomerization of 9 to 10, a 1,2-hydrogen
shift to produce carbene 11 which inserts into a
methyl C-H bond to produce silaoxetane 12,
decomposition of which affords both butadiene and
silanone. Indeed FVP of 9 at 650°C does produce
10, and FVP of either 10 or its propargyl isomer 13
affords both butadiene and D_3. Carbene 11 can also
be generated by a 1,2-silyl shift in acylsilane 14,
and indeed we find that FVP of 14 produces
butadiene as the major product along with lesser
amounts of 9, 10, and $D_{3,4}$.

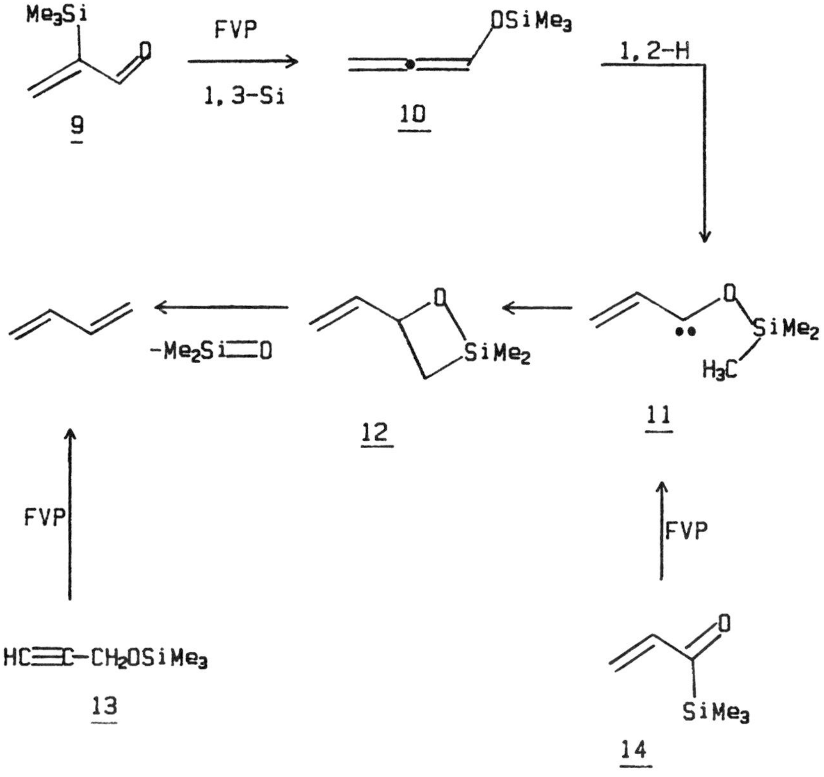

Another unexpected generation of silanone was
serendipitously discovered when we pyrolyzed a
mixture of dihydrofuran 15 and bis(dimethylsilyl)-
ketene (16). The synthesis of ketene 16 occurred
during the construction of 15 via metallation of

dihydrofuran and quenching with Me_2HSiCl. We have
now established that this process proceeds through
the intermediacy of the dilithioalkynolate 17,
trapping to form the acetylenic ether 18 which
rearranges (salt promoted) in solution to the
bis(silyl)ketene. We have found this to be an
excellent synthetic source of a wide variety of
bis(silyl)ketenes. Until recently the major synthetic problem was that the ketene was always
accompanied by the corresponding silylated dihydro-
furan (e.g. 15), but this problem has now been
taken care of through the use of 5-phenyl-4,5-
dihydro-furan which cleanly extrudes styrene to
afford uncontaminated bis(silyl)ketenes.

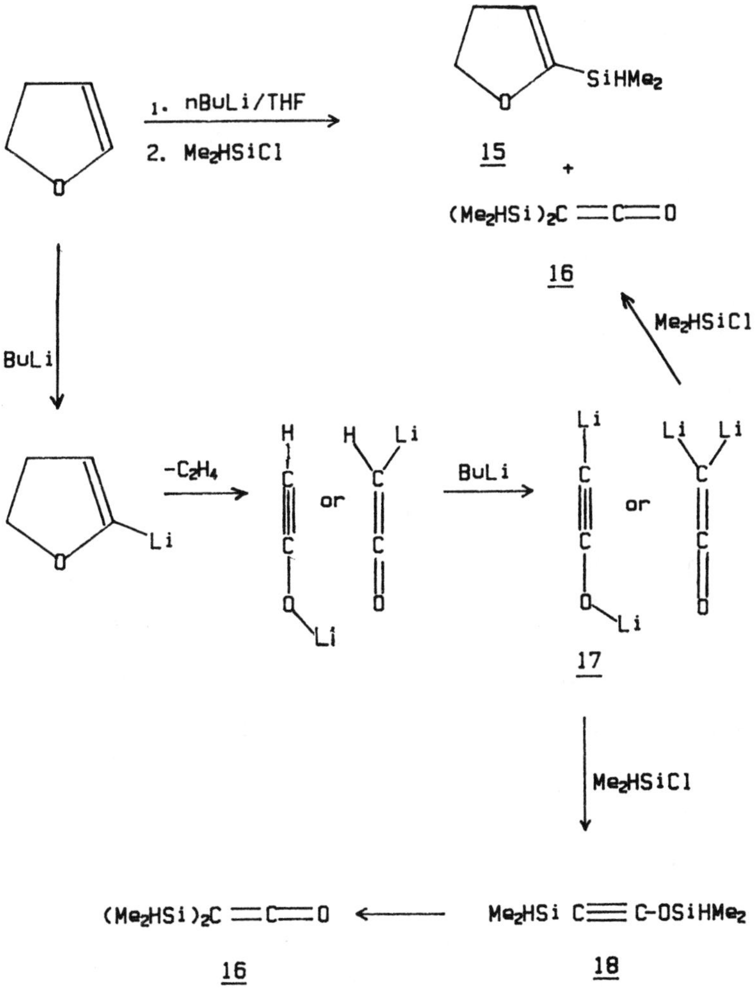

FVP of silylketenes <u>16</u> and <u>20</u> results in clean
extrusion of Me$_2$Si=O and formation of the corresponding silylacetylenes (<u>19</u> and <u>21</u>) (ref. 3). We
have recently found that the presence of an Si-H is
unnecessary since bis(trimethylsilyl)ketene (<u>22</u>)
also extrudes silanone to yield acetylene <u>23</u>.

$$(Me_2HSi)_2C{=}C{=}O \xrightarrow[-Me_2SiO]{FVP} Me_2HSi-C{\equiv}CH$$

<u>16</u>   <u>19</u>

$$(Me_3Si)(Me_2HSi)C{=}C{=}O \xrightarrow[-Me_2SiO]{FVP} Me_3Si-C{\equiv}CH$$

<u>20</u>   <u>21</u> (76%)

$$(Me_3Si)_2C{=}C{=}O \xrightarrow[-Me_2SiO]{FVP} Me_3Si-C{\equiv}C-Me$$

<u>22</u>   <u>23</u> (42%)

Although there are several mechanistic
possibilities, the mechanism which we currently
favor for this unique decomposition is a "retro-
Wolff" process which is initiated by a 1,2-silyl
shift to produce carbene <u>24</u>. The proposal of a
1,2-shift to produce a vinyl carbene has precedence
in the thermochemistry of allene which is now
thought to isomerize to propyne via an initial 1,2-
migration of hydrogen. The remainder of the steps
leading to silanone and acetylene, ultimately from
silaoxetene decomposition, are all precedented in
the work of Ando (ref. 4).

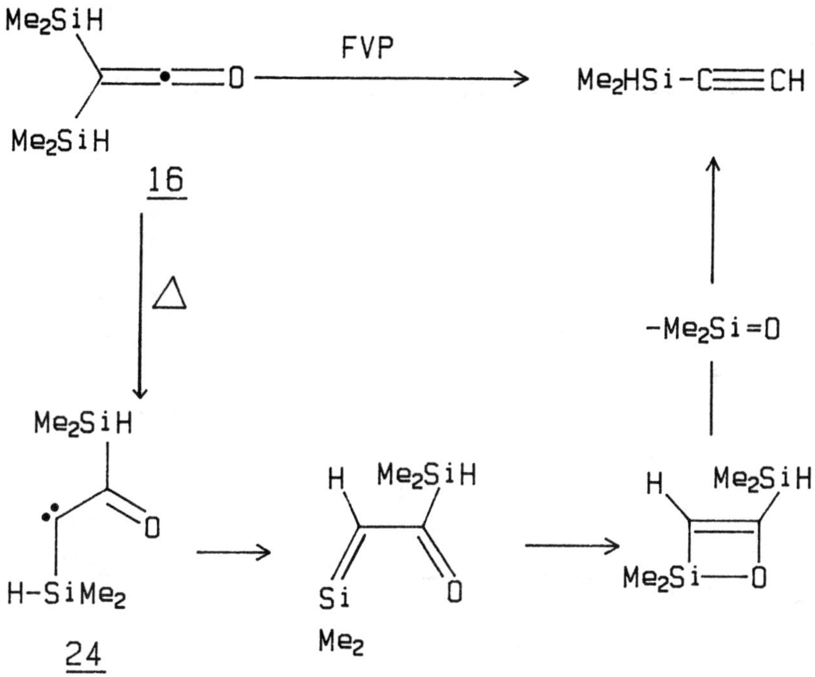

Me$_2$SiH

$\diagdown$C$=$C$=$O — FVP ————→ Me$_2$HSi-C$\equiv$CH

Me$_2$SiH

16

$\triangle$

$-$Me$_2$Si$=$O

Me$_2$SiH

C$=$O

H$-$SiMe$_2$

24

H   Me$_2$SiH

C$=$C$=$O
$|$
Si
Me$_2$

H   Me$_2$SiH

C$=$C
$|$ $|$
Me$_2$Si$-$O

     Most recently we have found that bis(silyl)-
thioketenes undergo an analogous extrusion of
dimethylsilathione (Me$_2$Si=S). Thus, for example,
thioketene **25** cleanly forms trimethylsilylacetylene
(92%) via extrusion of Me$_2$Si=S as evidenced by the
formation of ·its cyclic trimer **26**.

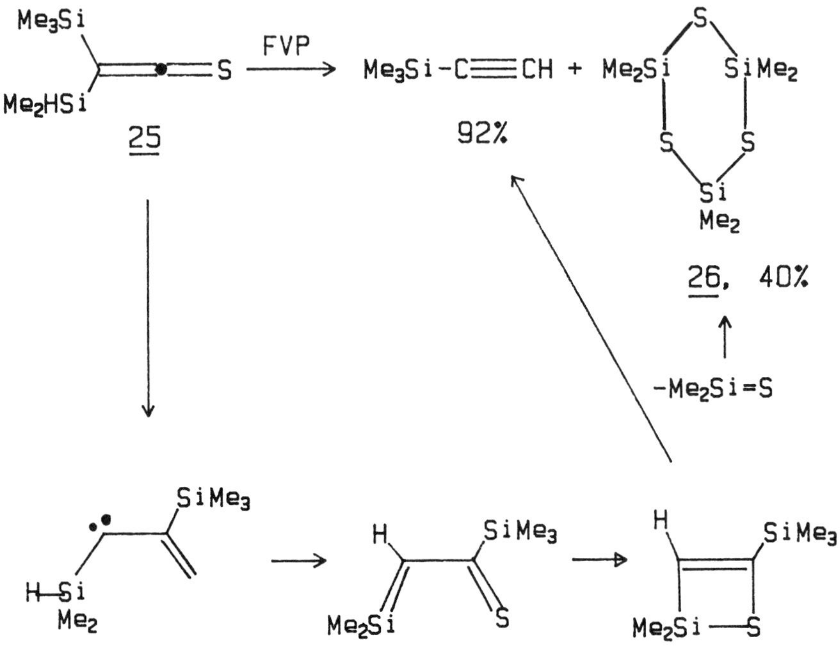

Finally we would make brief mention of some unexpected chemistry we have encountered in a study dedicated to the development of a general route to silaoxetanes as thermal precursors to silanones. Reaction of norbornanol 27 and KH was conducted in the hopes that intramolecular etherification would afford a silaoxetane (28) which would be stabilized by virtue of the fact that silanone extrusion would be accompanied by formation of the highly unstable, anti-Bredt olefin, 29. The actual product of this condensation, siloxane 30 undergoes a remarkable decomposition to tricyclane (31) and the cyclic disiloxane 32. At higher temperatures 32 also affords 31 along with D_3. The origin of tricyclane is almost certainly carbene 33, and we are intrigued by the possibility that 33 arises from isomerization of bicycloheptene 29. To our knowledge there only two literature suggestions of the thermal isomerization of an olefin to a carbene (refs. 5 & 6).

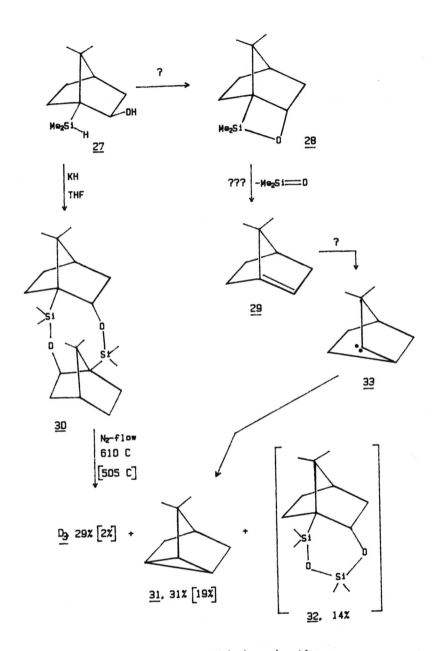

Perhaps even more surprising is the observation that in solution (at room temperature) reaction of 27 and I_2 in THF produces 31 in 80% yield, while this reaction in benzene affords the endo-iodide 34 in 73% yield. Since 34 is not formed by the reaction of HI and tricyclane, one must entertain the possibility that it arises from

reaction of HI and <u>29</u>. If indeed this is true, and
experiments are currently in progress to check
this, it implies that olefin <u>29</u> isomerizes to
carbene <u>33</u> at room temperature!

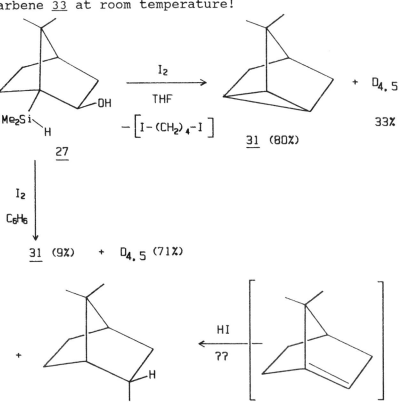

REFERENCES

1. N. S. Nametkin, L. E. Gusel'nikov, E. A. Volpina and V.
M. Vdovin, <u>Dokl</u>. <u>Akad</u>. <u>Nauk</u>. <u>SSSR</u>, (1975) **220**, 386.

2. G. T. Burns and T. J. Barton, <u>J</u>. <u>Organomet</u>. <u>Chem</u>.,
(1981), **216**, C5; T. J. Barton, G. T. Burns, E. V. Arnold, J.
Clardy, <u>Tetrahedron</u> <u>Lett</u>., (1981) **22**, 7.

3. T. J. Barton and B. L. Groh, <u>J</u>. <u>Am</u>. <u>Chem</u>. <u>Soc</u>., (1985)
107, 7221.

4. A. Sekiguchi and W. Ando, <u>J</u>. <u>Am</u>. <u>Chem</u>. <u>Soc</u>., (1984)
106, 1482.

5. T. H. Chan and D. Massuda, <u>J</u>. <u>Am</u>. <u>Chem</u>. <u>Soc</u>., (1977)
99, 936.

6. R. T. Conlin, H. B. Huffaker and Y.-W. Kwak, <u>J</u>. <u>Am</u>.
<u>Chem</u>. <u>Soc</u>., (1985) **107**, 731.

Chapter 21

Intermediates with Si=O multiple bond: generation, stabilization and direct spectroscopic study

A.K. Maltsev, V.N. Khabashesku, O.N. Nefedov and N.D. Zelinsky — Institute of Organic Chemistry, USSR Academy of Sciences, Moscow, USSR.

INTRODUCTION

The discovery of the property of silicon to form multiple bonds with carbon [1], for which spectroscopic evidence was obtained in 1976 [2,3], gave rise to the studies of stabilization of compounds with silicon-nitrogen, silicon-phosphorous, silicon-sulfur as well as silicon-oxygen multiple bonds. As far back as 1952 Andrianov and Sokolov proposed the involvement of silicon-oxygen doubly bonded species-silanones in the thermal redistribution of polysiloxanes [4]. Since that time a large body of information has been accumulated, suggesting formation of silanones in a variety of pyrolytic or photochemical reactions. These assumptions are based on the final products of reactions, chemical trapping data and in some cases on kinetic data. However, for unambiguous identification of silanones one should get their direct physical characteristics by instrumental methods. That is a difficult task as silanones, because of their instability at normal conditions, have not been isolated till now.

The theoretical [5] and experimental [6,7] values of Si=O multiple bond strength give the limits for π-component of that bond in the range of 38-62 kcal/mol, which is close to the energy of Si=C multiple bond estimated to be 39± kcal/mol

[1]. Thus, the instability of silanones as well as silaethylenes is caused by a kinetic factor rather than by a thermodynamic factor. This explains the high reactivity of silanones, for example, cyclo-oligomerization reactions. The high polarity of the Si=O bond is calculated to be 4.18 to 4.52 D for the $H_2Si=O$ molecule [5].

At the present time one may distinguish two general approaches to the problem of silanone stabilization. They consist of minimization of the disappearance of these intermediates during secondary reactions both by the thermal generation in the gas phase at low pressures and by photolysis of precursors in solid inert matrices at low temperatures.

Only two of the numerous physical methods can be successfully applied now for direct investigations of silanones. Namely, mass spectrometry and matrix isolation IR-spectroscopy. The data obtained by means of these methods will be discussed.

MASS-SPECTROSCOPIC STUDIES OF SILANONES IN THE GAS PHASE

One of the varieties of mass-spectroscopic methods, ion cyclotron resonance spectroscopy, was used for generation and investigation of the dimethyl-silanone enolate anion $^-CH_2(Me)Si=O$ (1) in the gas phase. This anion was produced by collision-induced dissociation of Me_3SiO^- during the ion-molecule reaction of $SiMe_4$ with hydroxyl anion. The proton affinity of 1 was estimated to be 366+3 kcal/mol. It was also characterized by ion-molecular reactions with CO_2 and SO_2, which led to the 2-methylsilaacetate anion [8].

In another work [7], 1 was obtained both from 2 and from dimethylsiloxide anion (3) by their infrared multiple photon (IRMP) irradiation using a CO_2 laser. The upper limit for Si=O bond strength in neutral silanone molecule was estimated to be 62 kcal/mol.

$$Me - \underset{\underset{Me}{|}}{\overset{\overset{O^-}{|}}{Si}} - H \xrightarrow{\text{IRMP}} \quad \underset{Me}{\overset{O}{\underset{\diagdown}{\diagup}}} \underset{1}{Si} \diagdown_{CH_2}^- + H_2 \quad (1)$$

3

Pyrolytic mass-spectrometry was applied by us for direct detection of dimethylsilanone (4) in the gas phase, using vacuum pyrolysis of epoxysilacyclopentanes (5,6). This work was done in collaboration with Dr. J. Tamas and his coworkers of the Central Research Institute of Hungarian Academy of Sciences in Budapest.

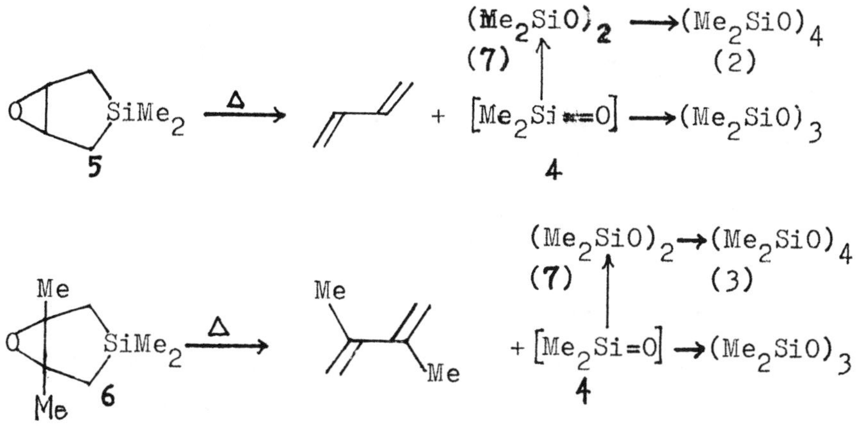

The formation of cyclosiloxanes as final products of thermolysis [9], and kinetic data [10] showed that 4 could be an intermediate in reactions 2 and 3. These reports also discussed the possibility of the formation of short-lived dimethylsilanone cyclodimer [7].

Butadiene, D_3 and D_4 as well as compound, characterized by a peak with m/e 133, were detected by us in the mass-spectra due to changes in the intensities of characteristic peaks of both initial compounds 5,6 and their pyrolysis products at temperatures of 700–1000°. At these conditions the peak intensity of the ion, m/e=133, increases about 10 times compared to the peak intensities of 5,6. Its measured exact mass, corresponds to the composition $C_3H_9Si_2O_2$ in good agreement with the calculated mass. The registration of a metastable ion with m/e=119.6 shows, that the daughter ion with m/e=133 is formed by loss of a methyl fragment from the ion with m/e=148 – the molecular ion of cyclodimer 7. That ion is unstable as well as the molecular ions of cyclosiloxanes D_3 and D_4, easily losing a CH_3 group under electron impact. Thus, the ion with m/e=133 characterizes 1,3-cyclodisiloxane 7 and we obtained the first direct spectroscopic evidence of its existence in the gas phase.

Unfortunately, the molecular ion of 4 with m/e=74 was not detected in the mass spectra with a decrease of ionization energy from 70 to 10–15 eV. Yet, a small increase of the intensity of peak with m/e=59 as well as the production of 7 leads us to assume, that 4 is present in the gas phase as a kineticly independent species. Its flow concentration is too low because of a high cyclodimerization rate, proceeding with zero activation barrier, according to calculations [11].

Only the application of matrix isolation techniques for freezing of products of vacuum pyrolytic

reactions allowed stablization and accumulation of
dimethylsilanone in an inert matrix in amounts sufficient for the IR-observation [17].

LOW-TEMPERATURE MATRIX ISOLATION OF THE INTERMEDIATES WITH Si=O MULTIPLE BOND

Among numerous variants of matrix isolation [12]
two techniques were applied to the studies of the
intermediates with Si=O multiple bond.

Photochemical reactions in low-temperature matrixes

H. Schnockel was the first to use this method for
generation and stabilization of inorganic silanones
$Cl_2Si=O$ and $F_2Si=O$. They were obtained by cocondensa-
tion of reactive silicon monoxide into an Ar-matrix
containing chlorine or fluorine and irradiated simultaneously by UV-light. A number of bands in the IR-
spectra of these silanones were recorded and a full
assignment to various vibration modes as well as to the
Si=O stretch was made [13,14].
    The simplest silanones $H_2Si=O$, (HO)HSi=O and
$(HO)_2Si=O$ were obtained by photolysis of a SiH_4 and
ozone mixture in an Ar-matrix at 17 K [15]. Unfortunately, each silanone was identified only by one IR-
band, observed in the Si=O stretch region. Recently,
the same technique was used [16] for generation of
MeHSi=O (8) and 4. In that case only the bands, assigned to the Si=O stretch were indicated at 1207.6
cm^{-1} for 8 and 1209.6 cm^{-1} for 4. It should be noted,
that the last value is in a good agreement with the
band at 1210 cm^{-1}, recorded by us for 4 first stabilized from the gas phase. We published these data in
May 1986 [17] - seven months earlier than the above
work [16].
    An attempt to stabilize 4 by photo irradiation of
dodecamethylcyclohexasilane (9) - source of dimethylsilylene in Ar matrix, containing 0.5-1% N_2O, used as
oxygen donor, was made by J. Michl and R. West [18].

$$Me_{12}Si_6 + N_2O \xrightarrow{h\nu} Me_{10}Si_5 + N_2O + Me_2Si=O$$
$$\quad \mathbf{9} \qquad\qquad\qquad\qquad\qquad\qquad\qquad \mathbf{4} \quad (4)$$

A new weak band at 1204 cm^{-1}, disappearing on warming
the matrix was assigned to the Si=O stretch of 4.
However by using ethylene oxide as an oxygen donor this
new band was observed at 1193 cm^{-1}. The high sensitivity of this band position in the IR-spectra to the
composition of the matrix allows us to assume, that in
this work no monomer of 4, but a complex of that polar
molecule with other molecules, present in the matrix,
was detected.

To avoid unplanned formation of such complexes, generation of reactive intermediates in the gas phase at low pressures followed by freezing in inert matrix was attempted.

Vacuum pyrolysis of precursors and freezing of the intermediates with Si=O multiple bond in matrix.

This technique developed by us specifically for studies of organic pyrolytic reaction mechanisms [19] has been widely used since 1967. It was successfully used by us for the detection of the first intermediate with Si=C multiple bond by a physical method in 1976 [2].

By way of extension we studied the possibility of dimethylsilanone stablization from the gas phase. The vacuum pyrolysis of a series of silicon-oxygen containing compounds (10-15 as well as 5,6) was investigated by matrix isolation IR-spectroscopy. These compounds have been considered in the literature as potential sources of **4**.

$$(Me_2SiO)_4 \qquad CH_2{=}CHCH_2OSiHMe_2 \qquad CH_2{=}CHCH_2{\diagdown}SiMe_2$$
$$CH_2{=}CHCH_2{\diagup}O$$

10   **11**   **12**

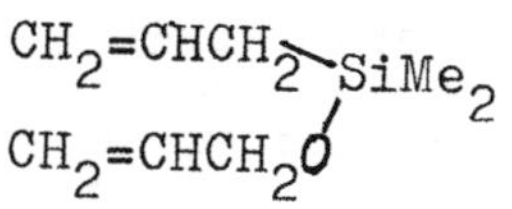

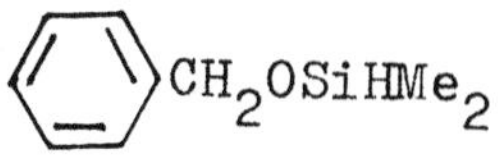

13 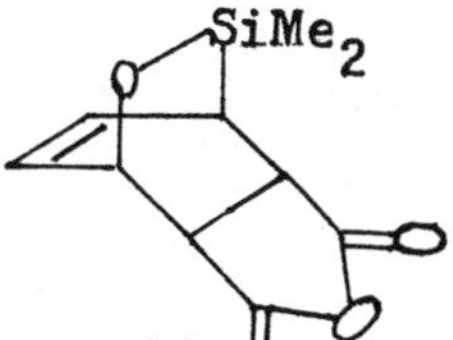   **14** 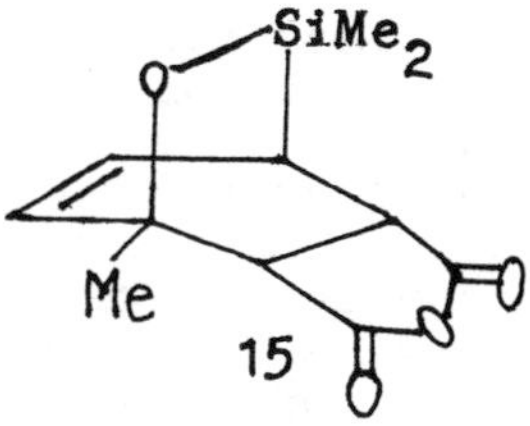   **15**

The gas phase pyrolysis of octamethylcyclotetrasiloxane was studied by Davidson at low pressures in a static system [6]. D_3 and D_5 were obtained as final products of the reaction. At small conversions, the decomposition kinetics corresponded to first order. These data have been considered as evidence for **4** as an intermediate, which could be trapped by D_4 forming D_5.

$$(Me_2SiO)_4 \xrightleftharpoons{\Delta} (Me_2SiO)_3 + [Me_2Si{=}O] \quad (5)$$

$$[Me_2Si{=}O] + (Me_2SiO)_4 \xrightleftharpoons{} (Me_2SiO)_5 \quad (6)$$

We studied the vacuum pyrolysis of **10** in a flow reactor at low pressures, 5×10^{-2}–5×10^{-4} torr and at temperatures between 700–1000°. A detectable decomposition of **10** (~ 30%) was observed at temperatures higher than when formation of the intense bands of D_3, CH_4, C_2H_2 as well as the bands of the unstable species – CH_3 radical and SiO (1225 cm^{-1} [20]) were detected. We didn't observed any bands, which could be assigned to **4** expected according to equation 5. However, because of the high decomposition temperature of **10** in vacuum, **4** is probably thermally unstable at these conditions and decomposes into SiO and CH_3 radical.

$$Me_2Si=O \xrightarrow{\Delta} SiO + \cdot CH_3 \qquad (7)$$

The choice of allyloxydimethylsilane (**11**) as a subject of investigation was based on the available data on allylsilane pyrolysis, which exhibited the capability of such compounds for retroene reactions resulting in the formation of double bonded Si=C species [1]. Moreover, a similar compound – (dimethylallyloxy)dimethylsilane – was considered as a potential source of **4** on the basis of a kinetic study [21]. Therefore we could also expect the formation of silanone during a retroene reaction.

$$CH_2=CHCH_2OSiHMe_2 \xrightarrow{\Delta} CH_2=CHCH_3 + [Me_2Si=O]$$
$$\mathbf{11} \qquad\qquad\qquad\qquad\qquad \mathbf{4} \quad (8)$$

According to matrix IR-spectroscopy data, vacuum pyrolysis of **11** proceeded at temperatures higher then 850°. At 940° and 5×10^{-4} torr conversion of the initial compound reached 75% and intense bands of propylene were observed in the IR-spectra of the pyrolysis products as well as the bonds of SiO, CH_3 radical, CH_4, C_2H_2, C_2H_4 – probably the decomposition products of **4**. Therefore, we detected the bands of allyl radical, which is probably formed by a parallel pathway.

$$CH_2=CHCH_2OSiHMe_2 \xrightarrow{\Delta} CH_2 \cdots CH \cdots CH_2 + [\cdot OSiHMe_2] \quad (9)$$
$$\mathbf{11} \qquad\qquad\qquad\qquad\qquad\qquad \downarrow \; \mathbf{16}$$
$$SiO + \cdot CH_3 + CH_4$$

Thermal instability of the siloxyl radical (**16**) and its fragmentation above 800° in vacuum was found recently in our mass spectrometric work [22].

Allyl(allyloxy)dimethylsilane (12) was suggested
by T.J. Barton as source of 4 in the gas phase. During
pyrolysis of 12, 4 was to be formed with consecutive
loss of two allyl radicals and formation of diallyl and
siloxane D_3 as final products shown in Eq. 10. [23].

$$
\begin{array}{c}
CH_2=CHCH_2 \\
{}_{SiMe_2} \xrightarrow{\Delta} 2CH_2{\cdots}\overset{\cdot}{C}H{\cdots}CH_2 + [Me_2Si{=}O] \quad (10) \\
CH_2=CHCH_2O \qquad\qquad\qquad\qquad\qquad\qquad \downarrow \qquad\qquad\quad \llcorner\!\!\rightarrow D_3 \\
\mathbf{12} \qquad\qquad\qquad\qquad\qquad \wedge\!\!\wedge\!\!\wedge\!\!/
\end{array}
$$

Bands of the allyl radical [24] were the most
intense in the matrix IR-spectra of the pyrolysis
products. Detectable conversion of the initial compound
was observed above 900° and at pressures 5×10^{-2}-5×10^{-3}
torr in the reaction zone. Bands of cyclosiloxane D_3
under these conditions were not detected, which shows
the absence of intermolecular collisions in the gas
phase, facilitating trimerization of the expected
silanone 4. However, we failed to stabilize 4 by this
method. Only the products of its thermodecomposition
were detected in the IR-spectra.

Vacuum pyrolysis of benzyloxydimethylsilane (13)
was studied in a collaboration with E.A. Chernishev and
T.L. Krasnova. Our hopes to generate 4 from 13 were
based on composition of the final products of its
pyrolysis at atmospheric pressure and temperatures
between 350-550°, where the quantitative formation of
toluene and D_3 was observed.

$$
C_6H_5CH_2OSiHMe_2 \xrightarrow{\Delta} C_6H_5CH_3 + [Me_2Si{=}O] \qquad (11)
$$
$$
\mathbf{13} \qquad\qquad\qquad\qquad\qquad\qquad \llcorner\!\!\rightarrow D_3
$$

However, decomposition of 13 in vacuum proceeded at too
high a temperature (950°-1100°). Besides the bands of
SiO, CH_3, CH_4 and C_2H_2 there were two intense bands at
762 and 667 cm^{-1}, recorded in matrix IR-spectra of
pyrolysis products of 13. These bands disappeared on
warming the matrix to diffusion temperatures (40-45 K)
with simultaneous appearance of toluene bands. The same
changes were observed on increasing the pressure in the
pyrolysis zone from 10^{-2} to 1 torr. According to these
data we assigned the two above bands to the benzyl
radical. Additional evidence for that conclusion was
obtained in a special study of the IR-spectrum of
benzyl radicals generated by pyrolysis from several
independent sources [25]. The lack of toluene bands in
the IR-spectra of vacuum pyrolysis products of 13
suggests that this reaction does not proceed in vacuum
by a molecular mechanism. The radical mechanism with

formation of benzyl radical and thermally unstable
siloxyl radical **16** is the most probable at these conditions.

$$C_6H_5CH_2OSiHMe_2 \xrightarrow[10^{-2}-10^{-4}torr]{950-1100^o} C_6H_5\overset{\cdot}{C}H_2 \; + \; \begin{bmatrix} O-SiMe_2 \\ | \\ H \end{bmatrix}$$

$$\textbf{13} \qquad\qquad\qquad SiO + C_2H_2 + CH_4 + \overset{\cdot}{C}H_3 \quad (12)$$

   After numerous failures to generate **4**, connected
probably with the high decomposition temperature of the
investigated compounds in vacuum, we managed our aim by
using 3,4-epoxysilacyclopentanes **5** and **6** as sources.
Their decomposition in vacuum proceeds under milder
conditions. Bands of butadiene, D_3 and two new weak
bands at 1067 and 1054 cm^{-1} were present in the IR-
spectrum of pyrolys products of **5** at pressures higher
than 1×10^{-2} torr and temperatures between 730-900°.
These two new bands disappeared on warming the matrix
to 35-40 K. They also were observed in the IR-spectra
on freezing the pyrolyzate at 12 K without dilution
with argon. According to mass-spectroscopic data under
these conditions, silanone cyclodimer **7** should be
present. Probably, the above bands, 1067 and 1054 cm^{-1},
belong to that unstable molecule, which agrees with
calculations of the cyclodisiloxane vibrational
spectrum **11**.
   At lower pressure, $(1.10^{-2}$ to 5.10^{-4} torr) new
bands at 1244, 1240, 1210, 882, 798, 770 and 657 cm^{-1}
in the IR-spectrum of the pyrolyzate, were observed
with bands of D_3 completely absent. On warming the
matrix from 12 to 40 K the above bands disappeared
simultaneously, which suggests that they belong to the
same unstable species. The bands of stable molecules -
butadiene and **5** - remained in the spectrum. Freezing of
the reaction products at 12 K without argon resulted in
formation of butadiene, D_3 and small amounts of D_4. The
same seven bands were recorded in the IR-spectrum of
the vacuum pyrolysis products of another source of **4**
the epoxide **6**. The increase of the pyrolysis temperature from 850° to 950° led to weakening of these seven
bands and the bands of SiO, CH_3 radical, C_2H_2 and CH_3
appeared in the spectra instead.
   These results prove the assignment of the seven
found bands to dimethylsilanone, first observed by us
from the gas phase. The most important conditions for
the generation of silanone as monomer are the lowest
pyrolysis temperature of the precursor in vacuum and
the lowest pressure in the reaction zone. The experiments on pyrolysis of epoxides as well as the consideration of the above data for other sources suggest,
that the limit of thermal stability of dimethylsilanone
in the gas phase is about 850° and the limit of kinetic

stability is pressure about 5.10^{-4} torr. That conclusion is supported by IR-spectroscopic study of Diels-Alder adducts 14, 15. According to data [26], the temperature of thermolysis of 14 is as low as 180°. The formation of D_3 in this reaction was explained by participation of 4.

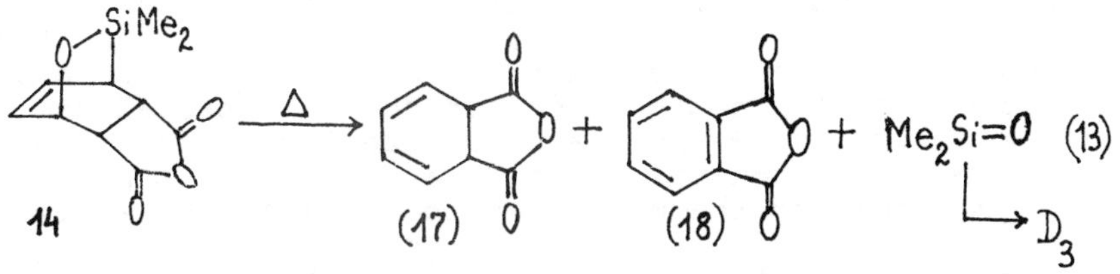

Vacuum pyrolysis of 14 at ~ 1×10^{-2} torr and 400-700° showed that full conversion of the initial compound was achieved at 600°. The intense bands of D_3, 17 as well as bands of 18, benzene, CO_2, CO and three bands at 1245, 1210 and 798 cm^{-1} were recorded in the IR-spectrum. These bands disappeared on warming the matrix and coincide with the most intense bands of 4, generated from 5 or 6. The observation of its weaker bands was complicated in this case by overlapping with the intense bands of 17. Similar frequencies were registered in the study of vacuum pyrolysis of another Diels-Alder adduct, 15.

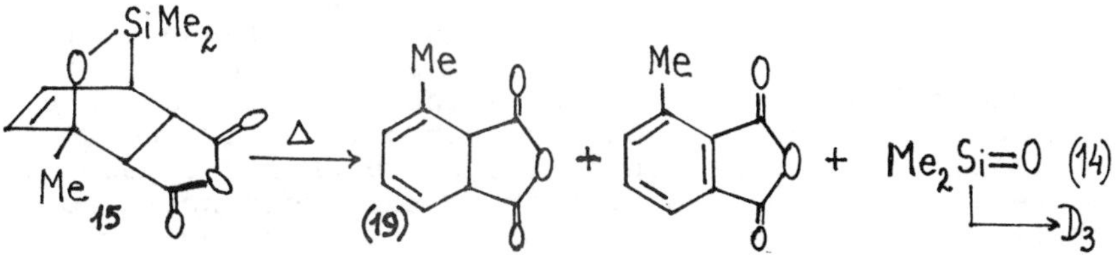

The full conversion of 15 in vacuum was observed at a lower temperature, ~ 500°. However, in these experiments a number of IR-bands of 4 also overlapped with the intense bands of 19, which allowed identification only of the most intense bands of silanone.
      Thus, vacuum pyrolysis of Diels-Alder adducts by taking into account the low thermostability of dimethylsilanone is the most suitable method for its generation. For IR-spectroscopic study of that intermediate the best sources are epoxysilacyclopentanes.
      Recently [27] in matrix the IR-spectrum of the pyrolysis products of benzyloxytrichlorosilane at pressures close to atmospheric, gave three bands which were assigned to dichlorosilanone, considered to be generated as a monomer at these conditions. That could lead to the assumption that dichlorosilanone is a much

less reactive intermediate than other earlier investigated silanones. However, this conclusion requires
additional experimental evidence.

VIBRATIONAL ASSIGNMENT IN IR-SPECTRA OF SILANONES.
VIBRATION FREQUENCY, FORCE CONSTANT AND ORDER OF
Si=O BOND

The frequencies of dimethylsilanone can be used as
an analytic characteristic for its detection among the
products of numerous organosilicon reactions. These
frequencies also can be used for the vibrational assignment in the IR-spectra of dimethylsilanone, involving necessary isotopic shift data. With this aim in
view we recorded the matrix IR-spectrum of $(CD_3)_2Si=O$
(d_6^{-4}), obtained by vacuum pyrolysis of
bis(trideuteromethyl)epoxysilacyclopentane (d_6^{-5}).

$$\text{Ar-matrix, 12 K}$$

$$d_6\text{-}5 \quad \xrightarrow{\;850°,5\cdot10^{-4}torr\;} \quad + (CD_3)_2Si=O \quad d_6\text{-}4$$

The bands 1215, 1032, 1007, 995, 712, 685 and 674 cm^{-1}
were assigned to d_6-4 in the matrix IR-spectra. These
bands disappeared simultaneously both on warming the
matrix to 35-40 K and on increase of the pressure from
5×10^{-4} to 1 torr in the reaction zone. Increase of the
reaction temperature to 900-950°C also brought about
the weakening of silanone bands and intensification of
the bands of the decomposition products – SiO and CD_3
radical. The information on frequency shifts caused by
deuteration allowed assignment of the frequencies to
different vibrational modes of the silanone molecule.
The planar structure of that molecule has 24 vibrations. Twenty of them are active in the IR-spectrum,
but only 16 vibrations are predicted in the 400-4000
cm^{-1} range. Most of the frequencies belonging to
stretching and deformation modes of methyl groups can
be predicted on the basis of methylsiloxane data [28].
The calculations have been done by the central
force field approximation with a program of
Schactschneider [29]. Bond lengths and valence angles
are borrowed from reference [30]. A final assignment is
shown in Table 1. The rather intense band at 1210 cm^{-1}
was assigned to Si=O stretching vibration [17] in **4**.
That band is located in spectra considerably higher
than the band of the Si-O single bond [28]. The same
kind of assignment was made later [16] for a single IR-
band, given for **4**. According to our calculation the
contribution of ν (Si=O) corresponding to the vibration

Table 1. Vibrational assignment in IR-spectra of silanones $(CH_3)_2Si=0$ and $(CD_3)_2Si=0$.

Vibration modes	Symmetry type	$(CH_3)_2Si=0$			$(CD_3)_2Si=0$		
		exp. ν,cm⁻¹	Calc. ν,cm⁻¹	Potential energy distribution (%)	exp.,cm⁻¹	calc.,cm⁻¹	Potential energy distribution (%)
δ CH₃	A₁		1397	96 δ CH₃	1032	1011	97 δ CD₃
δ CH₃	B₁		1395	96 δ CH₃		1008	97 δ CD₃
δ CH₃	B₂		1395	96 δ CH₃		1007	97 δ CD₃
δ CH₃	A₂		1395	96 δ CH₃		1007	97 δ CD₃
δ CH₃ ρ CH₃	A₁	1244	1247	54 δ CH₃,46 ρ CH₃	1007	976	47 δ CD₃,41 ρ CD₃
δ CH₃ ρ CH₃	B₁	1240	1235	56 δ CH₃,42 ρ CH₃	995	973	50 δ CD₃,44 ρ CD₃
ν Si=0	A₁	1210	1218	89 ν Si=0	1215	1217	90 ν Si=0
ρ CH₃	B₁	822	848	57 ν SiC.,41 ρ CH₃	685	638	63 ρ CD₃,16 δ CD₃
ρ CH₃	B₂		810	96 ρ CH₃	674	611	97 ρ CD₃
ρ CH₃	A₁	798	808	96 ρ CH₃		594	88 ρ CD₃
ρ CH₃	A₂		795	96 ρ CH₃		592	97 ρ CD₃
ν as SiC	B₁	770	772	55 ρ CH₃,44 ν SiC	712	773	58 ν SiC
ν s SiC	A₁	657	657	86 ν SiC		581	98 ρ CD₃

at 1210 cm^{-1} is 90%, which is witness to the very high
degree of specificity of the IR band of the Si=O
stretching vibration in dimethylsilanone. On deuteration of the methyl groups the Si=O bond frequency is
slightly increased and shifted in d$_6$-4 to 1215 cm^{-1}.
That agrees with a similar shift of the Si=C double
bond frequency from 1003 to 1016 cm^{-1} on transition
from $(CH_3)_2Si=CH_2$ to $(CD_3)_2Si=CH_2$ [31] and with the
increase of P=O frequency from 1161 cm^{-1} for $(CH_3)_3P=O$
to 1165 cm^{-1} for $(CD_3)_3P=O$ [32].

The best agreement of experimental and calculated
frequencies for silanone was achieved with a Si=O bond
force constant equal to 8.32 mdyn/Å, which agrees with
similar parameters for $Cl_2Si=O$ [13] and $F_2Si=O$ [14],
equal to 9 mdyn/Å. This value is considerably higher
than the force constant of the Si-O single bond in
hexamethyldisiloxane equal to 5,3 mdyn/Å [28]. According to the rule of Siebert the order of Si=O multiple
bond was calculated to be 1.45. The order of Si=C
multiple bond, calculated by us earlier for dimethyl-
silaethene, is equal to 1.62 [3].

Thus, the experimentally observed stretching
vibration frequency as well as that calculated from the
spectral force constant and order of Si=O bond are
direct evidence of considerable double bonding in
dimethylsilanone.

To sum up the sources, methods of generation and
direct spectroscopic studies of intermediates with Si=O
multiple bond it should be emphasized, that pyrolysis
at low pressures is most promising for obtaining short-
lived intermediates as monomers, and matrix isolation
IR-spectroscopy is the most efficient for stabilization
and study of the unstable molecules in question.

It is this very technique that allowed us for the
first time not only to stabilize dimethylsilanone from
the gas phase but also to observe seven rather intense
bands in the IR-spectrum. It enabled us to find both
the frequency of the Si=O stretching vibration and the
frequencies and modes of other vibrations in dimethylsilanone.

REFERENCES

1. G. Raab, J. Michl, Chem. Revs, 1985, 85, 416.
2. A.K. Maltsev, V.N. Khabashesku, O.M. Nefedov, Izv. Akad. Nauk SSSR, Ser. Khim., 1976, 5, 1193; Dokl. Akad. Nauk SSSR, 1977, 233, 421.
3. V.N. Khabashesku, Ph.D. Thesis, Moscow, 1979.
4. K.A. Andrianov, N.N. Sokolov, Dokl. Akad. Nauk SSSR, 1952, 82, 909.
5. R. Jaquet, W. Kutzelning, V. Staemmler, Theor. Chim. Acta, 1980, 54, 205.
6. I.M.T. Davidson, J.F. Thompson, J. Chem. Soc. Chem. Commun., 1971, 251.
7. W. Tumas, K.E. Saloman, J.E. Brauman, J. Am. Chem. Soc., 1986, 108, 2541.
8. S.W. Froelicher, B.S. Freiser, R.R. Squires, J. Am. Chem. Soc., 1984, 106, 6863.
9. C. Manuel, G. Bertrand, W.P. Weber, S.A. Kazoura, Organometallics, 1984, 3, 1340.
10. I.M.T. Davidson, A. Fenton, G. Manuel, G. Bertrand, Organometallics, 1985, 4, 1324.
11. T. Kudo, S. Nagase, J. Am. Chem. Soc., 1985, 107, 2589.
12. S. Cradock, A.J. Hinchcliffe, "Matrix Isolation", Cambridge University Press, 1975.
13. H. Schnockel, Angew. Chem., 1978, 90, 638; Z. Anorg. Allgem. Chem., 1980, 460, 37.
14. H. Schnockel, J. Mol. Struct., 1980, 65, 115.
15. R. Withnall, L. Andrews, J. Am. Chem. Soc., 1985, 107, 2567.
16. R. Withnall, L. Andrews, J. Am. Chem. Soc., 1986, 108, 8118.
17. V.N. Khabashesku, Z.A. Kerzina, A.K. Maltsev, O.N. Nefedov, Izv. Akad. Nauk SSSR, Ser. Khim., 1986, 1215.
18. C.A. Arrington, R. West, J. Michl, J. Am. Chem. Soc., 1983, 105, 6176.
19. A.K. Maltsev, R.G. Mikaelian, O.M. Nefedov, Dokl. Akad. Nauk SSSR, 1971, 201, 901.
20. J.S. Anderson, J.S. Ogden, J. Chem. Phys., 1969, 51, 4189.
21. I.T. Wood, Ph.D. Thesis, University of Leicester, 1983.
22. N.D. Kagramanov, I.O. Bragilevsky, V.A. Yablokov, A.V. Tomadze, S.V. Krasnodubskaya, A.K. Maltsev, Izv. Akad. Nauk SSSR, Ser. Khim., 1987, in press.
23. T.J. Barton, International Symposium on Organo-silicon Reactive Intermediates. Sendai, Japan, 1984.
24. A.K. Maltsev, V.A. Korolev, O.M. Nefedov, Izv. Akad. Nauk SSSR, Ser. Khim., 1982, 2415.
25. E.G. Baskir, V.N. Khabashesku, A.K. Maltsev, Third All Union Meeting on Low Temperature Chemistry, Moscow, 1985, Abstracts of papers, p. 223.

26. G. Hussmann, W.D. Wulf, T.J. Barton, J. Am. Chem. Soc., 1983, 105, 1263.
27. A.V. Golovkin, N.A. Mudrova, T.L. Krasnova, L.V. Serebrennikov, V.S. Nikitin, E.A. Cherneshov, J. Obsh. Khim., 1985, 55, 2802.
28. A.N. Lazarev, I.S. Ignatiev, T.F. Tenisheva "Vibrations of Simple Molecules with Si-O bonds", Leningrad, Nauka, 1980.
29. J.H. Schachtschneider, Vibrational Analysis of Polyatomic Molecules, Tech. Pept., 231-64, Shell Development Company, Emeryville, California, U.S.A., 1964.
30. M.S. Gordon, C. George, J. Am. Chem. Soc., 1984, 106, 609.
31. A.K. Maltsev, V.N. Khabashesku, O.M. Nefedov, J. Organomet. Chem., 1982, 226, 11.
32. F. Watari, E. Takayama, K. Aida, J. Mol. Structure, 1979, 55, 169.

Chapter 22

Some aspects of the reactivity of hypervalent species of silicon

R. Corriu – Institut de chimie Fine UA CNRS N° 1097, Universite' des Sciences et techniques du Languedoc, Montpellier, France.

The nucleophilic activation for nucleophilic displacements is a process extensively used at silicon mainly in the uses of organosilanes in synthesis: cleavage and activation of Si-O bonds (1), activation of Si-H (2), Si-C (3), Si-N (4) bonds, all processes performed using F⁻ or HMPA or more generally using nucleophiles having an high affinity towards silicon, as catalysts. The same process was also observed at Phosphorus (5). Since such a process is not known in carbon chemistry, we have proposed a mechanism involving the formation, in a preequilibrium, of a pentacoordinated intermediate formed with the nucleophilic catalyst. This step is followed by the nucleophilic attack of the incoming nucleophile in the rate determining step.

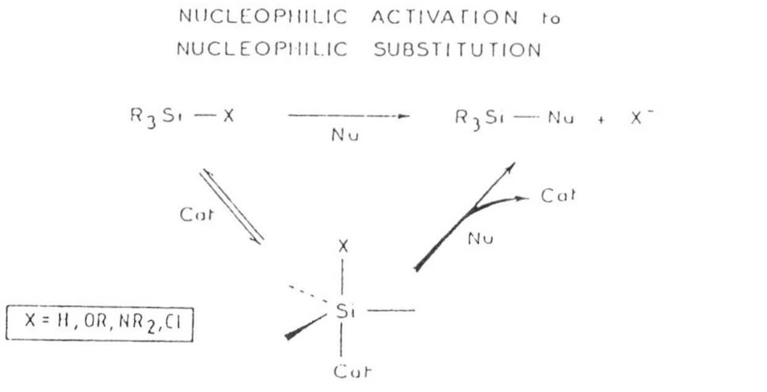

NUCLEOPHILIC ACTIVATION to
NUCLEOPHILIC SUBSTITUTION

$$R_3Si-X \xrightarrow[Nu]{} R_3Si-Nu + X^-$$

X = H, OR, NR$_2$, Cl

Cat = NON SUBSTITUING NUCLEOPHILE (HMPA, F$^-$, RCO$_2^-$, X)

In this process the nucleophilic reaction takes place on a pentacoordinated silicon and it could appear difficult to explain the nucleophilic activation, which corresponds to an acceleration, using an intermediate which can be considered as more crowded and less electrophilic at silicon.

In order to prove our assumption, it was necessary to compare the reactivity of tetra – and pentacoordinated species with very comparable struc- tures. In the first part of this lecture, the reactivity of such species is reported using Si-H, Si-Cl, Si-F, and Si-OMe as leaving groups.

The Si-H bond is a very interesting one for comparison since we have obtained a very good evi- dence for the formation of pentacoordinated struc- tures in solid state (XR) (6) and in solution (Si^{29} NMR) (7). The Si-H bond has a good ability to form pentacoordinated structures and the hydrogen was always in equatorial position.

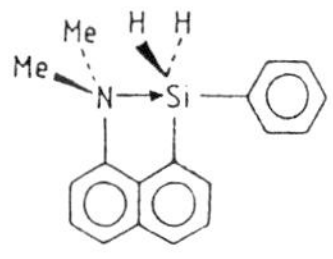

I

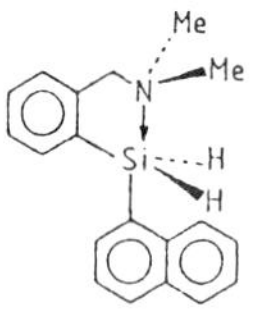

II

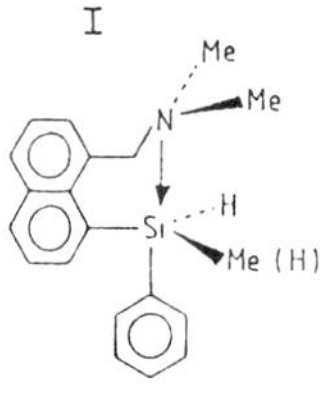

III

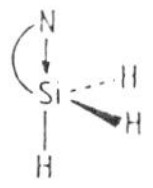

It is well known that the tetravalent dihydrogenosilanes do not react with alcohols or carboxylic acids, or carbonyl groups. At the opposite the same reagents are able to substitute the Si-H bond when the silicon is in a pentacoordinated state. We have even found the possible hydrosilylation of carbonyl groups (8).

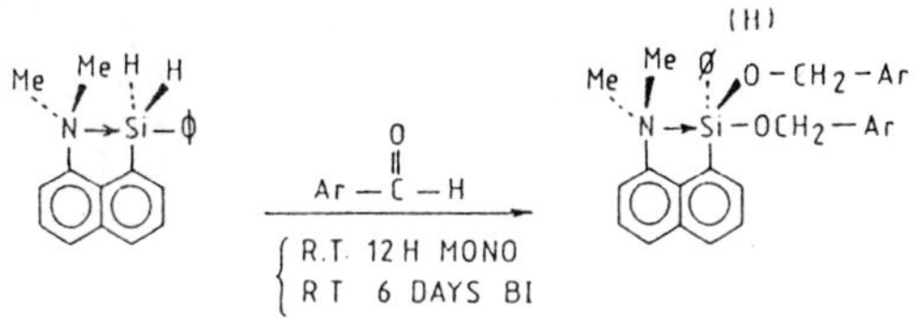

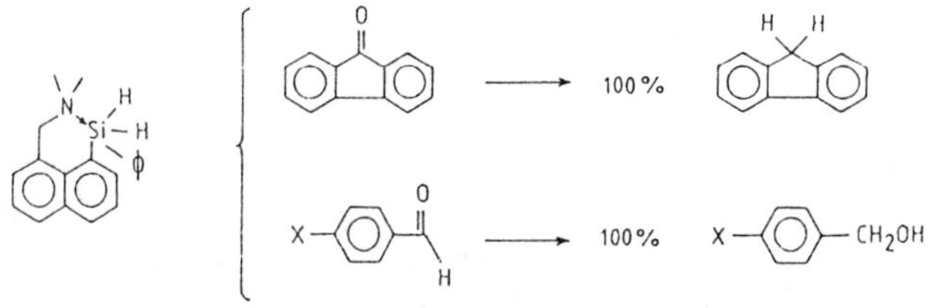

Recently DAMRAUER described a good way to prepare pure hypervalent fluorosilanes (9). It is now possible to study the reactivity of these species and to compare it directly to the reactivity of the corresponding tetravalent fluorosilanes.

We have observed a very high reactivity for the trifluorosilicates.

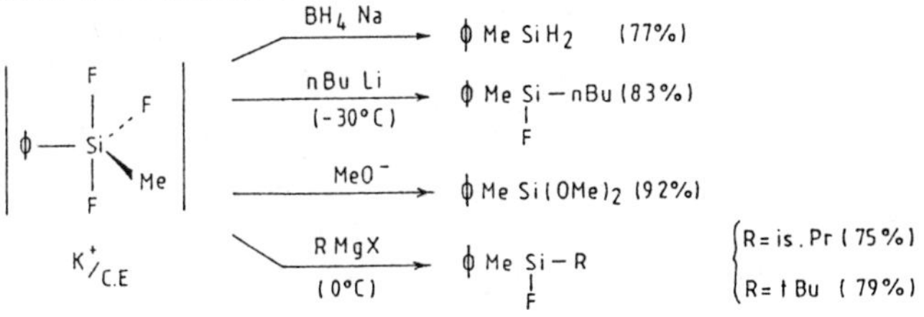

and a faster reactivity of the pentacoordinated towards the tetracoordinated ones.

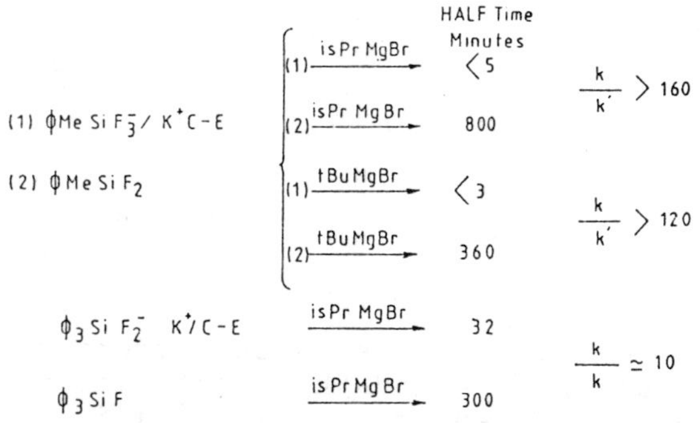

The same increase of reactivity was also
observed in the case of neutral pentacoordinated
species. The case of (IV) is particularly illustrative since the tri, bi, and monofluorosilanes are
pentacoordinated. The tri substitution takes place
much faster than in the case of trifluoronaph-
thylsilane.

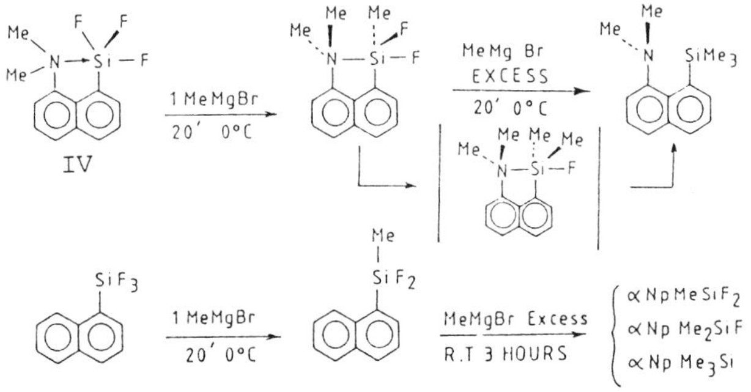

We have extended successfully the DAMRAUER
procedures to the preparation of hypervalent methoxysilanes. These compounds are very reactive. For
instance the pentamethoxy siliconate is hydrolysed
very quickly giving a gel :

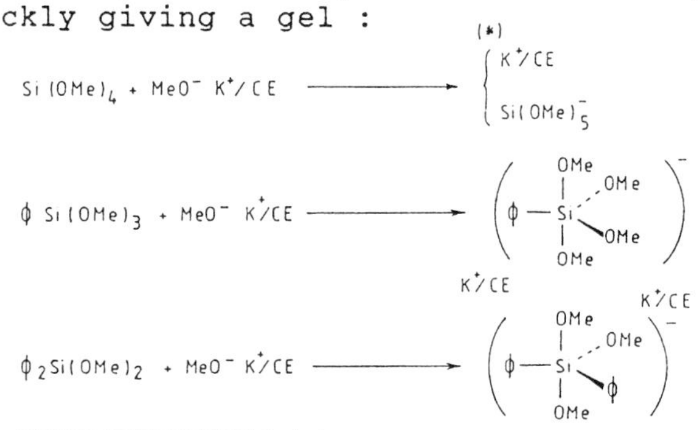

HIGHLY HYDROLYSABLE (∗)

We have observed once again a higher reactivity for the pentacoordinated silicon

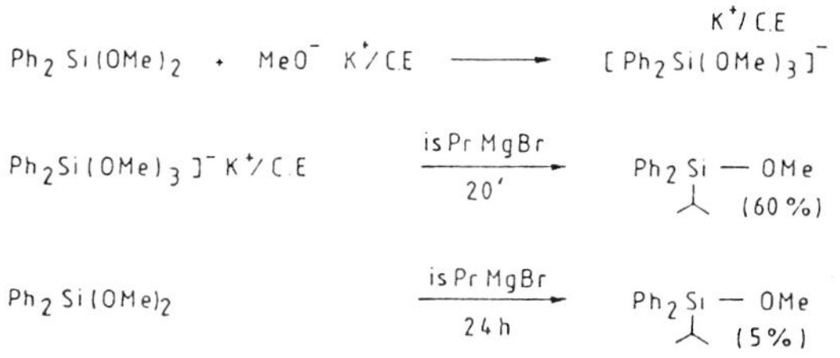

The comparison was also performed in the
case of neutral species and the reactivity was found
always in favour of pentacoordinated silicon :

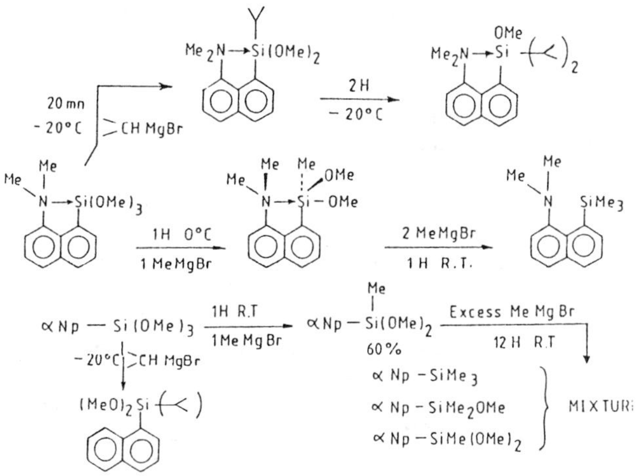

The higher reactivity of pentacoordinated Si-Cl bond was also observed in the reaction of the organolithium (V) towards Me_2SiCl_2. This reaction gives the exclusive formation of disubstituted product (VII) whatever is the ratio between V and Me_2SiCl_2 (even with a fixe time excess of Me_2SiCl_2). The conclusion is that the mono substituted pentacoordinated chlorosilanes VI react faster with (V) than Me_2SiCl_2. This high reactivity of VI was checked by separate preparation and reaction :

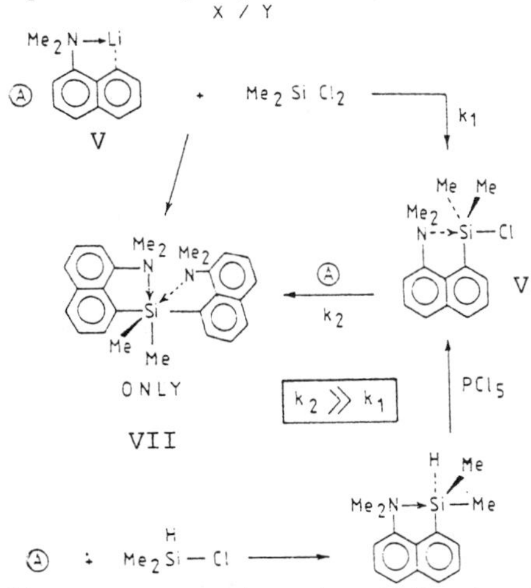

All these experiments illustrate very well the great reactivity of hypervalent species. They confirm the possibility of pentacoordinated intermediates in the nucleophilic activation, of nucleophilic substitution at silicon. This possibility cannot be ruled out only on the basis of the argument of a more crowded and less electrophilic species than tetracoordinated silicon. Fur-

Silicon Reactive Intermediates

thermore after these results, it becomes interesting
to understand why these hypercoordinated species
react faster than the tetracoordinated ones. The two
possible explanations are one part the increase of
the length of Si-X bonds which corresponds to an
higher lability and other part the increase of the
electrophilicity of the central Si atom (10).

In the second part of the lecture the very
uncommon reactivity of hypervalent species will be
illustrated. These species appear to have a very
unexpected chemical behaviour. The illustration of
that is the reactivity observed between dihydrogenosilanes (III) and CS_2 which gives a rearrangement
never reported in silicon chemistry. Interestingly
the process started with a double hydrosilylation of
CS_2

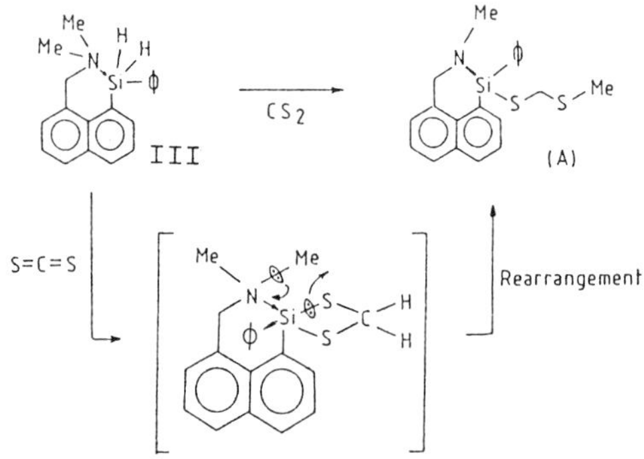

The reaction with CO_2 gives at first a
monoaddition product VIII which can be prepared by
reaction of HCO_2H on (III). Heating (VIII) at 120°C,
we obtained, under vacuum line, in very dry conditions and using degased solvents, the formation of
the trisiloxane X and we observe the evolution of
formaldehyde.

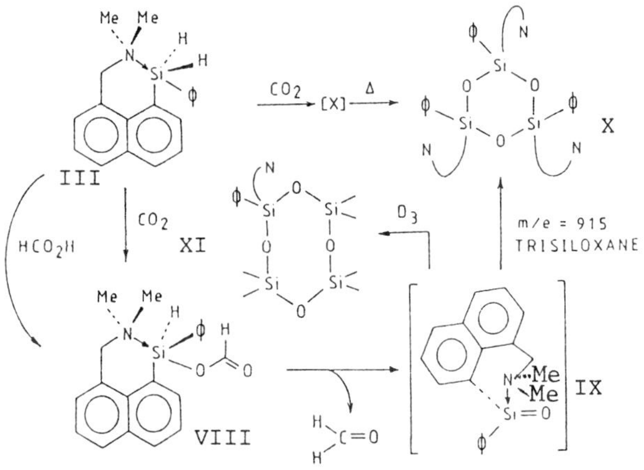

The mechanism of this process involve the elimination of $H_2C=O$ from VIII with formation of silanone (IX) which trimerizes giving X. The sila-none can be also trapped with D_3 with formation of D'$_4$ (XI).

In order to explain the reaction of (III) with CS_2 we have treated it with molecular S and we have obtained a product (XII) identified as a sila-thione stabilized by intramolecular coordination with the NMe_2 group. This product is oxidized easily; it is stable in anaerobic conditions and reacts instantaneously with air, with formation of the trisiloxane (X).

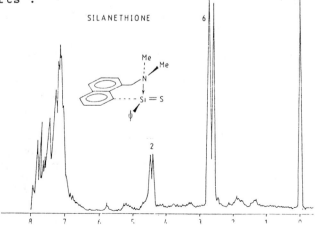

Besides the microanalysis, the spectroscopic arguments supporting the structure of the silathione are : 1) the mass spectrum obtained at two different electronic impacts

2) the H^1 NMR showing a clean diastereo-topism of $-CH_2-$ and NMe_2 groups due to the strong coordination of the nitrogen atom to silicon

3) the Si^{29} chemical shift which is just between the low valent and hypervalent species chemical shifts .

Finally the synthesis and the structure of an heptacoordinated specie is reported. Both Ge and Si compounds have been obtained by the following way

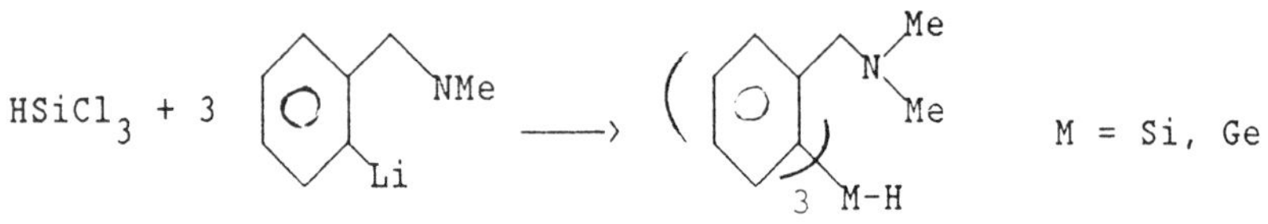

$$HSiCl_3 + 3 \quad \text{[structure]} \longrightarrow \left(\text{[structure]} \right)_3 \quad M = Si, Ge$$

The XR structure of the Ge compound was determined. The three NMe₂ groupes interact with the central atom as shown on the following scheme.

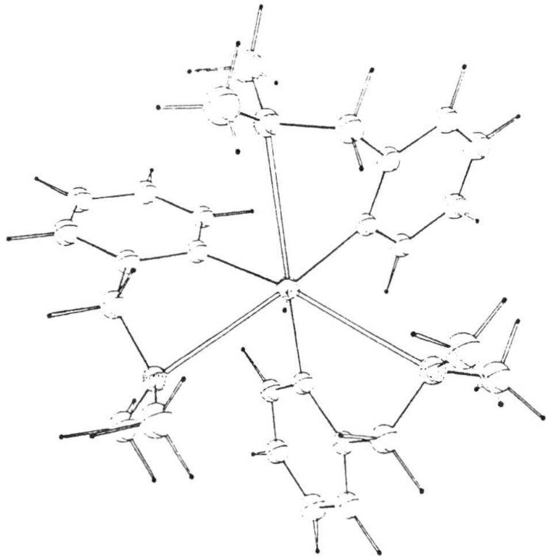

The angle between the Ge-H bond and each NMe₂ coordination is about 90°. The isomorphism of the silicon compound was proved by the phase diagram which shows a very nice solid solution instead of the expected eutectic, and also by the XR diffraction diagram of the two powders which are almost identical.

The reactivity of this Si-H bond is very different from the reactivity observed with pentacoordinated species. We observe no reaction with methanol, as observed in the case of pentacoordinated compounds. At the opposite, the abstraction of the hydride ion is performed easily by Ph₃C⁺ BF4⁻ giving an hypercoordinated siliconium ion as a salt

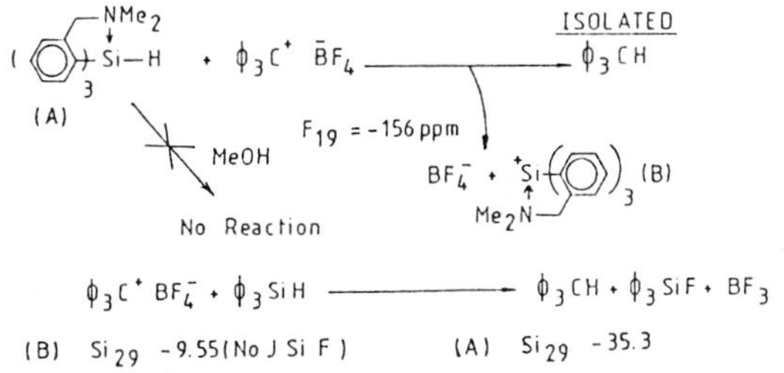

$$(\underset{(A)}{\boxed{\bigcirc}} \!-\! Si\!-\!H)_3 \;\; NMe_2 \;\; + \;\; \phi_3 C^+ \; \bar{B}F_4 \longrightarrow \; \phi_3 CH \quad \text{ISOLATED}$$

$$F_{19} = -156\,ppm$$

MeOH

No Reaction

$$BF_4^- + \; (^+Si \!-\! \boxed{\bigcirc})_3 \;\; (B) \quad Me_2N\!-\!$$

$$\phi_3 C^+ \; BF_4^- + \phi_3 SiH \longrightarrow \phi_3 CH + \phi_3 SiF + BF_3$$

(B) Si_{29} -9.55 (No J Si F)   (A) Si_{29} -35.3

The chemical shift NMR of F^{19} and Si^{29} demonstrate the formation of the salt. The normal expected reaction with Ph_3SiH is the formation of Ph_3SiF.

$$Ph_3SiH + BF_4^- CPh_3^+ \longrightarrow Ph_3SiF + BF_3 + Ph_3CH$$

Acknowledgements : R.CORRIU indebted his gratitude to his coworkers involved in this chemistry of hypercoordinated silicon compounds.
G. ROYO, C. BRELIERE, M. POIRIER, G. LANNEAU, F. CARRE, C. GUERIN, B. HENNER, C. CHUIT, G.CERVEAU, C. REYE, A. KPOTON, M. PERROT, A. BOUDIN, L. GERBIER, M. WONG CHI MAN.

REFERENCES

1.a) R.J.P. CORRIU, R. PERZ, C. REYE. Tetrahedron (1983) Vol 39 999-1009 and ref. therein.
  b) R. NOYORI, K. YOKOYAMA, J. SAKATA, I. KUWAJIMA, E. NAKAMURA, M. SHIMIZU, J. Am.   Chem. Soc. (1977) 99 1265.

2.   Ref 1a ; J. BOYER, R.J.P. CORRIU, R. PERZ, C. CARRE, C. REYE, Tetrahedron (1981) vol 37 2165.

3.   Ref 1a ; A. HOSOMI, A.SHIRAHATA, H. SAKURAI, Tetrahedron Letters (1978) 3043.

4.   Ref 1a ; R.J.P. CORRIU, J.J.E. MOREAU, Some uses of Si-N bonds organic synthesis in selectivity a goal for synthetic efficiency Verlag. Chemie (1984) 14 21 ;   R.J.P. CORRIU, J.J.E. MOREAU, M. PATAUD-SAT, Organometallics (1985) 4 623.

5.   G.F. LANNEAU Phosphorus and Sulfur (1986) 27 43-54

6. C. BRELIERE, F. CARRE, R.J.P. CORRIU, M. POIRIER, G. ROYO, Organometallics (1986) <u>5</u> 388.

7. B.J. HELMER, R. WEST, R.J.P. CORRIU, M. POIRIER, G. ROYO, J. Organomet. Chem. (1983) <u>251</u> 295.

8. J. BOYER, C. BRELIERE, R.J.P. CORRIU, A. KPOTON, M. POIRIER, G. ROYO. J. Organometal. Chem. (1986) 311 C39.

9. R. DAMRAUER, DANAHEY. J.E, Organometallics (1986) <u>5</u> 1490

10. S.N. TANDURA, M.G. VORONKOV, N.V. ALEKSEEV. Topics in current chemistry Springer- Verlag (1986) <u>131</u> p 99-189.

PART V

SILICON-SILICON CHEMISTRY

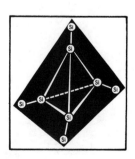

Chapter 23

Polysilanes

E. Hengge — Institute of Inorganic Chemistry, Technical University of Graz, Austria.

POLYSILANES

Polysilanes are silanes with more than three Si-
atoms, directly bonded together. Much is known in
this field of chemistry, published in about two
thousand papers and some books. Polysilanes are
known in linear and branched chains, but also in
cycles and high polymer frameworks. The Si-Si-
linkage is quite different from the C-C-bond. The
bond energy is lower, therefore, the thermal
stability is lower. Nucleophilic attack occurs much
easier in comparison to carbon, hydrolysis, halo-
genations and similar reactions are very common. In
carbon chemistry, the simple hydrocarbons, but also
fully or partially halogenated compounds are old and
very well known. The sensivity of a silicon framework, together with the very sensitive silicon-
hydrogen and silicon halogen bonds changes this
picture in silicon chemistry. Higher hydrosilane
chains were isolated up to about 8 Si-atoms. By
means of chromatographic methodes, chains up to
15 Si-atoms have been detected. Besides linear
chains, also branched chains were isolated, for
example i-Si_4H_{10} or the neopentane analogue structure of Si_5H_{12}. The first cyclic silicon hydrides
Si_5H_{10} and Si_6H_{12} can be prepared starting from perphenylated silicon rings by cleaving off the phenyl
groups with HX/AlX_3 and reacting the resulting per-

halocyclosilanes with $LiAlH_4$ (1,2). The cyclic
compounds exhibit a higher stability in comparison
to the chain compounds, which decompose slowly at
room temperature. All silicon hydrides ignite
spontanousely in air. All these compounds show a
high decomposition rate at relativly low temperature.

A similar situation is given with polysilanes
bearing halogens, mainly chlorine. The usual preparation route is the reaction of chlorine with
silicon or silicides. The separation of the
different polysilanes in the reaction mixture is
difficult, caused by cracking reactions during the
distillation and small differences in the boiling
points. It is impossible, for instance, to separate
n- and $i-Si_4H_{10}$, a distillation becomes impossible
for compounds with more than five Si-atoms in the
chain. Therefore, we prepared a large number of per-
chlorinated polysilanes, but also of mixed hydrogen
and chlorine substituted compounds by other ways.
The principial routes are cleavage of cyclic
silanes, removing of phenyl groups from Si-phenyl
bonds by use of hydrogen halides, formation of
Si-Si-bonds by reductive dechlorination or recombination of silylradicals after photolytic cleavage of
(bis)-polysilanylmercury compounds. Starting from
$c-Si_4Ph_8$ it was possible under mild reaction
conditions to split off the phenylgroups without
ring cleavage and to isolate Si_4Cl_8 (3). Subsequent
cleavage of one Si-Si-bond of the cyclic chloride
with tetrachloroethane yields $n-Si_4Cl_{10}$ (4). Pure
$iso-Si_4Cl_{10}$ could also be prepared by Höfler in our
group in Graz in 1975 (5). Another way yields
chlorinated polysilanes with hydrogens at the ends
of the chain. Starting from the perphenylated
cycles, ring-cleavage is possible with HCl, PCl_5 or
Li. The resulting chain products can further be
reacted to the α, ω -hydrogenated compounds (6).

Another example is the synthesis of chlorinated
silicon-chains with hydrogen substituents in the
middle of the chain by use of silylpotassium
compounds. Triphenylsilylpotassium reacts with
different chlorosilanes forming polysilanes. After
replacing the phenyl substituents, the desired
compounds are formed (7).

$$\underset{Ph}{\overset{H}{ClSiCl}} + 2\ KSiPh_3 \rightarrow \underset{Ph}{\overset{H}{(Ph_3Si)_2Si}} \xrightarrow[AlCl_3]{HCl} \underset{Cl}{\overset{H}{Cl_3Si-Si-SiCl_3}}$$

The way via the silyl-potassium compounds enables
the formation of polysilanes step by step, one
silicon atom after another. Starting from a polysilane with one hydrogen substituent, one can make

the corresponding silylmercurial, which can be
cleaved by sodium/potassium alloy yielding the poly-
silylpotassium-compound. This can be reacted with
a chloromonosilane yielding a polysilane with one
Si-atom more in the chain. Returning to the hydrogen
compound with $LiAlH_4$, this cyclic reaction course
can be started again. We did this up to four-
Si-atoms (8).

$$2\,R_{2n+1}Si_nH + {}^tBu_2Hg \xrightarrow{-2BuH} (R_{2n+1}Si_n)_2Hg \xrightarrow{h\nu} R_{2n+1}Si_n\text{-}Si_nR_{2n+1}$$

$$2\,R_{2n+3}Si_{n+1}H \qquad\qquad Na/K$$

$$LiAlH_4$$

$$2\,R_{2n+1}Si_n\text{-}Si(R_2)Cl \xrightarrow{+2R_2SiCl_2} 2\,R_{2n+1}Si_nK$$

Many papers investigated the physico chemical properties of polysilanes. Summarizing the spectroscopic results on polysilanes, NMR- and mass-spectra,
measurements of vibrations and normal coordinate
analyses, X-ray-structures and others, one can see
the following rules for the Si-Si-linkage (9):

1. The strength of the Si-Si-bond is dependent on
   the substituents. Substituents with high electronegativity and such with free electron pairs
   cause an increase of the strength.

2. The strength of the Si-Si-bond is also dependent
   on the structure and the size of the Si-frame-
   work in polysilanes. In cycles the Si-Si-bond
   becomes weaker in strained ring systems,
   expressed by longer bond distances and lower
   force constants. In chain systems, internal
   Si-Si-bonds are weaker than the Si-Si-bonds at
   the end of the chains. This also can be seen in
   bond distances, force constants and in mass
   fragmentation.

The properties of Si-Si-bonds not only depend on
electronic effects, but also on the size of
substituents.

Long Si-Si-distances were found in three-membered
rings (10). It is remarkable, that not only ring
strain causes such long distances. Weidenbruch (11)
recently showed very long Si-Si-bonds in a trisilane. The longest Si-Si-distance was found in
hexakis-(t-butyl)-disilane by Wiberg (12). So, the
space requirement of substituent groups seems to be

very important.

In larger ring systems the Si-Si-distances approach normal values and are mainly dependent on the substituents attached to the ring. Cartledge (13) has postulated special steric substituent parameters for silicon compounds. Nagai and Watanabe proved these parameters for the formation of cyclic poly-silanes (14).

In four-membered rings with bulky substituents we find relatively long Si-Si-bonds, with smaller groups like the methylgroup, normal distances are usually observed. More interesting is the folding angle of the four-membered ring, which is strongly dependent on the substituents. $Me_4tBu_4Si_4$ exhibits a strong folding angle (15), the bulky phenylgroups in Si_4Ph_8 also give rise to a folding angle of about 13° (16). On the other hand, Gaspar (17) showed a planar Si_4-cycle with large trimethylsilylgroups as substituents. In case of the permethylated four-membered ring, we found a planar ring structure in the crystal (18).

The geometries of five- and six-membered rings are consistent with the geometries of carbon compounds, their properties exhibit no special features except the formation of radical ions. In general, silicon cycles are less stable than carbon cycles and show a higher flexibility. This is demonstrated by the very facile pseudorotation of Si_5H_{10} (19) but also in the mobility of Si_6H_{12} (20). Under the conditions of electron diffraction, this ring shows a perfect equilibrium between chair-, boat and twist con-formation, the energy differencies between these forms are very low.

The knowledge of polycyclic polysilanes are rare, only a few compounds were found in small yields or only in chromatographic diagrams as by-product in the reductive dechlorination of dialkyldichloro-silanes.

Now we found a possibility to obtain polycyclic polysilanes by a special synthesis, we started with a monofunctional cyclosilane, for instance $c-Si_6Me_{11}Cl$. This compound was first synthesized by Kumada (21) by reacting $HCl/AlCl_3$ with $c-Si_6Me_{12}$.

The success of the reaction depends on the quality of $AlCl_3$ and we found it necessary to control the course of the reaction by gas chromatography. The reduction to $c-Si_6Me_{11}H$ gave no problems and with di(t-butyl)-mercury, we made the new compound $(c-Si_6Me_{11})_2Hg$.

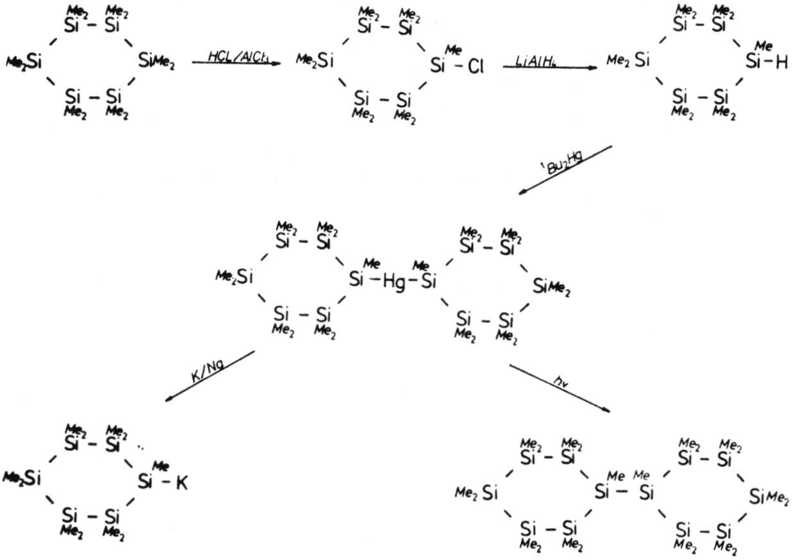

This mercury-compound can be split by action of sodium/potassium alloy and we got the first cyclo-silylpotassium compound. This is the key to poly-cyclic polysilanes and in some examples I want to show you very briefly some syntheses of such poly-cyclic silanes. By irradiation with light the mercury compound by itself reacts to a bicyclic system. The same compound can be obtained from the reaction of the potassium compound with the mono-brominated cycle (22). Other halogenated silanes like dichlorodimethylsilane or 1,2-dichlorotetra-methyldisilane also react with the potassium-compound (23).

One of the largest silicon frameworks, made by an aimed synthesis, can be obtained by the reaction of the potassium-compound with tris(chlorodimethyl-silyl)methylsilane:

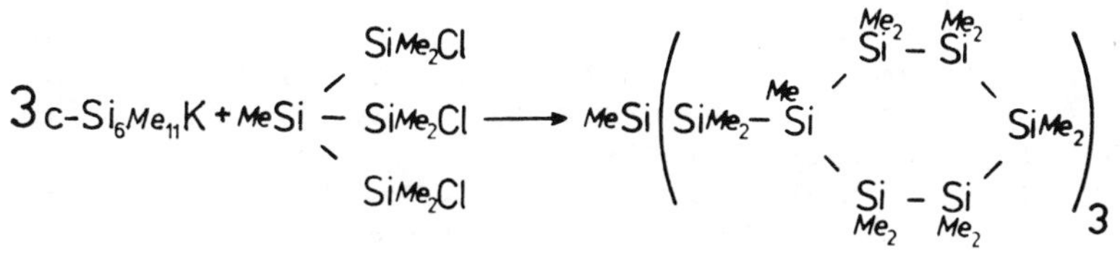

Remarkable in these systems of polysilanes are the
Si-Si-bonds connecting the cycles. X-ray investigations and force constants calculations show a
longer and weaker bond (24).

An interesting reaction product can be obtained from
the reaction of the potassium-compound with
1,4-dibromodecamethylcyclohexasilane yielding a
tricyclic polysilane (23):

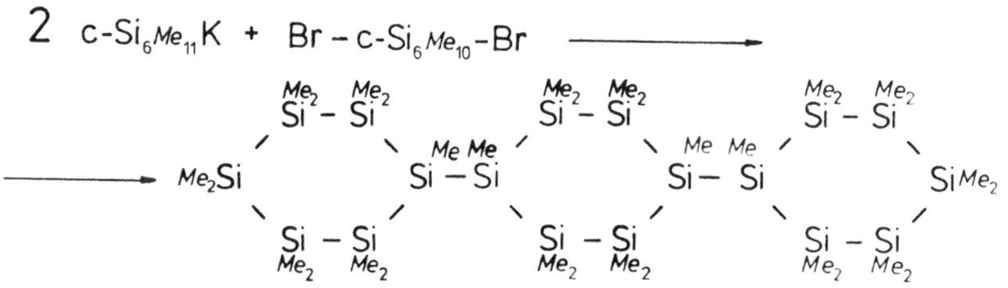

$$2 \ c\text{-}Si_6Me_{11}K \ + \ Br\text{-}c\text{-}Si_6Me_{10}\text{-}Br \longrightarrow$$

Similar investigations were also done with five-
membered rings. One of the outstanding properties of
polysilanes is the ability to form radical cations
and anions. Besides the first formation of anions
by addition of one electron (25), which is
completely delocalized in the cycle, the formation
of cations is also possible (27). Perphenylated
cyclosilanes are showing the same behaviour to form
radical anions (26).

Several derivatives of high molecular weight polysilane polymers are known for many years. First
investigations were done by Kipping, Burkhard and
others, but especially in the last years, research
on such polymers intensified, because their technological utility was discovered. Polysilanes are
useful in at least three applications (28):

1. as precursors to silicon carbide fibers and
   ceramics
2. as photoinitiators for vinyl-polymerization
3. as UV photo resists.

Let us just have a brief look to these applications.
The silicon carbide polymer in form of fibers, but
also as ceramics is of high technical interest. One
way to the SiC-frameworks starts from a polysilane
with organic substituents, which is converted at
high temperature to a Si-C-Si-framework. Other ways
starting from monosilanes to carbosilanes directly
are known, but cannot be dealt with in this lecture
on polysilanes.

What ways lead to the polysilanes needed? These are
in general the known ways for forming a Si-Si-bond.
Up to now, the most important way is the reductive
dechlorination. It is the classic way to form
Si-Si-bonds and is used also for the formation of
polysilanes in the Japanese Yajima-process. Many
details are known and investigated in many papers
about the best reaction conditions, about the choice
of solvents, of the alkali metals including
magnesium, temperature, reaction time and so on.

A second and less expensive way to polysilanes is
the disproportionation of disilane derivatives to
monosilane and polysilane. A lot of patents are
dealing with this polysilane synthesis, because the
chloromethyldisilane starting materials are
industrial by-products. Starting the disporportionation directly from the chloromethyldisilane,
yields a polymer containing some chlorine, which
effects the properties of the resulting SiC.

This can be avoided by changing the chlorine against
other groups or atoms before disproportionation.
This can be done by use of ammonia or amines,
yielding nitrogen containing polysilanes or with
methanol, yielding the methoxy-derivatives, which
can be converted into polysilanes by reacting them
with Si-H compounds and sodiummethoxylate as a
catalyst (29). Another way is the primary formation
of methyldisilanes, yielding polysilanes with good
properties. Another possibility would be the
preformation fo cyano- and cyanato-derivatives,
yielding nitrogen containing polysilanes (30).

At the pyrolysis of polysilanes, the formation of
volatile Si-compounds lowers the yield of SiC. To
avoid this side reaction crosslinking of the polysilane chains seems to be a possible way.Connections
with C-C-,Si-O-Si- and Si-Si-bonds are possible (31).

A second application was discovered by R. West (28)
and his group, the photoinitiation of polymerization
reactions by polysilanes. Irridiation with UV light
splits the Si-Si-bond forming both, radicals and
silylenes. These species initiate the polymerization
of olefins.

The photochemical behaviour of polysilanes leads us
to the third application. The formation of radicals
and silylenes upon irradiation by scission of
Si-Si-bonds generates reaction products, which are
generally more soluble than the starting material.
An application of this property is given by the use
of polysilanes as positive photoresists. Besides
this chain scission, polysilane cross-linking can

also be used in photoresist technology. Polysilanes
crosslink on irridiation with UV light in presence
of unsaturated organic compounds like vinyl
derivatives, which greatly reduces their solubility.
Therefore they are suitable to work as negative
photoresists.

Essential for all these applications is the knowledge about the absorption spectra of polysilanes
and their mechanism. Many years ago, Gilman (32)
found a bathochromic shift of the absorption maxima
with increasing chain length for methylated polysilanes. Recently, West (28) and his group investigated polysilanes with longer alkyl- and phenyl
groups and found a similar dependence.

All these absorptions appear in the UV region and
the compounds investigated are colorless. But also
colored polysilanes are known. Perhalogenated
polysilanes exhibit yellow, red or brown colors.

A class of structurally defined high polymer polysilanes are two dimensional polysilanes of the
general formula $(SiX)_n$. They are synthesized from
the preformed two dimensional silicon sheets in the
layer lattice of $CaSi_2$ (33). With ICl one can
isolate these Si layers from the lattice, yielding
yellow $(SiCl)_n$. The Si-Cl bonds attached to the
layers are reactive and many derivatives can be
formed in solid state reactions. All these compounds
are colored and the color depends on the
substituents.

The conclusion of all investigations is: the
electrons are strongly delocalized in the Si-sheets
and the electron density can be influenced by the
substituents. In all these derivatives including
the two dimensional Si-modification, sometimes
radical states are present arising from unsaturated
Si-valencies, being resonance stabilized in the
sheet.

We were interested to see more about the influence
of the substituents and the mechanismn of light
absorption. Therefore we turned our interest to
soluble cyclosilanes with halogen and other
substituents. Here just a brief overlook of the
results: The UV spectra of cyclosilanes with
halogens or methoxy groups exhibit additional low
intensity bands at the low energy side. They are
responsible for the color of some of these
compounds. These bands always can be found in the
presence of substituents with free electron pairs on
the Si-cycle. Position and intensity of the bands
are depending on the substituents and on the ring

size. Iodine shifts stronger than bromine and
chlorine. It was surprising, that the band in
question is shifted to longer wavelength in the
order $Si_4 > Si_6 > Si_5$ and did not follow the ring
size. Together with the results of photoelectron-
spectroscopy and in respect to the results of
Pitt (34), we came to the result, that this band in
question might be a charge transfer band.

In larger polycyclic ring systems or polymeric
compounds with larger Si-Si-frameworks, this band
also appears in the visible region and therefore
is responsible for the color of polymeric compounds.
This charge-transfer band not only seems to be
present in cyclic silanes, but also in long chains.

Nefedov (35) at al in 1968 showed, that linear polysilanes are also able to form radical anions.
Together with the fact that long linear Si-chains
are also colored, these observations suggest, that
Si-chains have similar properties like Si-cycles,
being able to add or to lose electrons. This
behaviour directly leads us to the properties of
semiconducting elementary silicon. But more
investigations are necessary to elucitate the
properties of this very interesting field of silicon
chemistry, the change of the properties going from
simple monosilanes to cyclic and catenated molecules
up to high polymeric structures and elementary
silicon.

REFERENCES

1. E.Hengge and G.Bauer, Monatsh.Chem.
   106 (1975) 503
2. E.Hengge and D.Kovar, Z.Anorg.Allg.Chem.
   459 (1979) 123
3. E.Hengge and D.Kovar, Z.Anorg.Allg.Chem.
   458 (1979) 163
4. W.Raml and E.Hengge, Z.Naturforsch.
   34 b (1979) 1457
5. F.Höfler and R.Jannach, Inorg.Nucl.Chem.Lett.
   11 (1975) 743
6. E.Hengge and G.Miklau, Z.Anorg.Allg.Chem.
   508 (1984) 33
7. E.Hengge and F.K.Mitter, Z.Anorg.Allg.Chem.
   529 (1985) 22
8. E.Hengge and F.K.Mitter, Monatsh.Chem.
   117 (1986) 721
9. E.Hengge, Rev.Inorg.Chem. 2 (1980) 139
10. M.Weidenbruch, Comments Inorg.Chem. 5 (1986) 247

11. M.Weidenbruch, B.Flintjer, K.Peters and
    H.G. von Schnering, Angew.Chem. 98 (1986) 1090
12. N.Wiberg, H.Schuster, A.Simon and K.Peters,
    Angew.Chem. 98 (1986) 100
13. F.K.Cartledge, Organometallics 2 (1983) 425
14. H.Watanabe, T.Muraoka, M.Kageyama, K.Yoshizumi
    and Y.Nagai, Organometallics 3 (1984) 141
15. C.J.Hurt, J.C.Calabrese and R.West, J.Organomet.
    Chem. 91 (1975) 273
16. L.Párkányi, K.Sasvári and I.Barta,
    Acta Crystallogr. B 34 (1978) 883
17. Y.-S.Chen and P.P.Gaspar, Organometallics
    1 (1982) 1410
18. C.Kratky, H.G.Schuster and E.Hengge,
    J.Organomet.Chem. 247 (1983) 253
19. Z.Smith, H.M.Seip, E.Hengge and G.Bauer,
    Acta Chem.Scand. A 30 (1976) 697
20. Z.Smith, A.Almenningen, E.Hengge and D.Kovar,
    J.Am.Chem.Soc. 104 (1982) 4362
21. M.Ishikawa and M.Kumada, Synth.Inorg.Met.-
    Org.Chem. 1 (1971) 191
22. F.K.Mitter, G.I.Pollhammer and E.Hengge,
    J.Organomet.Chem. 314 (1986) 1
23. F.K.Mitter and E.Hengge, J.Organomet.Chem.
    (in press)
24. K.Hassler, F.K.Mitter, E.Hengge, C.Kratky and
    U.G.Wagner, J.Organomet.Chem. (in press)
25. E.Carberry, R.West and G.E.Glass,
    J.Am.Chem.Soc. 91 (1969) 5446
26. M.Kira, H.Bock and E.Hengge, J.Organomet.Chem.
    164 (1979) 277
27. H.Bock, W.Kaim, M.Kira and R.West,
    J.Am.Chem.Soc. 101 (1979) 7667
28. R.West, J.Organomet.Chem. 300 (1986) 327
29. B.Pachaly, V.Frey and N.Zeller (Wacker-Chemie
    G.m.b.H.) DE 35 32 128 A1 (Cl. C08G77/60)
30. W.Kalchauer, Doktor Thesis, TU Graz, 1986
31. H.Stüger and R.West, Macromolecules 18 (1985)
    2349
32. H.Gilman, W.H.Atwell and G.L.Schwebke,
    J.Organomet.Chem. 2 (1964) 369
33. E.Hengge, Fortschr.Chem.Forsch. 9 (1967) 145
34. C.G.Pitt, J.Am.Chem.Soc. 91 (1969) 6613
35. V.V.Bukhtiyarov, S.P.Solodovnikov, O.M.Nefedov
    and V.I.Shiryaev, Izv.Akad.Nauk SSSR, Ser.Khim.
    (1968) 1012, engl. Ed. 967.

Chapter 24

Synthesis and properties of strained cyclopolysilanes

Y. Nagai, H. Watanabe and **H. Matsumoto** – Department of Chemistry, Gunma University, Kiryu, Gunma 376, Japan.

INTRODUCTION

The chemistry of strained silacycles provides an intriguing subject because they are expected to show
   unique physical and chemical properties.   The
synthesis of a cyclotetrasilane goes way back to 1921
when Kipping isolated octaphenylcyclotetrasilane [1].
After more than 50 years, West et al. reported the
successful isolation of 1,2,3,4-tetra-tert-butyl-
tetramethylcyclotetrasilane with Na/K in THF [2].
The first cyclotrisilane, hexakis(2,6-dimethylphen-
yl)cyclotrisilane, was obtained in 1982 by Masamune
and co-workers [3], whereas the first isolation of a
cyclodisilane, tetramesityldisilene, was reported in
1981 by West and co-workers [4].
We have been interested in the electronic properties
of the strained silicon frameworks and commenced an
investigation on the synthesis of cyclotetra-, tri-,
and di-silanes without having any aryl groups.   This
choice seemed reasonable, since aryl substituents
might significantly perturb the intrinsic electronic
structure of the silicon frameworks in question [5].

PERALKYLCYCLOTETRA-, TRI-, AND DI-SILANES

The cyclotetra- and tri-silanes can be most conveniently obtained as colorless crystals by the reductive

coupling of dialkyldichlorosilanes, 1,2-dichloro-
tetraalkyldisilanes, and 1,3-dichlorohexaalkyltri-
silanes bearing sufficiently bulky substituents with
lithium in THF.

$$R^1R^2SiCl_2 \xrightarrow{\text{Li}} (R^1R^2Si)_4$$

$\underline{1}$ $R^1=R^2={}^iBu$

$\underline{2}$ $R^1=R^2={}^iPr$

$\underline{3}$ $R^1={}^nPr, R^2={}^tBu$

$\underline{4}$ $R^1=R^2={}^sBu$

$\underline{5}$ $R^1=R^2=Me_3SiCH_2$

or

$$(R^1R^2Si)_3$$

$\underline{6}$ $R^1=R^2={}^tBuCH_2$

$$Cl(R^1R^2Si)_2Cl \xrightarrow{\text{Li}} (R^1R^2Si)_4$$

$\underline{2}$

$\underline{7}$ $R^1=Me, R^2={}^tBu$

$\underline{8}$ $R^1=Me, R^2={}^tBuCH_2$

$\underline{9}$ $R^1=R^2={}^tBuCH_2$

$$Cl(R^1R^2Si)_2Cl + Cl(R^3R^4Si)_2Cl$$

$$\xrightarrow{\text{Li}} (R^1R^2Si)_2(R^3R^4Si)_2$$

$\underline{10}$ $R^1=R^2={}^iPr,$
$R^3=R^4={}^tBuCH_2$

$$Cl(R^1R^2Si)_3Cl \xrightarrow{\text{Li}} (R^1R^2Si)_3$$

$\underline{6}$

Masamune and co-workers have shown that photolysis of
hexakis(2,6-dimethylphenyl)cyclotrisilane gave a
stable crystalline disilene, tetrakis(2,6-dimethyl-
phenyl)disilene [3].   Similar photolysis of hexaneo-
pentylcyclotrisilane was found also to give somewhat
less stable tetraneopentyldisilene, $\underline{11}$ [5c].   Tetra-
alkyldisilenes can be more conveniently obtained by
the photolysis of the corresponding octaalkylcyclo-
tetrasilanes which give successively cyclotrisilanes

and disilenes with extrusion of the dialkylsilylenes.

$$(R^1R^2Si)_n \longrightarrow R^1R^2Si=SiR^1R^2$$
$$n=3,4$$

11 $R^1=R^2={}^tBuCH_2$
12 $R^1=R^2={}^iBu$
13 $R^1=R^1={}^iPr$
14 $R^1={}^nPr,\ R^2={}^tBu$
15 $R^1=R^2={}^sBu$
16 $R^1=R^2=Me_3SiCH_2$
17 $R^1=Me,\ R^2={}^tBu$

Structures for peralkylcyclotetrasilanes 2, 5, 9, and 10 have been determined by x-ray crystallography. The silicon skeletons in these molecules all assume folded structures with large dihedral angles around 37-39° [5i,5v]. The Si-Si bond lengths range between 238 and 241 pm, the values being sizably larger than those for the normal Si-Si single bonds.

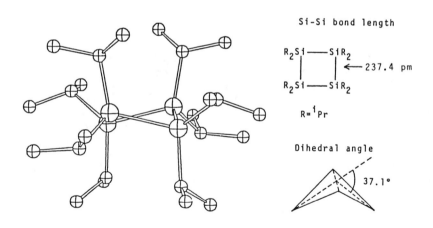

Si-Si bond length

$$R_2Si\text{———}SiR_2$$
$$|\qquad\quad| \leftarrow 237.4\ \text{pm}$$
$$R_2Si\text{———}SiR_2$$

$R={}^iPr$

Dihedral angle

37.1°

Fig. 1. ORTEP View of $({}^iPr_2Si)_4$.

In the crystal structure of cyclotrisilane 6 shown in Figure 2, the three silicon atoms form an isosceles triangle with Si-Si bond lengths near 237 and 240 pm which are very close to those found in the above cyclotetrasilanes.

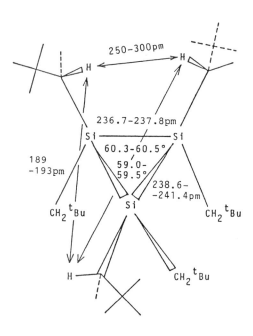

Fig. 2. The structure of $[(^tBuCH_2)_2Si]_3$

ELECTRONIC SPECTRA AND INTERNAL STRAIN

The electronic properties of peralkylpolysilanes are
generally recognized as being associated with σ-
electron delocalization, and thus in the linear polysilanes the energy transition of the bonding electrons can be observed in the uv region.   In keeping
with this interpretation there are found considerable
bathochromic shifts with increasing silicon catenation in the lowest transition absorptions for a
series of linear permethylpolysilanes, $Me(Me_2Si)_nMe$
where n = 2, 3, 4, 5, and 6, which appear at 194,
216, 235, 250, and 260 nm, respectively [6].   By contrast, in the lowest transition absorptions for a
series of peralkylcyclopolysilanes there is found an
exactly opposite trend.   Thus, longest wavelength
absorptions for these cyclopolysilanes, $(R_2Si)_n$
where n = 2, 3, 4, 5, and 6, occur in average at 400,
315, 290, 265, and 260 nm, respectively [5s].   The
smaller transition energies observed for the smaller
rings could be most simply explained in terms of
higher HOMO levels due to higher internal strain
energy.   Knowing that this explanation holds only to
the first approximation, one can define the ring

strain energy of the cyclopolysilanes as the difference in the transition energy between each set of
polysilanes with the same number of silicon atoms in
the two series of polysilanes.   Table 1 lists the
lowest transition energies for the linear and cyclic
polysilanes (TE) and the ring strain energies (SE)
which are set to be equal to TE(linear) - TE(cyclic).
Interestingly enough, the ring strain energies so
calculated for the cyclopolysilanes are in good agreement with those of the corresponding cycloalkanes,
$(CH_2)_n$, except for the case of n = 2.

Table 1. Ring strain energies (Kcal/mol) of peralkylcyclopolysilanes.

Polysilane n:	2	3	4	5	6
TE(lin)	148	132	122	115	110
TE(cyc)	72	89	99	109	110
SE(Si)	76	43	23	6	0
SE(C)	38	46	27	7	0

The HOMO levels for the peralkylsilacycles were
directly measured by photoelectron spectroscopy and
the IP values were thus determined for 2 and 6 to be
7.35 and 6.90 eV, respectively (Figures 3 and 4) [7].
It can been seen that the IP values for the cyclotrisilane is lower than that for the cyclotetrasilane,
which is in turn lower than the reported value for
permethylcyclopentasilane, 7.92 eV [8].

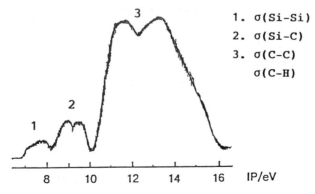

1. σ(Si-Si)
2. σ(Si-C)
3. σ(C-C)
   σ(C-H)

Fig. 3. PE spectrum of $(^iPr_2Si)_4$.

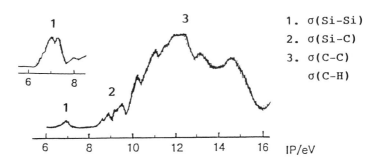

1. σ(Si–Si)
2. σ(Si–C)
3. σ(C–C)
   σ(C–H)

Fig. 4. PE spectrum of $[(^t BuCH_2)_2 Si]_3$.

The trends observed both in the lowest energy transitions and in the first ionization potentials may imply that the reactivity of the series of cyclopolysilanes toward electrophilic reagents should increase with decreasing ring size.   In accord with this expectation, rates for the ring-opening reaction with iodine as well as for one electron transfer reaction with tetracyanoethylene of the cyclopolysilanes were found to proceed faster with smaller rings [5u].

PERSILYLCYCLOPOLYSILANES

Gaspar and Chen obtained first a persilylsilacycle, octakis(trimethylsilyl)cyclotetrasilane, by pyrolysis of neat tris(trimethylsilyl)methoxysilane [9].   This class of compounds are of interest, since silyl substitution on the silicon framework is expected to show the electronic stabilizing effect by σ-conjugation.   In this view, we have undertaken a study along this line and found that reductive coupling of bis(trialkylsilyl)dibromosilanes with sodium in toluene at 110°C gave 18, 19, and 20 as colorless crystals in workable yields [5p,5q].   Obviously, the trimethylsilyl and dimethylethylsilyl substituents favor the four-membered ring, whereas the bulkier triethylsilyl substituent does the smaller three-membered ring.

$$(R^1 R^2 R^3 Si)_2 SiBr_2 \xrightarrow{\text{Na}} [(R^1 R^2 R^3 Si)_2 Si]_4$$

$$\underline{18} \quad R^1 = R^2 = R^3 = Me$$
$$\underline{19} \quad R^1 = R^2 = Me, \ R^3 = Et$$

or

$$[(R^1 R^2 R^3 Si)_2 Si]_3$$

$$\underline{20} \quad R^1 = R^2 = R^3 = Et$$

The corresponding tetrasilyldisilenes which are all stable in solution can be easily obtained by photolysis of these persilylcyclopolysilanes.

$$[(R^1R^2R^3Si)_2Si]_n \xrightarrow{h\nu} (R^1R^2R^3Si)_2Si=Si(SiR^1R^2R^3)_2$$

$$n=3,4$$

$$\underline{21} \quad R^1=R^2=R^3=Me$$
$$\underline{22} \quad R^1=R^2=Me, \quad R^3=Et$$
$$\underline{23} \quad R^1=R^2=R^3=Et$$

The crystal strcutures of the persilylcyclotetrasilanes desrve special comment, since it has been reported that the four-membered ring in 18 is **planar** [9]   in spite of the folded structures with large dihedral angles determined for the peralkylcyclotetrasilanes, 2, 5, 9, and 10 [5].   Our X-ray crystallographic analysis for 19 (Figure 5) also confirmed the planar structure for the persilyl substituted four-membered ring [10].   It is thus clear that the silicon rings assume planar structures in the persilylcyclotetrasilanes, while they display folded structures in the peralkylcyclotetrasilanes.   The structural difference might be the result of the difference in bond length between the Si-Si and Si-C bonds.

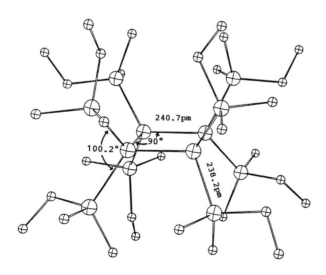

Fig. 5. ORTEP view of $[(Et_2MeSi)_2Si]_4$.

In the uv region, the lowest energy transitions for the persilylcyclotetra-, tri-, and di-silanes were found to occur at 300, 335, and 430 nm, respectively. The bathochromic shifts observed for the persilyl-silacycles relative to the corresponding peralkyl rings can be attributed to the electronic perturbation caused by the complete silyl substitution.

STRAINED POLYCYCLOPOLYSILANES

Polycyclopolysilanes consisting of strained rings provide another point of interest. Quite recently, Masamune and co-workers have isolated 1,3-di-tert-butyl-2,2,4,4-tetrakis(2,6-dimethylphenyl)bicyclo-[1.1.0]tetrasilane [11]. However, no single example of polycyclopolysilanes in which four-membered rings are fused each other has been reported to date. Now it was found by us that two new persilapolycyclo-alkanes, decaisopropylhexasilabicyclo[2.2.0]hexane, 24, and dodecaisopropyloctasilabicyclo[4.2.0.0^{2,5}]-octane, 25, can be isolated as air-stable colorless crystals by the co-condensation of 1,2-diisopropyl-tetrachloro- and 1,1,2,2-tetraisopropyldichloro-disilane with lithium in THF in rather low yields[12]. In HPLC charts of the reaction mixture there are a number of peaks due to unidentified products in addition to peaks due to 2, 24, and 25. Most of these compounds are air-stable but light-sensitive and might include compounds of unusual structures.

$$Cl_2RSiSiRCl_2 \quad + \quad ClR_2SiSiR_2Cl \xrightarrow{\text{Li}}$$

$$
\begin{array}{cc}
R_2Si\!-\!\overset{R}{\underset{|}{Si}}\!-\!SiR_2 & R_2Si\!-\!\overset{R}{\underset{|}{Si}}\!-\!\overset{R}{\underset{|}{Si}}\!-\!SiR_2 \\
R_2Si\!-\!\underset{R}{\underset{|}{Si}}\!-\!SiR_2 & R_2Si\!-\!\underset{R}{\underset{|}{Si}}\!-\!\underset{R}{\underset{|}{Si}}\!-\!SiR_2
\end{array} \quad (R = {}^{i}Pr)
$$

<u>24</u>   <u>25</u>

Both 24 and 25 absorb uv light at 310 and 315 nm, respectively, and exhibit marked color change in the condensed phase upon heating. Thus, crystals of 24 change in color from colorless to yellow at 410°C irreversibly, while crystalline 25 shows reversible color change in the region between room temperature

and 215°C from colorless to light yellow and then the
color of molten liquid turns bright yellow at 330°C
irreversibly.

CONCLUSION

Strained silacycles protected by bulky alkyl substituents constitute a unique class of compounds which are
stable toward air and yet sensitive toward electro-
magnetic interaction.   It seems clear that the
synthesis of newer polycyclopolysilanes will open an
entirely new area of organosilicon chemistry.

ACKNOWLEDGEMENT

A significant contribution to this work was made in
collaboration with Professors Y. Harada and K. Ohno
at the University of Tokyo as well as with Dr. M. Goto
at the National Chemical Laboratory for Industry.

REFERENCES

1.   F. S. Kipping, J. Chem. Soc. (1921), 830.
2.   M. Biernbaum and R. West, J. Organomet. Chem.
     (1974), 77, C13.
3.   S. Masamune, Y. Hanzawa, S. Murakami, T. Bally,
     and J. F. Blount, J. Amer. Chem. Soc. (1982),
     104, 1150.
4.   R. West, M. J. Fink, and J. Michl, Science
     (1981), 214, 1343.
5.   (a) H. Watanabe, T. Muraoka, Y. Kohara, and Y.
     Nagai, Chem. Lett. (1980), 735; (b) H. Watanabe,
     T. Muraoka, M. Kageyama, and Y. Nagai, J.
     Organomet. Chem. (1981), 216, C45; (c) H.
     Watanabe, T. Okawa, M. Kato, and Y. Nagai, J.
     Chem. Soc., Chem. Commun. (1983), 781; (d) H.
     Watanabe, J. Inose, K. Fukushima, Y. Kougo, and
     Y. Nagai, Chem. Lett. (1983), 1711; (e) H.
     Watanabe, Y. Kougo, and Y. Nagai, J. Chem. Soc.,
     Chem. Commun. (1984), 66; (f) H. Watanabe, Y.
     Kougo, M. Kato, and Y. Nagai, Bull. Chem. Soc.
     JPN (1984), 57, 3019; (g) H. Watanabe, T.
     Muraoka, M. Kageyama, K. Yoshizumi, and Y.
     Nagai, Organometallics (1984), 3, 141; (h) H.
     Watanabe, M. Kato, T. Okawa, and Y. Nagai, J.
     Organomet. Chem. (1984), 271, 225; (i) H.
     Matsumoto, M. Minemura, T. Takasuna, Y. Nagai,
     and M. Goto, Chem. Lett. (1985), 1005; (j) H.
     Matsumoto, T. Arai, H. Watanabe, and Y. Nagai,
     J. Chem. Soc., Chem. Commun. (1984), 724; (k) C.
     L. Wadsworth, R. West, Y. Nagai, H. Watanabe, H.
     Matsumoto, and T. Muraoka, Chem. Lett. (1985),

1525; (l) H. Watanabe, K. Yoshizumi, T. Muraoka,
M. Kato, Y. Nagai, and T. Sato, ibid. (1985),
1683; (m) H. Matsumoto, K. Takatsuna, M.
Minemura, Y. Nagai, and M. Goto, J. Chem. Soc.,
Chem. Commun. (1985), 1366; (n) C. L. Wadsworth,
R. West, Y. Nagai, H. Watanabe, and T. Muraoka,
Organometallics (1985), 4, 1659; (o) H.
Matsumoto, S. Sakamoto, M. Minemura, K.
Sugaya, and Y. Nagai, Bull. Chem. Soc. JPN
(1986), 59, 3314; (p) H. Matsumoto, N. Yokoyama,
A. Sakamoto, Y. Sramaki, R. Endo, and Y. Nagai,
Chem. Lett. (1986), 1643; (q) H. Matsumoto, A.
Sakamoto, and Y. Nagai, J. Chem. Soc., Chem.
Commun. (1986), 1768; (r) H. Watanabe, M. Kato,
E. Tabei, N. Kuwabara, N. Hirai, T. Sato, and Y.
Nagai, ibid. (1986), 1662; (s) H. Watanabe, H.
Shimoyama, T. Muraoka, T. Okawa, M. Kato, and Y.
Nagai, Chem. Lett. (1986), 1057; (t) H. Shizuka,
T. Tanaka, K. Okazaki, M. Kato, H. Watanabe, Y.
Nagai, and M. Ishikawa, J. Chem. Soc., Chem.
Commun. (1986), 748; (u) H. Watanabe, H.
Shimoyama, T. Muraoka, Y. Kougo, M. Kato, and Y.
Nagai, Bull. Chem. Soc. JPN (1987), 60, 769;
(v) H. Watanabe, M. Kato, T. Okawa, Y. Kougo, Y.
Nagai, and M. Goto, Appl. Organomet. Chem.
(1987), 1, 157.

6. B. G. Ramsay, "Electronic Transitions in Organo-
    metalloids", Academic Press, NY (1969).
7. H. Bock and W. Ensslin, Angew. Chem. (1971), 83,
    435.
8. H. Matsumoto, Y. Yagihashi, Y. Nagai, T. Takami,
    T. Ishida, S. Masuda, K. Ohno, and Y. Harada,
    54th National Meeting of Chem. Soc. Jpn., Tokyo,
    April (1987), Abstr., No. 1IIIJ03.
9. Y. Chen and P. P. Gaspar, Organometallics
    (1982), 1, 1410.
10. H. Matsumoto, A. Sakamoto, Y. Nagai, and M. Goto,
    54th National Meeting of Chem. Soc. JPN., Tokyo,
    April (1987), Abstr., No. 1IIIJ01.
11. S. Masamune, Y. Kabe, S. Collins, D. J.
    Williams, and R. Jones, J. Amer. Chem. Soc.
    (1985), 107, 5552.
12. H. Matsumoto, H. Miyamoto, N. Kojima, Y. Nagai,
    and M. Goto, 8th Intl. Symp. Organosilicon
    Chem., St. Louis, June (1987); J. Chem. Soc.,
    Chem. Commun., in press.

Chapter 25

Silicon and germanium double bond and polycyclic ring systems

Satoru Masamune – Department of Chemistry, Massachusetts Institute of Technology.

ABSTRACT

This article describes some important physical and chemical properties of silicon and germanium double bond and polycyclic compounds; specifically, 1) the stability of tetrakis[bis-(trimethylsilyl)methyl]disilene relative to that of the corresponding silylene, 2) the E-Z isomerization of disilene and digermene derivatives, and 3) substituent effects on the reactivity of tetrasilabicyclo[1.1.0]butanes and pentasilabicyclo-[1.1.1]pentanes as well as our efforts toward the synthesis of a tetrasilatetrahedrane derivative.

INTRODUCTION

It was in the fall of 1981 that we succeeded in synthesizing the first cyclotrisilane derivative, hexakis(2,6-dimethyl-phenyl)cyclotrisilane (A-1-a)* (see footnote* on next page) which upon photolysis yielded tetrakis(2,6-dimethylphenyl)di-silene (B-1-a) as shown in Scheme 1 (1). The discovery of this synthetic pathway laid a solid foundation for the project aimed at developing the chemistry of strained-ring and multiple-bond systems consisting of Si, Ge, and Sn. The project had undergone expansion and diversification in the usual, logical manner, and by the middle of 1986 we were in possession of a set of three-membered rings [cyclotrimetallanes, $(R_2M)_3$: M=Si, Ge, Sn] (A) and a set of formal double bond derivatives [dimetallenes, $(R_2M)_2$: M=Si, Ge, and Sn] (B) for comparative

studies as well as a tetrasilabicyclo[1.1.0]butane 1 (2) and a
pentasilabicyclo[1.1.1]pentane derivative 2 (3). This article
presents, after a brief discussion on the known chemistry of A
and B, several episodes chosen from the work we have carried

Scheme 1

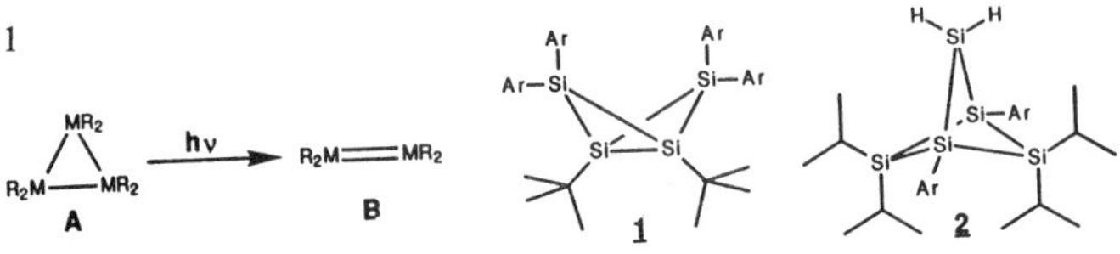

out in this area for the past one year.

OUTLINE OF CYCLOTRIMETALLANE AND DIMETALLENE CHEMISTRY

Hexaarylcyclotrimetallane (A. M=Si, Ge, Sn). Treatment of
bis(2,6-dialkylphenyl)dichlorometallanes [Ar$_2$MCl$_2$: Ar=2,6-
disubstituted phenyl, M=Si, Ge, and Sn] (C) with a reductant
provides the corresponding cyclotrimetallanes (A) (Scheme 2).
This successful construction of A is attributed mainly to the
advantageous use of bulky ligands which suppress the high
reactivity of product A, the strained three-membered ring
system, and also terminate the oligomerization of C at the
stage of the trimer [Cl-M(R$_2$)M(R$_2$)M(R$_2$)Cl] (D), the probable
direct precursor of A. Indeed, for M=Si and Ge the trimer D's

Scheme 2

have been found to undergo reductive cyclization with lithium
naphthalenide, rather than further chain elongation, even in
the presence of excess C's, (Thorpe-Ingold effect). We have
prepared A-1-a and A-1-b (see footnote* for the structure),
A-2-a and A-2-b, and also A-3-b, all of which are fully characterized spectroscopically (1, 4, 5, 6, 7). These compounds
are unique in that their strong electronic UV absorptions tail
off in the visible region, indicating the electrons in the M-M
bond are substantially delocalized over the three-membered
ring. The crystal structures of A-1-a, A-2-a, and A-3-b show
equilateral or near equilateral triangular Si, Ge, and Sn
skeletons, respectively, with an unusually long M-M bond.

---

*Several compounds appearing in this Article are expressed in
this way, with the capital letter, number, and lower case
letter referring to the basic structure indicated in a scheme,
metal M, and ligand R, respectively. Thus, 1, 2, or 3
indicates M=Si, Ge, or Sn and a, a', b, c, or d, indicates
R=2,6-dimethylphenyl, mesityl, 2,6-diethylphenyl, 2,4,6-tri-
isopropylphenyl, or bis(trimethylsilyl)methyl=disyl.

In contrast with the above cyclotrimetallanes including A-3-b, the way A-3-c is formed is unusual. As shown in Scheme 2, reaction of C-3-c (R=2,4,6-triisopropylphenyl) with lithium naphthalenide provides dimeric E-3-c (R=2,4,6-triisopropylphenyl) which upon further treatment with the same reductant leads to the formation of A-3-c (in the absence of C-3-c) (8). The mechanism of the latter transformation and the equilibration between A-3-c and the corresponding distannene B-3-c are described in the next Section.

Tetraaryldimetallenes (B. M=Si, Ge, Sn). Almost all of our dimetallenes B (1.5 equiv) have been prepared through the photolysis of the corresponding cyclotrimetallanes A (1.0 equiv), and both the photo-conversion of B into the metallylene F and the thermal dimerization of F into B have been well documented (9). The yield of photo-conversion of A

Scheme 3

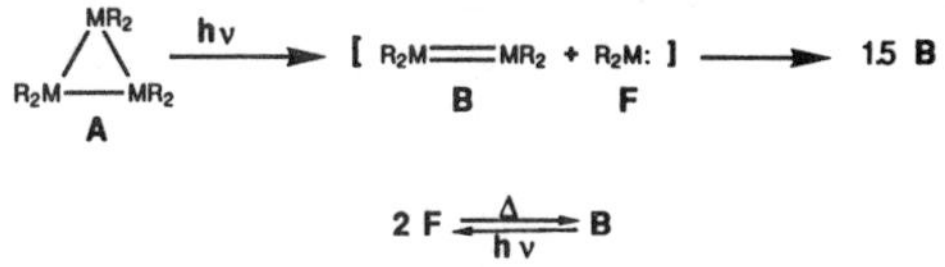

$$\text{A} \quad \xrightarrow{\text{hv}} \quad [\ R_2M\!=\!\!=\!\!MR_2 + R_2M: \ ] \quad \longrightarrow \quad 1.5\ \textbf{B}$$
(A: MR_2 / R_2M—MR_2; B; F)

$$2\ \text{F} \ \underset{hv}{\overset{\Delta}{\rightleftharpoons}} \ \text{B}$$

into B depends upon the relative "cleanness" of each of the above conversions (A→B, B→F, F→B) which is greatly influenced by both the kind of M and the ligands attached to M. For instance, in the case of M=Ge, conversion A→B is quantitative with R=2,6-diethylphenyl (A-2-b → B-2-b) (7), but is at most 65% with R=2,6-dimethylphenyl (A-2-a → B-2-a) (4). Of particular interest is M=Sn. While photolysis of A-3-b fails to provide B-3-b, (R=2,6-diethylphenyl) (5), upon irradiation at -78°C A-3-c (R=2,4,6-triisopropylphenyl) provides a 100% yield of B-3-c (8). It is noted that B-3-c is converted, at 0°C or below, to A-3-c, which upon further warming to 90°C is partially reconverted to B-3-c. This unusual thermal equilibrium between the tricyclostannane A-3-c and distannene B-3-c, thus far unprecedented in the Si and Ge series, indicates that both A-3-c and B-3-c generate the corresponding stannylene F-3-c even with heat and the equilibrium is attained in the following manner: 2(A-3-c)⇌2(B-3-c) + 2(F-3-c)⇌3(B-3-c) (8). In the preceding Section the direct conversion of E-3-c to A-3-c is mentioned (Scheme 2). This reaction is understood by the rapid formation of A-3-c from B-3-c which is derived from E-3-c through reductive dehalogenation.

The molecular structures of the disilene B-1-b and digermene B-2-b have been established (6, 7), and their salient structure features are compared with those of dimetallenes recorded by West and Michl (10) and Lappert (11) (Table 1).

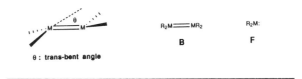

θ : trans-bent angle

Table 1.

Entry	Compound	θ (°)*	M=M Bond Length
1	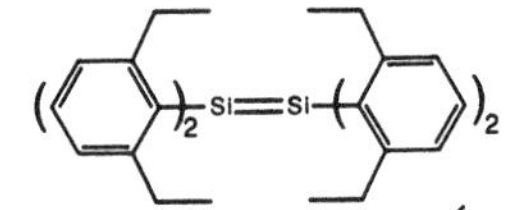 $(\text{...})_2$Si≡Si$(\text{...})_2$	0	2.14 (0.91)**
2	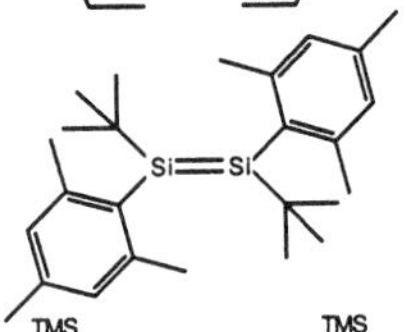 Si≡Si	0	2.14 (0.91)
3	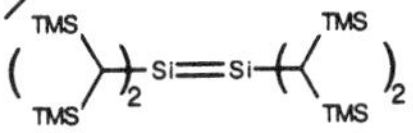 $(\text{...})_2$Si≡Si$(\text{...})_2$ TMS		
4	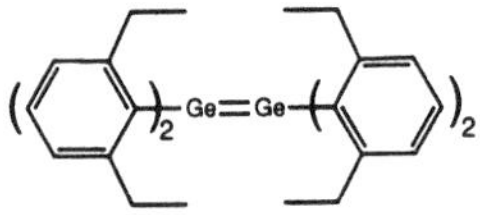 $(\text{...})_2$Ge≡Ge$(\text{...})_2$	13	2.21 (0.92)
5	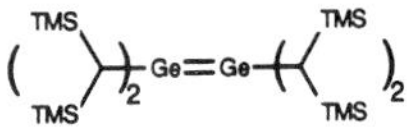 $(\text{...})_2$Ge≡Ge$(\text{...})_2$ TMS	32	2.35 (0.93)
6	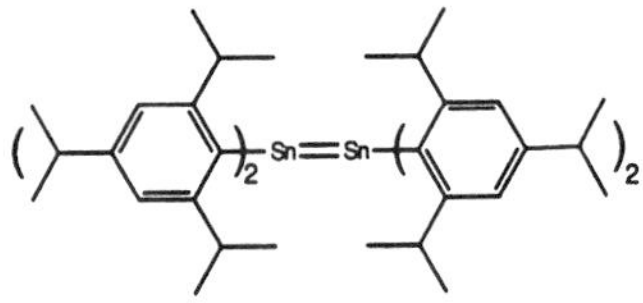 $(\text{...})_2$Sn≡Sn$(\text{...})_2$		
7	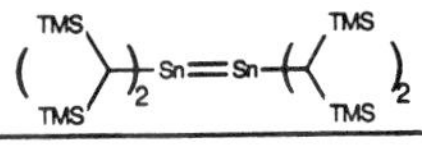 $(\text{...})_2$Sn≡Sn$(\text{...})_2$ TMS	41	2.76 (0.98)

*   θ : trans bent angle

**   the M=M bond length relative to the
      M-M bond length

It is now clear from accumulated data on dimetallene B and
monomeric metallylene F that the stability of F relative to B
increases in the order of Si, Ge, and Sn and is reflected in
both the degree of pyramidization at the metal (bend angle θ
observed in the dimetallenes) and the bond length. The higher
the stability is, the larger θ is. Thus, in Lappert's

compound B-3-d which represents an extreme example, the Sn=Sn double bond shows a bent angle of 41° and is found to be as long as the average single bond. The metallylene-dimetallene relative stability also varies with the kind of ligands attached to each metal. Thus, while both digermene B-2-b (entry 4) and distannene B-3-c (entry 6) retain their structural integrity (of double bonding) in solution, Lappert's tetrakis(disyl)digermene (B-2-d, entry 5) and tetrakis(disyl)-stannene (B-3-d, entry 7) exist as dimers only in crystalline form and completely dissociate into the corresponding metallylenes in solution. In this connection our attention was directed some time ago to the then unknown compound tetrakis(disyl)disilene (B-1-d, entry 3). This compound has now yielded to synthesis and its behavior is described in the next Section.

TETRAKIS(DISYL)DISILENE (B-1-d)

This target disilene has defied many synthetic attempts in these and other laboratories, but a short and direct route shown below has turned out, unexpectedly, to be successful and

Scheme 4

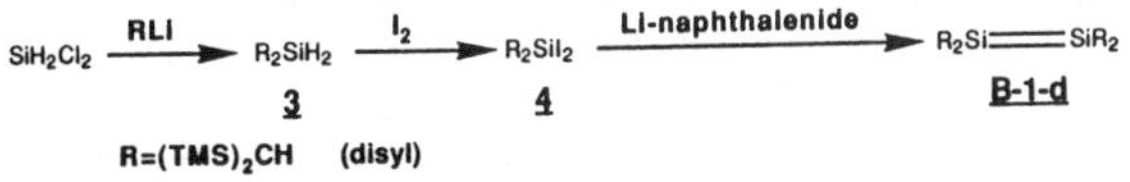

B-1-d represents the first disilene derivative that has been prepared under reductive conditions (12). Treatment of di-chlorosilane with disyllithium provides bis(disyl)silane (3) which in turn is converted to the corresponding diiodide (4). The final step of reductive coupling adopts the procedure Weidenbruch used for the synthesis of hexa-t-butylcyclotri-silane (13). Thus, reduction of 4 with lithium naphthalenide in THF provides a ca. 40% yield of air and moisture sensitive B-1-d.

The physical and chemical properties of B-1-d are fully consistent with its formulation as the disilene structure and show no indication of its dissociation into the corresponding silylene in solution. 1) Mass spectrum (EI, 70eV), m/z calcd for $C_{28}H_{76}Si_{10}$ 692.36398; found 692.36415. 2) Electronic spectrum (methylcyclohexane, r.t.) λ_{max} 393 nm (ϵ 12,600), 357 (8,600). The extinction coefficients remains constant with samples whose concentrations vary widely. In comparison with the spectra reported for all known disilenes, the 393 nm ab-sorption maximum and intensity of B-1-d is in the region ex-pected for this disilene chromophore (1, 6, 14). 3) The di-silene B-1-d reacts with water to provide the corresponding hydrate 5 (Scheme 5). Rather unexpectedly, B-1-d is inert to methanol and remains unchanged even after a MeOH-THF solution

Scheme 5

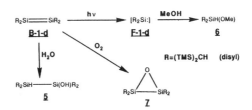

of B-1-d is heated at 80°C for 24 h. However photolysis
(Hanovia medium pressure, Pyrex filter, MeOH-methylcyclohexane
solution) immediately causes dissociation of B-1-d into bis-
(disyl)silylene (F-1-d) which subsequently reacts with methanol to afford the methoxysilane (6). 4) Bubbling dioxygen
through a methylcyclohexane solution of B-1-d produces the
epoxide 7 as the major product.

A ^{1}H NMR spectrum (250 MHz, benzene-d$_6$, r.t.) of B-1-d is simple: δ 0.18 (s, 2H, CH), 0.26 (s, 2H, CH), 0.37 (s, 36H, Si-
CH$_3$), and 0.39 (s, 36H, SiCH$_3$). Although the chemical shifts
of these signals are slightly temperature-dependent over the
range -50° to 100°C, all spectra show only one set of signals,
indicating that the sample solution contains a single species.
While the appearance of more than one signal for the Si-CH$_3$
groups rules out the possibility of this species being (monomeric) [(TMS)$_2$CH]$_2$Si, the presence of two signals for the CH-
(TMS)$_2$ groups and also for the Si-CH$_3$ groups in the spectra is
consistent with the interpretation that B-1-d is undergoing
rapid inversion between the two trans-bent conformers (S$_2$) via
the planar conformer (C$_{2h}$) as shown below, or B-1-d might be
frozen in the C$_{2h}$ conformation (Scheme 6).

Scheme 6

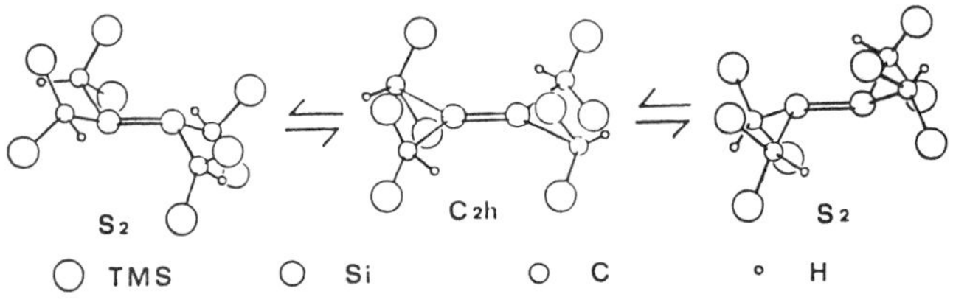

S$_2$ C$_{2h}$ S$_2$

○ TMS   ○ Si   ○ C   ∘ H

Unlike its Ge and Sn counterparts B-2-d and B-3-d compound
B-1-d shows no indication of dissociation into F-1-d in solution to a detectable extent. This is obviously due to the expected high thermodynamic stability of the Si, Si double bond
compared to the Ge, Ge and Sn, Sn formal double bonds (15).
The stability of the dimetallene double bond should also be
reflected in its geometry in terms of the bond length and bent
angle. For this reason, the crystal structure of B-3-d
attracts special interest and an x-ray analysis is underway.

THE *E-Z* ISOMERIZATION OF *E* AND *Z* 1,2-(2,6-DIETHYLPHENYL)-1,2-DIMESITYLDIGERMENES

It is noted above that tetraaryldigermenes retain their structural integrity. We now find that an E and Z pair of the titled digermenes (8Z and 8E) undergo facile interconversion without breaking the digermene double bond (16). Treatment of the dichlorogermane 9 provides the all-cis cyclotrigermane 10 in addition to its cis, trans isomer 11 (Scheme 7). When photolysis of 10 at -78°C is monitored by ¹H NMR spectroscopy, there appear, at the expense of the signals due to 10, two sets of new signals which are ascribable to 8Z and 8E. The singlets at δ 6.66 and 6.61 which are assigned to the aromatic protons indicated in 8Z and 8E, respectively, serve conveniently to determine the ratio of these two compounds at regular intervals. Thus, at 30% conversion of 10 the ratio of 8Z to 8E is ca. 4:1 and beyond 80% conversion it remains to be 1:1, indicating a photostationary state has been established.

Scheme 7

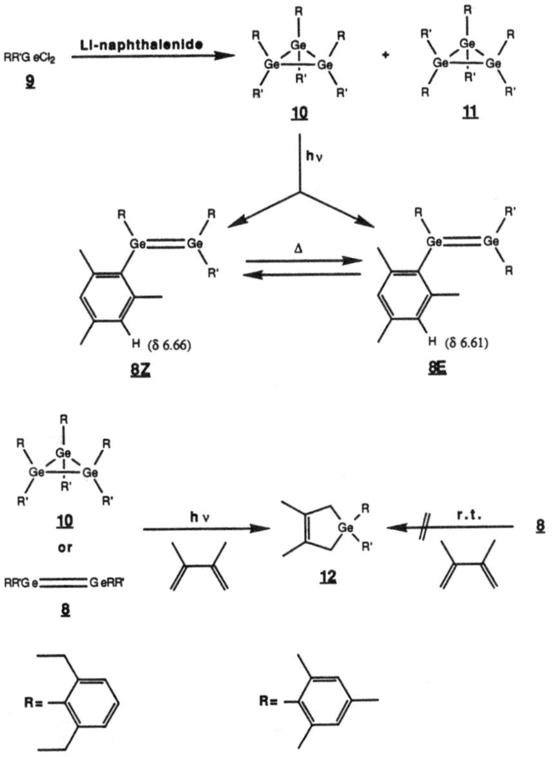

Upon warming a cold photolysis sample of 30% conversion (see above) to room temperature, thermal conversion of 8Z to 8E is observed (equilibrium ratio of Z to E, 0.64 at 7°C). Although we have not secured accurate kinetic parameters, enthalpy for

this thermal conversion is estimated to be ca. 17 kcal/mol
with a half life of 84 min at 7°C. The mechanism of this isomerization solely involves rotation about the Ge-Ge double
bond, since in the presence of the efficient germylene trap
2,3-dimethylbutadiene, the rates of isomerization are not affected and no trapped product 12 is formed, thus precluding a
thermal dissociation recombination pathway via the germylene.
Product 12 is readily obtained through photolysis of 8 or 10
in the presence of 2,3-dimethylbutadiene.

For comparative studies, the silicon analogues (13Z and 13E)
of 8Z and 8E have been prepared. Kinetic parameters for the Z
to E thermal interconversion of 8Z to 8E obtained in much the
same manner as that described for 8Z and 8E are : $\Delta H^{\ddagger}$ = 27.0
± 0.8 kcal mol^{-1}, $\Delta S^{\ddagger}$ = 1.8 ± 3.1 at cal K^{-1}mol^{-1} (Scheme 8)

Scheme 8

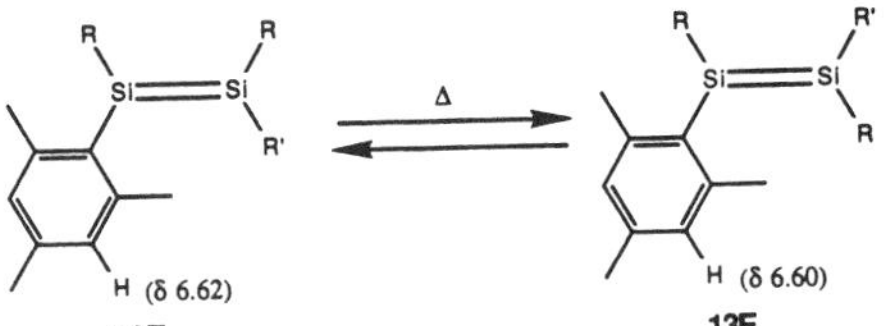

(17). The enthalpy of activation of the germanium double bond
isomerization is considerably lower than that of the corresponding silicon isomerization, as expected from the known
chemistry of these two dimetallenes.

BI- AND TRICYCLIC RING SYSTEMS

After several cyclotrisilane derivatives yielded to synthesis,
our obvious target molecules were tetrasilabicyclo[1.1.0]bu-
tanes. After many attempts, the pursuit of Scheme 9 has led
to the synthesis of 1,3-di-t-butyl-2,2,4,4-tetrakis(2,6-di-
ethylphenyl)tetrasilabicyclo[1.1.0]butane (14). Treatment of
trichlorosilane 15, prepared in a standard fashion, with
lithium naphthalenide gives rise to the cyclotetrasilane 16,
which after purification is subjected to a second reductive
cyclization with the same reductant to provide 14 quantitatively (2).

The bicyclo compound 14 is extremely reactive and its exposure
to air, water (atmospheric moisture), or chlorine immediately
converts 14 to 17, 18, or 16. The ring undergoes facile inversion (14 ⇌ 14a). Thus, while three distinct ethyl signals
in ^1H NMR spectra taken at -30°C or below, these signals are
replaced by a single quartet-triplet pair at 115°C, indicating

Scheme 9

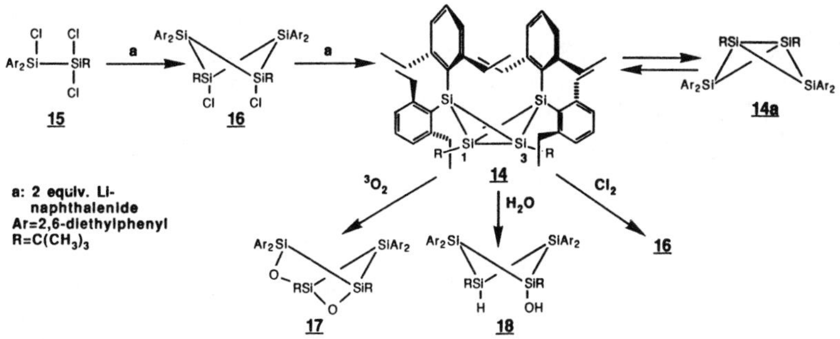

a: 2 equiv. Li-
naphthalenide
Ar=2,6-diethylphenyl
R=C(CH$_3$)$_3$

that all the ethyl groups have attained equivalence on the NMR
time scale. The barrier to this inversion is estimated to be
ca. 15 kcal/mol.

The crystal structure of 14 has been disclosed (18). The
geometry of the ring system as well as the barrier to inversion are in excellent agreement with the results obtained from
an ab initio (6-31G*) calculation (19) and we conclude that
the critical central Si(1)-Si(3) bond (see structure 14) is
highly bent and rich in p-character.

The reactivity of this Si(1)-Si(3) bond was expected to be
further enhanced by replacing the t-butyl substituents attached to Si(1) and Si(3) with aryl groups. We have now found
that reductive cyclization of trichlorodisilane 19 under similar conditions to those applied to 14 leads to products which
result from the reductive cleavage of the Si(1)-Si(3) bond
(20). Thus, treatment of 19 with 5 equiv of lithium naphthalenide provides, after H$_2$O quenching, 20 (33%) as the sole
isolable product (Scheme 10). The use of 3.5 equivalent of

Scheme 10

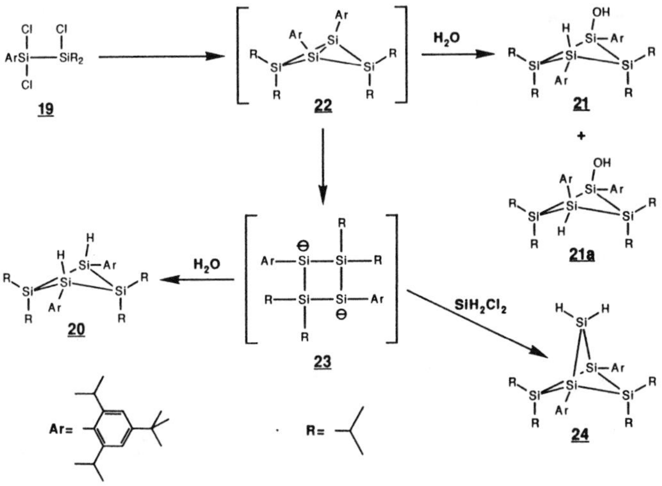

the reductant leads to a 4:1 mixture (37%) of 21 and 21a. In
view of the earlier observation with 14, the precursor of 21
must be 22, and a set of experiments have demonstrated that 22
is overreduced to 23 with the excess reductant. In fact, 23
is trapped with dichlorosilane to yield a pentasilabicyclo-
[1.1.1]pentane derivative 24 (3). This reactivity of the
Si(1)-Si(3) bond in 22 is in contrast with that in 14 which
resists reductive cleavage with the same reductant even at
room temperature.

Reductive coupling of trichlorodisilane 25 with an excess reductant appears to proceed in a manner similar to that described for 19. Thus, reaction of 25 with 5 equiv of lithium
dispersion at -78°C provides, after H₂0 workup, a mixture of
three cyclotetrasilane derivatives to which the stereostruc-
tures 26a, b and c are assigned. All of these products undoubtedly have formed from the corresponding dianion 27. In

Scheme 11

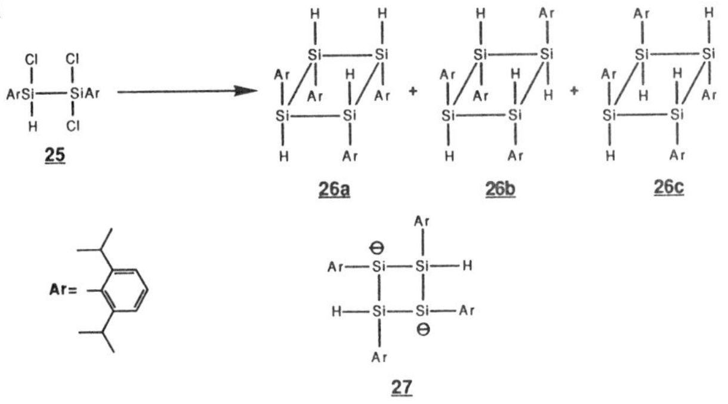

this connection, similar reductive coupling of the tetrachlor-
odisilane 28 is of some interest. This reaction leads to the
formation of numerous products, with only a minor product to
which the structure 29 may be assigned (NMR and mass spectrum). That intermediate 30 is involved in the formation of
29 is a distinct possibility.

Scheme 12

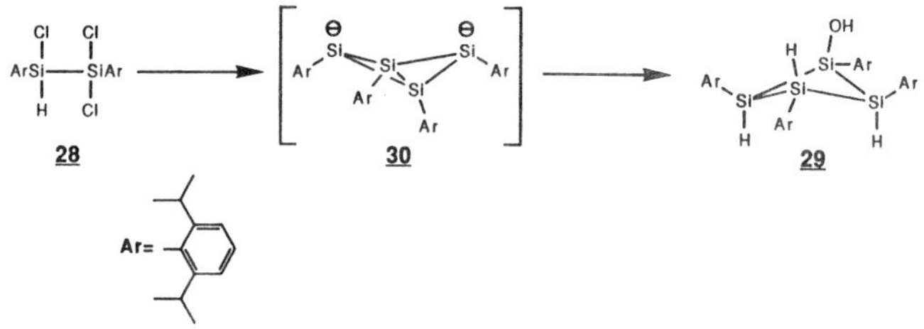

From the results outlined above it was increasingly clear that a tetrahedrane derivative 31, if successfully synthesized under reducing conditions, would have bulky alkyl (rather than aryl) substituents attached to the ring system. Thus, a great deal of effort has been expended to synthesize stable synthetic intermediates including disilanes $(RSiX_2)_2$ and cyclotetrasilanes $(RSiX_2)_4$ with R's and X's indicated below. Reductive coupling of some of these intermediates appears promising, but

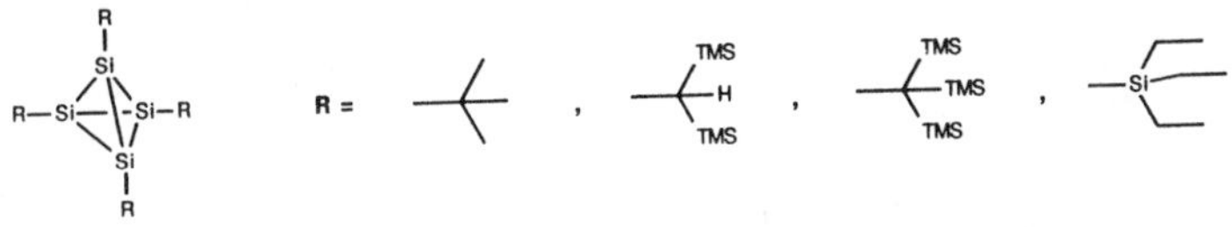

it is, at the present moment, too early to disclose definite results. Over the past five years several major achievements have been witnessed in this area of strained ring systems. In view of this progress one wonders when a silatetrahedrane derivative will yield to synthesis.

Acknowledgments: The author is grateful to his collaborators whose names appear in the references. They have contributed experimentally as well as conceptually to the development of this project. The work has been supported by grants from the National Science Foundation (USA) as well as from the Kao Corporation (Japan).

REFERENCES

1. Masamune, S.; Hanzawa, Y.; Murakami, S.; Bally, T.; Blount, J.F. J. Am. Chem. Soc. 1982, 104, 1150.

2. Masamune, S.; Kabe, Y.; Collins, S.; Williams, D.J.; Jones, R. J. Am. Chem. Soc. 1985, 107, 5552.

3. Masamune, S.; Kawase, T. unpublished results.

4. Masamune, S.; Hanzawa, Y.; Williams, D.J. J. Am. Chem. Soc. 1982, 104, 6136.

5. Masamune, S.; Sita, L.R.; Williams, D.J. J. Am. Chem. Soc. 1983, 105, 630.

6. Masamune, S.; Murakami, S.; Snow, J.T.; Tobita, H.; Williams, D.J. Organometallics 1984, 3, 333.

7. Snow, J.T.; Murakami, S.; Masamune, S.; Williams, D.J. Tetrahedron Lett. 1984, 25, 4191.

8. Masamune, S.; Sita, L.R. J. Am. Chem. Soc. 1985, 107, 6390.

9. Masamune, S.; Murakami, S.; Tobita, H.; Williams, D.J. J. Am. Chem. Soc. 1983, 105, 7776.

10. Fink, M.J.; Michalczyk, M.J.; West, R.; Michl, J. Organometallics 1984, 3, 793.

11. Goldberg, D.E.; Hitchcock, P.B.; Lappert, M.F.; Thomas, K.M.; Thorne, A.J. J. Chem. Soc., Dalton Trans 1986, 2387.

12. Masamune, S.; Eriyama, Y.; Kawase, T. Angew. Chem. Int. Ed. Engl. in press.

13. Schäfer, A.; Weidenbruck, M.; Peters, K.; von Schnering, H.-G. Angew. Chem. Int. Ed. Engl. 1984, 23, 302.

14. Masamune, S.; Tobita, H.; Murakami, S. J. Am. Chem. Soc. 1983, 105, 6524.

15. For a theoretical evaluation, see for instance Luke, B.T. et al. J. Am. Chem. Soc. 1986, 108, 260 and 270.

16. Batcheller, S.A.; Masamune, S. unpublished results.

17. Cf. (a) Michalczyk, M.J.; West, R.; Michl, J. Organometallics 1985, 4, 826. (b) Murakami, S.; Collins, S.; Masamune, S. Tetrahedron Lett. 1984, 25, 2131.

18. Jones, R.; Williams, D.J.; Kabe, Y.; Masamune, S. Angew. Chem. Int. Ed. Engl. 1986, 25, 173.

19. Collins, S.; Dutley, R.; Rauk, A. J. Am. Chem. Soc. in press.

20. Kawase, T.; Batcheller, S.A.; Masamune, S. Chem. Lett. 1987, 227.

Chapter 26

Oxidation of disilenes with atmospheric oxygen: a status report

OXIDATION OF DISILENES WITH ATMOSPHERIC OXYGEN:

A STATUS REPORT

Robert West*, Howard B. Yokelson, Gregory R. Gillette and Anthony J. Millevolte

Robert West*, Howard B. Yokelson, Gregory R. Gillette and Anthony J. Millevolte – Department of Chemistry, University of Wisconsin-Madison, Madison, WI 53706 USA.

INTRODUCTION

In 1981 the first isolation of a stable disilene was reported, [1,2] simultaneously with the discovery of a stable silene.[3] In the intervening six years, organosilicon chemists have learned a number of things about the spectroscopy, structure, and chemical reactions of these remarkable new substances,[4,5] but for every question answered several new ones have arisen. Nowhere has this been more true than in studies of the simple oxidation of disilenes with triplet oxygen of the atmosphere.

In this chapter, current knowledge and understanding of the oxygen oxidation of disilenes will be summarized. The various oxidation products, many of which have unusual structures, will be described, and possible models for chemical bonding in these novel compounds will be considered. Finally, mention will be made of the many unsolved problems in this surprisingly complex area of disilene chemistry.

THE 1,3-CYCLODISILOXANES

As solids, the disilenes take up oxygen when exposed to air. The reaction is not a vigorous one. Its rate depends on the nature of the disilene and its state of subdivision (surface area). Powdered samples of the most reactive stable disilenes become decolorized within minutes, indicating complete oxidation. Examples are tetramesityldisilene (1), Z- and E-1,2-bis(trimethylsilylamino)dimesityl disilene (2), and

tetrakis(2,6-xylyl)disilene, (**3**). The same process
requires hours for **E**-1,2-di-<u>tert</u>-butyl-1,2-dimesityl-
disilene (**4**), and days for **E**-1,2-di-1-adamantyldi-
mesityldisilene (**5**)[6] as well as for the tetrakis-
(2,4,6-tri-<u>iso</u>-propylphenyl)disilene, (**6**).[7]

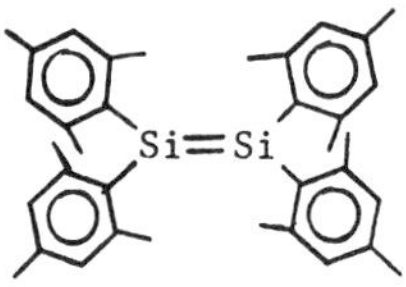

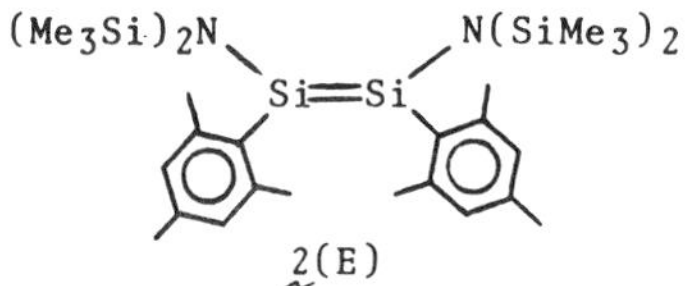

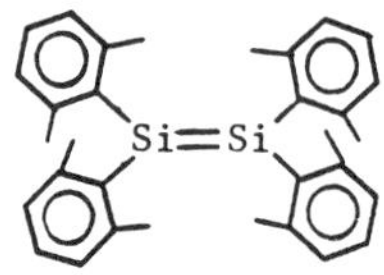

Oxygenation of the solid disilenes leads to
colorless crystalline compounds. Quite early it was
found by analysis that the products from **1**, **2** and **4**
each contain two oxygen atoms per two silicons.[8]
At this point two structures, **7** and **8** were considered for the oxidation products. The structural
question was resolved by x-ray crystallography,

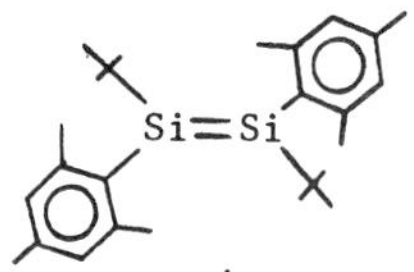

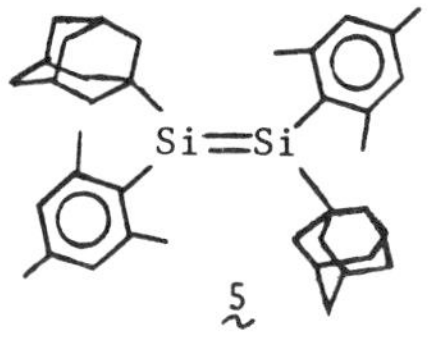

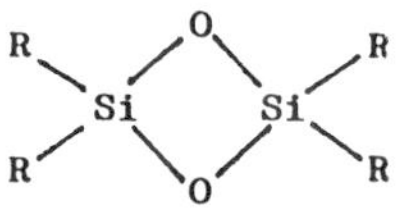

[9,10] which showed that the oxidation products of
solid **1**, **E**-**2**, and **E**-**4** all have the 1,3-cyclodi-
siloxane structure **1** (Figure 1). And although
cyclic siloxanes are among the best-known of
organosilicon compounds, manufactured in tonnage
quantities, all previously known compounds of this
class had six, eight, or larger-membered rings.

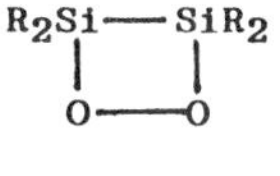

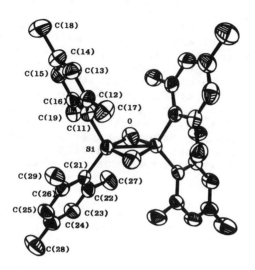

Fig. 1. The molecular structure of tetramesitylcyclodisiloxane (*7a*).

The x-ray crystal structures of these cyclodisiloxanes revealed some surprising features (Table 1). The four-membered rings deviate only slightly from planarity, as expected. However the Si-O-Si angles are quite acute, only 86° in the most extreme case, cyclodisiloxane **7a** obtained from **1**. And the silicon-silicon distance in **7a** is surprisingly short, only 231 pm. This is actually less than the normal Si-Si single-bonded distance of about 234-235 pm.

Table 1. Structural parameters for cyclodisiloxanes, RMesSi=SiMesR.

R	SiOSi,°	Si-Si,pm
Mesityl	86.2	230.6
t-Butyl	91.1	239.6
(Me₃Si)₂N	84.4,89.7	234.9

How can this short silicon-silicon separation
be rationalized? In a model invoking sigma bonding
between the Si atoms, the Si-O bonds would have to
be electron deficient, which seems unlikely (A).
Another view is that repulsion between the oxygen
atoms dominates the structure, forcing the silicon
atoms into antibonding contact (B). This problem
has attracted the attention of several theoreti-
cians.[11] Most of the ab initio treatments of
cyclodisiloxanes favor model B, although for rather
different reasons. However recently Grev and
Schaefer have proposed still a third model (C) for
the bonding in these compounds, in which there is

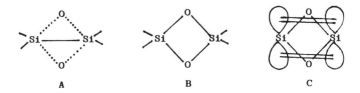

pi-bonding but no sigma bonding between the sili-
cons--an "unsupported pi bond".[12]

    We reasoned that the NMR coupling constant,
J_{SiSi}, in cyclodisiloxanes might provide important
evidence about the bonding. However, it was
necessary to synthesize a cyclodisiloxane in which
the silicon atoms were different, and this, in turn,
required synthesis of a disilene of type
$A_2Si=SiB_2$. To make such an unsymmetrically-
substituted disilene, we used the now-classic
photolytic method, employing an equimolar mixture of
trisilanes 9 and 10.[13] Our hope that these
compounds might photolyze at similar rates was
realized, in that we obtained an almost exactly
statistical mixture of the three possible disilenes.

$$Mes_2Si(SiMe_3)_2 + Xyl_2Si(SiMe_3)_2 \xrightarrow[C_5H_{12}, -60°C]{h\nu,\ 254\ nm} 1 + Mes_2Si=SiXyl_2 + 3$$

                                              25%   11, 50%   25%

However, the solution of the mixed disilene 11
slowly changed with time, undergoing transformation
to the Z-isomer Z-12. Thus this experiment led to
the discovery of the 1,2-sigmatropic rearrangement

of tetraaryldisilenes, first announced at the 7th
ISOC in Kyoto, Japan in 1984.[14] With time, the
Z-isomer of **12** undergoes the known[15] <u>cis-trans</u>
isomerization to give **E-12**:

Mes   Xyl   Mes   Mes   Mes   Xyl
  \\Si=Si/   —O→   \\Si=Si/   —O→   \\Si=Si/
  /   \\   slow   /   \\   slower   /   \\
Mes   Xyl   Xyl   Xyl   Xyl   Mes

  11   **Z-12**   **E-12**

These and similar experiments are easily
followed by ^{29}Si NMR spectroscopy, which also
provided the opportunity to determine for the first
time $^{1}J_{SiSi}$ for disilenes. Oxidation of the
disilenes led to the mixed 1,3-cyclodisiloxanes, for
which J_{SiSi} was also measured. The results for
several mixed disilenes are given in Table 2, along
with typical values of J_{SiSi} for Si-Si single
bonds and for open siloxanes Si-O-Si:

Table 2. Si-Si NMR coupling constants, Hz.

Disilene	$^{1}J_{SiSi}$	Cyclodisiloxane	J_{SiSi}
$Mes_2Si=Si(Mes)Xyl$	158		
$Mes_2Si=SiXyl_2$	155	$Mes_2SiO_2SiXyl_2$	3.85
$Xyl_2Si=Si(Xyl)Mes$	156	$Xyl_2SiO_2Si(Xyl)Mes$	4.02
$Mes_2Si=SiDmt_2$*	154		
Si-Si	80-90	Si-O-Si	0-5

*Dmt = 2,6-dimethyl-4-(<u>tert</u>-butyl)phenyl

For the disilenes, $^{1}J_{SiSi}$ is near 155 Hz,
much larger than the typical value (for aryldisilanes) of 80-90 Hz.[16] This indicates greatly
increased s character in the Si-Si sigma bond in the
disilenes, consistent with a pi bond which removes
one of the p orbitals from sigma bonding. Similar
increases in ^{1}J are found in ^{13}C and ^{31}P NMR,
on going from singly to doubly-bonded carbon and
phosphorus compounds.

For the cyclodisiloxanes, J_{SiSi} is only ~4 Hz
(Figure 2). This value is similar to that found for
open disiloxanes with no Si-Si bonding, and indicates that there is <u>no s character</u> contributing

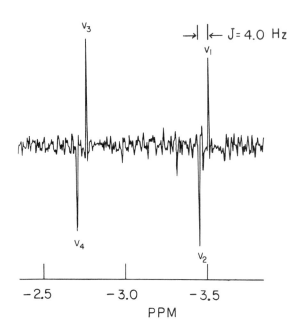

Fig. 2. INEPT-INADEQUATE 29-silicon NMR spectrum of disiloxane, Xyl$_2$SiO$_2$Si(Xyl)Mes.

to bonding between the silicon atoms.[16,17] This
experiment shows clearly that there is no sigma bond
between the silicon atoms in cyclodisiloxanes,
ruling out model A.   However it is consistent with
either of models B and C. The question of the exact
nature of bonding in 1,3-cyclodisiloxanes remains
open.

OXIDATION IN SOLUTION

     When disilenes are dissolved in organic
solvents and dry air or oxygen are bubbled through
the solution, the results are quite different from
those obtained upon complete air oxidation of the
solids in that no cyclodisiloxane is formed
initially. Two oxidation products are immediately
produced, having compositions (R$_2$Si)$_2$O and
(R$_2$Si)$_2$O$_2$. The first of these is a three-
membered ring disilaoxirane[18] (disilene epoxide,
13) and the second we believe to have the
1,2-dioxetane structure considered earlier, **8**.
These products are easily observed by ^{29}Si NMR

spectroscopy. For tetraaryldisilenes like <u>1</u>, the
ratio of <u>8</u>:<u>13</u> at 25°C is often about 4:1, but it
depends on the relative concentration of oxygen and
disilene, with higher concentrations of O_2 leading
to more <u>8</u>.

Further changes take place with time, however.
[19] The 1,2 dioxetanes slowly, but completely,
rearrange to cyclodisiloxanes, <u>7</u>. And if sufficient
oxygen is present, the disilaoxiranes undergo slow
further oxidation to the same product, <u>7</u>. Rearrangement of <u>8</u> also takes place in the solid

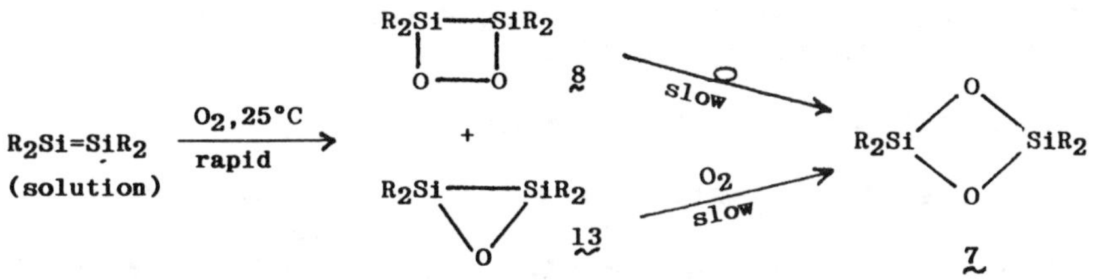

phase, so it is possible that <u>8</u>, or <u>8</u> and <u>13</u>, are
also intermediates in the solid-state oxidation
leading to <u>7</u>, the evident "thermodynamic sink" for
disilene oxygenation.

The two pathways to <u>7</u> appear to be quite
distinct. Oxidation of <u>13</u> does not lead to <u>8</u>, and
attempts to partially reduce <u>8</u> to give <u>13</u> (for
example, with excess added disilene) were
unsuccessful. It is possible that <u>8</u> and <u>13</u> may be
formed by partitioning from a common intermediate,
for which a likely candidate is the perepoxide,
<u>14</u>.[20]

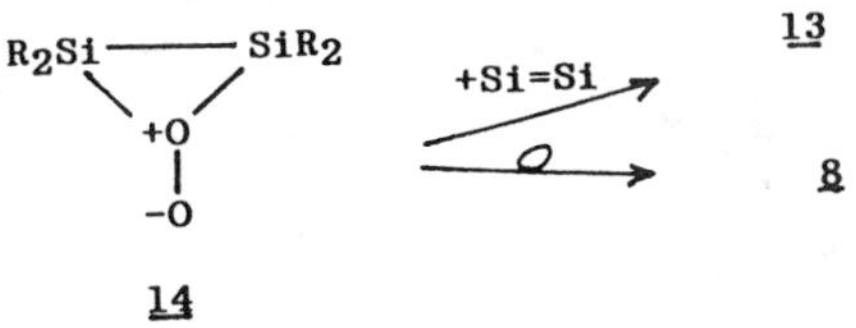

For Z and E isomers of disilenes <u>2</u> and <u>4</u> the
oxidation reactions are highly stereospecific. Both
in the gas-solid reaction and by either pathway in
solution, Z-disilene gives Z-cyclodisiloxane and E
gives E. For <u>4</u> this point was established by x-ray

crystal structures for the E isomers, combined with
NMR studies.[10]

These observations imply that in the conversion
of disilene to **8** and then **7**, either both steps proceed with retention of configuration or both steps
go with inversion. Stereospecific rearrangement of
8 to **7** most likely involves breaking of the O-O
bond, rotation and re-closing to the cyclodisiloxane, which should invert stereochemistry.[21]
This would require that the initial oxidation of
disilene to **8** also go with inversion, which is
possible but unexpected. The alternate explanation,
that both steps proceed with retention of configuration, would seem to be possible only if the structure of **8** is rather unusual.

THE STRUCTURE OF DISILENE EPOXIDE

The crystal structure for the disilaoxirane **13**
where R = mesityl has recently been determined, with
results which are quite striking.[22] An ORTEP
diagram and a Newman projection of **13** are shown in
Figures 3 and 4, respectively.

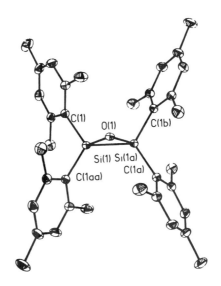

Fig. 3. Molecular structure of (*13*) (R = mesityl). Bond lengths (pm): Si(1)–Si(1a) 222.7(2),
Si(1)–C(1) 187.6(2), Si(1)–C(1a) 187.7(2), Si(1)–O(1) 173.2 (3). Bond angles (deg);
Si(1)–Si(1a)–O(1) 50.0 (1), Si(1)–o(1)–si(1a) 80.0(2).

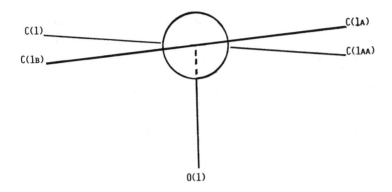

Fig. 4. Newman projecting of (R=mesityl) along the Si-Si axis showing the twist angle of 10°. The numbering scheme is shown in Fig. 3; Si(1) is in the back of Si(1a).

The surprising features are the <u>very short Si-Si distance of only 223 pm</u>, and the planar arrangement of carbon and silicon atoms attached to each of the silicons.

The Si-Si separation in <u>**13**</u> is closer to the usual Si-Si <u>double</u> bond distance of 215 pm than to the single bond distance, ~235 pm. Both this fact and the planar conformation at silicon suggest that much of the double bond character in the disilene is retained in the epoxide.

Further support for this view is provided by the silicon-silicon NMR coupling constant, $^{1}J_{Si_{1}Si_{1}}$, for the unsymmetrical disilene epoxide, $Xyl_{2}Si-O-SiDmt_{2}$, where Dmt = 2,6-dimethyl-4-(<u>tert</u>-butyl)phenyl. The observed value of 98 Hz is again intermediate between that for disilenes (~155 Hz) and disilanes (~85 Hz).

All these facts suggest that the disilene expoxide might be better regarded as an oxygen

atom-pi complex of the disilene (D) rather than as
an oxetane (E). Data for other three-membered ring
compounds with related structures are given in Table
3. The rings containing two silicons and a CH_2 or

Table 3. Si-Si distances and bond angle sums* for three-membered ring compounds.

Compound	Si-Si, pm	$\leq \angle$'s,°	Ref
$(Xyl_2Si)_2SiXyl_2$	237.5-242.5	346.5-349.4	3
$(Mes_2Si)_2SiS$	228.9	357.4	22
$(Xyl_2Si)SiCH_2$	227.2	357.4,357.7	23
$(Mes_2Si)Si_2O$	222.8	359.9	21

*Sum of $\angle$'s C-Si-Si', C'-Si-Si', C-Si-C'

S group are intermediate cases, showing shortening
of the Si-Si bond below the single-bond distance and
flattening of the configuration about silicon, but
less extreme than in 13.

The bonding in disilene epoxide can be understood in terms of a theoretical model presented by
Cremer.[25] Although his treatment was developed
for carbon-containing three-membered rings, C-C-X,
it should apply equally well to Si-Si-X rings. In
this treatment one-electron density distribution
functions are used to examine the continuous
variation between three-membered rings and olefin
pi-complexes. Cremer shows that the pi-complex
character of a C-C-X structure increases as the
acceptor ability of the bridging atom toward the
olefin pi-bond increases.

In the disilene epoxide, the oxygen atom should
be a powerful acceptor from the Si-Si pi-bond, which
in turn should be a very good pi-electron donor.
These factors may operate synergistically to produce
a particularly large amount of pi-complex character
in 13. In the Si-Si-CH_2 and Si-Si-S rings, the
acceptor power of the bridging groups is less, so
these compounds are intermediate cases, showing some
pi-complex character but less than in the disilene
epoxide.

UNANSWERED QUESTIONS

In spite of substantial progress in the past few years, the amount yet unknown about oxidation of disilenes almost surely exceeds that which is understood. For example, nothing is known of the mechanisms of reactions leading to the initial oxidation products, **8** and **13**. Do these products arise separately or via a common intermediate? The puzzling stereochemistry of the reaction of disilene → **8** → **7**, as well as the lack of understanding of the chemical bonding in **8** and **13**, have been discussed earlier in this chapter.

The structures of the 1,2-dioxetanes **8** are not yet known, and may be unusual. Like the corresponding epoxide **13**, the 1,2 dioxetane $Mes_2 Si\text{---}SiXyl_2$ shows an unusually

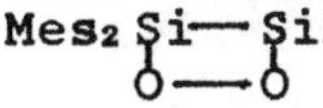

high value (98 Hz) for the $^1J_{SiSi}$ NMR coupling constant suggesting Si-Si double bond character in **8**. The chemical reactions of the oxidation products are very little known, since they have been investigated in only the most superficial way to date.[10] And, nothing is yet known of the fate of disilenes in the presence of other forms of oxygen: singlet oxygen or ozone.

The oxygen oxidation of disilenes is, it seems, a deceptively simple reaction which contains numerous hidden complexities. Further research in this minor area of disilene chemistry should be fruitful for some years to come.

ACKNOWLEDGEMENT This work was supported by the Air Force Office of Scientific Research, Air Force Systems Command, USAF, under Contract F49620-86-C-0010 and the National Science Foundation, Grant CHE-8318810-02. The United States Government is authorized to reproduce and distribute reprints for governmental purposes notwithstanding any copyright notation thereon.

REFERENCES

1.   West, R.; Fink, M. J.; Michl, J., 14th Organosilicon Symposium, Duke University, Durham, North Carolina, March 1981; West, R.; Fink, M. J.; Michl. J. Science (Washington, D.C.) 1981 **214**, 1343.

2. Masamune, S.; Hanazawa, Y.; Murakami, S.;
    Bally, T.; Blount, J. F. J. Am. Chem. Soc.,
    1982, 104, 1150.

3. Brook, A. G., 14th Organosilicon Symposium,
    Duke University, Durham, North Carolina, March
    1981; Brook, A. G.; Abdesaken, F.; Gutekunst,
    B.; Gutekunst, G.; Kallury, R. K. J. Chem.
    Soc., Chem. Commun., 1981, 191.

4. For reviews on disilenes see West, R., Angew.
    Chem. Int. Ed. Engl. 1987, 26, xxx ; West, R.,
    Science (Washington, D.C.) 1984, 225, 1109;
    West, R., Pure Appl. Chem., 1984, 56, 1.

5. For a recent review of silenes see Brook, A.
    G.; Baines, K. M. in Adv. Organomet. Chem.,
    Academic Press, Inc., New York, N. Y., 1986,
    Ch. 2, pp 1-44.

6. Shepherd, B.; West, R., unpublished studies.

7. Watanabe, H., private communication.

8. Fink, M. J.; De Young, D. J.; West R.; Michl,
    J. J. Am. Chem. Soc., 1983, 105, 1070-1071.

9. Fink, M. J.; Haller, K. J.; West, R.; Michl, J.
    J. Am. Chem. Soc., 1984, 106, 822-823.

10. Michalczyk, M. J.; Fink, M. J.; Haller, K. J.;
    West, R.; Michl, J. Organometallics, 1986, 5,
    531-538.

11. Kirichenko, E. A.; Ermarkov, A. I.; Samsonova,
    I. N.; Russ. J. Phys. Chem. 1977, 51, 146;
    Kudo, T.; Nagase, S. J. Am. Chem. Soc., 1985,
    107, 2589; Barharach, S. M.; Streitweiser, A.
    J. Am. Chem. Soc., 1985, 107, 1197; Jemmis, E.
    D.; Kumar, R.V.V.P.; Kumar, N.R.S. J. Chem.
    Soc., Dalton Trans., 1987, 271; Silaghi-
    Dimitrescu, I.; Haiduc, I. Inorg. Chim. Acta,
    1986, 112, 159; Brenstein, R. J.; Scheiner, S.
    Int. J. Quantum Chem. 1986, 29, 1191.

12. Grev, R. S.; Schaefer III, H. F. J. Am. Chem.
    Soc. in press.

13. West, R.; Gillette, G. R.; Yokelson, H. B.,
    Inorg. Syn. in press.

14. Yokelson, H. B.; Maxka, J.; Siegel, D. A.;
     West, R. J. Am. Chem. Soc. 1986, 108,
     4239-4242.

15. Michalczyk, M. J.; West, R.; Michl, J. Organo-
     metallics, 1985, 4, 826-829.

16. Yokelson, H. B.; Millevolte, A. J.; Adams, B.
     R.; West, R. J. Am. Chem. Soc., 1987, xxx,
     xxxx.

17. Maciel, G. E. in "Nuclear Magnetic Resonance
     Spectroscopy of Nuclei Other than Protons"
     (Axenrod, T.; Webb, G. A. eds.), Ch. 12, Wiley,
     New York, 1974.

18. (a) Michalczyk, M. J.; West, R.; Michl, J.
     J.C.S. Chem. Comm., 1984, 1525-1526. (b)
     Watanabe, H.; Tabei, E.; Goto, M.; Nagai, Y.
     ibid., 1987, 522-523.

19. Yokelson, H. B.; Millevolte, A. J.; Gillette,
     G. R.; West. R., J. Am. Chem. Soc., to be
     submitted (full paper).

20. Another conceivable intermediate is a
     pi-complex between the disilene and an oxygen
     molecule.

21. Another possible pathway, the cleavage of 8 to
     R₂Si=O and recombination, has been excluded
     by trapping experiments. See ref. 10.

22. Yokelson, H. B.; Millevolte, A. J.; Gillette,
     G. R.; West, R. J. Am. Chem. Soc., to be
     submitted (communication).

23. West, R.; De Young, D. J.; Haller, K. J. J. Am.
     Chem. Soc., 1985, 107, 4942-4946.

24. Masamune, S.; Murakami, S.; Tobita, H. J. Am.
     Chem. Soc., 1983, 105, 7776-7778.

25. Cremer, D.; Kafke, E. J. Am. Chem. Soc.,1985,
     107, 3800.

PART VI

SILICON-OXYGEN POLYMERS AND MATERIALS

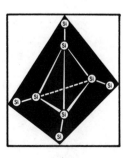

Chapter 27

"Silicones — past, present and future"

Donald R. Weyenberg — Dow Corning Corporation, Midland, Michigan 48640.

Polysiloxanes, commonly known in the trade as "silicones", are the major commercial products based on organosilicon chemistry. Since their introduction into commerce in 1943 with the formation of Dow Corning Corporation, these unique polymers have become a major new class of performance materials with a range of applications which span both our everyday experience and every corner of our economy.

This growth to the current multibillion dollar industry has been driven by a steadily expanding array of applications for an ever more versatile range of materials - - a range of materials made possible by an expanding base of materials science. The science and technology have grown simultaneously and synergistically for this field with that interactive cycle of science, invention and innovation spurring progress in both.

The exponential growth in scientific publications from less than 20 per year in the 1930's to 4000 per year today is matched by a very parallel growth in U. S. patents from less than 10 per year in the 1930's to 3000 per year today. Data from chemical abstracts for the last 10 years suggest that silicon chemistry is even more popular with the inventor than with the scientist. Silicon is mentioned in over 3% of their patent abstracts, but in somewhat less than 1% of their

other abstracts. The best measure of innovation is value in
the marketplace. Although comprehensive figures on the market
value of silicones are not available, it has clearly enjoyed
an exponential growth.

My focus is on the recent expansions of this broad frontier of
materials technology. However, a brief recap of the first
three and a half decades of industrial growth will place these
latter advances in perspective. The first era which extended
into 1960 concentrated on the classic silicones -- the polydimethylsiloxane fluids and heat-cured elastomers and the
methyl and phenylsiloxane resins for coatings and binders. It
was the era of basic monomer processes development. It was
the era in which the unique physical and chemical behavior of
silicones, particularly polydimethylsiloxanes, were being
detailed, and this understanding was leading to that broad
pattern of applications. These uses ranged from mechanical
application where stability in hostile environments was the
attribute, to surface treatments where low surface energy was
the predominate feature, to cosmetic and biomedical
applications taking advantage of the biological inertness of
these materials. But of more fundamental significance, the
science was being developed for an even broader expansion.

The next era, those years up through the mid-1970's, was
dominated by the second generation silicones. The explosive
growth during this period was fueled by several significant
advances including fluorosilicones and silicone-polyether
surfactants. But two separate and very fundamental expansions
of the science were to have the dominate impact.

It was the era for tailoring siloxane structures. The control
and chemistry of polymer end-groups was being added, and this
was to have a dramatic impact on the industry. Tying low
molecular weight and therefore flowable polymeric molecules
together into strong elastomers requires very precise ways of
connecting these polymer ends. Siloxane polymerization methods
provide access to a variety of end-groups and the chemistry
for tying those ends into networks was developed. The
chemistry was generally one of three types -- a terminal
silanol displacing acyloxy, amide, or alkoxy group from a
polyfunctional silane; a polyfunctional acyloxy, amide or
alkoxy siloxy ended polymer reacting with water; or a
platinum-catalyzed coupling of hydride and vinyl functional
polysiloxanes. The result being that liquid prepolymers could
be converted in place to strong elastomers by any one of
several techniques -- by adding a catalyst, or by heating, or
by exposing to moisture. This technology had a dramatic
impact· on the industry, because it offered the product
designer an opportunity to simultaneously design for the

ultimate material property profile as well as for the
application technique.

The second major advance with great impact on the industry was
organofunctional silicon chemistry -- the selection rules on
how to place both organic and silicon functionality and
reactivity into a molecule and realize the separate
chemistries of each. These molecules literally bridge the
domains of organic and inorganic chemistry and their major
industrial use which emerged during the 1960's was to do just
that -- to act as coupling agents to adhere inorganic fillers
and fibers into organic plastic matrices. Silicones with
organofunctional side-groups also appeared which offered the
designer one more degree of freedom. Many new surface
treatments -- polishes or textile finishes or cosmetic
ingredients -- took advantage of these functional sites. Much
of the explosive growth of the 1960's and 1970's came from new
product lines resting on these new technologies.

A survey of the newer commercial thrusts should begin with
some important new applications for that workhorse and still
the backbone of the industry, polydimethylsiloxane. These new
uses build from a steadily increasing knowledge of the
physical and chemical properties of this unique polymer.
Silicones are increasingly the fluid of choice to replace
PCB's in fluid-filled transformers. In addition to the many
evaluations of dielectric behavior, the flammability studies
summarized by Lipowitz(1) and Orbeck(2) were critical, because
it is the lower fire hazard that differentiate the silicone
from other replacement candidates. The inorganic crust which
forms on a burning silicone liquid -- or a silicone solid --
reduces the burning rate and instantaneous heat release and
significantly lowers the hazard associated with a given fire.
The service and safety experience has been summarized by
McClintick & Orbeck(3).

This increased understanding of flammability and the fire
hazards of polydimethylsiloxane fluids and elastomers is key
to an array of applications where fire safety is an issue.
These range from fabric coatings to cable insulation to these
key fluid applications. Also, recent studies by Frye(4) have
demonstrated that certain silicone formulations are very
effective flame retardants for polyolefins, particularly
polypropylene. This may represent another case where the
attributes of a silicone can be realized in an organic system.

Heat transfer is another growing application for polydimethylsiloxanes. Syltherm® heat transfer liquid, which is
specifically designed for industrial systems, is finding wide
acceptance for demanding applications above 600°F. Of
particular importance in this application is the ultimate chain

scission mode of decomposition at the upper temperature limits. Excursions above recommended temperatures do not result in film formation and fouling of the heat transfer surface as is common with organic systems(5).

Silicones, particularly the fluid polymers, have enjoyed dramatic growth in personal care products during the 1980's following the early publications by Todd and Byers(6) and by Scott and Turney(7). The volatile permethylcyclosiloxanes are the carriers of choice in many applications, particularly antiperspirants. The conventional polydimethylsiloxanes as well as polymers with amino or polyol side-groups are used widely in skin and hair care products for their unique surface characteristics. The in vivo behavior of polydimethylsiloxane on skin has been characterized by Klimisch and Chandra(8).

All of these broad industrial and consumer applications rest on an improving knowledge of the environmental consequence and fate of these fluid polydimethylsiloxanes. In fact, the silicones are replacing fluids -- PCB's -- which are very efficacious in transformers but unacceptable in the environment. These studies have been summarized by Frye(9).

The health-related uses of the classical silicones continue to grow, with many new applications based on unique designs which take advantage of the basic characteristics of polysiloxanes. Drug delivery systems take advantage of both permeability and biocompatibility. Since implantable devices were first described by Folkman and Long(10), this concept of steady, zero-order delivery of a drug from a reservoir through a permeable membrane has expanded into innumerable device forms. The variety of these systems along with the underlying principles have been summarized by Chien(11) and the more practical aspects of selecting a silicone system are summarized by Rankin(12). A most popular and convenient device form is transdermal where the drug is delivered through the skin from a reservoir contained in a skin patch.

Tough, biocompatible elastomers allow new device designs; tissue expanders are a good example. Our skin expands to accommodate internal pressure, as in reaction to a bruise or to pregnancy. The purpose of this device, a hollow implantable sack with a valve through which saline can be injected, is to generate new skin for skin grafts. Implanted adjacent to the wound and expanded over several weeks by periodic injection of saline, the surgeon has a new crop of skin to cover the wound. Commercialization followed a basic patent in the late 1970's by Radovan and Shulte(13), and has become an accepted technique for the plastic and reconstructive surgeon. Recent experience has been summarized by Austad(14).

Silicone foams and gels are finding application in improved wound management. Hughes and Wood(15) first described the use of a foamed-in-place silicone as a superior dressing for deep, open, granulating wounds. The catalyzed liquid silicone is poured into the open wound and the resulting foam provides a nearly ideal dressing. It allows oxygen to reach the wound surface while protecting that surface with a minimum of pressure. It can be temporarily removed for wound medication or irrigation, or can be easily replaced by foaming a new dressing as the wound shrinks in size. The treatment is available in Europe and clinical experience has been summarized by Harding(16).

Another promising application, also in use in Europe, uses a very soft silicone gel in the management of hypertrophic scarring which is often associated with thermal injury. This phenomenon was first observed by Perkins, et al(17). Recent studies by Quinn, et al(18) show rather dramatic results including the reversal of even well developed hypertrophic scars. Although silicone fluids have been used in various burn treatments for many years, the mechanism by which they aid the burn healing process is not understood.

New commercial thrusts also include several of the most recent second generation silicones; this process of designing silicones for both their ultimate properties and ease of application remains as one of the important growth vectors of the industry.

The significant advances for silicones in the construction industry have been summarized by Klosowski(19). Buildings have been sealed against the weather with silicones since these sealants appeared in the early 1960's. The newer sealants have been addressing those difficult building joints with high movement as well as the even tougher issue of our highways where joint failure is a major problem. High elongation, low modulus, high adhesion, cure-in-place sealants are finding widespread application in both new highway construction and in the repair of failed joints(20). This application is typical of many in the construction industry where a very sophisticated and expensive silicone sealant is replacing a less expensive material, in this case asphalts, because the total cost over the lifetime of the structure is dramatically reduced.

A most dramatic application is structural glazing. Beginning with a few demonstrations in the 1960's, silicone sealant adhesives slowly became the method of choice for securing the exterior glass windows to high-rise structures. This technology, in fact, allowed the all-glass exterior appearance which is increasingly dominating our city skylines. The

fascinating history of this development is covered by
Klosowski(19).

Silicone coatings are also gaining acceptance as a premium
roofing membrane. Applied either as a moisture curing
one-part or as a two-part system, these elastomeric silicone
coatings are often the final outer weatherproofing membrane in
a roofing system which incorporates layers of polyurethane
foam for insulation. Silicones are also playing an important
role in the increasingly popular fabric roof systems(21).
Silicone resin coated glass cloth has some distinct advantages
over the more commonly used fluorocarbons. The silicones can
transmit more light, thus allowing better aesthetics with more
natural vegetation within the enclosure.

Liquid silicone rubber represents a major advance in second
generation silicones for the industrial market. They are
pumpable liquids which are processable by fast cycle, liquid,
injection molding to high performance silicone elastomers.
They use the platinum-catalyzed coupling of a siloxane hydride
with a vinylsiloxane for the network forming chemistry; a
chemistry which has been the subject of some very excellent
recent studies. Chandra, Lo, Lappert and Hitchcock(22) have
characterized the siloxane soluble catalysts of the patent
literature as platinum zero complexes of vinylsiloxanes.
This chemistry along with silanol crosslinking systems has
been used by Macosko(23) and by Mark(24) to design networks
with precise three dimensional geometry to determine how model
networks behave under load, both to test and to extend the
theories of rubber elasticity. Mark has further examined
networks from mixtures of short and long chains which give a
bimodal distribution of chain length between crosslinks. As
first demonstrated by the technologist, these bimodal networks
can rather dramatically change the dynamic mechanical behavior
of the network. Not only are these studies very relevant to
the practical issues of product design, but they again
demonstrate the very real synergy between the science and
technology. The practical application of this technology has
been described by Cush(25) and by Macosko and Lee(25).

Foamed silicones represent another growing application.
Silicone foams based on the organotin-catalyzed reaction of a
siloxane hydride and silanol have been in use since the very
early 1960's; but the recent growth is based on the
platinum-catalyzed version of this hydrogen elimination
chemistry described by Smith(26) in the 1970's. Subsequent
studies by Lee and coworkers(27) have demonstrated the great
versatility of this chemistry and the ability to control both
the cure rates and foam characteristics such as cell geometry,
density, and modulus. The major application has been in
fire-resistant penetration seals. Fire rated walls contain

penetrations for passing cables and pipes and ducts; sealing
these penetrations so that they do not become the carriers of
fire and smoke through a structure is a continual problem.
Fire stop silicone foam has become the standard seal for
nuclear plants and its use in other power or manufacturing
plants and in commercial structures is growing. Liquid
silicone foams are becoming a valuable companion product line
to the various liquid silicone elastomers.

Protection of electronic devices has been a key application
for specialty silicones and these applications continue to
keep pace with the rate of device development. The silicones
range from resinous circuit board coatings to encapsulants
with the silicone gels representing a unique solution. These
very lightly crosslinked polydimethylsiloxanes, cured in place
via the platinum-catalyzed coupling of siloxane hydrides and
vinylsiloxanes, are elastomeric but with an extremely low
modulus. They provide the stress-relief characteristics of a
fluid with the nonflow of an elastomer for circuit
protection(28) over an extremely wide temperature range, from
-80°C to 200°C. Optical fibers require a soft coating for
protection against a variety of loss mechanisms, and a UV
curing silicone is providing an answer(29).

The development of new siloxane structural forms is an equally
important thrust for this industry, and several new materials
illustrate this expanding dimension. A most promising new
form for polydimethylsiloxane is a latex -- a latex which
deposits a coherent and fully cured elastomer on removal of
the water. This development builds on the emulsion
polymerization of polydimethylsiloxane, a unique process where
a surfactant catalyst, like dodecylbenzenesulfonic acid,
serves the dual role of siloxane polymerization catalyst and
supporting surfactant. This polymerization technique gives
hydroxyl-ended polydimethylsiloxanes of high molecular weight
in aqueous emulsion(30). Recent studies by Saam and
coworkers(31) describe the interaction of these emulsion
polymers with aqueous silicates to form networks within each
colloidal particle. Furthermore, they describe the unique
phenomenon which occurs on removal of the water and
coalescence of the particles; these elastomeric colloidal
particles interact with one another and with added colloidal
silica, presumably by silanol condensation, to deposit
continuous, crosslinked, reinforced, elastomeric films of
polydimethylsiloxane. These systems are appearing as
easy-to-use, water-based, silicone caulks and as new high
performance exterior coating systems. This technology
represents a new dimension for conventional silicones.

One of the fascinating new technologies of the 1980's is an
antimicrobial surface treatment, first described in the early

1970's by Isquith and coworkers(32). Surfaces treated with quaternary ammonium functional silanes destroy a wide range of microorganisms on contact. This is not a slow release phenomenon; bacteria are destroyed on contact with the surface. Trimethoxysilylpropyldimethyldodecylammonium chloride was selected for commercial development and became the first organosilicon product designed for control of microbial growth. The mechanism by which this treated surface destroys microorganisms has been studied by Speier and Malek(33) who show that this effect is due to a particular density of charge on the surface and the disruption that this charge has on the biomembranes; specifically, the microbial cell walls. The function of the silane is to permanently alter the surface charge. A recent review by White and Gettings(34) illustrates the breadth of this technology for controlling microbial growth. The treatment is applied to a variety of surfaces ranging from articles of clothing to carpet for home or institutional use.

Silica-siloxane resins, commonly known as abrasion resistant coatings, certainly deserve a key spot on any list of recent silicone expansions. They can be described in many ways -- as silicone inorganic hybrids, as silicone-silica block copolymers, or as a modified sol-gel. All are descriptive. They are silica-silsesquioxane copolymers prepared by the hydrolysis of an alkytrialkoxysilane in the presence of colloidal silica in an acidic aqueous alcoholic media. The resulting highly functional resin with up to 70% silica is stable in solution, but cures rapidly via silanol condensation on removal of solvent. The resulting film is an exceedingly hard and scratch resistant coating. The 10 to 30 nanometer silica particles are uniformly dispersed in the cured film. These resins were first described in a landmark patent by one of the pioneers of silicone resins, Hal Clark(35) in 1976. Abrasion resistant coatings for plastics are still the major applications and are described in a more recent article by Vincent, Kimball and Boundy(36). These resins and dielectric gels illustrate the range of useful, problem-solving structures from polymethylsiloxanes.

Silicone-organic hybrids are a fourth growth vector for polysiloxanes. Silicone-polyether surfactants and silicone-polyester coatings are well established and enduring applications. The adaptation of silicone moisture curing chemistry to silane-functional organic polymer systems is also in use. The Sioplas® process for grafting vinyltrialkoxy- silane to polyethylene is rather widely used to provide a moisture crosslinking polyethylene(37). Polyether sealants which cure via alkoxysilyl end-groups are in commerce in Japan(38), and the patent literature is describing an increasing array of organic resins and coatings and sealants which use these silicone curing chemistries. The use of silicones in a variety of blends with organic polymers is also a growing commercial trend. Covulcanizing blends of silicone and ethylene-propylene rubber were discussed by Wada(39) at the last Organosilicon Symposium and reviewed more recently by Mitchell(40). Navitove, Crosby, and DeAngelis(41) have described interpenetrating networks by adding a liquid, heat-curable silicone to a thermoplastic and curing the silicone during injection molding of the plastic. A unique copolymer, described at the Fifth Organosilicon Symposium in Karlsruhe, is also finding increased application(42).

The future commercial thrusts of organosilicon chemistry will come from today's scientific frontiers, all of which are well represented at this Symposium. Silicone-organic composites seem poised for major impact, based mainly on the block copolymers of polydimethylsiloxane and a wide range of organic systems. Polydimethylsiloxane is an ideal soft phase and can confer many of the beneficial attributes of silicones on these microphase composites. A very large number of these systems are described in the recent literature(43). McGrath will provide a comprehensive review of this frontier at this Symposium.

Synthons, the use of organosilicon chemistry in organic synthesis, may be the surest prediction for new commercial opportunities. Approximately half of the current papers on organosilicon chemistry deal with this topic. It is well represented at this Symposium. A process with significant industrial promise, group transfer polymerization, will be described by Webster.

Silylated surfaces represent another very active frontier of this science with significant commercial potential. This technology has been the topic of a series of symposia organized by Professor Leyden(44), the third to be held at Colorado State University in June, 1987. The exciting opportunities included in this rapidly moving science and rapidly evolving technology are catalysis, biosurfaces, modified electrodes and electronic surfaces, and separations. Molecules tethered to a surface open a range of opportunities, and functional silanes offer a very convenient way of tethering a variety of structures to an equally broad variety of surfaces.

The use of this chemistry for the preparation of new inorganic structures will certainly be a future commercial driving force for this field. Preceramic polymers -- polymers which can be formed by conventional polymer processing like spinning or molding and then converted via pyrolysis into a ceramic -- are excellent examples. Although isolated examples of this

technique date back to the 1960's, the current extensive
activity in this field flows directly from the pioneering work
of the late Professor Yajima and coworkers(45). He described
his process for silicon carbide fiber at the Fifth
Organosilicon Symposium in 1978 and this fiber is available in
Japan. Many complementary approaches to preceramic polymers
based on polysilanes or polysilazanes have been
developed(46,47,48,49,50) and are being applied to the
preparation of silicon carbide and silicon nitride ceramics.
Ceramic-in-ceramic composites -- materials with the strength
and stiffness and heat resistance of a ceramic like silicon
carbide but with the fracture toughness of fiber reinforced
composites -- are a tantalizing and realistic goal from this
technology(51).

The linear polydiorganosilanes are also of interest in several
other applications(50). As close relatives of amorphous
silicon, they show unique electronic behavior like
photoconductivity. Also, their rapid photodepolymerization
makes them ideal candidates for positive photoresists, a topic
to be covered by Miller at this Symposium. These new
materials thrusts will certainly provide commercial driving
forces for organosilicon chemistry over the next several
years; and polysilanes which are central to both the
preceramic and electronic applications will be very much at
the forefront of that thrust.

In this brief review of industrial applications, I have tried
to show how this array of specialty siloxane polymers known as
silicones continues to expand and why they continue to be the
materials of choice for demanding applications. Also, a quick
overview of current scientific frontiers illustrates our
optimism for future industrial growth. Finally, I have tried
to illustrate the critical role which this interactive cycle
of science, invention, and innovation has had both on our
industry and on this exciting and expanding field of
chemistry.

REFERENCES

1. Lipowtiz, J.; Ziemelis, M. J.; Journal of Fire and
   Flammability, 7, 504 (1976).
2. Orbeck, T.; Ninth International Conference on Fire
   Safety, (Jan. 16, 1984).
3. McClintick, B. R.; Orbeck T.; IEEE Conference, (September
   28, 1986).
4. Frye, R. B., U. S. Pat. 4,387,176 (1983).
5. Brooks, K., Chemical Week, 56 (Jan. 7, 1987).
6. Todd, C., Byers, T.; Cosmetics and Toiletries, 91, 29-32
   (1976).
7. Scott, R. J.; Turney, M. E.; J. Soc. Cosmet. Chem., 30,
   137-156 (1979).

8. Klimisch, H. M.; Chandra, G.; J. Soc. Cosmet. Chem.,
    37, 73-87 (1986).
9. Frye, C. L.; Soap/Cosmetics/Chemical Specialities,
    32-35 (Aug., 1983).
10. Long, D. M., Jr.; Folkman, M. J., U. S. Pat. 3,279,996
    (1966).
11. Chien, Y.; Pharmaceutical Technology, 9(5), 50-56
    (1985).
12. Rankin, F. S.; Interphex 86 Conference, Cahners
    Exhibitions Ltd., UK, Chapter 4, 7-16 (1985).
13. Radovan, C.; Schulte, R. R.; U. S. Pat. 4,217,889 (1980).
14. Austad, E. D., Thomas, S. B., Pasyk, K.; Plastic &
    Reconstructive Surgery, 78(1), 63-67 (1985).
15. Wood, R. A. B.; Hughes, L. E.; Abstract British Journal
    Surgery, 61(11), 921 (1974).
16. Harding, K.; Advances in Wound Management, Turner, T. D.;
    Schmidt, R. J.; Harding, K. G., Ed.; John Wiley and Sons,
    Ltd., 41-51 (1986).
17. Perkins, K., Davey, R. B.; Wallis, K. A.; Burns, 9,
    201-204 (1983).
18. Quinn, K. J.; Evans, J. H.; Courtney, J. M.; Gaylor, J.
    D. S.; Burns, 12, 102-108 (1985).
19. Klosowski, J. M.; "Construction Sealants", Marcel Dekker,
    Inc., in press.
20. Spells, S.; Klosowski, J. M.; American Concrete
    Institute, 1(1), SP-70, 217-235 (1981).
21. Melnick, S.; Building Design & Construction, 140-144
    (May, 1986).
22. Chandra, G; Lo, P. Y.; Hitchcock, P. B.; Lappert, M. F.;
    Organometallics, 6, 191 (1987).
23. Macosko, C. W.; Benjamin, G. S.; Pure & Applied
    Chemistry, 53(8), 1505-1518 (1981).
24. Mark, J. E.; Ning, Y. P.; Polymer Engineering and
    Science, 25(13), 824-827 (1985).
25. Cush, R. J.; Plastic Rubber Int., 9(3), 14-17 (1984);
    Macosko, C. W.; Lee, L. J.; Rubber Chem. and Tech. 58,
    436-448 (1984).
26. Smith, S. B.; U. S. Pat. 3,923,705 (1975).
27. Lee, C. L.; Ronk, G. M.; Spells, S.; Journal of Cellular
    Plastics, 29-33 (1983).
28. Otsuka, K., Takeo, Y.; Tachi, H.; Ishida, H.; Yamada, T.;
    Kuroda, S., International Electronics Packaging Society
    (Nov., 1986).
29. Dehli, G. L.; Lee, C. L.; Lutz, M. A., Suzuki, T.;
    Optical Fiber Material and Properties, S. R. Nagel,
    G. Sigel, J. W. Fleming and D. A. Thompson, Ed., In
    press.
30. Saam, J. C.; Huebner, D. J.; Journal of Polymer Science,
    20, 3351-3368 (1982).
31. Saam, J. C., Graiver, D.; Baile, M.; Rubber Chemistry
    and Technology, 54(5), 976-987 (1981).

32. Isquith, A. J.; Abbott, E. A.; Walters, P. A.;
     Applied Microbiology, 24(6), 859-863 (1972).
33. Speier, J. L.; Malek, J. R.; Journal of Colloid and
     Interface Science, 89(1), 68-76 (1982).
34. White, W. C.; Gettings, R. L.; Silanes, Surfaces and
     Interfaces, Leyden, D. E., Ed., 107-140, Gordon and
     Breach Science Publishers, New York (1986).
35. Clark, H. A.; U. S. Pat. 3,986,997 (1976).
36. Vincent, H. L.; Kimball, D. J.; Boundy, R. R.;
     Polymer Wear and Its Control, ACS Symposium Series 287,
     Lee, L. H. Ed., 129-134 (1985).
37. Scott, H.; Humphries, J.; Modern Plastics, 50, 82 (1973).
38. Isayama, K.; Hatano, I.; U. S. Pat. 3,971,751 (1976).
39. Wada, T.; Organosilicon and Bioorganosilicon Chemistry,
     Saburai, Ed., Ellis Horwood Ltd., 281-292 (1985).
40. Mitchell, J. M.; SAE International Congress and
     Exposition (Feb., 1987).
41. Navitove, M. H.; Crosby, J. M.; DeAngelis, P. J.;
     Plastics Technology, 57-61 (Aug., 1985).
42. Bruner, L. B.; Chemical Week, 29 (June 2, 1976);
     Marquart, K.; Kreuzer, F. H.; Wick, M.; Angew. Macromol.
     Chem., 58, 243-57 (1977).
43. Kilic, S.; Summers, J. D.; Elsbernd, C. S.; Arnold,
     C. A.; Pullockaran, J; McGrath, J. E.; Polymer
     Preprints, 28(1), 398 (1987).
44. Leyden, D. E.; Silanes, Surfaces and Interfaces,
     Gordon and Breach Science Publishers, New York (1986).
45. Yajima, S.; Hayashi, J.; Omori, M; Chem. Lett., 9, 931-4
     (1975).
46. LeGrow, G. E.; Lim, T. F.; Lipowitz, J.; Reaoch, R. S.;
     Ceramic Bulletin, 66(2), 363-367 (1987).
47. Seyferth, D.; Wiseman, G. H.; Poutasse, C. A.; Schwark,
     J. M.; Yu, Y. F.; Polymer Preprints, 28(1), 389 (1987).
48. Laine, R. M., Blum, Y. D.; Chow, A.; Hamlin, R.;
     Schwartz, K. B., Rowecliffe, D. J.; Polymer Preprints,
     28(1), 393 (1987).
49. Baney, R. H.; Chandra, G.; Encyclopedia of Polymer
     Science and Engineering, John Wiley & Sons, Inc., New
     York, Preceramic Polymers, In press.
50. West, R.; Maxka, J.; Sinclair, R.; Cotts, P. M.;
     Polymer Preprints, 28(1), 387 (1987).
51. Mah, T-I; Mendiratta, M. G.; Katz, A. P.; Mazdiyasni, K.
     S.; Ceramic Bulletin, 66(2), 304-317 (1987).

Chapter 28

Mechanistic features of processes leading to linear siloxane polymers

Julian Chojnowski − Center of Molecular and Macromolecular Studies of the Polish Academy of Sciences, Boczna 5, 90-362 Lodz, Poland.

INTRODUCTION

Chemical processes leading to linear siloxane polymers could be classified as either polycondensation of bifunctional silanes or ring opening polymerization of cyclic siloxanes according to the way in which the polymer chain is formed. However, it is more convenient here to define these processes as either anionic or cationic depending on the structure of intermediates which appear in these reactions.

ANTONIC PROCESSES

Silanolate ion appears to be a common intermediate in the anionic processes used in synthesis of polysiloxanes [1]:

$$(1)\ \text{wSiO}^- +\ \begin{cases} \overset{O}{>Si\diagdown Si<} \longrightarrow \text{wSiOSi SiO}^-; \text{ polymerization} \\ \text{XSiw} \longrightarrow \text{wSiOSiw} + X^-; \text{ polycondensation} \end{cases}$$

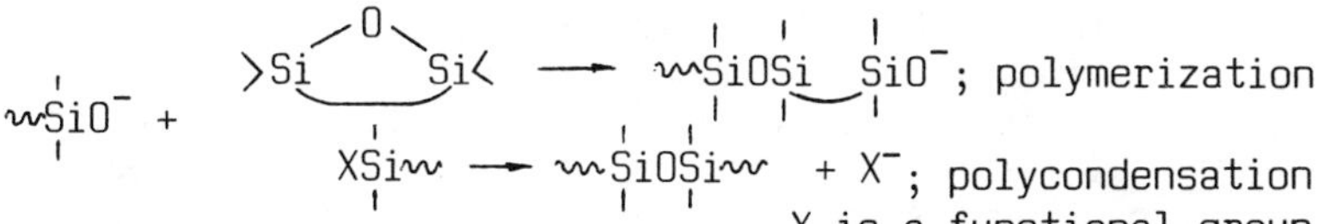

X is a functional group

In the above reaction systems the silanolate ion shows a considerable tendency to associate with electron deficient species, therefore various association phenomena dominate the reaction pattern. Two types of interaction involving the silanolate will be considered, interaction between ions, and association of ionic centers with siloxane chains and rings.

Ion-ion association

Synthesis of polysiloxanes is usually performed in a medium of low polarity, which also shows a low ion solvation ability. Two types of ion-ion interaction are important in these processes:

(2)
$$\text{ion pair formation; } \sim\!\!\overset{|}{\text{SiO}^-} + M^+ \rightleftharpoons \sim\!\!\overset{|}{\text{SiO}^-}M^+; \quad M^+ \text{ is alkali metal or}$$
$$\text{ion pair association; } n\sim\!\!\overset{|}{\text{SiO}^-}M^+ \rightleftharpoons \sim\!\!(\overset{|}{\text{SiO}^-}M^+)_n \; ; \quad \text{q.ammonium}$$
$$\text{(aggregation)} \qquad\qquad\qquad\qquad\qquad\qquad \text{ion}$$

Although free ions show the highest reactivity their concentration is most often kinetically insignificant. Silanolate ion pairs are active propagation centers in the polymerization or are intermediates in polycondensation. However, they appear in equilibrium with their aggregates which, being unreactive, form dormant centers. Since the equilibrium lies usually well to the side of the aggregates, the kinetic law and the rate depend to a considerable extent on the aggregation state and on the strength of these associates, which are different for various counter-ions, solvents, additives, and polysiloxane structures. For example, in PDMS (polydimethylsiloxane) bulk, dimer of chain end potassium silanolate is the dominating species [2] (scheme 3), while tetramer was found to be the most important for sodium polysilethylenesiloxane in n-heptane-dioxane 95:5 v/v [3].

(3)
$$\sim\!\!\overset{Me}{\underset{Me}{\text{SiO}}}\overset{\cdot\cdot K^+}{\underset{K^{+\cdot\cdot}}{\diagdown}}\overset{Me}{\underset{Me}{\text{O Si}}}\!\!\sim$$

Kinetics of the anionic polymerization of cyclosiloxanes has been studied for a long time since the pioneering works of Grubb and Osthoff [4]. Results of these studies were comprehensively reviewed by Wright [2]. Although the importance of the ion pair aggregation was pointed out only by few research groups [3,5-8], many observations are in full accord with the aggregation role [2]. In particular, the polymerization usually shows a fractional order with respect to the initiator and the rate increases strongly with the size of the counter-ion. The addition of small amounts of nucleophiles strongly increases the rate, which may be partly due to the loosening of the aggregates. Some physical properties of the system, in particular, very low electrical conductivity and dipole moment, as well as high viscosity of living oligomeric and polymeric species indicate also a high association state of the silanolate ion pairs. The ion pair association introduces many complications to kinetics of systems in which different ionic species exist, because mixed complexes (cross-aggregates) are formed. This is the case for initiation and copolymerization as well as for propagation in the presence of ionic additives. A theory predicting the effect of the cross-aggregation on the kinetics has been developed [3,8].

The phenomenon of the ion pair aggregation has also some implications in thermodynamics of the polymerization of cyclic siloxanes. Since chain scrambling during the process occurs, in principle, at random, the normal Flory distribution of the chain size should be achieved at the equilibrium [9]. However, such is not a case for silanolate ended living PDMS polymer. Shorter chains are thermodynamically favored, because the formation of the silanolate aggregates involving both ends of the same chain is possible, which makes lower the thermodynamic potential of the over-all living polymer system [10].

(4)

Interaction of silandlate group with siloxane ring and chain

Polydimethylsiloxane chains and larger rings are flexible enough to acquire a conformation making possible their multi-center interaction with the silanolate counter ion [11]. These interactions are much weaker than those of corresponding cations with polydentate ether ligands. Cyclic siloxanes are not efficient as crown ligands because of the unfavorable polyoxymethylene-like structure of siloxane chain, and also because of hindrance caused by methyl substituents and low basicity of siloxane oxygen. Nonetheless the cation-siloxane interaction has a pronounced effect on the kinetics of the polymerization. It makes lower the energy barrier of siloxane bond cleavage by the silanolate ion pair (scheme 5). Therefore

(5)

a considerable increase of the rate of cleavage of larger rings and also chains is observed. It is purely kinetic effect i.e. it accelerates to the same extent the polymerization of larger rings and their formation by back-biting [12]. Rate of the chain scrambling is effectively increased as well.

This effect has important practical implications since it is an undesired phenomenon in the anionic polymerization of cyclic siloxanes performed under kinetically controlled regime.

In this mode of polymerization strained highly reactive six-
-membered ring monomer (D3) is used to carry out the polymerization in the absence of back-biting and chain scrambling.
Such a process, if quenched at a proper moment, gives a high
yield of the polymer of narrow molecular weight distribution
and is free of cyclic oligomers [2]. Certainly back-biting and
chain scrambling would be too fast to produce such a polymer
unless the silanolate-siloxane association was suppressed.
There are two ways of doing it. Introduction of nucleophiles
which interact more strongly than siloxane with the counter
ion is a common way. The role of the nucleophile is not only
to enhance the rate, but also to increase specificity of the
processes, since the polymerization of D3 is accelerated to a
much higher extent than other siloxane opening processes.
Using a hard counter-ion like Li+ in connection with a hard
nucleophile may be also profitable, since the rate enhancement
effect is likely to be more differentiated. The other way of
reducing the role of the silanolate-siloxane interaction is to
use a weakly interacting counter-ion like tetraorganoammonium
[13] or a cryptand [7].

Selectivity in siloxane bond making and breaking

The cleavage of polysiloxanes with a base occurs sometimes
with an amazing regioselectivity. It was shown earlier [14,15]
that siloxane bond in the neighborhood of lithium silanolate
groups is readily cleaved by butyllithium. More recently, when
attempting to study the base catalyzed polycondensation of
oligodimethylsiloxanediols in dioxane containing 2% w/w of water,
much to our surprise we found that the process was dominated
by intermolecular last unit migration [16]:

$$
\begin{array}{c}
\text{NaOH} \\
+ \text{ I}
\end{array}
$$

$$
(6) \quad HO[Si(Me)_2O]_5H \longrightarrow
\begin{cases}
HO[Si(Me)_2O]_{10}H + H_2O \\
[Si(Me)_2O]_5 + H_2O \\
HO[Si(Me)_2O]_4H + HO[Si(Me)_2O]_6H
\end{cases}
$$

At the beginning of the process the disproportionation of the
diol into its oligohomologues having one dimethylsiloxane unit
more or less was the exclusive reaction. Further inter-chain
exchange led to other oligohomologues. Condensation processes
proceeded at a rate lower at least by two orders of magnitude.
Thus the cleavage of the $\equiv Si-O-Si\equiv$ is strongly preferred
over the cleavage of the $\equiv SiOH$ group, although the former is
known to be thermodynamically much more stable.

The enhancement of the $\equiv SiOSi\equiv$ reactivity by adjacent
hydroxyl group is to some extent a kinetic phenomenon, because
this bond is also easier formed. Dimethylsilanediol is much

more reactive in the condensation than oligosiloxanediols,
which leads to an unusual mechanism for the polycondensation of
oligosiloxanediols. First a series of interchain exchange leads
to the formation of the reactive monomeric silanediol which
then is added by condensation to an oligosiloxanediol [17].
For example, the scheme of the formation of the tetramer-diol
from the dimer is the following:

(7) $2HO[Si(Me)_2O]_2H \longrightarrow HO[Si(Me)_2O]_3H + HOSi(Me)_2OH$

$$HO[Si(Me)_2O]_4H + H_2O$$

This pathway leaves no room for the formation of cyclic products. This is perhaps the explanation for a low yield of
cyclic oligomers in base catalyzed hydrolysis and polycondensation of bifunctional silicone monomers.

The kinetics of the last unit migration was recently
studied in isopropanol or THF + 2 w % H_2O [18]. Results did
not fit the elimination-addition scheme involving silanone
intermediate, nor did they correspond to classic SN2 substitution. An interesting observation was that the rate decreases
in the series of counter-ions; Li+ > Na+ > K+ > Cs+, which
indicates that the silanolate ion pair is the intermediate and
that the counter-ion plays a substantial role in the process.
It is supposed that a cyclic complex of the siloxanol and
siloxanolate is formed, which is then transformed into the
migration products. A contribution from the mesomeric structure
B of the complex would explain weakening of the bond to be
cleaved.

(8)

Although the transformation is perhaps a one step process, we
believe that siloxane bond cleavage is so advanced in the
transition state and the extent of the formation of new
siloxane bond so small that the transformation may be considered to proceed via a quasi elimination-addition pathway.

Some evidence for the transient formation of tricoordinate
silicon species having silicon doubly bonded to oxygen has
recently been provided in a solvolytic process in basic media
[19].

CATIONIC PROCESSES

Cationic processes leading to linear polysiloxanes most often involve primary or secondary oxonium ions $\equiv SiO(H+)H$, $\equiv SiO(H+)Si\equiv$ as intermediates and these processes are at least as important as the anionic ones.

Ring opening polymerization

It was about forty years ago, when the first contribution to the mechanism of the cationic polymerization of cyclic siloxanes was made [20]. Since that time the reaction has been extensively studied and comprehensively reviewed up to 1980 [2]. Two general schemes for this process have been considered: (i) acidolytic polymerization comprising alternation of opening of the monomer ring with a strong protonic acid and condensation in which the catalyst is recovered, and (ii) addition polymerization involving direct addition of the monomer to an active propagation center. Two research groups including ourselves proposed a dual mechanism comprising both above schemes [21,22].

Oligomer formation, classical kinetics, and component reactions have been the subject of studies in our laboratory. D_3 was preferred for this study, since its use made possible the investigation of the polymerization unobscured by back biting and chain scrambling. The polymerization was studied in methylene chloride or n-hexane solution using CF_3SO_3H or CF_3COOH as the initiator. The polymerization of D_4 was recently showed [23] to exhibit many common features with that of D_3, although it was much more complex. Results of kinetics of D_3 conversion [21] and oligomer formation [24] were in agreement with scheme 9.

a) Acidolytic and hydrolytic ring opening:

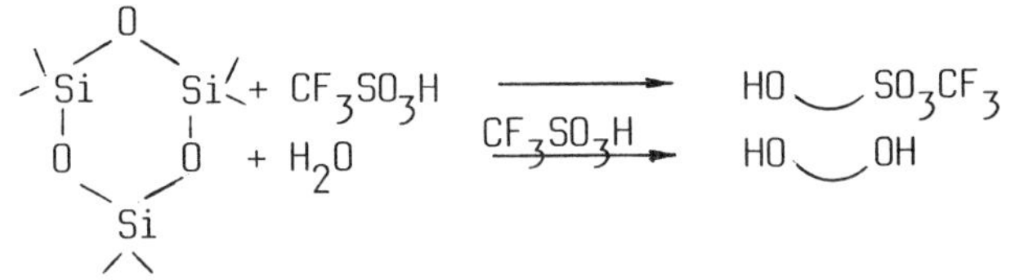

(9)   b) Homo and heterocondensation:(cyclization and chain coupling)

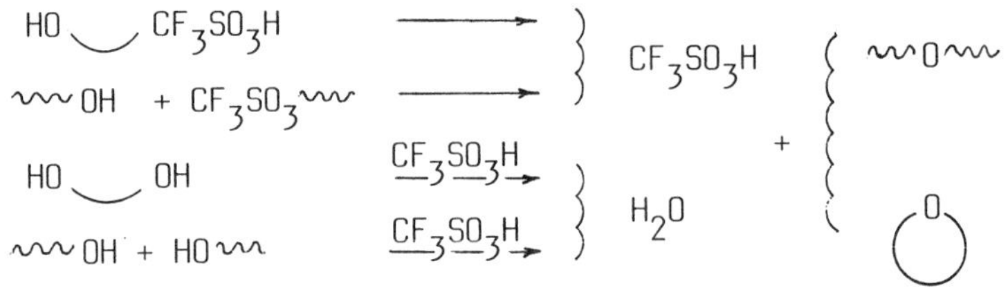

c) End group interconversion: $\sim SO_3CF_3 + H_2O \longrightarrow \sim OH + CF_3SO_3H$

d) Direct addition:

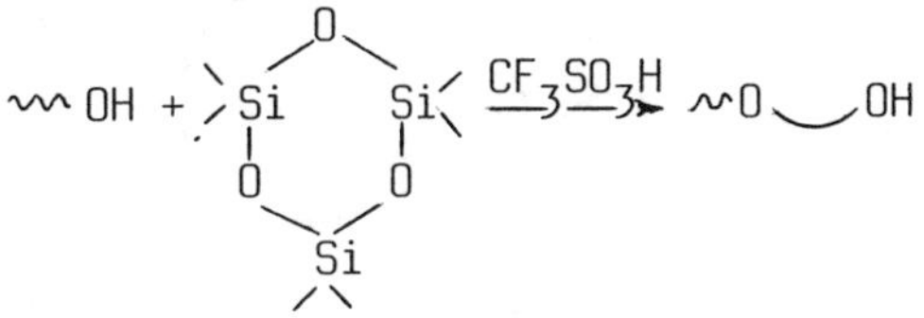

Part of the water effect is also due to hydrogen bonding association. Water changes the state of association which affects the kinetics [27].

Polycondensation reactions of model oligosiloxanes in the presence of acids

Studies of acid-catalyzed condensation processes of silanol end groups in polysiloxanes are important for a better understanding of the cationic polymerization of cyclic siloxanes. These reactions are also important as principal processes in hydrolytic polycondensation of bifunctional silicon monomers. Although much is known about the polycondensation of monomeric silanediols mostly due to the extensive research of Lasocki et al.[29,30] knowledge of the behavior of hydroxyl end groups of the polysiloxane chain is scarce in spite of some recent activity in studies of oligosiloxanols [31-33].

Cyclization versus chain coupling competition

Studying the behavior of model decamethylpentasiloxanediol. I in the presence of protonic acids (CF_3SO_3H, CH_3SO_3H, CF_3COOH) in a homogenous media (solution in CH_2Cl_2 or dioxane) we observed all three reactions of scheme 6. The competition of linear chain growth and cyclization was of our main interest. This competition was such as might have been expected when the polycondensation was carried out in dioxane solution with CF_3SO_3H. The kinetics were observed to be first order with respect to the diol for unimolecular cyclization and second order for bimolecular chain coupling. With dilution the contribution of the coupling was strongly reduced. The cyclics were almost exclusive products for initial concentration of the diol of 0.01 M.

Quite different behavior was exhibited by the same system in methylene chloride. True order with respect to the diol in both cases was second order. Thus rates of both processes were reduced to the same extent with dilution. Even at low concentration of the substrate a considerable portion of linear products appear. This unusual result is due to a phenomenon which we called intra-inter catalysis.

(10)   General scheme of intra-inter catalysis:

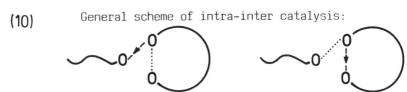

intra molecular catalysis   inter molecular catalysis
inter molecular reaction   intra molecular reaction

An intermolecular reaction catalyzed intramolecularly competes
with the analogous intramolecular reaction catalyzed inter-
molecularly. In such a case both reactions show formally the
second order in the substrate, and no preference for cyclization is observed with dilution of the system.

This general scheme applies here in a following way.
Although the reaction is catalyzed by acid, it requires some
base assistance for withdrawal of proton from the attacking
silanol group. In dioxane this assistance is provided by the
solvent molecule, however, in methylene chloride, which is an
acid-base inert solvent the assistance is provided by the
substrate molecule. Since the OH group is the most basic, this
is the group which is the cocatalyst in the reaction. This
base assistance in intermolecular condensation is illustrated
below:

(11) in methylene chloride   in dioxane

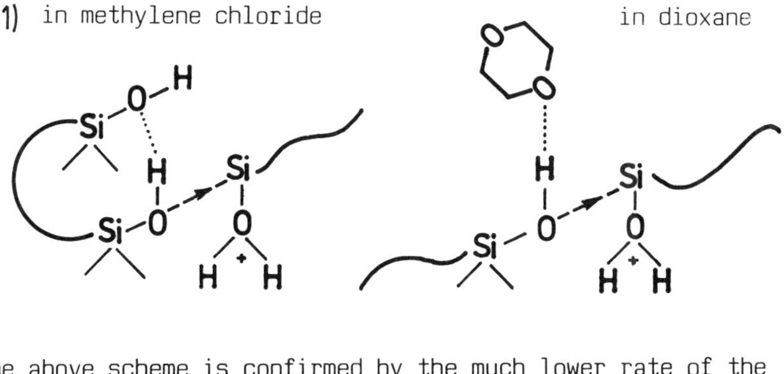

The above scheme is confirmed by the much lower rate of the
intermolecular condensation of the $Me_3SiO[Si(Me)_2O]_4H$ i.e. the
analogue of the oligomer I having only one hydroxyl group in the
chain, thus demonstrating the involvement of the other hydroxyl
group in I.

Competition of homo and heterocondensation

Polycondensation of oligosiloxanols in dioxane occurs
according to the classical mechanism proposed by Lasocki and
Chrzczonowicz [29] for the condensation of the monomeric
dialkylsilanediols, which involves two silanol groups, one of

which is protonated (as in the scheme 11). In methylene chloride, another variant should be taken into consideration, i.e. one involving esterification of the acid with the siloxanol followed by the rate limiting heterocondensation of siloxanol with siloxy ester [27]. The mechanism of intra-inter catalysis would operate in a similar manner with a leaving ester group. The competition of homo and hetero condensations is a general feature of the cationic processes leading to siloxane polymers.

The competition has also some practical implications. Polycondensation of bifunctional monomers or oligomers is sometimes used for synthesis of polymers with a regular chain microstructure e.g. chains with regularly arranged carbofunctional side groups. The polycondensation which involves reaction between ≡SiOH groups is not suitable for this purpose because of the last unit migration processes. More promising are heterocondensation processes involving the ≡SiOH group. However, they may be seriously disturbed by the homocondensation. A good way of suppressing the homocondensation is the nucleophilic catalysis which may very actively accelerate the cross-condensation of the ≡SiOH with ≡SiO(O)CR or ≡SiCl [34]. The nucleophilic catalysis leads therefore to a higher selectivity of the process.

The last unit migration in cationic systems

The last unit migration process occurs not only in basic systems, but also in the presence of acids. Although condensation is generally more important, under some conditions the migration may be the dominating process. Such is the case in some heterogeneous systems. We observed that the disproportionation of oligosiloxanols takes place during the contact of oligomer solution with the surface of aluminosilicates. Since the domination of the last unit migration over the polycondensation may mean a lower content of cyclic products, perhaps it is a general method for decreasing the cyclic content in the cationic polymerization products [17].

REFERENCES

1. Hurd, D.T.; Osthoff, R.C.; Corrin, M.L. J.Am.Chem.Soc. 1954, 76, 41.
2. Wright, P.W. in "Ring-Opening Polymerization", Volume 2, Ivin, K.J.; Saegusa, T. editors, Elsevier Applied Science Publishers, London and New York, 1984, p.1055.
3. Mazurek, M.; Chojnowski, J. Macromolecules 1978, 11, 347.
4. Grubb, W.T.; Osthoff, R.C. J.Am.Chem.Soc. 1955, 77, 1405.
5. Ostrożynski, R.L. Polymer Preprint 1967, 8, 474
6. Yuzhelevskii, Yu.A.; Kagan, A.G.; Fedoseeva, N.N. Dokl. Akad.Nauk SSSR 1970, 190, 647.
7. Boileau, S. ACS Symposium Series 1981, 166, 283.
8. Chojnowski, J.; Mazurek, M. Makromol.Chem. 1975, 176, 2999.
9. Carmichael, J.B.; Heffel, J. J.Phys.Chem. 1965, 69, 2218.

10. Mazurek, M.; Ścibiorek, M.; Chojnowski, J.; Zavin, B.G.;
    Zhdanov, A.A. Europ.Polym.J. 1980, 16, 57.
11. Mazurek, M.; Chojnowski, J. Makromol.Chem. 1977, 178, 1005.
12. Laita, Z.; Jelinek, M. Vysokomol.Soedin. 1962, 4, 1939.
13. Lee, C.L.; Johannson, O.K. J.Polym.Sci.Polym.Chem. 1976, 17,
    729.
14. Frye, C.L.; Salinger, R.M.; Fearon, F.W.G.; Klosowski, J.M.;
    Young, T.D. J.Org.Chem. 1970, 17, 729.
15. Juliano, P.C.; Fessler, W.A.; Cargioli, J.D. Macromol.Prep.
    1971, 2, 1211.
16. Chojnowski, J.; Rubinsztajn, S.; Stańczyk, W.; Ścibiorek, M.
    Makromol.Chem.Rapid Commun. 1983, 4, 703.
17. Rubinsztajn, S.; Chojnowski, J. in preparation.
18. Chojnowski, J.; Kaźmierski, K.; Rubinsztajn, S.; Stańczyk,W.
    Makromol.Chem. 1986, 187, 2039.
19. Chmielecka, J.; Chojnowski, J.; Eaborn, C.; Stańczyk, W.
    J.Chem.Soc.Chem.Commun. in press.
20. Patnode, W.; Wilcock, D.F. J.Am.Chem.Soc. 1946, 68, 358
21. Chojnowski, J.; Wilczek, L. Makromol.Chem. 1979, 180, 117.
22. Sauvet, G.; Lebrun, J.J.; Sigwalt, P. in "Cationic Polymeri-
    zation and Related Processes" Goethels, E.J. editor Academic
    Press, Orlando (Florida) 1984 p.237.
23. Wilczek, L.; Rubinsztajn, S.; Chojnowski, J. Makromol.Chem.
    1986, 187, 39.
24. Chojnowski, J.; Ścibiorek, M.; Kowalski, J. Makromol.Chem.
    1977, 178, 1351.
25. Wilczek, L.; Chojnowski, J. Macromolecules 1981, 14, 9.
26. Chojnowski, J.; Rubinsztajn, S.; Wilczek, L. J.Chem.Soc.
    Chem.Commun. 1984, 69.
27. Chojnowski, J.; Rubinsztajn, S.; Wilczek, L. Macromolecules
    in press.
28. Chojnowski, J.; Wilczek, L.; Rubinsztajn, S. in "Cationic
    Polymerization and Related Processes" Goethels, E.J. editor
    Academic Press, Orlando (Florida) 1984 p.253.
29. Lasocki, Z.; Chrzczonowicz, S. Polym.Sci. 1962, 59, 259.
30. Lasocki, Z.; Dejak, B. Polimery 1970, 15, 391.
31. Rühlmann, K. Phosphorus and Sulfur 1986, 27, 139.
32. Rutz, W.; Lange , D.; Popowski, E.; Kelling, H. Z.Anorg.
    Allg.Chem. 1986, 536, 197.
33. Graiwer, D.; Huebner, D.J.; Saam, J.C. Rubber Chemistry and
    Technology 1983, 56, 918.
34. Rubinsztajn, S.; Cypryk, M.; Chojnowski, J. in preparation.

Chapter 29

Synthesis of porous tectosilicates: parameters controlling the pore geometry

Friedrich Liebau — Mineralogisches Institut der Universität Kiel.

INTRODUCTION

A tectosilicate (synonym: framework silicate) is a silicate in which $[(Si,T)O_n]$ polyhedra are linked via common oxygen atoms to a framework that has infinite expansion in three dimensions. The $[(Si,T)O_n]$ polyhedra are almost exclusively $[(Si,T)O_4]$ tetrahedra. There are only a few tectosilicates which contain $[(Si,T)O_6]$ octahedra; they shall not be considered in this review. T is usually Al^{3+}, and less commonly Be^{2+}, Mg^{2+}, B^{3+}, Fe^{3+}, and Ti^{4+} in minerals and Ga^{3+} and Ge^{4+} in synthetic phases.

The general formula for tectosilicates is

$$A_a[Si_{1-y}T_yO_{2+z}]X_v \cdot wM, \qquad\qquad 0 \leq z < 1 \qquad\qquad (1)$$

where A are cations not replacing Si, X anions, and M neutral molecules, all three being located in voids of the tetrahedral framework. Ignoring the few silicates with interrupted frameworks $(z \neq 0)$[1], this formula reduces to

$$A_a[Si_{1-y}T_yO_2]X_v \cdot wM, \qquad\qquad \text{and to} \qquad\qquad (2)$$

$$A_a[SiO_2]X_v \cdot wM \qquad\qquad \text{if } y = 0. \qquad\qquad (3)$$

CLASSIFICATION OF TECTOSILICATES

For crystal chemical as well as for purposes of technical applications it seems most sensible to have a classification of tectosilicates which is based on the size and shape of voids within the tetrahedral frameworks and on the chemical character of the tetrahedrally coordinated framework cations.

Dense tectosilicates

In this group the interstices between the oxygen atoms of the tetrahedral frameworks are either too small to house atoms or ions, or the interstices are only large enough to contain small atoms (He) or monatomic cations such as Li^+, Na^+, K^+, Mg^{2+}, Ca^{2+} etc. Examples for such dense tectosilicates are the silica polymorphs quartz, cristobalite and coesite and the tectosilicates proper nepheline, $(Na,K)[SiAlO_4]$, and the feldspars $Na_{1-x}Ca_x[Si_{2+x}Al_{2-x}O_8]$ respectively.

Porous tectosilicates

In these silicates at least some of the voids are large enough to house larger species which may either be cations (A), anions (X) or uncharged molecules (M). Fig.1 presents schematic drawings of a number of such voids which can, in general, be described as polyhedral voids

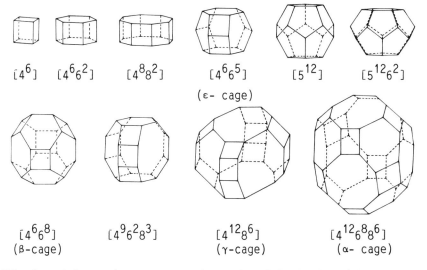

Fig.1   Schematic presentation of polyhedral voids observed in the tetrahedral framework of tectosilicates

or pores. Here and in the other Figures (Si,T) atoms are
located at the corners of the polyhedra and the framework
oxygen atoms near the midpoints of the lines connecting
adjacent Si atoms. Faces of·the polyhedra are [-(Si,T)-
O]$_n$- rings. A short-hand notation [$n_i{}^{m_i}$] is used where m_i
is the number of faces having n_i corners.

The tetrahedral framework is considered to be the host
and the ionic or uncharged species enclosed in the pores
to be the guests.

If the free apertures of the faces of a polyhedral
void are too small to let the enclosed guest species pass
then such a pore acts as a cage for the guest. The resulting tectosilicate is called a **clathrate compound**. If a
polyhedral void shares a face with each of two other
voids and the apertures of these faces are wide enough to
let the enclosed guest molecule pass then the polyhedra
can form channel-like pores in which the guest species
can more or less readily migrate. For medium sized guest

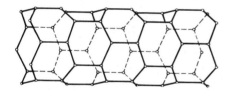

Fig. 2. Schematic diagram of a channel-
like pore in silica-ZSM-48 [13].

molecules such migration is possible if the ring size is
n$\geq$7 (Figs.1, 2). Tectosilicates which contain such channels are **zeolite-like compounds**. The channels may either
be separated from each other or intersect each other. The
corresponding pore systems have **pore dimensionalities**,
D_{pore}=1, 2, and 3, compared with D_{pore}=0 for cages of
clathrate compounds.

Chemical differentiation

Depending on the chemical character of the framework
cations, the tectosilicates are further subdivided as
described in Table 1 [2]. Although a pure SiO_2 framework
could formally contain ionic species A and/or X in its
pores, in agreement with the general formula (3), no
porosil is known for which the existence of ionic guest
species has been proven (see e.g. [3]). As a consequence,
formula (3) reduces further to

$$[SiO_2]·wM. \tag{4}$$

Table 1. Classification of tectosilicates based on porosity, pore dimensionality and chemical character of framework cations.

		Si	Al	
Dense tecto-silicates	T e c t o s i l s	Pyknosils	T e c t o l i t e s	Pyknolites
Porous tecto-silicates		Porosils clathrasils zeosils		Porolites clathralites zeolites

STRUCTURES AND THERMAL PROPERTIES OF POROSILS

At present there are 22 topologically distinct cristalline silica structures known. Some of these exist in several crystallographically distinct phases (e.g. high- and low-temperature quartz, high- and low-temperature cristobalite etc.). In Table 2 the phases with topologically distinct silica frameworks are divided into the three groups: dense silicates (pyknosils), clathrate-like silicates (clathrasils) and zeolite-like silicates (zeosils), and arranged in the order of decreasing framework density, d_f, which is measured in $n(SiO_2)/1000Å^3$.

Pyknosils

Pyknosils with framework densities between 43.0 and 22.9 $SiO_2/1000Å^3$ are known. If the very minor amounts of Al^{3+}, A^+ and A^{2+} which seem necessary to stabilize tridymite were taken into account tridymite would be considered as pyknolite.

Clathrasils

To date, nine clathrasil families with topologically distinct frameworks have been recognized (Table 2). Most of them have been synthesized with one or more combinations of guest molecules [5, 20 - 22]. In Fig.3 and Table 3 a survey of the 13 cage-like pores which have been observed in clathrasils is given. They vary considerably in size and maximum topological symmetry. The size of the guest molecules successfully incorporated ranges from that of Kr, N_2, CO_2, and CH_4 to that of 1-aminomethyl adamantane $C_{11}NH_{19}$. A typical example of a clathrasil structure is shown in Fig.4a, and additional clathrasil structures may be obtained from the references given in Table 2.

Zeosils

The topologically distinct zeosils known are listed in Table 2. As an example, the structure of silica-ferrierite (Fig.4c) clearly shows wide channel-like pores per-

Table 2. Survey of crystalline phases which have topologically distinct SiO_2 frameworks. Abbr.: abbreviated name, D_{pore}: pore dimensionality, d_f: framework density in $n(SiO_2)/1000Å^3$. M^f: guest molecule located in a capge that has $f=\Sigma m_i$ faces. Reference is made to the most recent publication only. For systematic nomenclature see [2].

Class	Name	Abbr. (D_{pore})	d_f	Formula, unit cell content	Ref.
P					
y	Stishovite	Sti	43.0	SiO_2	
k	Coesite	Coe	29.3	SiO_2	
n	Quartz	Qu	26.6	SiO_2	
o	Moganite	Mog	26.3	SiO_2	[4]
s	Keatite	Kea	25.1	SiO_2	
i	Cristobalite	Cr	23.2	SiO_2	
l	Tridymite	Tr	22.9	SiO_2	
s					
C					
l	Nonasils	Non	19.2	$88SiO_2 \cdot 8M^8 \cdot 8M^9 \cdot 4M^{20}$	[5]
a	Melanophlogites	Mel	19.0	$46SiO_2 \cdot 2M^{12} \cdot 6M^{14}$	[6]
t	Dodecasils 3C	D3C	18.6	$136SiO_2 \cdot 16M^{12} \cdot 8M^{16}$	[7]
h	(Silica-ZSM-39)				
P r	Dodecasils 1H	D1H	18.5	$34SiO_2 \cdot 3M^{12} \cdot 2M'^{12} \cdot 1M^{20}$	[8]
o a	Silica-sodalites	Sod	17.4	$12SiO_2 \cdot 2M^{14}$	[9]
r s Z	Deca-	DD3R	17.6	$120SiO_2 \cdot 6M^{10} \cdot 9M^{12} \cdot 6M^{19}$	[10]
o i e	dodecasils 3R	(3)			
s l o	Deca-	DD3H	17.6	$120SiO_2 \cdot 6M^{10} \cdot 9M^{12} \cdot 1M^{15} \cdot 4M^{19} \cdot 1M^{23}$	[11]
i s s	dodecasils 3H	(3)			
l i	Silica-ZSM-23	(1)	20.0	$24SiO_2 \cdot (CH_3)_2N(CH_2)_7N(CH_3)_2$	[12]
s l	Silica-ZSM-48	(1)	19.9	$48SiO_2 \cdot H_2N(CH_2)_8NH_2$	[13]
s	Silica-ZSM-22	(1)	19.7	$24SiO_2 \cdot HN(C_2H_5)_2$	[14]
	Silica-ferrierite	(2)	19.3	$36SiO_2 \cdot 2H_2N(CH_2)_2NH_2$ (en)	[15]
	Silica-ZSM-12	(1)	18.5	$28SiO_2 \cdot N(C_2H_5)_3$	[16]
	Silicalite II (ZSM-11)	(3)	17.9	$96SiO_2 \cdot n[N(C_4H_9)_4]OH$	[17]
	Silicalite I (ZSM-5)	(3)	17.8	$96SiO_2 \cdot 4[N(C_3H_7)_4]F$	[18]
Other	Fibrous silica		19.6	SiO_2	[19]

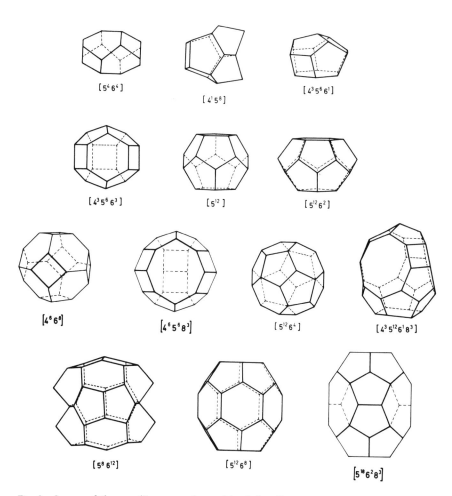

Fig. 3. Survey of the cage-like pores observed in clathrasils.

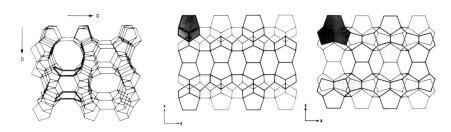

Fig. 4. Schematic presentation of the tetrahedral frameworks of a, b) nonasils [5] c) silica-ferrierite [15].

Table 3. Size and maximum topological symmetry of cages observed in clathrasils $f=\Sigma m_i$: number of faces of the cage.

Cage	f	Free volume [Å³]	Max. topol. symmetry	Number of cages per unit cell in						
				Non	Sod	Mel	D3C	D1H	DD3R	DD3H
$[5^4 6^4]$	8	25	222 o'rh.	8						
$[4^1 5^8]$	9	30	mm o'rh.	9						
$[4^3 5^6 6^1]$	10	35	3m trig.						6	6
$[5^{12}]$	12	80	m3 cub.			2	16		9	9
$[4^3 5^6 6^3]$	12	100	$\bar{6}$m2 hex.					2		
$[4^6 6^8]$	14	130	$\bar{4}$3m cub.		2			3		
$[5^{12} 6^2]$	14	160	$\bar{4}$2m tetr.			6				
$[4^6 5^6 8^3]$	15	ca. 200	$\bar{6}$m2 hex.							1
$[5^{12} 6^4]$	16	250	m3 cub.				8			
$[4^3 5^{12} 6^1 8^3]$	19	350	3m trig.						6	4
$[5^8 6^{12}]$	20	290	mmm o'rh.	4						
$[5^{12} 6^8]$	20	430	6/mmm hex.					1		
$[5^{18} 6^2 8^3]$	23	540	$\bar{6}$m2 hex.							1

pendicular to the plane of projection. They are inter-
sected by narrower channels parallel to the a axis. Two
8-membered rings which are common to two intersecting
channels are shown in bold lines. For other zeosil
structures see the references in Table 2.

Thermal properties of porosils

Zeosils lose guest molecules which are located in the
channels above ca. 400 K. In contrast, clathrasils lose
their guest molecules at higher temperatures, melanophlogite above ca. 750 K, the others above ca. 1000 K. Prolonged heating between 1250 and 1350 K releases all guest
species and leads to the formation of pure silica polymorphs without change of the framework topology. Their
density is 1.8 gcm^{-3} compared with 2.65 gcm^{-3} for quartz.

There is no sharp demarcation between clathrasils and
zeosils. For example, as synthesized deca-dodecasils
containing 1-aminoadamantane as a guest molecule are
clathrasils. After complete degassing their cages could
hypothetically be refilled with smaller guest molecules,
which may necessitate reclassification of the compounds
as zeosils.

PARAMETERS CONTROLLING THE FRAMEWORK TYPE DURING SYNTHESIS

Simultaneous presence of Al^{3+} and A^+, A^{2+} cations

With the exception of fibrous silica, which is extremely hygroscopic, all other tectosils can be formed in
aqueous solutions, provided that Al^{3+} and alkali and
alkaline earth cations are not simultaneously present in
substantial amounts. Otherwise zeolites and clathralites
are formed at low temperatures, and pyknolites are formed
at higher temperatures.

Presence of potential guest molecules (M) Therman

Crystallization of silica from aqueous solution takes
place by condensation of monomeric or low-molecular silicic acids. In the absence of suitable guest molecules,
pyknosils are formed which have dense frameworks.

In the presence of suitable guest molecules, however,
silicic acid condensates around these molecules thus
forming porous silica frameworks. The molecules are then
enclosed in the pores. Given this formation mechanism, it

is apparent that the molecules' physical properties, such as size and shape critically influence the type of pore and, therefore, of framework that is formed.

Thermal and hydrothermal stability of potential guests

A suitable guest molecule should be stable at the temperature of synthesis. In addition, it should neither decompose nor hydrate under synthesis conditions.

Lewis basicity of guest molecules

The tendency to form a porosil increases with higher Lewis basicity of the guest molecules. For example, the rate of crystallization and the yield of dodecasils 3C in the presence of cyclic guest molecules of equal size and shape increase with increasing pK_A value (Table 4). In fact, most of the molecules that have been successfully used to synthesize porosils are amines which interact with the weak Lewis acid SiO_2 via their lone pair electrons.

Shape of guest molecules

Globular and ellipsoidal guest molecules form clathrasils with polyhedral pores. Such molecules are either acyclic with up to four non-hydrogen atoms or cyclic. Acyclic chain molecules with more than seven non-hydrogen atoms form zeolites with channel-like pores instead of clathrasils. Acyclic pores with five or six non-hydrogen atoms form either clathrasils or zeosils or both (Tables 5 and 6).

Table 4. Influence of Lewis basicity of guest molecules on crystallization rate and yield of dodecasil 3C.

Guest molecule	 ⬠	 ⬠S	 ⬠O	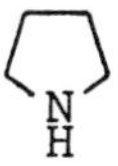 ⬠NH
pK_A value at 300K	> -10	-4.5	-2.1	11.1
Rate of D3C crystallization	extremely slow	very slow	slow	fast
Yield of D3C	very low	low	good	high

Table 5. Correlation between shape, size and symmetry of guest molecules and the type of clathrasil formed. Only those cages are given which contain the guest molecules indicated.

Typical guest molecules	Kr, CH_3NH_2, $CH_4 + N_2 + CO_2$	SF_6, $N(CH_3)_3$, 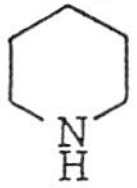 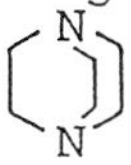 (piperidine, piperazine)	$CH_3CH(NH_2)C_2H_5$, 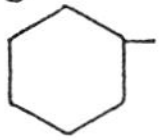 (cyclohexyl)$-N(CH_3)_2$	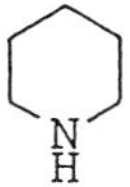 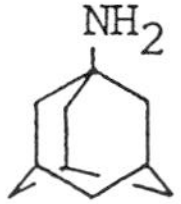 (piperidine), (adamantyl)$-NH_2$
Porosil	M e l	D 3 C	N o n	D 1 H
Free cage volume [$Å^3$]	80 160	250	290	430
Maximum topological symmetry	m3 (cubic) $\bar{4}$2m (tetrag.)	m3 (cubic)	mmm (o'rhomb.)	6/mmm (hexag.)

Acyclic molecules with unbranched chains of at least 5 non-hydrogen atoms form zeosils with 1-dimensional pores (D_{pore}=1); ethylenediamine and diethylenetriamine in the presence of $B(OH)_3$ form silica-ferrierite with D_{pore}=2; branched acyclic amines form silica-ZSM-5 with D_{pore}=3 (Table 6) [23].

Size of guest molecules

It is obvious that a pore must be large enough to accommodate the guest molecule. The correlation between molecular size and free volume of the cage enclosing it can be seen for some typical guest molecules of clathrasils from Table 5.

Flexibility of guest molecules

A suitable guest molecule is one which does not change its conformation with time but is still flexible enough to adjust to the inner walls of a cage without distorting the framework considerably. These properties are typical of cyclic molecules which have only single bonds. If, however, a molecule contains double or even

Table 6. Influence of guest molecule shape on the formation of zeosils with different pore dimensionality.

Typical guest molecules	Zeosil	D_{pore}
$H_2NC_4H_9$, $HN(C_2H_5)_2$, $H_2N(CH_2)_3NH(CH_2)_2NH(CH_2)_2NH_2$, $H_2N(CH_2)_3NH(CH_2)_3NH(CH_2)_3NH_2$	Silica-ZSM-22	1
$H_2N(CH_2)_2NH(CH_2)_2NH_2$, $H_2N(CH_2)_2NH(CH_2)_2NH(CH_2)_2NH_2$, $H_2N(CH_2)_3NH(CH_2)_3NH(CH_2)_3NH_2$	Silica-ZSM-48	1
$H_2N(CH_2)_2NH_2$ + $B(OH)_3$, $H_2N(CH_2)_2NH(CH_2)_2NH_2$ + $B(OH)_3$	Silica-ferrierite	2
$N(C_2H_5)_3$, $N(C_3H_7)_3$, $N(C_4H_9)_3$	Silica-ZSM-5	3

triple bonds it will be too rigid to contribute to the neccessary mutual adjustment between host framework and guest molecule. Such molecules like benzene and alkenes do not form clathrasils.

If a molecule changes its conformation easily by free rotation about single bonds C-C, C-N, etc. then at a given time only some of the molecules of a particular species will have a suitable conformation for the formation of clathrasils. Consequently, acyclic molecules and cyclic molecules with longer sidechains show no or little tendency to form clathrasils. This is demonstrated for the rate of crystallization of the dodecasils 3C and 1H in Table 7.

Symmetry of guest molecules

The increased symmetry of the guest molecules correlate with the increased symmetry of the corresponding pores (Table 5). This is particularly true at low synthesis temperature.

Usually, the symmetry of the cage is higher than that of the enclathrated molecule. As a consequence, the guest molecules can have several energetically equivalent orientations leading to orientational disorder [7,8].

Synthesis temperature

Porosils have been synthesized between 120 and 250°C. Some guest molecules can form two topologically different porosils. These have to be considered to be polymorphs although no solid state phase transition between them has been observed. Table 8 lists some pairs of such polymorphs (the arrow indicates the high-temperature phase [22]). Three different temperature effects can be distinguished.

Change of cage volume   As temperature increases, thermal movements of the atoms increase so that the molecules require more space at higher temperatures than at lower temperatures. Consequently, if a particular guest molecule forms two polymorphic porosils, the high-temperature phase is expected to have the larger cages.

1- and 2-aminobutane, which are the smallest molecules found to be located in the 290Å^3 large [$5^8 6^{12}$] cage of nonasil, are the only guest molecules known to contradict the temperature - volume rule. At higher synthesis temperature they still fit into the slightly smaller (250Å^3),

Table 7. Influence of guest molecule flexibility on the crystallization rate of clathrasils.

Guest molecules	$HN(C_2H_5)_2$, [piperidine], [quinuclidine]	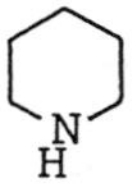  $CH_3CH(NH_2)C_3H_7$, [N-propylpiperidine], [adamantaneamine NH_2] 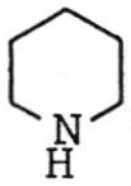 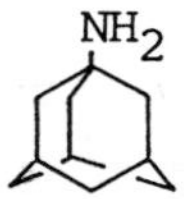
Phase	D3C	D1H
Rate of crystallization	slow medium fast	slow slow fast

Table 8. Polymorphism in clathrasils.

Cage housing characteristic guest M	V[Å³]	80	160	250	290	350	430
Guest molecule M	Symm.	m3 cub. $[5^{12}]$	$\bar{4}2m$ tetr. $[5^{12}6^{2}]$	m3 cub. $[5^{12}6^{4}]$	mmm o'rh. $[5^{8}6^{12}]$	$\bar{6}m2$ hex. $[4^{3}5^{12}6_{1}8^{3}]$	6/mmm hex. $[5^{12}6^{8}]$
Krypton		D3C ——170°C——▶ Me1					
Methylamine			Me1 ——185°C——▶ D3C				
2-Butylamine				D3C ◀——180°C—— Non			
Piperidine					D3C ——220°C——▶ D1H		
2-Methylpiperazine					Non ——150°C——▶ D1H		
1-Aminoadamantane						DD3R ——170°C——▶ D1H	

but more symmetrical $[5^{12}6^4]$ cage of dodecasil 3C (Table 8).

Change of cage symmetry   There is a general tendency that the space occupied by a molecule becomes more symmetrical as temperature increases. The temperature - symmetry rule, therefore, states that, if a particular molecule forms two polymorphic porosils, the high-temperature phase is expected to enclose this molecule in a pore of higher symmetry.

Only Kr and Xe are known to contradict this rule by going into the slightly less symmetric $[5^{12}6^2]$ cage of melanophlogite rather than into the excessively large $[5^86^{12}]$ cage of nonasil (Table 8).

Change of pore dimensionality   If a particular guest molecule forms a zeosil as well as a clathrasil, there is a tendency that it prefers the zeosil structure at lower temperature.

DISCUSSION

Rules for the synthesis of clathrasils and zeosils have been deduced from experimental data. In doing so, a number of parameters controlling the framework type formed had to be considered. The key of a successful elucidation and interpretation of these rules is the fact that the interaction between the uncharged SiO_2 framework and the guest molecules is only of the weak van der Waals and Lewis acid/base type.

For zeolites and clathralites, the situation is much more complicated. Due to partial Si^{4+}/Al^{3+} replacement, the tetrahedral framework is negatively charged. Consequently, both ionic species and neutral molecules are possible guest species. Interaction between host and guests, therefore, is mainly of the stronger ionic type and to a lesser extent of the van der Waals and Lewis acid/base type.

As a result, the general properties of guest molecules in clathralites and zeolites (porolites) are more poorly understood than those of the guest molecules in clathrasils and zeosils (porosils). Establishing corresponding rules for the incorporation of ionic and/or neutral guest species in porolites will require systematic exploration of systems in which the content of Al^{3+} and other cations are incrementally varied. Hopefully, this will lead to a better understanding of the important interactive parameters of charged and uncharged guest molecules in porolites, and ultimately allow us to design experiments through which zeolites can be engineered for specific applications.

Acknowledgments. I thank B. Marler, R.P. Gunawardane, H. Gerke, N. Dehnbostel, H.-H- Eulert and in particular H. Gies for their insight and enthusiasm which made this cooperative research so enjoyable. Financial support from the Deutsche Forschungsgemeinschaft and from the Humboldt-Stiftung is gratefully acknowledged.

REFERENCES

1 Liebau, F.: Structural Chemistry of Silicates. Springer-Verlag, Heidelberg 1985
2 Liebau, F., Gies, H., Gunawardane, R.P., Marler, B.: Zeolites 6, 373-377, 1986
3 Fyfe, C.A., Gobbi, G.C., Klinowski, J., Thomas, J.M., Ramdas, S.: Nature 296, 530-533, 1982
4 Miehe, G., Flörke, O.W., Graetsch, H.: Fortschr. Mineral. 64, Beiheft 1, 117, 1986
5 Marler, B., Dehnbostel, N., Eulert, H.-H., Gies, H., Liebau, F.: J. Inclusion Phenom. 4, 339-349, 1986
6 Gies, H.: Z. Kristallogr. 164, 247-257, 1983
7 Gies, H.: Z. Kristallogr. 167, 73-82, 1984
8 Gies, H.: J. Inclusion Phenom. 4, 85-91, 1986
9 Bibby, D.M., Dale, M.P.: Nature 317, 157-158, 1985
10 Gies, H.: Z. Kristallogr. 175, 93-104, 1986
11 Gies, H.: J. Inclusion Phenom. 2, 275-278, 1984
12 Rohrman, A.C., LaPierre, R.B., Schlenker, J.L., Wood, J.D., Valyocsik, E.W., Rubin, M.K., Higgins, J.B., Rohrbaugh, W.J.: Zeolites 5, 352-354, 1985
13 Schlenker, J.L., Rohrbaugh, W.J., Chu, P., Valyocsik, E.W., Kokotailo, G.T.: Zeolites 5, 355-358, 1985
14 Marler, B.: Zeolites, in press
15 Gies, H., Gunawardane, R.P.: Zeolites, in press

16 LaPierre, R.B., Rohrman, A.C., Schlenker, J.L., Rubin, M.K., Rohrbaugh, W.J.: Zeolites 5, 346-348, 1985

17 Kokotailo, G.T., Chu, P., Lawton, S.L.: Nature 275, 119-120, 1978; Bibby, D.M., Milestone, N.B., Aldridge, L.P.: Nature 280, 664-665, 1979

18 Baerlocher, C.: Proc. 6th Int. Zeolite Conf., Reno, USA, 1984, p. 823-833; Chao, K.-J., Lin, J.-C., Wang, Y., Lee, G.H.: Zeolites 6, 35 - 38, 1986

19 Weiss, A., Weiss, A.: Z. anorg. allg. Chem. 276, 95-112, 1954

20 Gies, H.: Nachr. Chem. Tech. Lab. 33, 387-391, 1985

21 Gunawardane, R.P., Gies, H., Liebau, F.: Z. anorg. allg. Chem. 546, 189-198, 1987

22 Gies, H.: Habilitationsschrift, submitted to Universität Kiel, 1987

23 Gunawardane, R.P., Gies, H., Marler, B.: submitted to Zeolites

PART VII

INORGANIC CHEMISTRY
OF SILICON

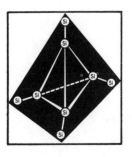

Chapter 30

Transition metal substituted silanes: ligand exchange at the silicon and the metal centre

W. Malisch, G. Thum, D. Wilson, P. Lorz, U. Wachtler and W. Seelbach — Institut für Anorganische Chemie der Universität Würzburg, Am Hubland, 8700 Würzburg (Federal Republic of Germany).

Since the first report of a silicon transition metal compound $Cp(CO)_2Fe-SiMe_3$ by Piper and Wilkinson in 1956 [19] a vast series of compounds with nearly all transition metal fragments and different ligands at the silicon have been described [2]. However no systematic studies have been undertaken to synthesize derivatives with silicon in a highly electron rich state, which might be suitable precursors for the generation of transition metal substituted silicenium ions according to eq. (1).

$$L_nM-SiR_2-X \xrightarrow{-X^-} [L_nM-SiR_2]^+ \qquad (1)$$

Despite numerous attempts, a silicenium ion has not been isolated to date [3, 7]. Recent experiments however give strong evidence for the existence of organo substituted derivatives R_3Si^+ in solution [13, 14].
According to theoretical calculations donor groups like NR_2, PR_2, SR should stabilize a three coordinated silicon atom [1, 10]. For that reason we tried to combine functional silicon moieties of the type SiR_2X (X = H, Hal) containing these groups with a strongly electron donating metal fragment.

Cp(L)$_2$Fe-substituted silanes appeared to be rather attractive candidates for this project since the iron silicon bond is of reasonable stability and allows substitution at the silicon without Fe-Si bond cleavage [11, 12, 15-18]. Moreover the organoiron fragment is characterized by its strong electron releasing capacity [17, 20], especially when it contains good donor ligands, e.g. phosphines. In this case activation of a silicon bonded halogen or hydrogen atom with respect to its anionic abstraction can be expected.

LIGAND EXCHANGE AT SILICON

Ferrio-aminosilanes

Although the silicon chemistry is rich in compounds having a Si-N unit [24] only a few transition metal substituted aminosilanes have been prepared [4, 8, 9, 11, 21]. We have found, that aminolysis of metal coordinated halosilyl groups can be extensively used for the synthesis of ferrio-aminosilanes. Starting with the ferrio-halosilanes 1a-i the ferrio-dialkylaminosilanes 2a-h are obtained from 1a-c,g and Me$_2$NH, 1a-c,e,h and Et$_2$NH or 1d and iPr$_2$NH respectively (scheme 1, (i)). Introduction of two amino groups is achieved for the combinations 1d/Me$_2$NH, 1d/Et$_2$NH and 1e,f,h/Me$_2$NH to give 3a-d (ii). 2-Ferrio-1.3.2-diazasilapentanes 4a-d are generated from 1d-f,h,i and N,N'-dimethylethylendiamine (iii). Interaction of 1i, having the more reactive silicon bromine bond with Me$_2$NH yields the ferrio-tris(amino)silane 5 (iv). Analoguously the C$_5$Me$_5$-substituted ferrio-aminosilanes 7a-d can be prepared from Me$_2$NH and 6a-d, generated via the nucleophilic metallation [15] of the corresponding halosilane with Na[Fe(CO)$_2$C$_5$Me$_5$].

	6a	b	c	d
SiR$_3$	SiMe(H)Cl	SiMe$_2$Cl	SiHCl$_2$	SiBr$_3$

	7a	b	c	d
SiR$_3$	SiMe(H)NMe$_2$	SiMe$_2$NMe$_2$	SiH(NMe$_2$)$_2$	Si(NMe$_2$)$_3$

The ferrio-aminosilanes of scheme 1 are surprisingly thermolabile and decompose with Si-Fe bond cleavage. This fact and the unsusual low activity of the silicon towards nucleophiles can be referred to a (N-Si) -interaction. Exchange of the amino groups is only accomplished by the reaction of 4b with MeOH to give Cp(CO)$_2$Fe-Si(OMe)$_3$ (9). On the other hand H/Cl exchange with CX$_4$ (X = Cl, Br), which involves a radical chain mechanism [15] is extremely favored. From 2a, b,e,3a,b under mild conditions the corresponding ferrio-aminohalosilanes Cp(CO)$_2$Fe-SiR$_2$Hal (SiMe(Cl)NMe$_2$ (8a), SiMe(Cl)NEt$_2$ (8b), SiCl$_2$NiPr$_2$ (8c), SiMe(Br)(NMe$_2$)$_2$ (8d), SiCl(NEt$_2$)$_2$ (8e)) are accessible in quantitative yield.

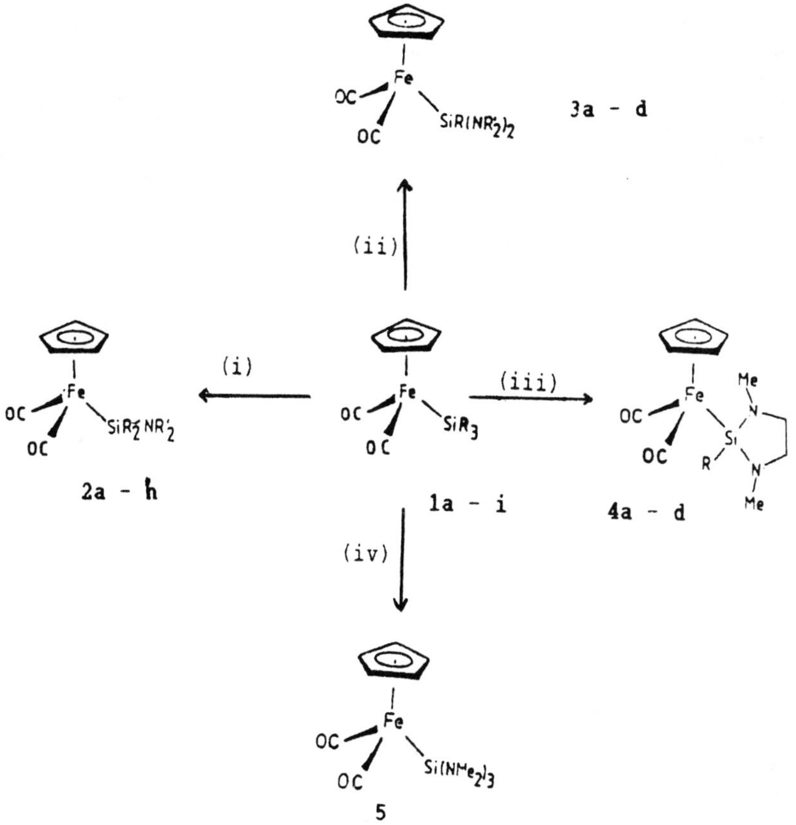

1: SiR_3 = SiMe(H)Cl (**a**), $SiMe_2Cl$ (**b**), $SiMe_2Br$ (**c**), $SiHCl_2$ (**d**), $SiMeCl_2$ (**e**), $SiMeBr_2$ (**f**), Si(iPr)Cl_2 (**g**), $SiCl_3$ (**h**), $SiBr_3$ (**i**), SiH(iPr)Cl (**j**).

2: $SiR_2(NR'_2)$ = SiMe(H)NMe_2 (**a**), SiMe(H)NEt_2 (**b**), $SiMe_2NMe_2$ (**c**), $SiMe_2NEt_2$ (**d**), SiH(Cl)$NiPr_2$ (**e**), SiMe(Cl)NEt_2 (**f**), Si(iPr)(Cl)NMe_2 (**g**), $SiCl_2NEt_2$ (**h**).

3: $SiR(NR'_2)_2$ = $SiH(NMe_2)_2$ (**a**), $SiH(NEt_2)_2$ (**b**), $SiMe(NMe_2)_2$ (**c**), $SiCl(NMe_2)_2$ (**d**).

4: R = H (**a**) Me (**b**) Cl (**c**) Br (**d**)

Scheme 1: Reaction of diverse ferrio-halosilanes with secondary amines; excess R_2NH (i,ii,iv), $[Me(H)N(CH_2)]_2$ (iii).

As the experiments with iPr_2NH demonstrate, which are successful only in the case of **1g**, aminolysis is not a productive route to introduce bulky amino ligands at silicon, necessary for the kinetic stabilization of a silicenium ion. When the more nucelophilic amide iPr_2N is employed, the lithium iron salt $[(iPr_2N)_2HSiC_5H_4Fe(CO)_2]Li$ is obtained from **1g**, which can be converted by MeI to the neutral methyl iron complex **11f**, having a bis(amino)silyl group fixed to the cyclopentadienyl ring [22].

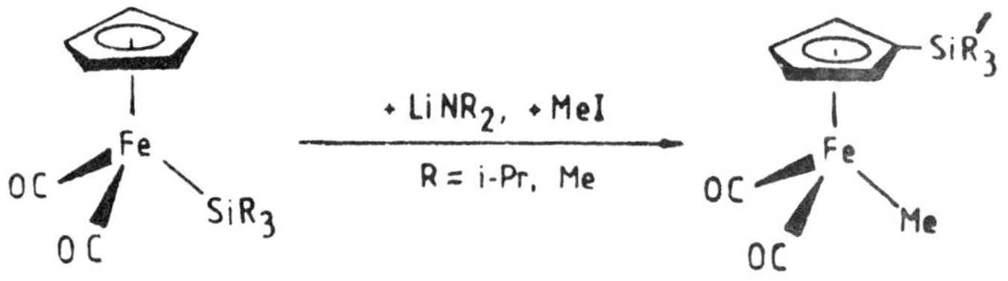

1a,c-e, 2c,h, 3d, 9, 10a-c   11a-i

11: SiR'_3 = $SiMe_3$ (a), $SiMe_2H$ (b), $SiMe(H)N^iPr_2$ (c),
$SiMe_2N^iPr_2$ (d), $Si(NMe_2)_2N^iPr_2$ (e), $SiH(N^iPr_2)_2$ (f),
$SiMe(NMe_2)_2$ (g), $SiNEt_2(NMe_2)_2$ (h), $Si(OMe)_3$ (i)

Starting from **1a**, **c**, **3d** and $Cp(CO)_2Fe-SiMe_2H$ (**10c**) the η^5-diisopropylaminosilylcyclopentadienyl derivatives **11c** – **e** are obtained in the same manner, indicating that silyl group transfer is combined either with chloride/amide or hydrido/amide exchange. This reaction sequence ligand exchange/silyl shift can be performed in the case of **1e**, **2h** with $LiNMe_2$ to give **11g**, **h**.
Exclusive transfer of the silyl groups Me_3Si, Me_2HSi, $(MeO)_3Si$, comparable to that described by Berryhill [6, 5] is involved in the course of the conversion of **10a**, **b**, **c** to **11a**, **b**, **i-h** with $LiNR_2$ (R = Me, iPr). In contrast $Na[N(SiMe_3)_2]$ preferentially attacks the silicon to produce the ferrio-bis(silyl)aminosilane **12**.
The observation, that reaction of **5** with $LiNiPr_2$ causes no silyl migration, while in the case of **3d** the $Si(NMe_2)_2NiPr_2$ group appears at the cyclopentadienyl ligand provides evidence that the third amino group is introduced after migration of the $Si(NMe_2)_2Cl$ group. Eq. (2) offers the opportunity to shift functional silyl groups from the metal to the cyclopentadienyl ring, which can be further modified by established procedures.

The lithium iron salt [R$_3$SiC$_5$H$_4$(CO)$_2$Fe]Li primarily formed
in the amide reaction (2), can in addition be transformed to
the antimonido, silyl and hydrido iron complexes 13a-b.

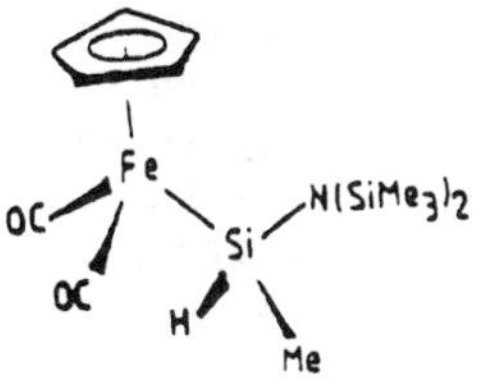

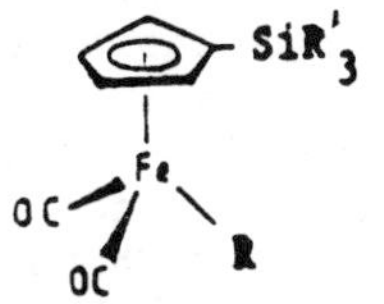

	13a	b	c
SiR'$_3$	SiMe$_3$	SiMe$_3$	SiMe(NMe$_2$)$_2$
R	SbMe$_2$	SiH(Me)Cl	H

12

13c yields the dinuclear species 14 on photolysis, the
iron hydride 15 on treatment with Me$_3$P and the phosphonium
iron salt 16, structurally related to the intermediate of
the transfer reaction, via deprotonation with the phosphorus ylide Me$_3$P=CH$_2$.

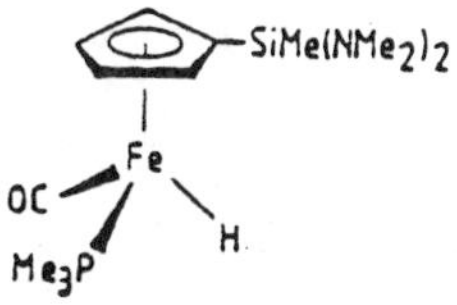

15

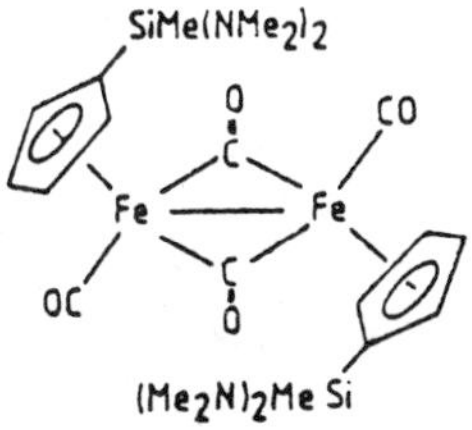

14

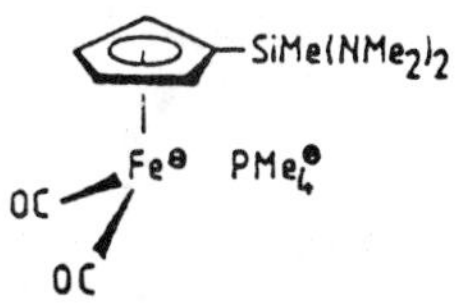

16

The reaction of the ferrio-chlorosilanes 1a, d, e, j and 6a
with the primary amines MeNH$_2$ or tBuNH$_2$ leads to the formation of the bifunctional ferrio-aminosilanes C$_5$R$_5$(CO)$_2$Fe-
SiH(R')NHMe (R = H, R' = Me (17a), iPr (17b); R = Me: R' =
Me (18)) Cp(CO)$_2$Fe-SiMe(R)NHtBu (R = H (19a), Cl (19b)) and
Cp(CO)$_2$FeSIH(NHtBu)$_2$ (20). Thermal stability of these species strongly depend on the substituents at silicon and
iron. While the 17a converts at room temperature to Cp(CO)$_2$-
FeH with formal elimination of the MeHSi=NMe fragment, the
presence of an isopropyl ligand at silicon of 17b, a tert.
butyl group at nitrogen (19 a,b) or methyl groups at the
cyclopentadienyl ligand (18) guarantees compounds of
reasonable stablility.

Ferrio-alkoxy-, -alkylthio-, -phosphino- and-η^1-
pentamethylcyclopentadienylsilanes.

The reactions (i), (ii) illustrated in scheme 2 prove, that
chlorine exchange at the silicon of ferrio-chlorosilanes
opens up an easy access to ferriosilanes with Si-bonded
alkylthio groups. Reactions of 1a with isopropane- or tert.-
butanethiol in the presence of triethylamine, yields the
ferrio-alkylthiosilanes 22a, b. 1d undergoes with butanethiol monosubstitution to give 22c, with isopropanethiol
substitution of both chlorine atoms occurs to yield 21.
Succesive treatment of 1d with tert.-butanol and isopropanol
leads to the formation of the ferrio-dialkoxysilane 23
(iii).
The introduction of phosphino groups is realized in the case
1a, d with alkali phosphides NaPR$_2$ (R = Ph, SiMe$_3$). The
resulting ferrio-phosphinosilanes 24a, b , 25 decompose
within 2 days with Ph$_4$P$_2$ formation. 1a, d reacts with
LiC$_5$Me$_5$ to the ferriosilanes 26a,b containing a η^1-bonded
"mobile" C$_5$Me$_5$ group, which is easily split off by CCl$_4$.

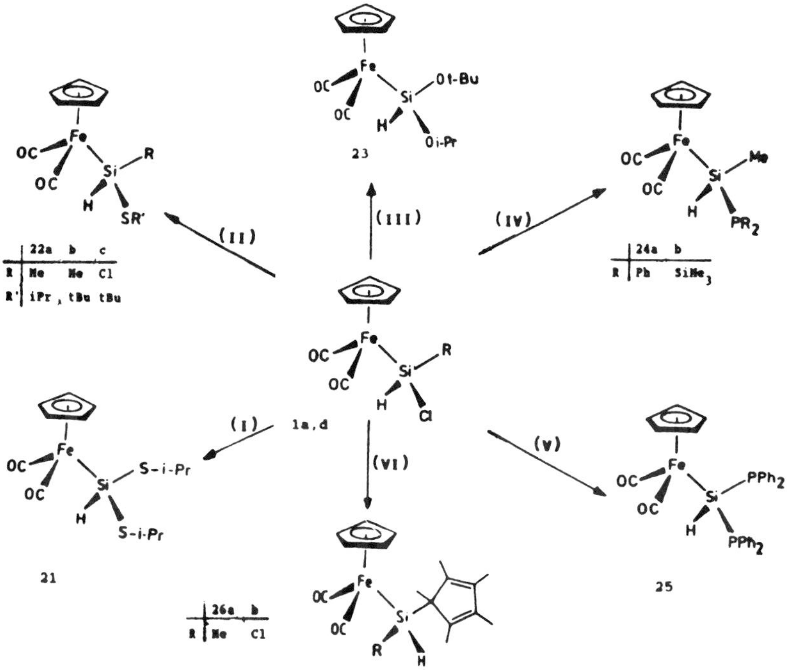

Scheme 2: (i) iPrSH/Et$_3$N; (ii) R'SH/Et$_3$N; (iii) tBuOH, iPrOH; (iv) NaPPh$_2$, LiP(siMe$_3$)$_2$;
(v) NaPPh$_2$ (vi) NaC$_5$Me$_5$.

CARBONMONOXIDE/PHOSPHINE EXCHANGE AT IRON

Photochemical treatment of the ferrio-silanes of scheme 1 and 2 as well as $Cp(CO)_2Fe-SiR_3$ (SiR_3 = SiF_3, SiH_3, $SiPh_2Cl$, $Si(o-Tol_2)H$) in benzene in the presence of an excess of Me_3P results in the displacement of both CO ligands by the phosphorus donor [23] to yield 27a – p.

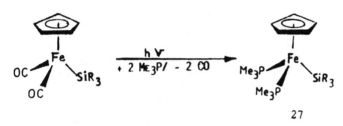

27

27:- SiR_3 = $SiMe_2H$ (a), $SiMe(H)Cl$ (b), $SiHCl_2$ (c), SiF_3 (d),
SiH_3 (e), $SiPh_2Cl$ (f), $Si(o-Tol)_2H$ (g), $SiMe(H)NEt_2$ (h),
$SiH(NEt_2)_2$ (i), $SiMe_2(NMe_2)$ (j), $SiMe(NMe_2)_2$ (k),
$SiCl(NMe_2)NEt_2$ (l), $Si(NMe_2)_3$ (m), $SiCl(Me)NHtBu$ (n),
$SiH(OiPr)_2$ (o), $SiH(Me)C_5Me_5$ (p).

The monosubstitution product is detectable by nmr and ir spectroscopy, but can be isolated only in special cases (SiR_3 = $SiMe_3$, SiF_3). In comparison to the $Cp(CO)_2Fe$ derivatives the values of J(SiH) and v(SiH) are strongly reduced, indicating that a high degree of electron density is accumulated on silicon. In several cases the values are below 150 Hz or 2000 cm^{-1} respectively and are among the lowest so far reported 2 . As the structure of 27c reveals, the silyl- and iron fragment adopt a nearly ideal staggered conformation. The FeSi bonding distance is relatively small and proves Fe-Si bond shortening by good donor ligands. In accordance with the electronic situation the silicon atom is rather inactive towards nucleophiles. For instance 27b exhibits only reactivity towards $Na[Fe(CO)_2Cp]$ leading to the generation of the phosphine substituted bis(ferrio)silane $Cp(Me_3P)_2Fe-SiMe(H)-Fe(CO)_2Cp$ (28), in which two transition metal donor groups are interacting with silicon.

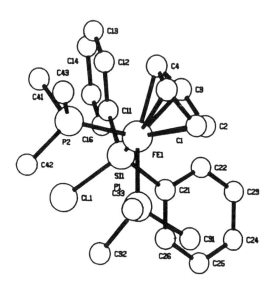

Fig. 1. Molecular structure of $Cp(Me_3P)_2Fe\text{-}SiPh_2Cl$: $d(Fe\text{-}Si) = 225$ pm; $d(Si\text{-}Cl) = 216$ pm.

LIGAND ABSTRACTION EXPERIMENTS

Experiments to reduce the coordination number of silicon in the ferrio-aminosilanes via protonation or alkylation have given so far the following results: **2a** reacts with $HOSO_2CF_3$ to yield $Cp(CO)_2Fe\text{-}SiH(Me)OSO_2CF_3$ (**29**), **7d** gives $C_5Me_5(CO)_2Fe\text{-}Si(NMe_2)_2I$ (**30**) with MeI, while **27m** is converted by $MeOSO_2CF_3$ to the trifluorsilyl complex $Cp(Me_3P)_2Fe\text{-}SiF_3$ (**27d**). In none of these cases the silicenium ion has been obtained, but might exist as a short lived intermediate. Hydride abstraction from $CpL_2Fe\text{-}SiR_2H$ (L = CO, Ph_3P, Me_3P; R = Me, Ph, o-Tol, NMe_2, NEt_2) with $[Ph_3C]ClO_4$ in donor solvents leads to the quantitative formation of the cationic solvent complex $[CpL_2FeS]ClO_4$ (S = THF, CH_3CN), Ph_3CH and polysilanes. In the case of $Cp(Ph_3P)(CO)Fe\text{-}SiR_2H$ (R = Ph, o-Tol) the intermediate existence of a cationic species can be deduced from a considerable increase of the νCO absorption band to higher frequencies, which might belong to the cationic silylene complex $[Cp(CO)(Ph_3P)FeSiR_2]ClO_4$. Experiments to trap the eliminated silylene with 1.4-di-(tert.-butyl)-o-benzoquinone are successful, when $Cp(CO)_2Fe\text{-}SiR_2H$ (R = Me, o-Tol) are employed in the hydride elimation reaction.

Acknowledgements: It is my particular pleasure to acknowledge the contribution of all my coworkers cited in the references. I would also like to thank Prof. W. S. Sheldrick (University of Kaiserslautern) for the crystal structure determination. The work was generously supported by the Deutsche Forschungsgemeinschaft and the Fonds der Chemischen Industrie.

REFERENCES

[1] Apeloig, Y.; Godleski, S. A.; Heacock, D. J.; McKelvey J. M.; Tetrahedron Lett. 22 (1981) 3297.

[2] Aylett, B. J.; Adv. Inorg. Chem. and Radiochem. 25 (1982) 1.

[3] Barton, T. J.; Hovland, A.K.; Tully, C. R.; J. Am. Chem. Soc. 98 (1976) 5695.

[4] Bentham, J. E.; Craddock, S.; Ebsworth, E. A. V.; J. Chem. Soc. A 1971, 587.

[5] Berryhill, S. R.; Sharenow, B.; J. Organomet. Chem. 221 (1981) 143.

[6] Berryhill, S. R.; Clevenger, G. L.; Burdurlu, F. Y.; Organometallics 4 (1985) 1509.

[7] Corey, J. Y.; Gust, D.; Mislow, K.; J. Organomet. Chem. 101 (1975) C7.

[8] Ebsworth, E. A. V.; Ferrier, H. M.; Fraser, T. E.; J. Chem. Soc., Dalton Trans. 1981, 836.

[9] Ebsworth, E. A. V.; Ferrier, H. M.; Rankin, D. W. H.; J. Chem. Soc., Dalton Trans. 1976, 1673.

[10] Godleski, S. A.; Heacock, D. J.; McKelvey, J. M.; Tetrahedron Lett. 1982, 4453.

[11] Höfler, M.; Scheuren J.; Weber, G.; J. Organomet. Chem. 78 (1974) 347.

[12] Höfler, M.; Scheuren, J.; Spilker, D.; J. Organomet. Chem 102 (1975) 205.

[13] Lambert, J. B.; Schulz Jr., W. J.; J. Am. Chem. Soc. 105 (1983) 1671.

[14] Lambert, J. B.; McConnel, J. A.; Schulz Jr.,W. J.; J. Am. Chem. Soc. 108 (1986) 2482.

[15] Malisch, W.; Chem. Ber. 107 (1974) 3835.

[16] Malisch, W.; Kuhn, M.; Chem. Ber. 107, (1974) 2835.

[17] Malisch, W.; Ries, W.; Chem. Ber. 112, (1979) 1304.

[18] Malisch, W.; Wekel, H.-U.; Grob, I.; Köhler, F. H.; Z. Naturforsch., B 37 (1982) 601.

[19] Piper, T. S.; Wilkinson, G.; J. Inorg. Nucl. Chem. 3 (1956) 104.

[20] Ries, W.; Malisch, W.; J. Organomet. Chem. 241 (1983) 321.

[21] Thum, G.; Malisch, W.; J. Organomet. Chem. 264 (1984) C5.

[22] Thum, G.; Ries, W.; Greißinger, D.; Malisch, W.; J. Organomet. Chem. 252 (1983) C67.

[23] Treichel, P.M.; Komar, D. A.; J. Organomet. Chem. 206 (1981) 77.

[24] Wannagat, U; Adv. Inorg. Chem. Radiochem. 6 (1964) 225.

Chapter 31

From aminofluorosilanes to iminosilanes

Uwe Klingebiel — Institute of Inorganic Chemistry, University of Goettingen, D-3400 Goettingen
Federal Republic of Germany.

ABSTRACT

Lithium salts of aminofluorosilanes are stable at room-temperature and
thus can be characterized fully by n.m.r. spectroscopy, crystal structure analysis, etc.
LiF-elimination leads to dimerisations and interconversions of the formed iminosilanes. The roles played by the substituents at nitrogen
and silicon in forming the iminosilane are discussed.
The result of a crystal structure analysis of a lithium salt proves the
iminosilane character of the compound.

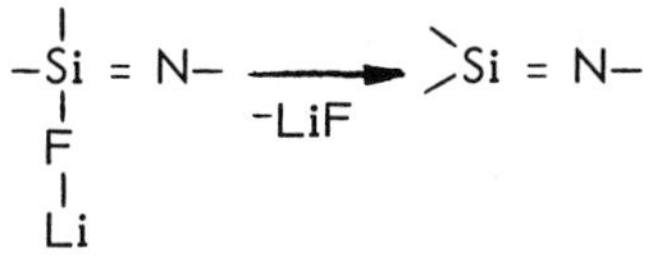

$$\underset{\underset{\underset{Li}{|}}{\overset{|}{\underset{F}{|}}}{-Si}} = N- \xrightarrow[-LiF]{} \overset{\diagdown}{\diagup}Si = N-$$

Reactions of free iminosilanes are presented.

In our research work about ten years ago we started to look for an iminosilane.

$$-\overset{|}{\underset{X}{Si}} - N\overset{/}{\underset{Y}{\diagdown}} \xrightarrow{-XY} \quad \overset{\diagdown}{\diagup}Si = \bar{N} - \quad \longleftrightarrow \quad \overset{\diagdown}{\diagup}\overset{\oplus}{Si} - \overset{\ominus}{\underset{}{\underline{N}}} -$$

A silicon-nitrogen ($p\pi$ - $p\pi$) bond should be very polar, because of the great differences in the electronegativity of the two elements Si (1.8), N(3.0). That means, a silicon-nitrogen double bond is ylidic and has the tendency to undergo oligomerisation reactions for example to dimerize. Therefore it is necessary to use bulky substituents which kinetically stabilize a double bond and prevent dimerisation.

On the other hand, substituents which reduce the polarity of the system should be useful.

Another important point to prepare an iminosilane is the use of a thermodynamically good leaving group, for example an electronegative substituent at the silicon atom, for instance halogeno atoms, and an electropositive ligand at the nitrogen atom, for instance alkali metals. We started to synthesize primary aminofluorosilanes and tried to lithiate them. We used fluorine compounds because SiF-NH-groups do not condense   as easily as the chlorine compounds.

$$-\overset{|}{\underset{F}{Si}} - N\overset{/}{\underset{H}{\diagdown}} \xrightarrow{+\ RLi} \begin{array}{c} \underset{-RH}{\overset{-LiF}{\longrightarrow}} \quad [\,\overset{\diagdown}{\diagup}Si{=}N{-}]\longrightarrow 1/2\ (\overset{\diagdown}{\diagup} Si{-}N{-})_2 \\[2ex] \underset{-LiF}{\longrightarrow} \quad \overset{\diagdown}{\diagup}\overset{|}{\underset{R}{Si}} - N\overset{/}{\underset{H}{\diagdown}} \end{array}$$

Substitution compounds and four-membered rings were the first results of these experiments. The formation of the four-membered ring was interpreted by us as (2+2) cycloaddition of an intermediary iminosilane. If the ligands at the aminofluorosilane are bulky enough no dimerisation and LiF-elimination takes place.

In 1976 we isolated the first lithium salts of aminofluorosilanes, for example, $(CMe_3)_2SiF\text{-}NLiCMe_3)$.

Unfortunately, rather high temperatures were necessary to eliminate LiF. Therefore we failed to synthesize an iminosilane.

We had to find a better leaving group, for example LiCl, and we had to reduce the polarity of the system $\overset{\oplus}{>}Si-\overset{\ominus}{N}-$ by better ligands which stabilize the system thermodynamically. We synthesized aminofluorosilanes with $-NSiMe_3$, $-OSiMe_3$ and $-CSiMe_3$ groups. It's known that these substituents stabilize elements in low coordination. Very often such systems are isoelectronic to stable π-systems in carbon chemistry. Because of the bulkiness of the substituents dimerisation of the intermediate iminosilanes does not occur easily, but in silicon chemistry novel cyclisation mechanisms were found.

Two examples:

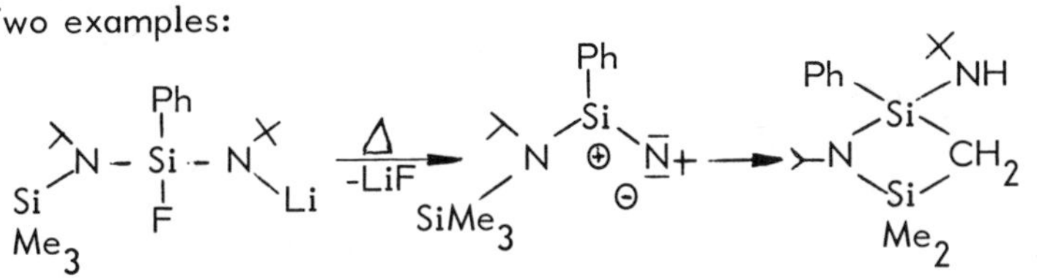

Using the isopropylsilylamino-substituent cyclisation occurs by cleavage of the polar SiC-H bond. If the size of the substituents is increased even more, for example by replacing the isopropyl-group by a second tert-butyl-group, as in the following molecule, another surprising ring closure occurs. Two unsymmetrically substituted Si_2N_2 rings are the products of this reaction.

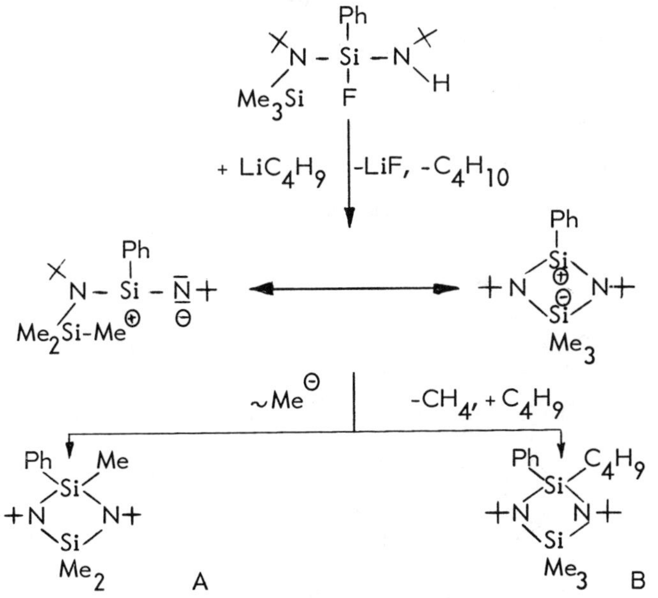

The formation of the cyclic system A can be attributed to a nucleophilic 1.3-rearrangement of a methanide-ion at silicon. The butyl-sub-
stituted ring B is isolated in low yields, if the lithium salt isn't isolated first. In this case the cyclic ylid is attacked by the butanide ion
which is in solution.

The cyclic ylid is formed because the two nitrogen atoms are symme-
trically substituted. The silyl-group doens't know to which nitrogen
atom it should go. But what happens if the central silicon carries one
tert.-butyl-amino-group and one bis(trimethylsilyl)amino-group.

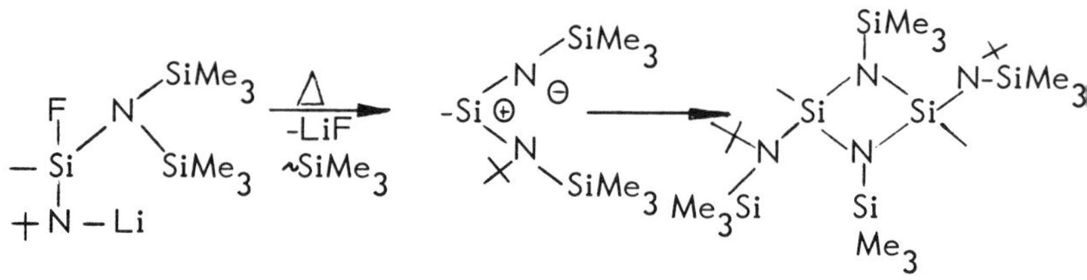

Now one of the silyl-groups notices the more negatively charged orga-
nyl nitrogen, and these lithium salts cyclisize by thermal LiF-elimination with simultaneous 1.3-migration of a silyl-group from one nitrogen
atom to the other.

The silyl-group migration can be explained by the propensity of silyl-
groups to be bonded to the more negatively charged atom. The iminosilane B dimerizes. The 1.3-silyl-group migration in such systems is
proved by crystal structure of a dimer isopropylaminoiminosilane.

A silyl-group migration would be without energy profit in a molecule
which is substituted with silylamino-groups on both sides.

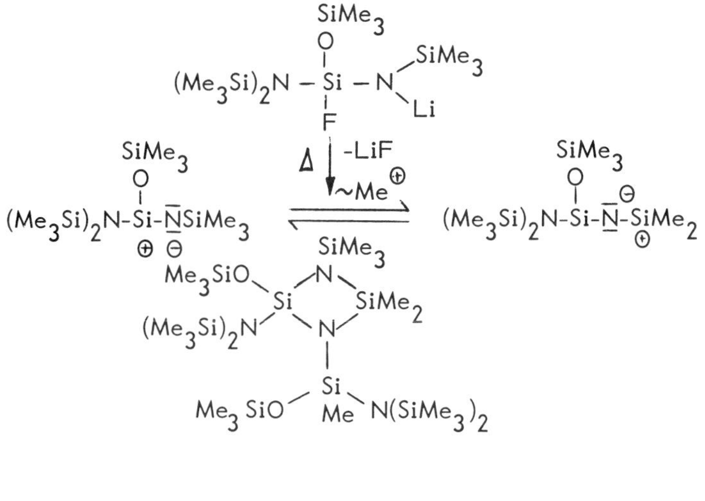

Now only half a mole of the ensuing iminosilane undergoes methanide ion migration. The result is that a crossdimer of the two ylides is formed in high yields. Two diastereoisomers could be separated by recrystallisation. So much for the mechanistic proof of intermediate iminosilanes. Now I come to the chemistry of the lithium salts of aminofluorosilanes.

$$-\overset{|}{\underset{\underset{F}{|}}{Si}} - N\overset{\diagdown}{\underset{\underset{Li}{\diagup}}{}} \xrightarrow[-LiF]{\Delta} \quad {>}Si = N- \longrightarrow \tfrac{1}{n} ({>}Si\text{-}N\text{-})_n$$

Insertion   a-b   2 x dimerisation a=b   cycloaddition

$$-\overset{|}{\underset{\underset{a}{|}}{Si}} - N\overset{\diagup}{\underset{\diagdown b}{}} \qquad {>}Si\overset{\diagup N \diagdown}{\underset{\diagdown N \diagup}{}}Si{<} \qquad {>}Si\overset{\diagup N \diagdown}{\underset{\diagdown a \diagup}{}}b$$

These lithium salts react like iminosilanes. Insertions in polar bonds occur, dimerisation reactions occur and cycloadditions are possible. Some examples:

$$>SiF - N\overset{\diagup}{\underset{\diagdown Li}{}} \xrightarrow{-\ LiF\ +RH} -\overset{|}{\underset{\underset{R}{|}}{Si}} - N\overset{\diagdown}{\underset{H}{}}$$

Insertion products are obtained in reactions with H-acid compounds and silanoles, silylethers, thioethers and acetylenes are formed. Of course a two-step mechanism can be discussed in these reactions. (2+2) Additions occur in reactions with iminophosphanes and SiN_2P-rings are formed. Carbodiimides, sulfurdiimides and azomethines are obtained with isocyanates, thionylamides and aldehydes. For example

$$-\overset{|}{\underset{\underset{F}{|}}{Si}} - N\overset{\diagup}{\underset{\diagdown Li}{}} \xrightarrow[-\ LiF]{O=C=N-} \left[-\overset{|}{\underset{\underset{O\ -\ C=N-}{|}}{Si}} - N\overset{\diagup}{}\right] \xrightarrow{(-\dot{S}iO)_n} -N=C=N-$$

These reactions are analogous to Wittig-reactions. The Si-N-bond is cleaved by metal carbonyls, and isonitril complexes are obtained.

If two different lithium salts are present cross dimerisations can take place.

Lithium salts of aminofluorosilanes react like an "ene" with "dienes".
That means Diels-Alder reactions may be carried out and six-membered rings are obtained.

So much for the proof that iminosilanes are indeed intermediates in
these reactions. Another point of interest can be seen in reactions of
lithium salts of aminofluorosilanes with halides of main group elements,
some of which I'll discuss here, namely with phosphorus, aluminium,
and silicon halides. The ratio, lithium salt: halide, is 2:1 and rings are
obtained. An example is shown in the reaction with phosphorus halides.
Initially formed silylaminophosphanes react under formation of iminophosphanes. It is always the difluorosilane that is cleaved   in these
reactions.

No bis(silyl)amino-organyl iminophosphan is isolated but always the
silyliminophosphan instead, which means that here also a silyl-group
migration to the more negatively charged nitrogen occurs. The isoelectronic silicon compounds dimerize as shown before. The results of
the reaction of aluminium halides with our lithium salts depend on the
reaction conditions. Three examples: 1. A dimer $AlCl_2$ compound is
obtained if $AlCl_3$ is added to a lithium salt of an aminofluorosilane
in unpolar solvents.   2. If the lithium salt is added to $AlCl_3$ a disubstituted product is obtained not a dimer but a monomer.

$\delta^{29}Si = 28$ pm

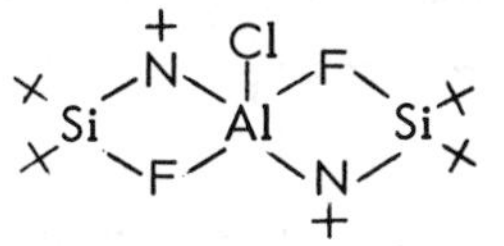

SiF: 170.8 pm

SiN: 169.7 pm

Aluminium is now five coordinate on account of its Lewis acid character. The Si-F bond is extremly long, the Si-N bond is correspondingly
short. The nitrogen is planar. So this compound can be described as
the two to one adduct of an iminosilane with an aluminum halide.

3. In ether as solvent however the one to one adduct is formed. The
compound is a monomer. Hydrolysis leads to an aminoalan with an
intact $AlCl_3$-group.

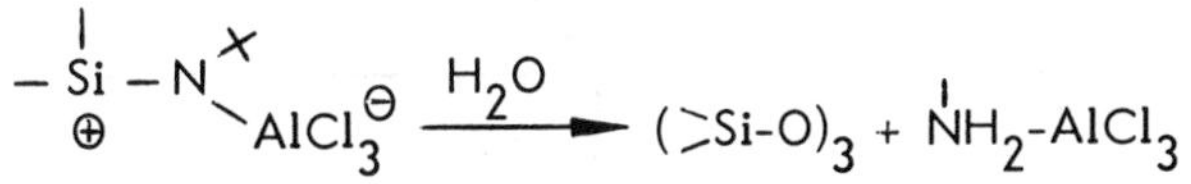

The ^{29}Si-n.m.r. resonance is below + 50 ppm.

To put two and two together these are the first silicon ylids with
three-coordinate silicon.

Experiments to synthesize such iminosilane adducts by cleavage of
Si-N-rings led to another interesting result.

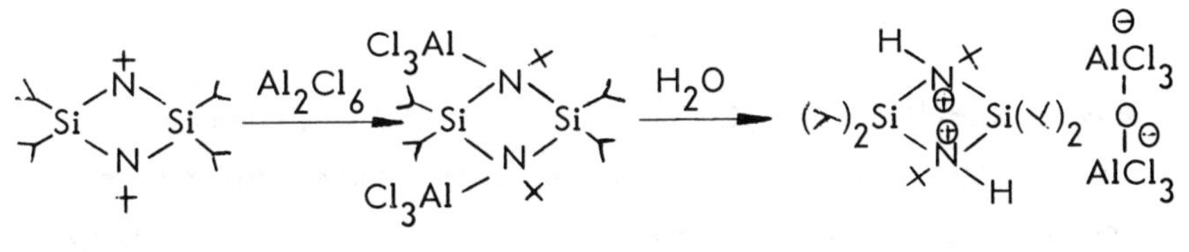

SiN = 174.7 pm

$\delta^{29}Si = 7$ ppm

Si-N = 179.3 pm

$\delta^{29}Si = +30$ ppm

Adduct formation was observed. Their hydrolysis products are the
first cationic Si-N-rings with protonated nitrogen and correspondingly
very long Si-N-single bonds.

The nitrogen is sp^3 hybridized. The ring is unsaturated.

Last but not least some reactions of lithium salts of aminofluorosilanes
with halogenosilanes: In reactions with fluorosilanes the high reactivi-

ty of the lithium salts allows the synthesis of very bulky molecules with
a frozen Si-N-bond at room-temperature.

Reactions of the salts of aminofluorosilanes with chlorosilanes give
very surprising results.

Until today we haven't been successful in isolating an $SiCl_3$-substitu-
ted compound in the reaction with $SiCl_4$.

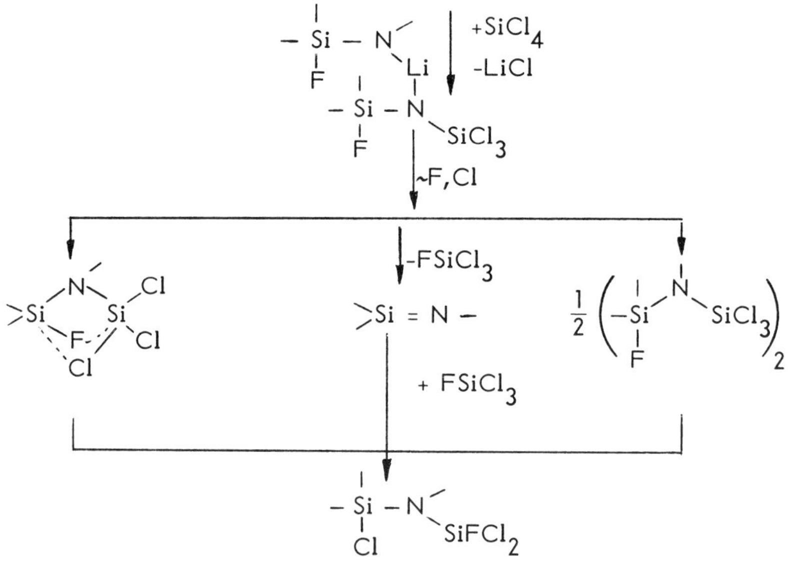

However first there is LiCl-formation and the $SiCl_3$-compound can be
identified by n.m.r. spectroscopy. Very soon - sometimes minutes sometimes hours later - a new signal appears in the ^{19}F-n.m.r. spectra.
It is the signal of a compound formed in a fluorine-chlorine exchange.
The possible exchange mechanisms are intra or inter molecular. The
exchange is fast, if the two [29]Si-n.m.r. shifts of the $SiFNSiCl_3$ compound differ a lot. In other words the fluorine atom looks for and
goes to the more positively charged silicon atom. Let's now look at
the result of a crystal structure analysis of a lithium salt of an aminofluorosilane.

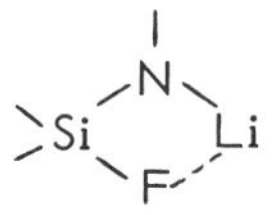

There is an electronegative atom and an electropositive atom in the
molecule and it is understandable that these two atoms try to be bonded to each other, a simple Coulomb effect. On the other hand the
nitrogen atom is also electronegative - more if an alkyl-group is bonded - less if there is an aryl-group bonded because of the plus M
effect. The tendency to be bonded to the lithium is therefore smaller.
If the lithium salt is in solution or crystallizes in an nonpolar solvent,
for instance in hexane, the lithium is bonded to the fluorine and the
nitrogen atom. The Si-F bond is lengthend. The molecule often is a
dimer: $(\geq SiFNLi-)_2$.
If the lithium salt is in solution or crystallizes in a polar solvent,
for instance in THF, the oxygen of THF is present. Now the lithium
is able to coordinate the oxygen of THF and to loosen the bond to
the nitrogen. The result often is the formation of a monomer.

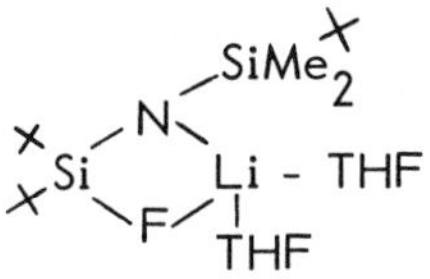

If the basicity of the nitrogen is lowered further the electropositively
charged lithium goes to the most negatively charged atom in the
molecule, which is the fluorine, not the nitrogen atom.

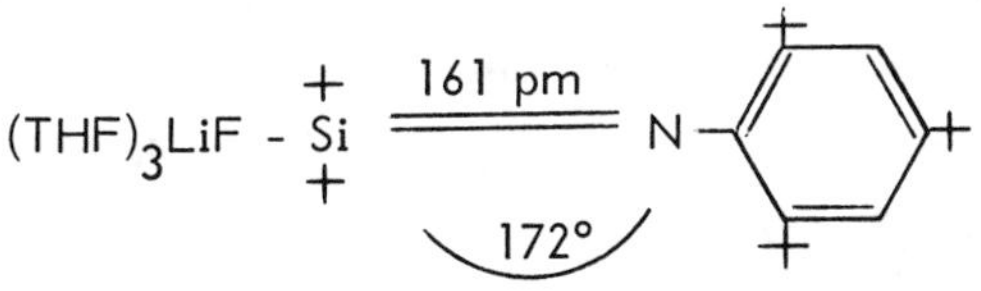

The lithium salt of the aminofluorosilane is really a LiF-adduct of
an iminosilane. Now it is understandable that these salts react like
iminosilanes. The crystal structure proves the imine character of the
nitrogen. Silicon has a tetrahedral environment and five bonds. The
Si-N bond is a double bond. Now it is also understandable that the
chemical shift appears at high field in the ^{29}Si-n.m.r. experiment.
The ^{19}F-n.m.r. signal shifts to lower field compared with the amino-

fluorosilane. Experiments to eliminate LiF lead to dimerisation. The cyclodisilazan is a strongly stretched four-membered ring with the longest Si-N bonds in a ring which are known so far. The molecule is totally asymmetric.

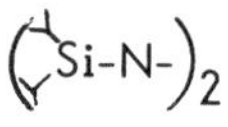

$$\left(\text{Si-N-}\right)_2$$

The CC_2-isopropyl–groups are twisted. All this shows that here the limits of dimerisation are reached.

Now what happens if trimethylchlorosilane is added to this lithium salt? To our astonishment fluorine was displaced by chlorine leading to a lithium aminochlorosilane. Now we had the leaving group we always wanted our efforts were rewarded.

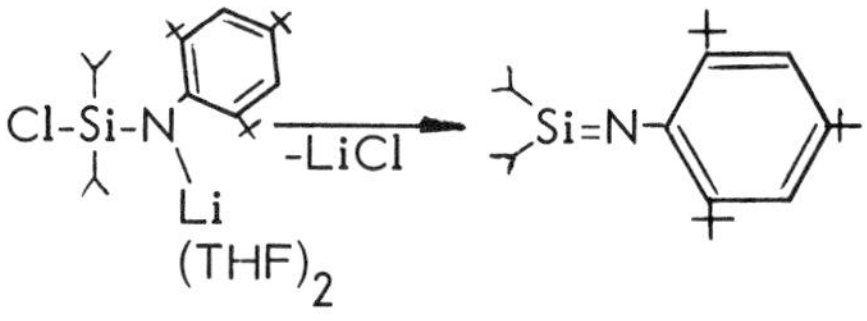

$$\text{Cl-Si-N} \xrightarrow{\text{-LiCl}} \text{Si=N-} \quad + \quad \delta^{29}\text{Si:} \quad 60.3 \text{ ppm}$$

Attempts to eliminate LiCl lead to the wished for iminosilane.

The iminosilane is orange, thermally rather stable and a chromophor. The ^{29}Si-n.m.r. shift is significant for three coordinate silicon compounds.

Remarkable reactions of iminosilanes are Wittig, Diels-Alder, $AlCl_3$ and cycloadditions, but remember it is not always necessary to isolate the iminosilane because the lithium salt if the iminosilane very often reacts the same way.

Chapter 32

The $HSiR_3/CO/Co_2(CO)_8$ catalytic reactions

Shinji Murai*, **Naoto Chatani** and **Toshiaki Murai** — Department of Applied Chemistry, Faculty of Engineering, Osaka University.

INTRODUCTION

The new catalytic reactions, $HSiR_3/CO/Co_2(CO)_8$, that uses a hydrosilane and carbon monoxide as a reagent and cobalt carbonyl as a catalyst precursor, enables incorporation of carbon monoxide into various oxygen-containing substrates [1, 2]. The cobalt carbonyl reacts with the hydrosilane to give trialkylsilylcobalt tetracarbonyl (1)[3-5], a key catalyst (1). The latter reacts, in turn, with the oxygen containing substrate to give an intermediate having a carbon-cobalt bond, that is essential for incorporation of carbon monoxide. In this paper, described are the selected examples of the new catalytic reactions based on the new methodology for the formation of a carbon-transition metal bond.

$$Co_2(CO)_8 \xrightarrow[-H_2]{R_3SiH} 2\ R_3SiCo(CO)_4 \qquad (1)$$

$$\mathbf{1}$$

CATALYTIC CARBONYLATION OF KETONES AT THE KETONIC CARBON ATOM

Incorporation of carbon monoxide in a catalytic manner into organic molecules at their carbonyl

group may be one of the most important but most
difficult reactions in the chemistry of carbon
monoxide. The difficulty lies in the unavailability
of the efficient methods for the bond formation
between the carbonyl carbon and the transition metal
in a catalyst species. We have reported a clean
catalytic carbonylation reaction of aldehydes
(2)[6]. The key catalytic step in this reaction is

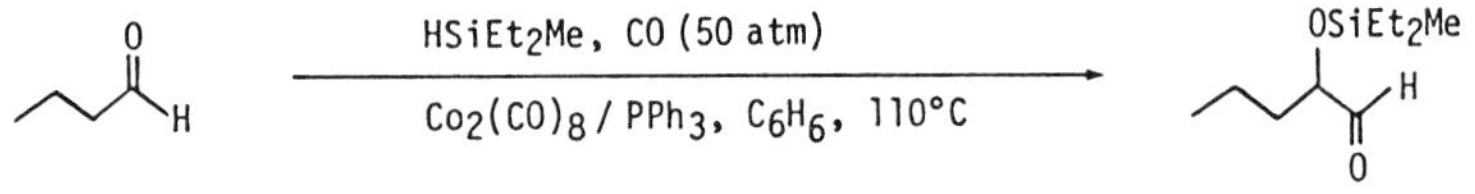

$$\text{(aldehyde)} \xrightarrow[\text{Co}_2(\text{CO})_8 / \text{PPh}_3, \text{C}_6\text{H}_6, 110°C]{\text{HSiEt}_2\text{Me}, \text{CO (50 atm)}} \text{(product OSiEt}_2\text{Me)} \quad (2)$$

the formation of the carbon-cobalt bond (3) that is
necessary for the incorporation of carbon monoxide.

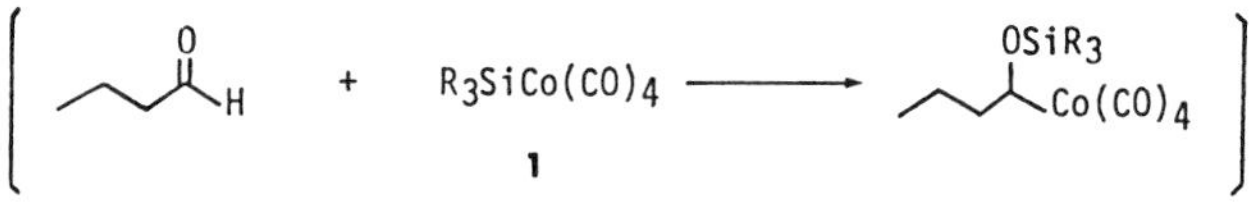

$$\left[\text{(aldehyde)} + \underset{\textbf{1}}{R_3\text{SiCo(CO)}_4} \longrightarrow \text{(product} \underset{}{\text{OSiR}_3}, \text{Co(CO)}_4) \right] \quad (3)$$

On the other hand, the catalyst species in the
$H_2/CO/M$ or $H_2O/CO/M$ system is generally a hydrido
metal carbonyl ($HM(CO)_n$) and this species tends to
add across a carbonyl group in a opposite way to
give an O-metal intermediate but not desired C-metal
intermediate (4). This is one of the reasons of
difficulty in realizing catalytic C-carbonylation.

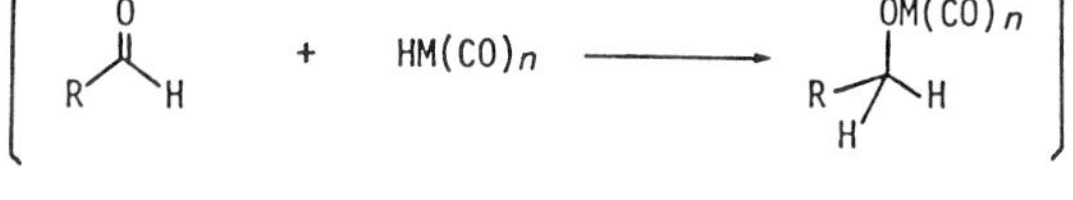

$$\left[\underset{R}{\overset{O}{\parallel}}\text{C}{}_H + HM(CO)_n \longrightarrow R\underset{H}{\overset{OM(CO)_n}{\mid}}H \right] \quad (4)$$

No example of catalytic C-carbonylation of
ketones has been reported. We have attempted the
carbonylation of cyclohexanone or acetophenone by
$HSiR_3/CO/Co_2(CO)_8$ reaction (5). The product

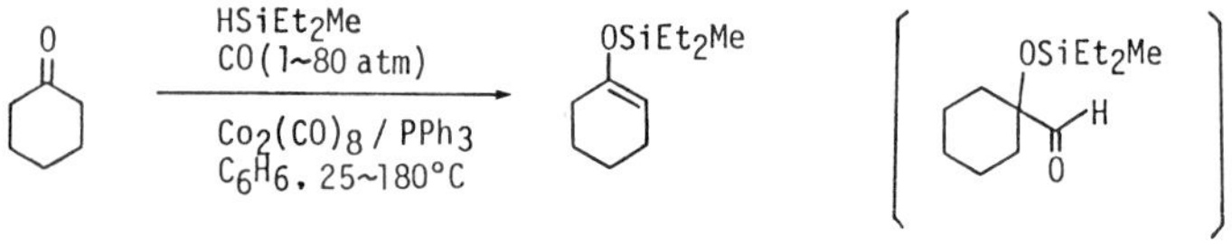

$$\underset{\text{cyclohexanone}}{\overset{O}{}} \xrightarrow[\text{Co}_2(\text{CO})_8 / \text{PPh}_3\; \text{C}_6\text{H}_6, 25\sim180°C]{\text{HSiEt}_2\text{Me}\; \text{CO}(1\sim80\;\text{atm})} \text{(OSiEt}_2\text{Me)} \quad \left[\text{(OSiEt}_2\text{Me, H, O)} \right] \quad (5)$$

(not formed)

obtained, however, was an enol silyl ether [7] and any product that might arise from incorporation of carbon monoxide was not detected (5).

This result suggested that an α-siloxyalkyl-cobalt complex (2), if formed, underwent β hydride elimination to afford the enol silyl ether instead of the desired CO insertion (6).

Therefore we examined the reaction of cyclo-

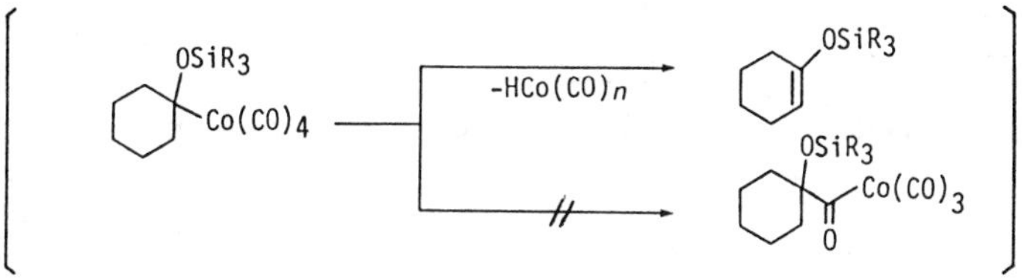

$$
\left[
\begin{array}{c}
\text{(6)}
\end{array}
\right]
$$

butanone as a ketone that might hardly afford the corresponding enol silyl ether because of ring strain. The clean incorporation of carbon monoxide has been observed for cyclobutanone (7)[8]. The reaction provides the first example of transition metal catalyzed C-carbonylation of a ketone, to the best of our knowledge.

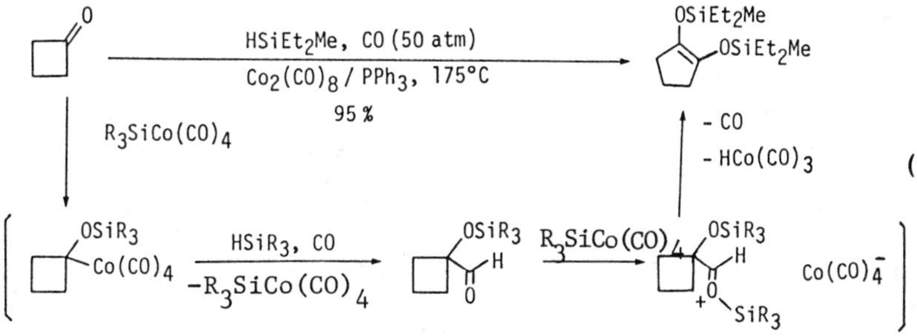

$$
\text{(7)}
$$

The new chemistry is applicable to cyclopenten-one annulation at an olefinic linkage [8], which constitutes a useful synthetic operation (8, 9).

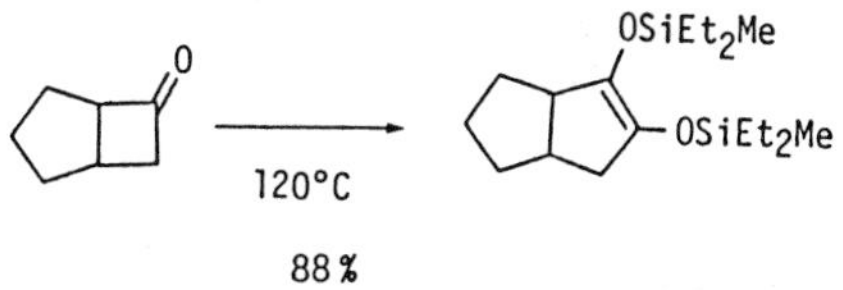

$$
\text{(8)}
$$

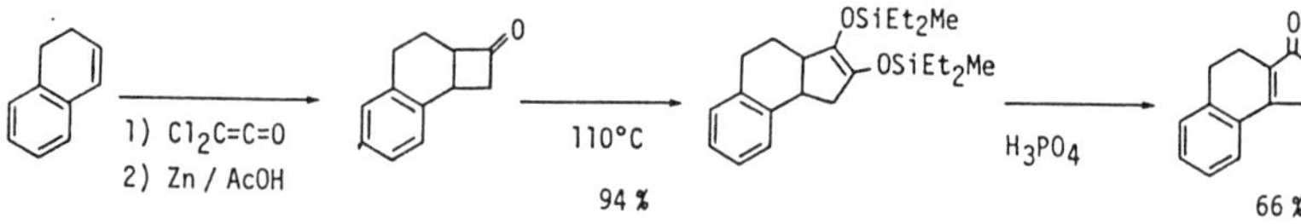

(9)

CATALYTIC SILOXYMETHYLIDENATION OF ESTERS

Silylcobalt carbonyl **1** has high reactivity towards various oxygen-containing compounds. The high affinity of the silicon in **1** (hard acid center) to the oxygen atom of a oxygen-containing substrate (hard base) initiates the reaction. For example, the reaction of **1** with an alkyl acetate provides an new entry to an alkylcobalt complex (10).

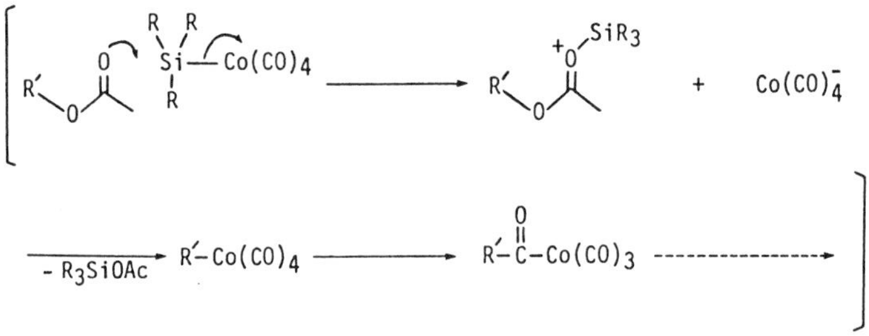

(10)

Thus, the ester group is activated into a better leaving group by silylation, which in turn is displaced by a good nucleophile $Co(CO)_4^-$ (soft base) leading to an alkylcobalt complex.

The $HSiR_3/CO/Co_2(CO)_8$ catalytic reaction of cyclohexyl acetate has been examined in detail [9, 10]. Discovered was a new type of transformation that enables the introduction of synthetically useful siloxymethylidene group upon cleavage of an ester linkage (11). The selectivity was best at 200°C for this particular reaction. The reaction did not take place below 100°C and it suffered from by-product formation at 150°C.

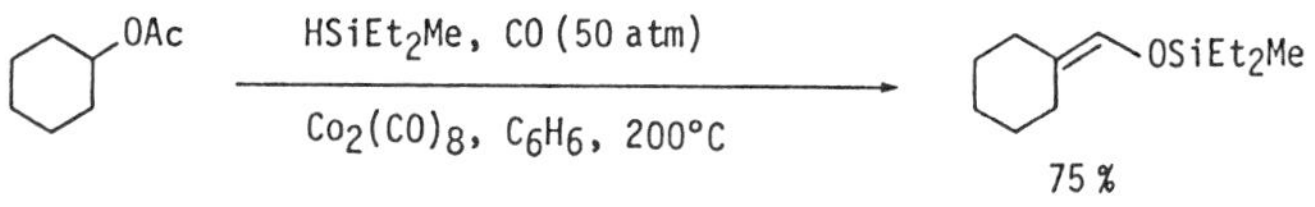

(11)

The catalytic steps leading to the product and the steps responsible for regeneration of the key catalyst **1** are suggested as follows, (12)(13).

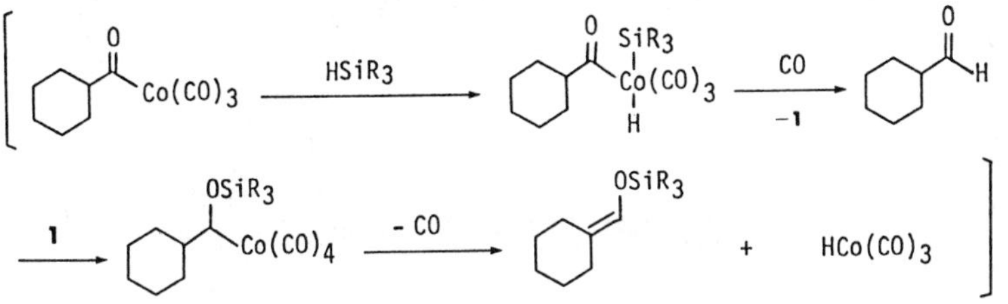

$$ \text{(12)} $$

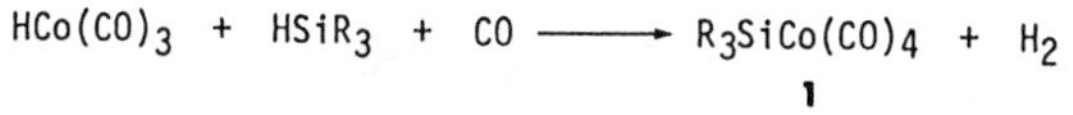

$$ HCo(CO)_3 + HSiR_3 + CO \longrightarrow R_3SiCo(CO)_4 + H_2 \qquad (13) $$
$$ \mathbf{1} $$

The relatively low yields (75 %) may be attributed to a side reaction from cyclohexylcobalt complex, an initial intermediate, to give cyclohexene. Indeed, 2-adamantyl acetate gives a good yield of the product, since the similar olefin formation is almost impossible with this molecule (14). A chlorine-containing ester (15) and an aliphatic ester (16) react similarly. The siloxymethylidene groups are known amenable to various synthetic transformations [11], for example, aldehyde formation (14).

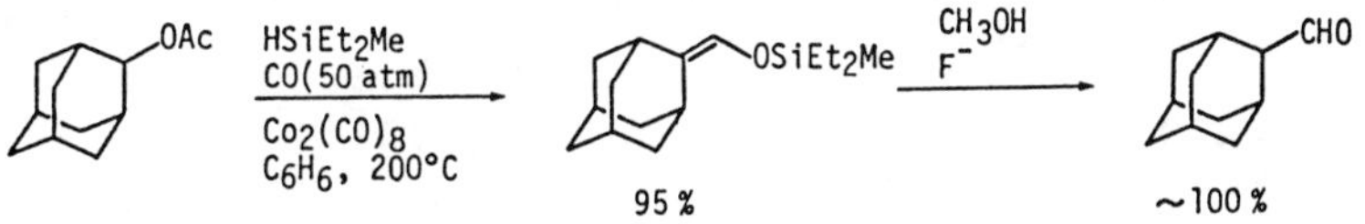

$$ \text{(14)} $$

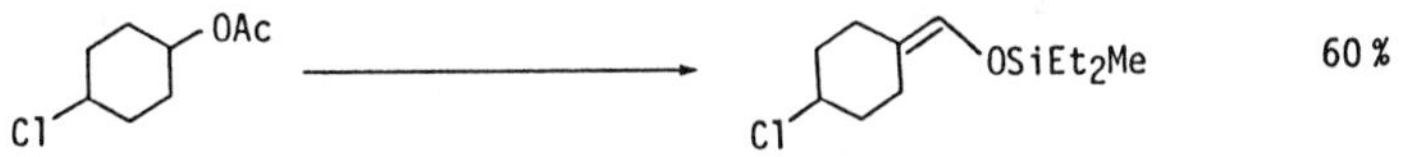

$$ \text{(15)} $$

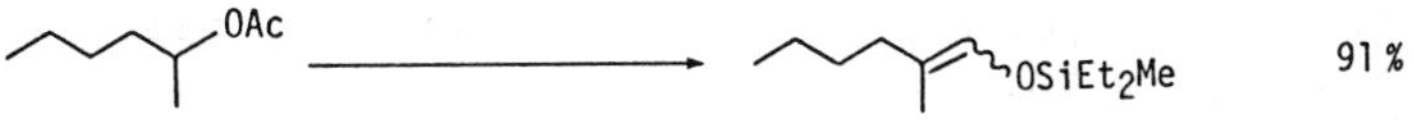

$$ \text{(16)} $$

Lactones react in the same manner except for that the silyl group remains in the products in the form of silyl esters. Examples given below are introduction of a carba-functional group to a cyclopentane ring (17) and the synthesis of a keto

aldehyde involving in situ protection/deprotection
of the ketone part followed by decarboxylation of
β-keto acid (18).

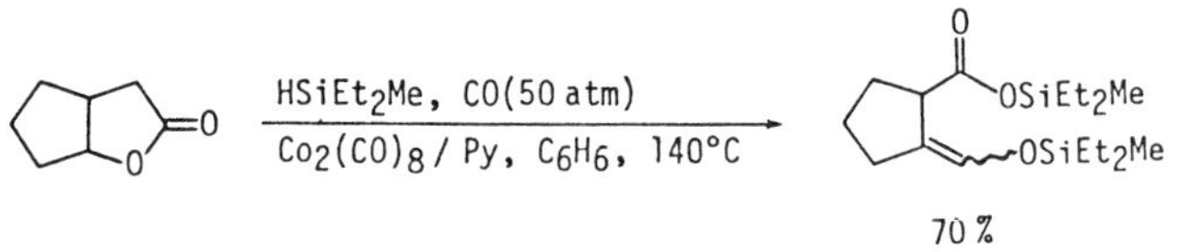

70 %   (17)

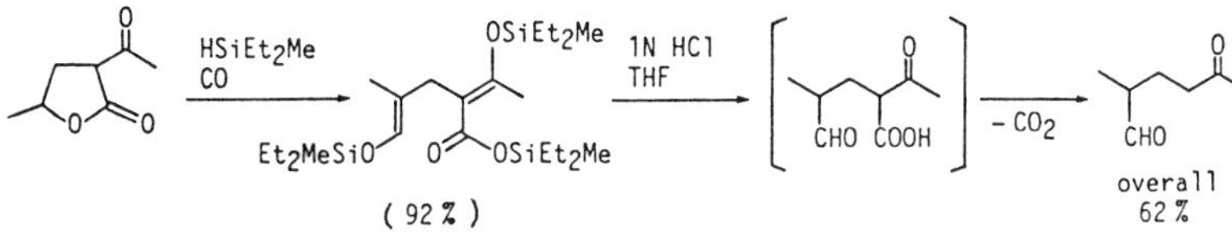

(92 %)   (18)

overall
62 %

CATALYTIC SILOXYMETHYLATIVE RING OPENING OF OXIRANES

The $HSiR_3/CO/Co_2(CO)_8$ catalytic reaction of
oxiranes has been found to proceed under exceptionally mild reaction conditions [12], an example being
given in eq 19.

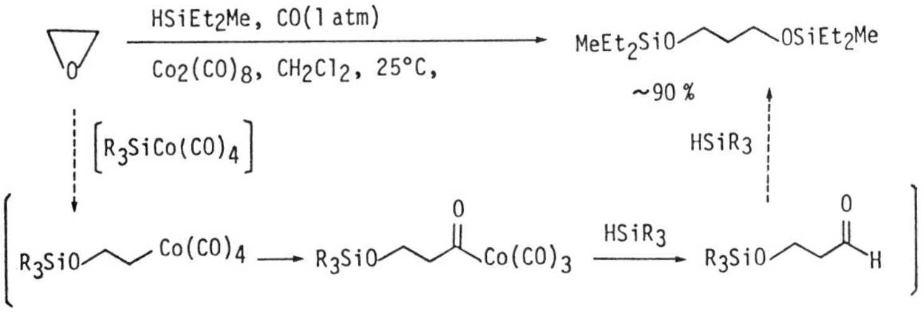

~90 %   (19)

Two novel features should be mentioned for this
reaction. Firstly, the reaction takes place at 25°C
and 1 atm. Secondly, carbon monoxide has been
incorporated in the form of siloxymethyl group, a
saturated form. It seems that, if an oxygen-
containing compound is sufficiently reactive, the catalytic reaction proceeds at room temperature and at
this temperature the intermediary aldehyde undergoes
hydrosilylation (simple addition of $HSiR_3$).

The catalytic reaction of oxiranes has been
revealed to be general, selective, and synthetically
useful.

The ring opening is stereoselective and gives trans products (20, 21). The stereochemistry is determined by back-side attack of Co(CO)$_4^-$ to an initially formed silyloxonium ion (22).

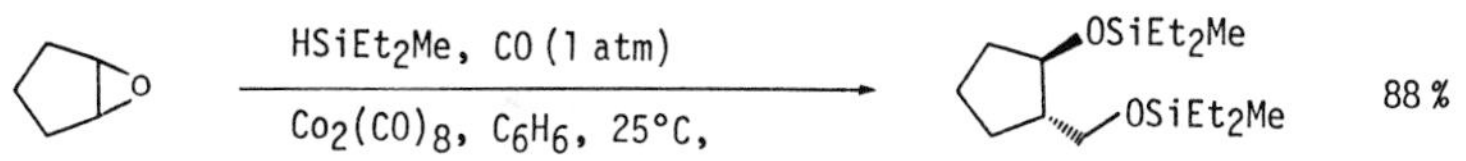

$$88\% \quad (20)$$

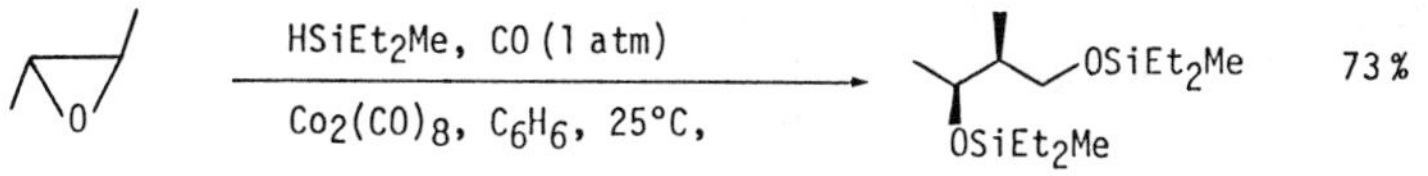

$$73\% \quad (21)$$

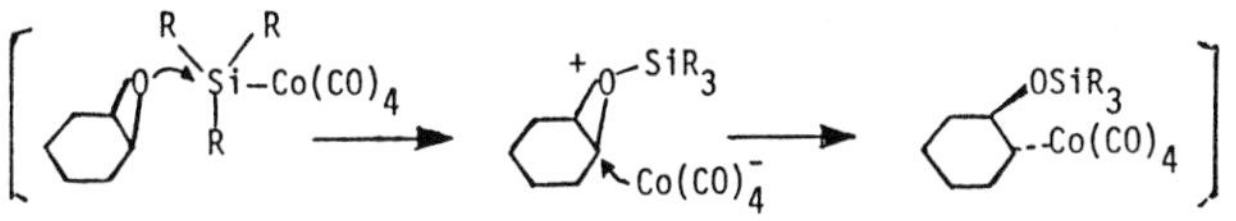

$$(22)$$

The catalytic reaction is regioselective in many cases studied. For oxiranes with one alkyl group, a mixture of regioisomers is obtained (23). Those having an electron-withdrawing substituent undergo ring opening selectively at the site away from the substituent (24-26).

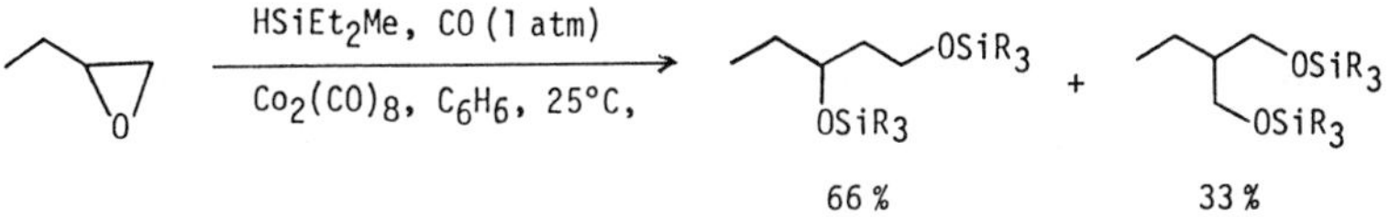

$$66\% \qquad\qquad 33\% \qquad (23)$$

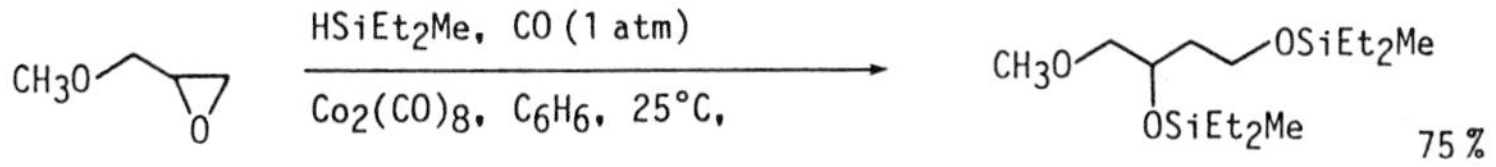

$$75\% \quad (24)$$

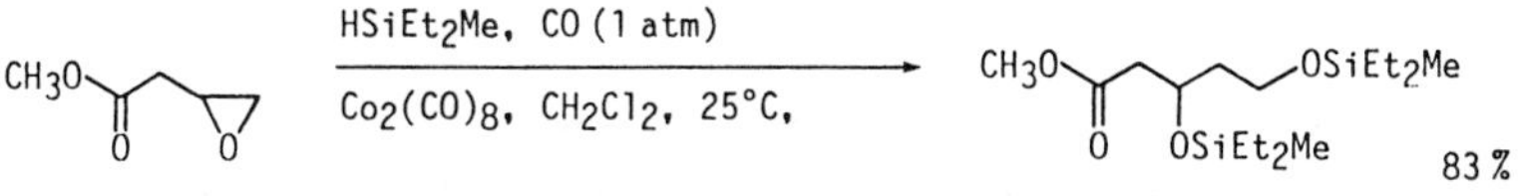

$$83\% \quad (25)$$

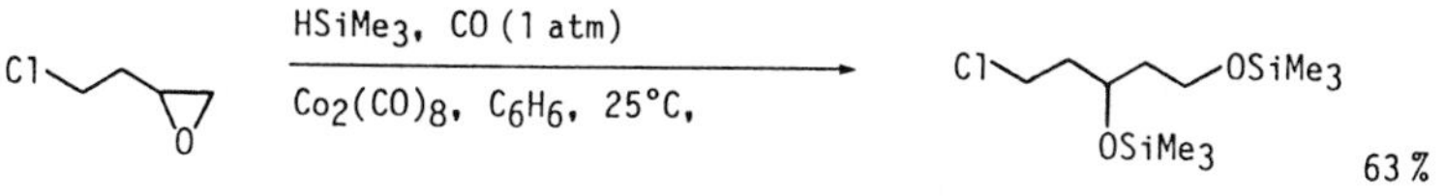

$$63\% \quad (26)$$

The electron-withdrawing group would have suppressed the positive charge development at the internal carbon atom in the transition state, thus directing the Co(CO)$_4^-$ to the terminal carbon (27).

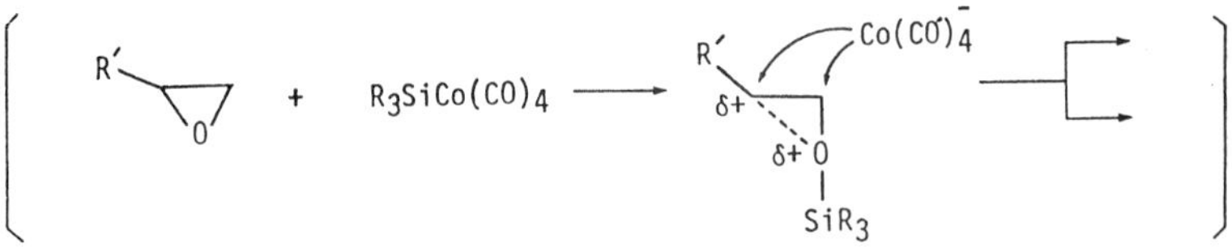

$$\left[\quad \overset{R'}{\underset{O}{\triangle}} \quad + \quad R_3SiCo(CO)_4 \quad \longrightarrow \quad \overset{R'}{\underset{\underset{SiR_3}{\delta^+ O}}{\underset{\delta^+}{\diagdown}}} \overset{Co(CO)_4^-}{\diagup} \quad \longrightarrow \quad \boxed{} \quad \right] \qquad (27)$$

From the synthetic point of view, the present transformation is formally a nucleophilic oxymethylation (28) that is difficult in a conventional manner.

$$\overset{R'}{\underset{O}{\triangle}} \quad + \quad {}^-CH_2OR \quad \longrightarrow \quad \overset{R'}{\underset{O^-}{\diagdown}} \diagup \diagdown OR \qquad (28)$$

Regioselective siloxymethylation is possible also in the case of some internal oxiranes when chloroacetic acid ester is employed as the directing group.

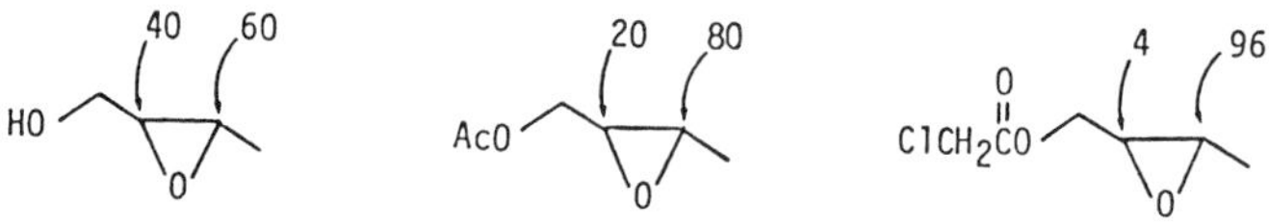

In the similar manner, four and five membered cyclic ethers also undergoes catalytic ring opening with incorporation of carbon monoxide (29, 30)[13-15].

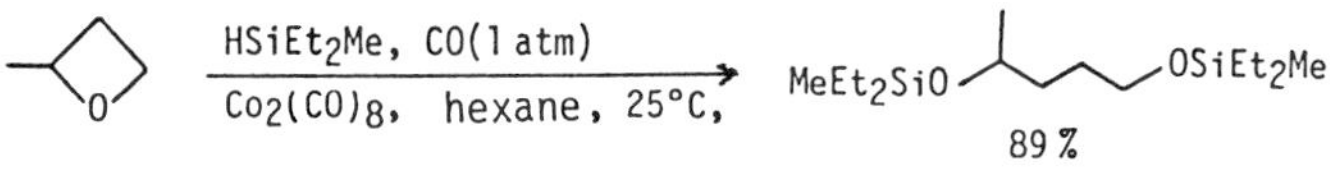

$$\underset{O}{\square} \quad \xrightarrow[\text{Co}_2(CO)_8, \text{ hexane, } 25°C,]{\text{HSiEt}_2\text{Me, CO(1 atm)}} \quad MeEt_2SiO\diagdown\diagup\diagdown\diagup OSiEt_2Me \qquad (29)$$
$$89\ \%$$

$$\underset{O}{\square}\diagdown \quad \xrightarrow[\text{Co}_2(CO)_8, \text{ C}_6\text{H}_6, 25°C,]{\text{HSiEt}_2\text{Me, CO(1 atm)}} \quad MeEt_2SiO\diagdown\diagup\diagdown\diagup OSiEt_2Me \qquad (30)$$
$$84\ \%$$

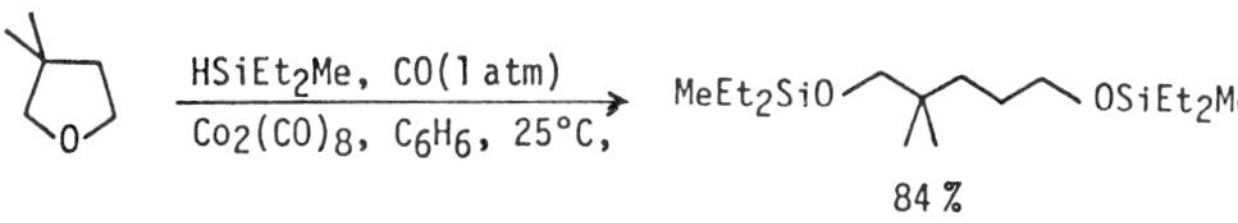

OUTLOOK

The $HSiR_3/CO/Co_2(CO)_8$ catalytic reaction involves new methodology for the formation of a carbon-transition metal bond. The high affinity of silicon in $R_3SiCo(CO)_4$ towards oxygen in the substrate accounts for the driving force of the bond formation. The discovery of the reaction proceeding under very mild conditions has expanded the scope of this catalytic reaction and incorporation of carbon monoxide into other substrates such as reactive esters, acetals, and other S_N1-reactive compounds are now becoming possible.

ACKNOWLEDGEMENT

We gratefully acknowledge Professor Noboru Sonoda of Osaka University and Professor Shinzi Kato of Gifu University for their discussion and support. Thanks are due to our able coworkers whose names are appered in the references. A generous gift of chlorosilanes by Shin-Etsu Chemical Industry Ltd. is gratefully acknowledged. This work was supported in part by grants by The Ministry of Education, Science, and Culture, Japan.

REFERENCES

[1] S. Murai, N. Sonoda, Angew. Chem., Int. Ed. Engl., **18**, 837(1979).
[2] S. Murai, Y. Seki, J. Mol. Catal., in press.
[3] A. J. Chalk, J. F. Harrod, J. Am. Chem. Soc., **89**, 1640(1967).
[4] Y. L. Baay, A. G. MacDiarmid, Inorg. Chem., **8**, 986(1969).
[5] A. Sisak, F. Ungray, L. Marko, Organometallics, **5**, 1019(1986).
[6] S. Murai, T. Kato, N. Sonoda, Y. Seki, K. Kawamoto, Angew. Chem., Int. Ed. Engl., **18**, 393 (1979).
[7] H. Sakurai, K. Miyoshi, Y. Nakadaira, Tetrahedron Lett., 2671(1977).
[8] N. Chatani, H. Furukawa, T. Kato, S. Murai, N. Sonoda, J. Am. Chem. Soc., **106**, 430(1984).
[9] N. Chatani, S. Murai, N. Sonoda, J. Am. Chem. Soc.,**105**, 1370(1983).
[10] N. Chatani, S. Fujii, Y. Yamasaki, S. Murai, N. Sonoda, J. Am. Chem. Soc., **108**, 7361(1986).
[11] K. Ruhlmann, Synthesis, 236(1971).
[12] T. Murai, S. Kato, S. Murai, T. Toki, S. Suzuki, N. Sonoda, J. Am. Chem. Soc., **106**, 6039 (1984).
[13] T. Murai, Y. Hatayama, S. Murai, N. Sonoda, Organometallics 2, 183(1983).
[14] T. Murai, S. Kato, S. Murai, Y. Hatayama, N. Sonoda, Tetrahedron Lett., **26**, 2683(1985).
[15] T. Murai, K. Furuta, S. Kato, S. Murai, N. Sonoda, J. Organomet. Chem., **302**, 249(1986).

Chapter 33

New enthusiasm for metal silicides: their relationship to zintl phases

Bernard J. Aylett – Department of Chemistry, Queen Mary College, London E1 4NS, UK.

B.A. Scott, R.D. Estes and D.B. Beach – IBM Thomas J. Watson Research Center, Yorktown Heights, NY 10598.

ABSTRACT

Examples of classical Zintl compounds are first given, and it is shown how metal silicides may exhibit related or differing structures. In these, silicon atoms (with partial negative charge) can be isolated or joined together in clusters, rings, chains or 3-d networks. It is shown how the useful properties of transition metal silicides lead to a variety of applications, particularly in electronic devices. Finally, the use of molecular precursors with silicon-metal bonds to provide thin films of metal silicides via prevenient chemical vapour deposition is described.

ZINTL PHASES [1,2]

Almost 100 years ago, A. Joannis (1891) reported that a Na/Pb alloy dissolved in liquid ammonia to give a dark green solution; analysis of the derived solid gave a composition of Na_4Pb_9, and it was suggested that the solution contained Pb_9^{4-} ions. Extensive later work by E. Zintl (1930 onwards) showed that many similar compounds between alkali or alkaline-earth metals and later main group elements could be produced, and species involving ions such as Sn_9^{4-}, Pb_7^{4-} and Sb_7^{3-} were proposed.

The first structural study of a Zintl-type solid by D. Kummer [3] on the rather unstable adduct Na_4Sn_9 . 7en (en = 1,2-diaminoethane) gave evidence for an anion of D_{3h} symmetry, i.e. a tri-capped trigonal prism. More recently, J. D. Corbett [4] has shown that much more thermally stable species may be isolated if the cation is firmly complexed by a polydentate cryptand ligand such as $N[(CH_2CH_2O)_2CH_2CH_2]_3N$ (crypt). Thus the dark red compounds $(crypt.K^+)_4 M_9^{4-}$ (M = Ge, Sn) are found by X-ray analysis to involve an anion of C_{4v} symmetry, viz. a capped square antiprism. This may be be described in terms of Wade's rules as a nido 9-atom 22-electron species, which is isoelectronic with Bi_9^{5+}. Similarly, the anion Ge_9^{2-} of D_{3h} symmetry is isoelectronic and isostructural with $B_9H_9^{2-}$. Thus Zintl compounds can be regarded as naked members of the large family of cluster compounds which includes in particular metal carbonyls, boranes, and carboranes.

METAL SILICIDES

In contrast to the compounds of Ge, Sn and Pb above, silicides of electropositive metals have not yet been shown to produce discrete anionic clusters in solution. There is plenty of evidence, however, to show that in the solid state these silicides contain a wide variety of species, some of which are analogous to Zintl anions. Table 1 shows a range of examples; it may be noted that the formal anions in entries (a) - (e) are analogous and sometimes isoelectronic to Ar, $C\ell_2$, B_4H_{10}, P_4, and S_n respectively. Similarly the formal anion in the mixed Zintl species $M_4^{II}SiAs_4$ is analogous to SiO_4^{4-}. Entry (g) has a particularly unusual structure, with stacked Si_5 rings and Si_4 stars, cemented together by Li cations [5].

Silicides of transition metals exhibit a correspondingly wide range of structural types (Table 2). As before, high metal : silicon ratios tend to produce alloy-like arrangements, but lower metal : silicon ratios often lead to chains or networks of silicon atoms, doubtless carrying a partial negative charge [6 - 8]. Non-stoicheiometric structures, such as silicon-rich MnSi or silicon-poor α-FeSi$_2$, are quite common; sometimes compounds with large complex unit cells such as $Mn_{27}Si_{47}$ are found.

PROPERTIES AND USES

In common with related transition metal compounds such as borides, carbides and nitrides, transition metal silicides possess a number of very useful properties [8]. Thus they have high melting points, are very resistant to chemical attack, and are hard, while at the same time their densities are relatively low. They have therefore found application as inert surface coatings on metals, particularly where resistance to oxidation at high temperatures is needed.

Table 1. Some silicides of electropositive metals.

Compound	Structural features	Formal anion
(a) Ca_2Si	isolated Si	Si^{4-}
(b) Li_7Si_2	Si_2 pairs + isolated Si	Si_2^{6-}, Si^{4-}
(c) Ba_3Si_4	Si_4 butterfly	Si_4^{6-}
(d) NaSi	Si_4 tetrahedra	Si_4^{4-}
(e) SrSi (form I)	zig-zag Si chain	Si_n^{2n-}
(f) SrSi (form II)	Si_6 rings with extra Si atoms at 1, 2, 4, 5 - positions	Si_{10}^{20-}
(g) $Li_{12}Si_7$	planar Si_5 rings + trigonal planar Si_4 stars	Si_5^{6-}, Si_4^{12-}
(h) $CaSi_2$	puckered layers of fused Si_6 rings	-
(i) Na_8Si_{46}	defect Si_n lattice, Na in holes	-
(j) $Li_{22}Si_5$	} alloy-like structures	-
(k) $Li_{10}Si_3$		-

Table 2. Examples of transition metal silicides.

Compound	Structural features
(a) V_3Si	isolated Si, 12-coordinated by V, in 3-d lattice
(b) Co_2Si	isolated Si, 3-d lattice like $PbCl_2$
(c) U_3Si_2	Si_2 pairs
(d) USi	zig-zag Si_n chain
(e) β-USi_2	puckered layers of fused Si_6 rings
(f) $ThSi_2$	3-d network of Si
(g) $CrSi_2$	Cr and Si in each close-packed layer

Most interest centres, however, on their use in the fabrication of microelectronic devices [9]. Transition metal silicides are generally quite good electrical conductors (in the "poor-metallic" range), although some, such as β-$FeSi_2$ and WSi_2 are semiconductors and can be doped to produce devices [10]. They are widely used as barrier layers to prevent mutual interdiffusion between silicon substrates and the metallic connections needed to link them to the outside world. Other actual or projected applications include the formation of interconnects in MOSFET devices, Schottky barriers, and multiple quantum well devices using silicon technology rather than the more developed III/V materials approach.

PRESENTDAY PREPARATIVE ROUTES

The simplest way to make transition metal silicides is to heat together the metal and silicon in appropriate proportions. This is the basis of the method routinely used to make thin films for electronic devices : a thin layer of metal is sputtered on to the silicon surface, the whole is heated at 800-900°C , and silicon is abstracted from the surface layer to produce one or more silicide phases [8]. This method has three disadvantages:

(i)   phase composition is hard to control
(ii)   the relatively high temperature needed may harm other circuit elements
(iii) abstraction of silicon may affect the device characteristics (e.g. by causing migration of dopants)

Consequently other methods have been proposed, particularly molecular beam epitaxy, plasma transport, and chemical vapour deposition (CVD) <u>via</u> gaseous mixtures of silicon and metal compounds. The last-mentioned, CVD, is attractive because it is potentially the most suitable for large-scale commercial application and can be linked with existing CVD production steps. However, .in practice, CVD reactions such as (1) and (2) can lead to mixtures of phases and contamination of the product by halide [11] :

$$WCl_5 + SiCl_4 \xrightarrow[600°C]{H_2} WSi_2 \quad etc. \quad (1)$$

$$TiCl_4 + SiH_4 \xrightarrow[600°C]{} TiSi_2 \quad etc. \quad (2)$$

PREPARATION BY MOCVD WITH MOLECULAR PRECURSORS

A different approach involves the use of metal-organic chemical vapour deposition (MOCVD), in which the strong silicon-metal bond is already present in the precursor. By thermolysis, pyrolysis, etc. of the precursor molecules, groups L and L′ attached respectively to Si and the metal M are removed, leaving Si-M units which rearrange to form the silicide (eqn. 3).

$$L_m Si - ML'_n \longrightarrow \{Si - M\} \longrightarrow Silicide \quad (3)$$

precursor 1

Because the silicon and metal have already come together in the synthesis of the precursor, we refer to this as the prevenient approach.

Some 10 years ago, we reported that flow pyrolysis of suitable SiH_3-metal compound in the presence of a large excess of inert gas can indeed lead to thin films of transition metal silicides [12]. Thus the volatile compound $SiH_3Co(CO)_4$ yielded CoSi (eqn. 4) :

$$SiH_3Co(CO)_4 \xrightarrow[excess Ar]{450°C} CoSi \quad (4)$$

b.p. 112°C

A number of other potential precursors 1 are known for which L = H, L′ = CO in eqn. (3); they are summarised in Table 3 [13, 14].

Table 3. Possible H-Si-M-CO precursors.

Si:M ratio	Compound	
1:1	$H_3SiM(CO)_n$	M = Co, Ir (n = 4) Mn, Re (n = 5) V (n = 6)
	$H_3SiFeH(CO)_4$	
2:1	$(H_3Si)_2Fe(CO)_4$	
1:2	$H_2Si[M(CO)_n]_2$	M = Co (n = 4) Mn, Re (n = 5)
1:3	$HSi[M(CO)_n]_3$	M = Mn, Re (n = 5)

In all cases so far studied, the overall resulting Si:M ratio in the product silicide is the same as that in the precursor, although product disproportionation may occur [12, 15], e.g.

$$H_3SiMn(CO)_5 \quad \xrightarrow{\quad 450^o \quad} \quad Mn_5Si_3 \quad + \quad MnSi_n \qquad (5)$$
$$(n \sim 1.25)$$

It will be noted that no compounds containing Ti, Zr, Hf, Cr, Mo, W, Ni, Pd or Pt appear in Table 3 : none of this type has yet been synthesised. However, a number of such compounds with other L and L' groups are known [13], e.g. $Me_3SiW(CO)_3(C_5H_5)$. Can they be used as MOCVD precursors? Experience has shown that alkyl and aryl groups attached to silicon (i.e. L = CH_3, C_6H_5, etc in $\underset{\sim}{1}$) are undesirable, in that they lead to incorporation of carbon in the final product. Similarly C_5H_5 attached to the metal can give the same result (e.g. when using $H_3SiCr(CO)_3(C_5H_5)$ as a precursor) [16]. Further synthetic work is needed to identify other ligands L and L' in $\underset{\sim}{1}$ which can be cleanly removed.

FUTURE OUTLOOK

The prevenient approach to the formation of binary metal silicides by MOCVD is clearly a very general one, and is capable of extension to ternary silicides, etc., and also to germanides, borides, etc. and to mixed species. Because of the generally low temperatures used in their formation, there is the possibility of forming new low-temperature phases, not readily accessible by other methods. Even low volatility of the molecular precursor species may not be an insuperable obstacle, since other deposition techniques such as spray deposition and photo- or plasma-assisted processes may prove suitable. The reward could be the realization of a cheap silicon-based technology, using thin films with almost infinite variety of composition, and yielding devices with novel characteristics.

The rate-determining step in all this will be the development of new synthetic routes to provide appropriate precursors, containing the same proportions of "core" elements as the desired end product. This presents a considerable challenge to silicon chemists.

REFERENCES

1. H. Schäfer, B. Eisenmann, and W. Müller, Angew. Chem. Int. Ed. Eng., 12 694 (1973); J. D. Corbett, Prog. Inorg. Chem. 21, 129 (1976).

2. H.-G. von Schnering, Angew. Chem. Int. Ed. Eng., 20, 33 (1981); idem, Amer. Chem. Soc. Symposium Series, 232, 69 (1983). See also D. M. P. Mingos, Chem. Soc. Rev., 15, 31 (1986).

3. D. Kummer and L. Diehl, Angew. Chem. Int. Ed. Eng., 9, 895 (1970); D. Kummer et al., Chem. Ber. 109, 3404 (1976).

4. J. D. Corbett, S. C. Critchlow and R. C. Burns, Amer. Chem. Soc. Symposium Series, 232, 95 (1983).

5. M. C. Böhm, R. Ramirez, R. Nesper, and H.-G. von Schnering, Phys. Rev. B, 30, 4870 (1984); idem, Ber. Bunsenges. Phys. Chem. 89, 465 (1985).

6. A. F. Wells, "Structural Inorganic Chemistry", Oxford University Press, 5th. edition (1984).

7. B. Aronsson, T. Lundström and S. Lundqvist, "Borides, Silicides and Phosphides", Methuen, London (1965).

8. B. J. Aylett, British Polymer J., 18, 359 (1986).

9. S. P. Murarka, "Silicides for VLSI Applications", Academic Press, New York, 1983.

10. B.-Y. Tsaur and C. H. Anderson, Appl. Physics Letters, 47, 527 (1985).

11. M. J. Cooke, Vacuum, 35, 67 (1985); P. K. Tedrow, V. Ilderem, and R. Reif, Appl. Physics Letters, 46, 189 (1985). But see also: A. Bouteville, A. Royer and J.-C. Remy, Proceedings 6th EUROCVD Conference, Jerusalem, Israel, March 1987, p.264.

12. B. J. Aylett and H. M. Colquhoun. J. Chem. Soc. Dalton Trans., 2058 (1977).

13. B. J. Aylett, Adv. Inorg. Chem. Radiochem., 25, 1 (1982).

14. B. J. Aylett and M. T. Taghipour, J. Organomet. Chem., 249, 55 (1983).

15. B. J. Aylett and A. A. Tannahill, Vacuum, 35, 435 (1985).

16. B. J. Aylett, M. J. Hampden-Smith and A. A. Tannahill, unpublished results.

PART VIII

SILICON IN
SOLID STATE TECHNOLOGY

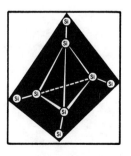

Chapter 34

The chemical vapor deposition
of silicon thin films

B.A. Scott, R.D. Estes and D.B. Beach – IBM Thomas J. Watson Research Center, Yorktown Heights, NY 10598.

ABSTRACT

The growth of silicon thin films by chemical vapor deposition (CVD) involves reactions that occur in the gas phase and at the gas-solid interface. The results of conventional CVD and HOMOCVD experiments using SiH_4/Ar mixtures are presented for temperatures between 400-650°C and pressures of 1-100 Torr. By varying the silane partial pressure at constant total pressure and the gas residence time, the deposition process can be separated into regimes in which either heterogeneous or homogeneous SiH_4 reactions dominate the kinetics of film growth. Using this approach, the contribution of each process can be assessed for a range of conditions encompassing various CVD and silane pyrolysis experiments.

INTRODUCTION

The chemical mechanistic aspects of silicon CVD still remain poorly understood despite the importance of this process to the semiconductor industry. In the case of silicon growth from SiH_4, for example, only a few of the simplest gas phase and surface processes have been delineated with any certainty. The state of understanding of important film growth issues have been considered from the physical chemistry and chemical engineering perspectives in two recent reviews [1,2].

In this paper we present an investigation of silicon thin film deposition from SiH_4 under conventional CVD (isothermal) and HOMOCVD (non-isothermal) conditions. Emphasis is placed upon experiments at relatively high pressures and low temperatures. This allows us to make a connection to previous studies of silane pyrolysis. Most of these investigations have generally ignored silicon, a major final product of the decomposition reaction. As a consequence, we are able to provide insight into SiH_4 pyrolysis itself, especially with respect to the role of heterogeneity in the reaction.

HOMOGENEOUS CHEMICAL VAPOUR DEPOSITION (HOMOCVD)

HOMOCVD [3] employs hot SiH_4 gas (typically, $T_g > 600°C$) to produce very reactive intermediates which deposit to form a thin film on a substrate held on a cooled pedestal at a much lower temperature ($T_s < 400°C$). A diagram of the apparatus used for the experiments is shown in Fig. 1. The method permits the growth of amorphous hydrogenated silicon, a-Si:H, containing as much as 40 atomic% hydrogen, at substrate temperatures too low for conventional SiH_4 or higher silane CVD.

Previous modelling of HOMOCVD attributed film growth at such low substrate temperatures to silylene (SiH_2) intermediates generated in the homogeneous decomposition which diffuse to the cold substrate to form a-Si:H [4]. Process simulations showed the buildup of SiH_2 and primary thermolysis products (H_2, Si_2H_6) in the hot gas. The SiH_2 concentration in the thermal gradient at the substrate was estimated to be large enough to explain the observed growth rates [4]. These simulations used kinetic parameters available from pyrolysis and copyrolysis experiments [5-7] to calculate the concentrations of gaseous products present in the reactor.

Recently, laser-induced fluorescence and laser absorption spectroscopy have been applied to the direct determination of the rate constants for the reaction of SiH_2 with H_2, SiH_4 and Si_2H_6 [8,9]. The new experiments definitively prove that the previous estimates [7] of these quantities are low by two to four orders of magnitude. In Fig. 2, we have recalculated the concentrations of SiH_2, Si_2H_6 and Si_3H_8 in the bulk gas employing the new rate constants in a simple gas kinetic model [10], including RRKM corrections [11]. The results are presented for a reaction time of 1 sec, which is characteristic of gas residence times in most CVD reactors. On the basis of previous considerations [4], it can be shown that the SiH_2 concentrations in Fig. 2 are now too low to account for the growth of HOMOCVD a-Si:H films.

Disilane is a primary homogeneous pyrolysis product (Fig. 2) earlier excluded from consideration as a film growth precursor because of the low substrate temperatures of HOMOCVD [12]. Recently, we have obtained more accurate estimates of T_s using IR pyrometry, and the results are shown in Fig. 3 for gas temperatures of 550 and 650°C, the range of most HOMOCVD experiments. The pyrometric measurements were made through a water-cooled NaCl window mounted directly on the reactor (Fig. 1). This experiment, however, does not measure temperature directly at the surface, but rather the temperature distribution within a very thin layer whose thickness is approximately equal to an absorption depth at the sensing wavelength of the pyrometer. Therefore, the results of Fig. 3 should be considered lower limits, and we conclude that Si_2H_6 and higher silanes are probably decomposing at pedestal temperatures $T_s > 200°C$. Under these conditions, the actual surface temperatures are probably in excess of 350°C. Films of a-Si:H have been grown from Si_2H_6 and Si_3H_8 in a conventional CVD reactor [13,14] at temperatures above 380°C. Moreover, recent measurements of the silicon hydride sticking probabilities under UHV dosing conditions [15] show trisilane to be reactive on hydrogen-covered Si(111) surfaces even at room temperature, and disilane is found to chemisorb strongly on bare Si(111). Both are a thousand times more surface-reactive than SiH_4 [15]. Higher silanes could therefore account for the growth of a-Si:H films in HOMOCVD at higher temperatures; however this does not fully explain how films containing large amounts of hydrogen deposit at $T_s < 200°C$. Experiments described in the following section suggest that other gas phase processes are involved. As discussed subsequently, these reactions also can play a role in film deposition under isothermal CVD conditions, and will prove to be a clue to the nature of silane pyrolysis beyond the early stages of decomposition.

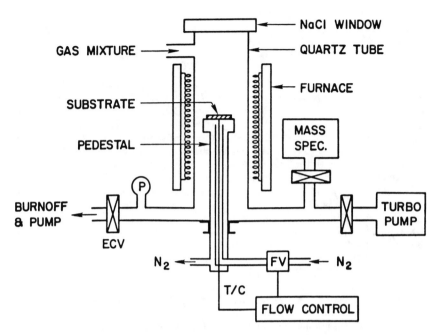

Fig. 1. Diagram of the CVD reactor. Constant pressure (P) is achieved with automatic valve ECV. The gas flows are controlled in a separate manifold. The substrate temperature in HOMOCVD is maintained with a flow of nitrogen to the sample pedestal.

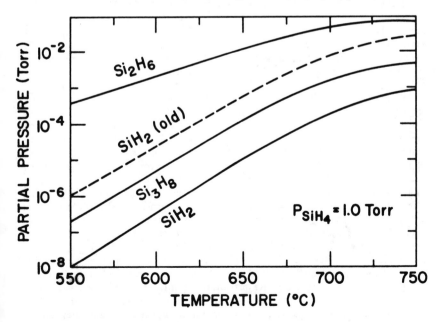

Fig. 2. Product concentrations after 1 sec of SiH_4 homogenous pyrolysis at 1 Torr calculated with the rate constants of refs. 8,9 (solid curves and ref. 7 (dashed curve).

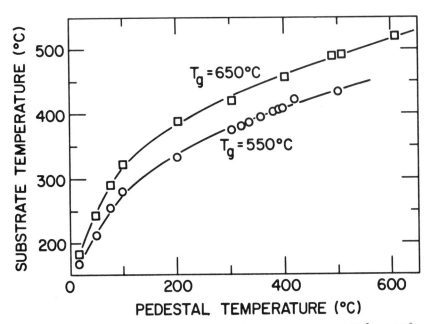

Fig. 3. HOMOCVD substrate (SiO_2) temperatures for gas temperatures of $550°$ and $650°C$ measured with an IR pyrometer at a total gas pressure of 3 Torr. The pedestal temperature is measured with a thermocouple (T/C) imbedded in the sample holder just below the substrate (see Fig. 1).

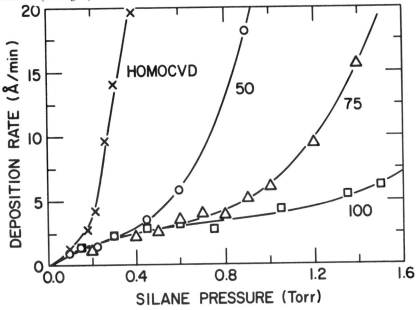

Fig. 4. Silicon film deposition rate vs. silane partial pressure at $465°C$ and a total pressure of 15 Torr (Ar diluent) for flow rates of 50, 75 and 100 sccm. HOMOCVD results are presented for $T_g=640°C$ and $T_s=300°C$ (pyrometrically measured).

SILICON FILM GROWTH UNDER ISOTHERMAL CONDITIONS

Silylene has been postulated as a film growth intermediate in many models of CVD silicon growth from SiH_4 [16]. From our previous results and discussion, it is clear that this assumption also must be seriously questioned, for conventional CVD is carried out isothermally at temperatures and pressures similar to those in HOMOCVD. While its high reactivity leads to a very low steady state concentration of silylene in the gas, the higher silane products of SiH_2 reactions (Fig. 2) can contribute significantly to film growth. Indeed, such intermediates could be transformed into even more reactive species at the high temperatures of most CVD processes. It is therefore very important to determine the conditions under which homogeneous or heterogeneous reactions dominate the kinetics of film formation.

The following series of isothermal ($T_s = T_g$) and HOMOCVD ($T_s \ll T_g$) experiments were therefore devised. They were carried out at relatively low temperatures (400°-650°C) and high pressures (P(tot) = 1−100 Torr), so as to overlap with previous SiH_4 pyrolysis studies. By maintaining constant total pressure and varying the SiH_4 partial pressure, we are able to observe two distinct growth regimes, as shown in Fig. 4 for isothermal CVD experiments at 464°C. At low SiH_4 partial pressure (region I) the growth rate (R_G) follows a Langmuir-Hinshelwood (LH) type mechanism in which $R_G \propto P(SiH_4)$ followed by a pressure dependence which appears close to saturation. This is most clearly depicted in the 100 sccm data of Fig. 4 for $P(SiH_4) \lesssim 1$ Torr. Measurements at 418°C are presented in Fig. 5, where R_G saturates rapidly and then remains constant to silane partial pressures beyond 10 Torr. At this temperature, region II is not observed at total gas flows down to 10 sccm.

The width of region I depends on gas residence time and temperature. It can persist to high $P(SiH_4)$ if the total flow is large. However, at low flow rates a second region (II) appears in which the film growth rate changes very rapidly with $P(SiH_4)$. This is evident for experiments at 50 and 75 sccm in Fig. 4. Region II can be made to completely dominate if the temperature is raised, although the total pressure must be decreased commensurately to avoid homogeneous nucleation.

In addition to its dependence on gas residence time, region II exhibits several other significant features. First, the film deposition rate is superlinear in SiH_4 partial pressure. Secondly, R_G does not continue upward unabated with $P(SiH_4)$, because the threshold for Si homogeneous nucleation is soon reached. For example, at 464°C, P(tot) = 15 Torr and a total flow rate of 50 sccm (Fig. 4), it occurs at $P(SiH_4)=1$ Torr. At the same total pressure and a temperature of 418°C, gas phase nucleation occurs at a silane partial pressure of 10 Torr and 10 sccm total flow. Also, as Fig. 6 clearly demonstrates, decreasing P(tot) will suppress region II, causing the reappearance of the LH behavior of region I over the same silane partial pressure range. An order of magnitude decrease in P(tot) almost completely eliminates II over the range of $P(SiH_4)$ shown in Fig. 6. Finally, HOMOCVD experiments also depicted in Fig. 4 clearly demonstrate that it operates entirely within region II, below the onset of gas phase nucleation.

On the basis of the above experiments, it is possible to ascribe region I to the heterogeneous decomposition of SiH_4. Hydrogen desorption from the growth interface has often been cited as the rate-limiting step in this case [16]. On the other hand, growth in region II is primarily due to the surface decomposition of intermediates created in the gas. From the previous discussion regarding the mechanism of HOMOCVD, higher silanes are the most likely film precursors in II because they exhibit much greater surface decomposition probabilities than does silane. Clearly, the suppression of II occurs with a decrease in P(tot) because the rate constant for the homogeneous pyrolysis [5,11] and the residence time of the gas are both pressure de-

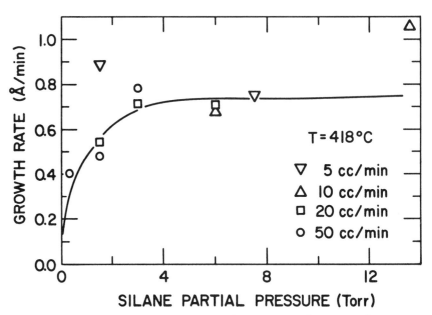

Fig. 5. Silicon film growth rate vs. silane partial pressure at 418°C and a total pressure of 15 Torr for 5, 10 and 50 sccm flow rates.

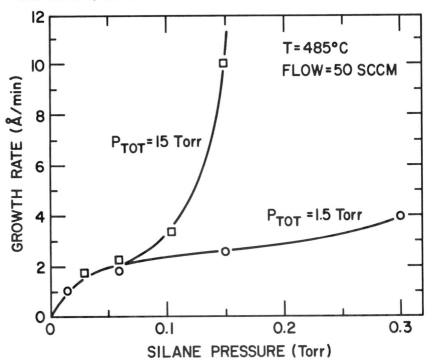

Fig. 6. Silicon film growth rate vs. silane partial pressure at total pressures of 1.5 and 15 Torr at a flow rate of 50 sccm.

pendent. Homogeneous channels to film precursors are repressed when $P(SiH_4)$ and/or P(tot) are low, and/or when gas residence times are too short.

As pointed out earlier, growth from higher silanes in the homogeneous regime is consistent with recent thermal desorption studies of SiH_4, Si_2H_6 and Si_3H_8, which demonstrate the high reactivity of the latter on hydrogen-saturated surfaces at temperatures as low as 25°C [15]. If hydrogen desorption is rate-limiting for SiH_4, it does not appear to be a bottleneck in higher silane surface decomposition (at least at surface temperatures >400 °C). Very different chemisorption and hydrogen elimination channels seem to be available to the higher hydrides [15].

The involvement of large oligomeric species or even "clusters" is also likely in the homogenous routes to film growth, particularly as conditions for gas phase nucleation are approached. The primary evidence for this process is the unusual dependence of R_G on $P(SiH_4)$ in region II, and in the HOMOCVD data (Fig. 4). The reason for this is discussed in the following section, and is related to the course of silane decomposition beyond the early time.

SILANE PYROLYSIS AND SILICON CVD

In a recent reinvestigation of silane pyrolysis, Robertson, Hils and Gallagher (RHG) concluded that silane decomposition even in its earliest stage is initiated on surfaces [17]. Their conclusions were based solely on the pressure change observed in static bulb experiments and silane consumption measured in a flow reactor. In the static case, initial slopes of the change in P(tot) scaled with surface area and exhibited a very weak pressure dependence. Silane consumption under flow conditions was consistent with these results. In addition, an activation energy of 56 Kcal/mole was obtained, nearly identical to the results of Purnell and Walsh (PW) for homogeneous pyrolysis in the early time [5]. Although the controversy created by these experiments has not been completely resolved, it is now generally recognized that gas and surface processes are both involved, but to relative degrees that depend on experimental conditions in an unknown manner [18-23]. Clearly this issue can be clarified from the CVD experiments we have described, for they overlap the P,T regimes of most pyrolysis studies.

In previous static tube measurements, total pressures [5,17] and product partial pressures [5] were measured as a function of time, whereas in our CVD experiments the residence time of the gas is a constant fixed by the flow rate. When we approach the "early time" by increasing the total flow, region I is obtained, and our observations of LH kinetics in this limit corresponds to RHG's result of an almost pressure-independent initial rate; i.e., $dP(tot)_i/dt \propto P(tot)^n$, with n = 0.2 ± 0.2 [17]. PW observed no change in P(tot) in the early time, but this has been attributed to the the sensitivity of their apparatus in the original study [19]. This is not surprising, for the pressure rise due to hydrogen release from the growing film is quite small. Figures 4-6 show that the degree of heterogeneity is not very great. Indeed, using RHG's reactor dimensions and pressure change data [17], we calculate film growth rates that are even lower than our CVD values under comparable conditions of pressure and temperature.

PW and RHG observed dP(tot)/dt to exhibit a maximum proportional to $P(SiH_4)$ in the later time [5,17] We believe that this corresponds to conditions well into region II of our CVD studies. However, R_G displays a greater than linear dependence on $P(SiH_4)$ in region II. This is because film growth rate, not pressure change, is being measured in our experiment. In this regime, deposition is occurring initially from higher silanes, but these can become transformed into species which are themselves highly reactive toward SiH_4. The precursors grow larger by reaction and aggregation until "homogeneous nucleation" occurs. Higher silanes are transformed into oligomeric species or clusters with increasingly larger molecular weights as the homogeneous nucleation point is approached. Consistent with this view, films grown under conditions close to the gas phase instability point are found to be porous and only weakly bonded to the substrate. In this interpretation of region II, the oligomers and clusters grow by creating their own highly reactive "surfaces" which can directly decompose SiH_4. Such an autocatalytic mechanism leading to homogeneous nucleation would explain why PW and RHG found $dP(tot)_{max}/dt$ independent of reactor surface-to-volume ratio [5,23,24]. Both groups report the final consequences of this process: powder formation in the later time, a condition eventually reached in our region II by increasing the gas residence time and/or raising the pressure.

Although the nature of the higher order processes occurring in the gas phase beyond the early time is still a mystery, it is at least possible to qualitatively explain some of the puzzling features of silane pyrolysis from the CVD experiments we have described. A more detailed analysis of the connection between the two experiments will be published separately.

ACKNOWLEDGEMENTS

We are indebted to Dr. Joseph Jasinski for providing us with the results of his RRKM calculations, and for many helpful discussions. We also wish to thank Dr. Bernard Meyerson for help with the kinetic simulations, and Dr. Robin Walsh for providing us with a copy of his Ph. D. thesis.

REFERENCES

1. J. M. Jasinski, B. S. Meyerson and B. A. Scott, *Ann. Rev. Phys. Chem.* **38**, 109 (1987).

2. D. W. Hess, K. F. Jensen and T. J. Anderson, *Rev. Chem. Eng.* **3**, 97 (1985). *J. Vac. Sci. Tech. A* (in press).

3. B. A. Scott, "Homogeneous Chemical Vapor Deposition," *Semicond. and Semimetals,* J. Pankove, ed., **21A**, pp. 123-149 (1984).

4. B. A. Scott, W. L. Olbricht, B. A. Meyerson, J. A. Reimer, and D. J. Wolford, *J. Vac. Sci. Tech. A* **2**(2), 450 (1984).

5. J. H. Purnell and R. Walsh, *Proc. Royal Soc. A* **293**, 543 (1966).

6. M. Bowery and J. H. Purnell, *Proc. Royal Soc. A* **321**, 341 (1971).

7. P. John and J. H. Purnell, *J. Chem. Soc.* Faraday Trans. I, **69**, 1455 (1973).

8. J. M. Jasinski, *J. Phys. Chem.* **90**, 555 (1986).

9. G. Inoue and M. Suzuki, *Chem. Phys. Lett.* **122**, 122 (1985).

10. B. S. Meyerson, B. A. Scott and R. Tsui, *Chemtronics* **1**, 155 (1986).

11. B. S. Meyerson and J. M. Jasinski, *J. Appl. Phys.* **61**, 8705 (1987).

12. B. A. Scott, R. M. Plecenik and E. E. Simonyi, *Appl. Phys. Lett.* ,**39** 73 (1981).

13. S. C. Gau, B. R. Weinberger, M. Akhtar, Z. Kiss and A. G. MacDiarmid, *Appl. Phys. Lett.* **39**, 436 (1981).

14. S. R. Kurtz, J. Proscia and R. G. Gordon, *J. Appl. Phys.* **59**, 249 (1986).

15. S. M. Gates, B. A. Scott, D. B. Beach, R. Imbihl and J. E. Demuth, J. Vac. Sci. Tech. A (in press).

16. Refs. 1,2 contain a discussion of many of the models.

17. R. Robertson, D. Hils, and A. Gallagher, *Chem. Phys. Lett.*, **103**, 397 (1984).

18. R. Robertson and A. Gallagher, *J. Chem. Phys.*, **85**, 3623 (1986).

19. J. H. Purnell and R. Walsh, *Chem. Phys. Lett.*, **110**, 330 (1984).

20. H. E. O'Neal and M. A. Ring, *Chem. Phys. Lett.*, **107**, 442 (1984).

21. J. M. Jasinski and R. D. Estes, *Chem. Phys. Lett.*, **117**, 495 (1985).

22. J. M. Jasinski, *J. Chem. Phys.*, **86**, 3057 (1987).

23. R. Robertson and A. Gallagher, *J. Chem. Phys.* **86**, 3059 (1987).

24. R. Walsh, Ph.D. Thesis, Cambridge University, 1965.

Chapter 35

Silicon containing photoresists

R.D. Miller – IBM Almaden Research Center, 650 Harry Road, San Jose, California 95120-6099.

Most of the patterning for current electronic circuitry is accomplished using photolithographic techniques [1]. Using this procedure, a substrate (usually silicon) is coated with a layer of photoresist and exposed through a mask to produce a latent image. The image is then developed with an appropriate solvent to generate the desired relief pattern. Photoresists are classified as either positive or negative functioning depending on whether the exposed area becomes more or less soluble in the developer. Negative resists are usually materials (often containing unsaturation) where polymer chain crosslinking or other types of interchain reactions occur upon exposure, thus greatly reducing the solubility of the exposed areas. These materials frequently show high sensitivity due to the gain inherent in photopolymerization processes, but are often resolution limited due to the swelling of the resist patterns during development. Recently, some negative resists which function by changes in the polarity of the material in the exposed areas rather than by crosslinking have been described [2-4]. These materials are less prone to swelling and are useful for the generation of high resolution patterns.

The vast majority of the commercial positive photoresists are comprised of a base soluble Novolac type resin (Cresol-formaldehyde condensation polymers) and a sensitizer which is usually a substituted derivative of 1,2-naphthoquinone-2-diazide [5]. The hydrophobic sensitizer inhibits the dissolution of the matrix resin in base prior to exposure. After exposure, the sensitizer is converted into a substituted indene carboxylic acid which greatly increases the rate of dissolution of the resin in the exposed regions. The combination of these effects results in a large difference in solubility between the exposed and unexposed regions which is responsible for the facile development of the latent mask image. Since the dissolution process in dilute base is a result of chemical reaction between the basic developer and the weakly acidic Novolac resin, it proceeds with very little swelling. Positive resists of this type are hence intrinsically capable of high resolution, although their sensitivity is often much lower than that observed for negative resists.

The current drive toward higher density circuitry requires smaller features and consequently higher resolution lithographic techniques. As the industry moves toward smaller lateral geometries, the aspect ratio (height/width) of the features increase because the resist thickness in a single layer process is limited by the necessary coverage of chip topography. The generation of small, high aspect ratio features often requires near vertical wall profiles. This sometimes creates problems using the classical single layer positive resists because the wet development process is intrinsically isotropic (i.e., there is some lateral as well as vertical development) which can erode the pattern profile. There are also additional problems with single layer resists which are associated with precise line width control for small features generated over significant chip topography [6]. This is caused by light scattering from edges and corners of the topographic features coupled with problems of exposure homogeneity exacerbated by the differing resist thicknesses resulting from the topography. In addition, for monochromatic radiation, standing waves in the variable thickness resist coating can also cause linewidth control problems.

In an effort to circumvent some of the problems caused by chip topography, a number of multilayer resist schemes have been introduced [6]. The simplest of these is the imagable bilayer illustrated in Figure 1. In this scheme, the chip containing any surface topography is coated with a thick planarizing polymer layer. Many materials can be used for this purpose (e.g., PMMA, polyimides, hardbaked photoresist, etc.) and the selection is dictated by specific process requirements. This layer is then overcoated with a thin (0.1-0.2 μm) layer of an imagable material. The top layer is imaged with high resolution in the normal fashion and the pattern developed usually by wet development techniques. The thinness of this layer allows the generation of high resolution patterns even using wet development processes. These patterns can then be transferred into the planarizing layer either by deep UV flood exposure utilizing the masking characteristics of the resist layer or by oxygen reactive ion etching (O_2-RIE). The former requires that the remaining resist layer effectively mask the underlying layer from exposure and that the planarizing layer is itself UV sensitive. This process also requires a subsequent wet development step for the imaged, planarizing layer with its attending drawbacks. Alternatively, the image can be transferred by oxygen plasma etching techniques. This procedure is attractive because it is dry (no solvents), alleviates adhesion problems and, most importantly, the etching process can be controlled so that it is highly anisotropic. In this manner, high aspect ratio patterns with near vertical wall profiles can be produced.

Transfer of patterns by O_2-RIE does, however, place some unique demands on the thin imaging layer. Not only must it be patternable with high resolution, but it must also be much more stable under the etching conditions than the underlying planarizing polymer. The latter criterion is admirably satisfied by organometallic polymers which contain elements (Si, Ti, B, Sn, Al, etc.) which form refractory oxides in an oxygen plasma [7]. Since a large number of silicon containing polymers are known and SiO_2

forms a good oxygen etch barrier, these materials have received considerable attention for multilayer applications [8].

The question of just how much silicon must be present to create a suitable oxygen etch barrier has been answered recently by a number of groups working on silicon containing copolymers [4,9]. The etch resistance of the investigated copolymers varies nonlinearly with the amount of silicon, but materials which contain at least 10% by weight seem to form effective oxygen etch barriers. The relative effectiveness of these materials seems also to depend somewhat on etching conditions such as power, voltage bias and oxygen pressure.

The silicon in a photoresist formulation may be simplistically viewed as playing either passive or active role. We define a passive role as one where a silicon containing monomer or polymer is utilized solely for the purpose of providing oxygen etch resistance and is not transformed during the exposure. In either role the silicon may be present as a pendant polymer substituent, be incorporated into the polymer backbone or it may simply be a monomer additive.

SILICON CONTAINING NEGATIVE RESISTS

Since silicon containing polymers are often prone to radiation induced crosslinking during irradiation, it is not surprising that the first reports of such materials for multilayer lithographic applications were as negative resists. In this regard, Hatzakis and co-workers [10] showed that the siloxane terpolymer 1 crosslinked upon exposure to light or ionizing radiation and etched very slowly under O_2-RIE conditions. The low glass transition temperature (Tg) of this material requires the use of very thin films to prevent thermal deformation of the imaged structures upon processing. Workers at Nippon Telephone and Telegraph (NTT) subsequently showed that poly(diphenylsiloxane) which has a much higher Tg could be partially chloromethylated to yield a polymer which readily crosslinked upon deep UV (λ 220-280 nm) exposure [11].

An "all dry" process (no wet development step) for X-ray exposure has been reported by Taylor et al. [12,13] who incorporated bifunctional silicon containing acrylates such as 2 into the radiation sensitive matrix polymer poly(2,3-dichloro-1-propylacrylate). X-ray exposure produces radical sites which subsequently polymerize and lock the silicon containing monomer into the polymer matrix. The exposed resist was then baked to volatilize the unreacted monomer in the unexposed regions and the patterns developed by O_2-RIE. Using this technique, high resolution submicron patterns were created. One drawback is that the resist has relatively low contrast (i.e., poor etch rate differential between the exposed and unexposed regions) because all or most of the unreacted monomer must be thermally removed from the unexposed regions prior to pattern transfer.

$$\left[\begin{array}{c} Me \\ | \\ -Si-O- \\ | \\ Me \end{array} \right]_x \left[\begin{array}{c} Me \\ | \\ -Si-O- \\ | \\ R \end{array} \right]_y \left[\begin{array}{c} Me \\ | \\ -Si-O- \\ | \\ R' \end{array} \right]_z \qquad (CH_2{=}CRCO_2(CH_2)_n\,Si(Me)_2)_2O$$

$$\underline{2} \quad R = H, Me$$
$$n = 4$$

R = Ph, R' = vinyl

$\underline{1}$

A number of negative resists where silicon is incorporated as a pendant substituent have also been reported. In this regard, since both p-chloro and p-chloromethyl polystyrene crosslink readily upon deep UV irradiation [5], it is not surprising that copolymers of these respective styrene monomers with p-trimethylsilylstyrene [14,15] also function as negative resists for bilayer applications. These systems offer the advantage that the oxygen etch stability of the copolymers can be varied by changes in the copolymer composition.

A somewhat different approach was adopted by Saigo and coworkers [16] who found that phenyl triallylsilane could be polymerized to yield a linear polymer 3 containing pendant allylic functionality. The unsaturated groups could then be utilized for crosslinking by irradiation in the presence of appropriate photocrosslinking reagents such as aryl bis azides.

R = vinyl, allyl

$\underline{3}$ $\underline{4}$

Recently, Ishikawa et al. [17] have reported a new class of negative resists 4 which contain both pendant and backbone silicon. These materials undergo crosslinking through silyl radicals produced by side chain scission of the disilanyl units.

SILICON CONTAINING POSITIVE RESISTS

Considerable effort has been focused recently on silicon containing materials which function as positive resists. Many of these contain pendant silicon functionality which often plays a passive role. The generation of silicon containing Novolac-type resins represents one such effort. Workers at AT&T Bell Labs have prepared a number of base soluble Novolacs by the condensation of formaldehyde with a mixture of either phenol-m-trimethylsilylphenol [18] or more recently with m-cresol-p-trimethylsilylmethyl phenol [18]. The monomer proportions were adjusted to produce resins with a silicon content of ~9% or more while maintaining the solubility in dilute base. Related resins have also been described by workers at NEC from m-trimethylsilylalkoxyphenol, 2-methyl resorcinol and formaldehyde [20]. The silylated Novolac type resins were designed for use with substituted 1,2-napthoquinone-2-diazide sensitizers employed in classical photoresist formulations. The silylated Novalac resists have the obvious advantage that they can be used with conventional mercury light sources available in commercial projection scanning and step-and-repeat lithographic tools. A potential drawback of these materials is, however, that the etch resistance is, in some cases, barely adequate and they are not suitable for use with shorter wavelength (e.g., deep UV) exposure sources. Workers at Hitachi [21] have recently described a related system composed of a Novolac-type resin containing a soluble, nonvolatile, base soluble silicon additive. The additive 5 was a mixture of the cis-1,3,5,7-tetrahydroxy 1,3,5,7-tetraphenylcyclotetrasiloxane and a commercially available polyphenylsilsesquioxane. An interesting variation of the silylated Novolac approach for the production of negative relief images has recently been reported by Coopmans et al. [22]. This procedure exploits the observation that certain Novolac-1,2-napthoquinone 2-diazide formulations are resistant to silylation by reagents such as hexamethyldisilazane (HMDS) prior to exposure. After irradiation, the phenolic hydroxyl functionality in the exposed regions is rapidly silylated under the processing conditions to produce an oxygen etch barrier. An "all dry" process capable of submicron resolution in a single layer of resist has been described [22].

Other positive, silicon containing resists where silicon plays a more active role have also been described. For example, an oxygen etch resistant, deep UV sensitive copolymer with pendant silicon composed of trimethylsilylmethyl methacrylate and 3-oximino-2-butanone methacrylate units has been reported (8). The former comonomer provides improved etch resistance while the latter is used to enhance the resist sensitivity to deep UV radiation.

A number of radiation sensitive positive resists which contain silicon in the backbone have also been reported. In this regard, workers at Hitachi [23,24] have reported that polymeric disilanes such as 6 generated by the condensation of substituted bis-chlorosilyl benzenes with alkali metals function in a positive fashion. Chain breaks results from the photoscission of the disilanyl units. These materials absorb strongly in the deep UV and also form excellent oxygen etch barriers.

$$5 \qquad\qquad 6$$

R = Me, Et, Ph

Recently, a new class of radiation sensitive materials which contain only silicon in the backbone has been described. The unusual spectroscopic properties and radiation sensitivity of these polysilanes have resulted in many new applications including a number in microlithography. The unusual characteristics of the polysilanes have created considerable interest recently and will be discussed in some detail.

POLYSILANES

Substituted silane polymers were probably first synthesized over 40 years ago [25]. The early materials were highly crystalline and insoluble because of the nature of pendant substituents. These undesirable properties are not characteristic of this general class of polymers and recent synthetic efforts have generated a large number of soluble, high molecular weight homo and copolymers [26]. The ready availability of soluble polymers from which high quality films can be cast has stimulated an intensive investigation of these materials. From these studies have emerged a variety of new applications including their use in multilayer microlithography [27-30].

High molecular weight linear polysilanes can be produced by a Wurtz type condensation of the respective substituted dichlorosilanes as shown below. In this regard sodium dispersion seems most effective for the production of the linear polymers. The yields in this process are dependent on structure and range from 5-70%. The use of lithium usually results in cyclic oligomers [26]

$$R_1R_2SiCl_2 + Na \rightarrow (R_1R_2Si)_n + 2NaCl$$

Despite the fact that the backbone bonding in these materials is predominantly sigma in nature, the extensively catenated silicon derivatives absorb strongly in the UV-visible region [26]. In this regard, both the λ_{max} and the $\epsilon/SiSi$ depend on molecular weight, both increasing rapid at first and approaching limiting values at a degree of polymerization of around 40-50 [31]. The molar absorptivities of these materials are very large and range from 4000-25000 per silicon-silicon bond depending on the nature and position of the substituents. Atatic alkyl derivatives absorb strongly around 300-325 nm with sterically demanding substituents causing significant red shifts (see Figure 3). Aryl substituents directly bonded to silicon cause strong red shifts of 25-40 nm because of the electronic interaction of the substituents with the sigma bonded framework [32]. Recently, we have demonstrated by studies in the solid state that the position and intensity of the absorption maxima also depend on the conformation of the polymer backbone [33-35]. In this regard, it was observed that a planar zigzag conformation causes a red shift of over 55 nm relative to observed solution values where the backbone is presumably disordered and the polymer has been shown by light scattering to adopt a random coil configuration [36]. In the solid state, deviation by as little as 30° from a trans coplanar conformation even in a regular structure results in blue shifts of ~60 nm in the absorption spectrum [37]. Furthermore, a recent prediction based on conformation calculations suggests that polysilanes with large substituents may actually prefer a helical configuration rather than planar zigzag [38,39] and that in most cases strong solid state intermolecular interactions are necessary before the latter conformation predominates.

Interestingly, soluble symmetrical diaryl polysilanes are the most red shifted of all known polysilane derivatives and absorb around 400 nm [40]. The magnitude of this shift far exceeds that anticipated by electronic substituent predictions which suggests that it might also be conformational in origin. In this regard, we have tentatively proposed that these materials may contain extended planar zigzag segments even in solution.

PHOTOCHEMISTRY

Since both the λ_{max} and ϵ_{SiSi} are dependent on molecular weight [31], processes which reduce the molecular weight should lead to a bleaching of the original absorption. We have observed this phenomenon for all of the polysilanes examined although the bleaching rate varies with structure. Photobleaching of a film of a typical polysilane is shown in Figure 3. Bleaching of the original absorption both in solution and in the solid state is associated with the formation of lower molecular weight fragments as determined by gpc analysis of the irradiated samples. In this regard, gpc analysis of irradiated samples of alkyl polysilanes show little indication of crosslinking while similar studies on irradiated samples of aromatic derivatives such as poly(phenyl methylsilane) indicate that a higher molecular weight portion remains. The quantum yields for scission (Φ_s) and crosslinking (Φ_x) for irradiated polymers can be determined routinely from plots of $1/\overline{M}_n$ and $1/\overline{M}_w$ versus dose [41]. Using this technique we have determined that the quantum yields for scission for alkyl substituted polysilanes in solution are high and range from ~0.5-1.0 depending on the nature of the substituents. Similarly high values are obtained for aryl derivatives, but in these cases a competitive crosslinking component is often observed. (Φ_x ~ 0.12-0.18). In spite of this, the aryl derivatives are still predominantly scissioning polymers in solution and Φ_s/Φ_x values of 5-7 are observed. In all cases, the quantum yields for both processes decrease significantly (>50 fold) in going from solution to the solid state presumably because solid state cage effects would be expected to promote chain repair (vide infra).

Exhaustive irradiation of high molecular weight polysilane derivatives at 254 nm in the presence of trapping reagents such as triethyl silane or alcohols lead to products which indicate that both substituted silylenes and silyl radicals are produced as intermediates [42]. The isolation of these products suggests that the photochemical pathways observed for the high polymers may be similar to those proposed in mechanistic studies of the photodecomposition of shorter acyclic silicon catenates [43,44]. Other authors have also suggested, on the basis of mass spectroscopic studies, that substituted silylenes are produced in the photovolatilization of some polysilane copolymers upon deep UV exposure [45]. On the basis of the accumulated evidence, we propose a basic mechanism shown below for the photodecomposition of substituted silane polymers which not only accommodates the proposed intermediates, but is consistent with the decreased efficiency in the solid state. While the proposed mechanism is consistent with the limited data, it has not been determined whether the silylenes arise from the photolysis of the polymer itself and/or are produced in a subsequent decomposition of the silyl radicals. In the case of the polymers is assumed that silyl radical disporportionation, which has been demonstrated to be a facile reaction for trimethyl silyl radicals themselves [46], is competitive with chain repair.

$$\underset{\substack{R_2\ R_2\ R_2}}{\overset{\substack{R_1\ R_1\ R_1}}{\wwww\ Si-Si-Si\ \wwww}} \overset{h\nu}{\longrightarrow} \underset{\substack{R_2\ \ \ R_2\ \ \ R_2}}{\overset{\substack{R_1\ \ \ R_1\ \ \ R_1}}{\wwww\ Si\cdot\ +\ Si\!:\ +\ \cdot Si\ \wwww}}$$

POLYSILANES IN PHOTOLITHOGRAPHY

The intense optical absorption over an extended spectral range, ready bleachability and oxygen reactive ion etch stability displayed by the polysilanes suggest a number of photomicrolithographic applications [27-31,45]. To date, we have exercised these materials as short wavelength contrast enhancing materials and as imagable etch barriers in bilevel processes. In the latter application, we have demonstrated feasibility both with processes utilizing wet development of the imaged layer as well as in an "all dry" process where the initial image is produced by excimer laser ablation of the polysilane [30].

Contrast enhancement lithography is a clever procedure which utilizes a thin, strongly absorbing but bleachable contrast enhancing material coated over a classical photoresist to resharpen the diffraction distorted mask image at photoresist surface. The result is greatly improved optical pattern resolution. The mechanism by which the image quality is improved is too complicated to discuss here and the interested reader is referred to the pertinent references [47,48].

Most commercial contrast enhancing materials are designed to operate in the near UV-visible range. Certain polysilane derivatives possess all of the necessary features to provide contrast enhancement in the mid (300-340 nm) and deep UV (220-280 nm) regions thus allowing the utilization of the improved resolution intrinsic to short wavelength exposure sources. Particularly significant in this regard is their strong absorption and excellent bleachability at short wavelengths which are essential for the process. Accordingly, we have demonstrated this principle for mid UV projection printing using a nonoptimized polysilane contrast enhancing layer coated over a commercial mid UV photoresist and the results are shown in Figure 4 [28]. The figure clearly demonstrates that there is less thinning of the central resist line for the contrast enhanced process relative to the conventional process.

We have also demonstrated the utility of polysilanes in a number of mid UV bilayer processes. Figure 5 shows submicron features printed in a thin layer of poly(cyclohexylmethylsilane) and transferred by O_2-RIE [29]. In this case, the polysilane image was wet developed with isopropanol prior to the image transfer. The image quality and vertical wall profiles clearly shows the utility of polysilane derivatives. In an extension of this work, we have also demonstrated submicron capability for an "all dry" bilayer process. In this case, the polysilane imaging layer was first photoablated using a KrF excimer laser (55 mJ/cm^2-pulse, 550 mJ total dose at 248 nm) [30,49]. The clean images produced in this step were subsequently transferred into the thick planarizing layer using O_2-RIE techniques.

In summary the demands for improved resolution over complex chip topography has resulted in studies of multilayer lithographic processes especially those using O_2-RIE techniques for image transfer. This has resulted in a greatly increased interest in imagable organometallic resists particularly those containing silicon. Although many of the initial efforts have centered on negative resists there has been recent progress in the development of new silicon containing positive resists. In this regard, the polysilane derivatives which represent a new class of radiation sensitive, O_2-RIE resistant polymers, show considerable promise and versatility.

ACKNOWLEDGEMENTS

R. D. Miller gratefully acknowledges partial financial support for this work from the Office of Naval Research.

REFERENCES

1. Elliot, D. J. "Integrated Circuit Fabrication Technology," McGraw-Hill Book Company, New York, 1982.
2. Hofer, D. C.; Kaufman, F. B.; Kramer, S. R.; Aviram, A. *Appl. Phys. Lett.* **1980** *37*, 314.
3. Ito, H.; Willson, C. G. *Polym. Eng. Sci.* **1983**, *23*, 1012.
4. MacDonald, S. A.; Ito, H.; Willson, C. G. *Microelectronic Engineering* **1983**, *1*, 269.

5.   Willson, C, G. in "Introduction to Microlithography," Thompson, L. F.; Willson, C. G.; Bowden, M. J., Eds., ACS Symposium Series, No. 219, American Chemical Society, Washington, DC, 1980, Chap. 3.
6.   Lin, B. J., in Ref. 5, Chap. 6.
7.   Taylor, G. N.; Wolf, T. M.; Stillwagon, L. E. *Solid State Technology* **1984**, *27*, 145.
8.   Reichmanis, E.; Smolinsky, G.; Wilkins, Jr., C. W. *Solid State Technology* **1985**, *28*, 130.
9.   Reichmanis, E.; Smolinsky, G. *Proc. SPIE* **1984**, *469* 38.
10. Hatzakis, M.; Paraszczak, J.; Shaw, J. *Proc. Intl. Conf. Microlith., Microcircuit Eng.* **1981**, *81* 386.
11. Tanaka, A.; Mouta, M.; Imamura, A.; Tamamura, T.; Koyure, O. *Polym. Preprints* **1984**, *25* 309.
12. Taylor, G. N.; Wolf, T. M. *J. Electrochem. Soc.* **1980**, *127*, 2665.
13. Taylor, G. N. *Solid State Technology* **1980**, *2*, 73.
14. Suzuki, M.; Saigo, H.; Gokan, H.; Ohnishi, Y. *J. Electrochem. Soc.* **1983**, *130* 1962.
15. MacDonald, S. A.; Steinmann, A. S.; Ito, H.; Hazaksis, M.; Lee, W.; Hiroaka, H., Abstr Int'l Symp. Electron, Ion; Photon Beams, **1983**, Los Angeles, California.
16. Saigo, K.; Ohnishi, Y.; Suzuki, M.; Gokan, H., Abstr. Int'l Sympos. Electron, Ion, Proton Beams 1984 Tarrytown, New York.
17. Ishikawa, M. *Polym. Preprints* **1987**, *28*, 426.
18. Wilkins, Jr., C. W.; Reichmanis, E.; Wolf, T. M.; Smith, B. C. *J. Vac. Sci. Technol.* **1985**, *3*, 306.
19. Tarascon, R. G.; Shugard, A.; Reichmanis, E. *Proc. SPIE* **1986**, *631*, 40.
20. Saotome, Y.; Gokan, H.; Saigo, K.; Suzuki, M.; Ohnishi, Y. *J. Electrochem. Soc.* **1985**, *132*, 909.
21. Hayashi, N.; Ueno, T.; Shiraishi, H.; Nishida, T. Toriumi, M.; Nonogaki, S., *Proc. Polym. Mat. Sci. Eng.* **1986**, *55*, 611.
22. Coopmans, F.; Roland, B., *Proc. SPIE* **1986**, *631*, 34.
23. Nate, K.; Suguyama, H.; Inoue, T., Abstr. Meeting Electrochem. Soc., No. 530, 1984.
24. Ishikawa, M.; Hongzhi, N.; Matsusaki, K.; Nate, K.; Inoue, T.; Yokomo, H. *J. Polym. Sci., Poly. Lett. Ed.* **1984**, *22*, 669.
25. Kipping, F. S. *J. Chem. Soc.* **1924**, *125*, 2291.
26. For a current review of the chemistry and properties of polysilanes, see, West, R. *J. Organomet. Chem.* **1986**, *300*, 327.
27. Miller, R. D.; McKean, D. R.; Hofer, D.; Willson, C. G. West, R.; Trefonas, III, P. T., in "Materials for Microlithography," Thompson, L. F.; Willson, C. G.; Frechét, J. M. J., Eds., ACS Symposium Series, No. 266, American Chemical Society, Washington, D.C., **1984**, Chap. 3.
28. Hofer, D. C.; Miller, R. D.; Willson, C. G.; Neureuther, A. R. *Proc. SPIE* **1984**, *469*, 108.
29. Hofer, D. C.; Miller, R. D.; Willson, C. G. *Proc. SPIE* **1984**, *469*, 16.
30. Miller, R. D.; Hofer, D.; Fickes, G. N.; Willson, C. G.; Marinero, E.; Trefonas, III, P.; West, R. *Polym. Eng. Sci.* **1986**, *26*, 1129.
31. Trefonas, III, P.; West, R.; Miller, R. D.; Hofer, D. *J. Polym. Sci., Polym. Lett. Ed.* **1983**, *21*, 823.
32. Pitt, C. G., in "Homoatomic Ring Chains and Macromolecules of Main Group Elements," Rheingold, A. L., Ed., Elsevier, Amsterdam, 1977, 203, and references cited therein.

33. Miller, R. D.; Hofer, D.; Rabolt, J.; Fickes, G. N. *J. Am. Chem. Soc.* **1985**, *107*, 2172.
34. Rabolt, J. F.; Hofer, D.; Miller, R. D.; Fickes, G. N. *Macromolecules* **1986**, *19*, 6114.
35. Kuzmany, H.; Rabolt, J. F.; Farmer, B. L.; Miller, R. D. *J. Chem. Phys.* **1986**, *85*, 7413.
36. Cotts, P. M.; Miller, R. D.; Trefonas, III, P. T.; West, R.; Fickes, G. N. *Macromolecules* **1987**, *20*, 1046.
37. Miller, R. D.; Farmer, B. L.; Fleming, W.; Sooriyakumaran, R.; Rabolt, J. *J. Am. Chem. Soc.* **1987**, *109*, 2509.
38. Farmer, B. L.; Miller, R. D.; Rabolt, J. F. *Macromolecules* **1987**, *20*, 1169.
39. For a recent contrasting theoretical study on poly(di-n-hexylsilane), see Damewood, Jr., J. R. *Macromolecules* **1985**, *18*, 1793.
40. Miller, R. D.; Sooriyakumaran, R. *J. Polym. Sci., Polym. Lett. Ed.* **1987** (in press).
41. Schnabel, W.; Kiwi, J. in "Aspects of Degradation and Stabilization of Polymers," Jellinek, H. H. G., Ed., Elsevier, New York, 1978, Chap. 4.
42. Trefonas, III, P.; West, R.; Miller, R. D. *J. Am. Chem. Soc.* **1985**, *107*, 2737.
43. Ishikawa, M.; Takaoka, T.; Kumada, M. *J. Organomet. Chem.* **1972**, *42*, 333.
44. Ishikawa, M.; Kumada, M., *Adv. Organomet. Chem.* **1981**, *19*, 51 and references cited therein.
45. Zeigler, J. M.; Harrah, L. A.; Johnson, A. W. *Proc. SPIE* **1985**, *539*, 166.
46. Raabe, G.; Michl, J. *Chem. Rev.* **1985**, *85*, 419.
47. Griffing, B. F.; West, P. R. *Polym. Eng. Sci.* **1983**, *23*, 947.
48. West, P. R.; Griffing, B. F. *Proc. SPIE* **1984**, *33*, 394.
49. Marinero, E. E.; Miller, R. D. *Appl. Phys. Lett.* **1987** *50*, 1041.

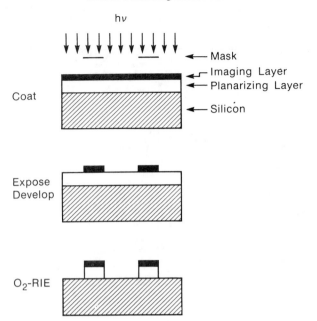

Fig. 1. A typical bilayer lithographic scheme utilizing oxygen reactive ion etching for image transfer.

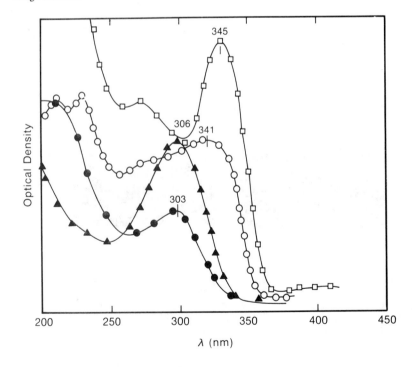

Fig. 2. UV spectra of some typical polysilane films. (●—●) poly(methyl phenethylsilane), (△—△) poly(methyl hexylsilane), (○—○) poly(phenylmethyl-co-dimethysilane), □—□ poly(phenyl methyl silane).

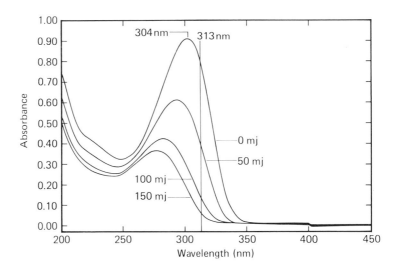

Fig. 3. Photochemical belaching of a film of poly(methyl hexylsilane) at 313 nm.

conventional   **CEL**

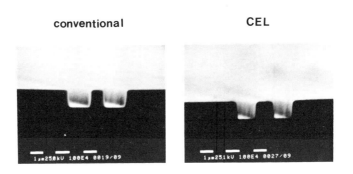

Fig. 4. Contrast enhanced resist profiles from projection lithography at 313 nm. 0.2 μm of poly(cyclohexylmethylsilane) over AZ-2400 photoresist, 110 mJ/cm² dose incident at photoresist surface.

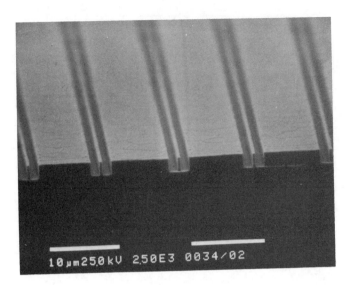

Fig. 5. 0.75 μm features generated in a bilayer of 0.2 μm of poly(cyclohexyl methylsilane) over 2.0 μm of a hardbaked Novolac-naphthoquinone-2-diazide photoresist; 100 mJ/cm^2, O$_2$/RIE image transfer.

Chapter 36

Plasma etching of silicon and silicon oxides

Daniel L. Flamm – AT&T Bell Laboratories, Murray Hill, New Jersey 07974, USA.

INTRODUCTION

Integrated circuits are fabricated by sequentially depositing and patterning a series of thin layers[1]. These layers are mostly silicon, silicon oxide, silicon nitride, metallizations and silicides. To pattern a layer, a light-sensitive organic polymer resist "film" is "spun on" the wafer from a viscous solution of the polymer[2]. The resist is exposed to light projected through a master mask and, depending on the type of resist, exposed areas become soluble (or insoluble) to a developer-solvent. The developer dissolves unwanted areas creating a surface masking layer. The resist mask then protects selected areas of the inorganic film from attack during plasma or wet etching processes. After etching, the resist stencil is stripped off using a strong solvent, or by etching in an oxygen plasma.

Plasma etching has largely replaced wet etching for microcircuit fabrication because of its finer resolution and adaptability to increased throughput and automation. The simplest reactors consist of opposed parallel plate electrodes in a chamber that can be maintained at low pressure, typically ranging from 0.01 to 1 Torr (1.33-133 Pa)[3]. A high frequency voltage is applied between the electrodes forming a plasma and semiconductor wafers or other substrate materials on the electrode surfaces are exposed to reactive neutral and charged species. Some of these species combine with the substrate and form volatile products which evaporate, thereby etching the substrate. For plasmas of interest to etching, the density of charged particles is low, about 10^9-10^{11} cm^{-3}, which corresponds to one charged particle per 10^4 to 10^6 neutrals at 0.1 Torr.

Electrons in the plasma attain high energy while the neutral gas remains relatively cool. This elevated electron temperature permits electron-molecule collisions to excite high-temperature reactions that form free radicals. The coexistence of a warm gas with high temperature species is an important distinction between the plasma reactor and conventional reaction systems, which permits the processing of sensitive materials.

Another important characteristic of plasmas is the presence of a negative-going electric sheath-field along the boundaries, which propels positive ions into surfaces at normal incidence at energies from a few eV to 100's of eV, depending on plasma conditions.

Electron energy is channelled into *inelastic* electron-neutral collisions which supply fresh species to the plasma. For instance, key dissociation reactions in CF_4 discharges (used to etch Si and SiO_2) are:

Ion and Electron Formation	**Atom and Radical Formation**
$e + CF_4 \rightarrow CF_3^+ + F + 2e$	$e + CF_4 \rightarrow CF_2 + 2F + e$
	$e + CF_4 \rightarrow CF_3 + F^-$
	$e + F^- \rightarrow F + 2e$
	$e + CF_4 \rightarrow CF_3 + F + e$

Plasma processes are often characterized according to rate, anisotropy, selectivity, the degree of loading effect and surface quality.[3] Anisotropy refers to preferential etching perpendicular to the surface of a wafer. Selectivity is the ratio of etching rates between two different materials immersed in the same plasma, for example Si and SiO_2. "Loading" is a term used to describe a measurable depletion of active etchant species from the gas phase, brought about by the consumption of this reactant in the etching process. In principle, the plasma etchant feed gas would be chosen according to the type of material to be etched, for selectivity over other substrates, to minimize any loading effect, and to avoid excessive surface degradation and polymer deposition.

Isotropic or anisotropic etching can be defined by reference to Figure 1.

Fig. 1. Isotropic chemical etching (left) has no preferred direction, which leads to circular undercut profiles. In anisotropic etching (right) ions impinging vertically induce straight-walled profiles.

Usually, the desired result is a transfer of the mask pattern to the substrate, with no distortion of critical dimensions. However lateral chemical attack (isotropic etching) underneath the mask will enlarge a feature. Mask openings may be made smaller to compensate for enlargement (undercutting), but the minimum size attainable with this technique is limited to the film thickness (typically a few μM), which is why anisotropic etching is required to make features smaller than 3-5 μ.

BASIC MECHANISMS OF PLASMA ETCHING

While a wide variety of phenomena may play a role, etching mechanisms can be grouped into the four categories depicted in Figure 2. At a basic level these categories encompass diverse elementary phenomena − many of which are not well-understood.

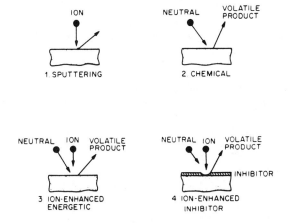

Fig. 2. The four generic mechanisms through which plasma etching takes place.
(1) sputtering; (2) chemical ("isotropic") etching; (3) *energetic* ion-enhanced etching
and (4) *inhibitor* ion-enhanced etching.

In sputtering, impinging particles (usually positive ions accelerated across the plasma
sheath) strike the surface with high kinetic energy. Some of the energy is transferred to
surface atoms which then are ejected, leading to a net removal of material. This process is
distinguished from other etching mechanisms in that the interaction is mechanical- only
the bonding forces between surface atoms and the ballistics of dislodging them are
important.

Chemical etching comes about when active species from the gas phase encounter a
surface and react with it to form a *volatile* product. Involatile products would coat the
surface and protect it from further attack. In this type of etching the plasma reactor
converts the feed into reactive chemical species, which are usually free radicals such as
fluorine atoms (F). Chemical etching shares the characteristics of other chemical reactions.
There is usually no directionality (as opposed to ion induced material removal) and the
etching can be specific (high selectivity) since it is governed by the relative chemical
affinities between the etchant species and exposed materials.

The negative-going electric sheath-field in plasmas propels positive ions into surfaces
at normal incidence. Ion bombardment in the presence of reactive neutral etchant species
often leads to a synergism in which fast *directional* material removal rates greatly exceed
the separate sum of chemical attack and sputtering rates. There are two general types of
mechanism for this ion-enhanced plasma etching: (a) energy-driven ion-enhanced etching
and (b) inhibitor-driven ion-assisted etching.

In the energy-driven mechanisms, there is usually little or no etching when the
substrate surface is exposed to neutral chemical species alone in the absence of ion
bombardment. Impinging ions "damage" the substrate material by virtue of their impact
energy, and thereby render the solid substrate more reactive toward incident neutral
radicals. The term "damage" is meant in a generic sense to include diverse mechanisms
such as the formation of reactive dangling bonds, disruption of lattice structure and
formation of dislocations, forcible injection of absorbed reactant into a lattice by the
collisional cascade, or bond-breaking in tightly-absorbed surface intermediates. There is
no evidence to support a single universal elementary mechanism for energy-driven ion-
enhanced etching. Diverse elementary mechanisms are active depending on the etchant,

surface material being etched and, perhaps, on the magnitudes of radical and ion fluxes to the surface (e.g. the pressure regime).

In inhibitor protected-sidewall ion enhanced etching, ion bombardment performs another function. Neutral etchant species from the plasma spontaneously gasify the substrate, and ions play a role by interacting with an another component – a "protective" inhibitor film. The role of ions in the surface-inhibitor mechanism is to "clear" the inhibitor from horizontal surfaces that are bombarded by the flux of ions impinging in the vertical direction. The protective film is not removed from the vertical walls of masked features because these surfaces only intercept those few ions that are scattered as they cross the sheath. This protective film may originate from involatile etching products or from film-forming precursors that adsorb during the etching process. Some fluorocarbons, for instance C_2F_6 or CHF_3, are a source of unsaturated monomeric species in plasmas. These species include the CF_2 radical (or CCl_2 radical) and derivatives (e.g. C_2F_4, C_3F_6, C_2Cl_4, etc.) which tend to polymerize and form thin films on surfaces (or sometimes thick films).

While inhibitor-producing compounds are usually a deliberate additive, films can originate from more subtle sources, for example from sputtered reactor material at low frequency where sheath potentials are high or from migration of material in the resist mask. It is generally believed that low energy ion impingement can prevent inhibitor films from growing on horizontal surfaces, in systems of this type. By contrast, *energetic* ions are usually required to cause energy-driven ion-enhanced etching.

VARIABLES CONTROLLING ETCHING

The instrumental variables: RF power, RF excitation frequency, pressure, temperature and feed chemistry are used to set the operation of a plasma process.[3]. Low pressure favors ion enhanced reactions. As pressure falls below about 0.1 Torr, the characteristic potentials of the sheaths rise sharply and the mean free path lengthens, increasing the energy of ions bombarding surfaces. Plasma density tends to be insensitive to total pressure, while the neutral radical density varies directly with pressure. Both factors-higher ion energies and a higher ion to reactive neutral ratio favor damage-induced ion-enhanced etching under low pressure conditions.

High pressure favors oligomer and film formation. Simple etching reactions such as the reaction of F atoms with Si to form SiF_4 are commonly rate-limited by a first-order step. By contrast, reactions that lead to inhibitor films and polymers involve chain growth reactions which vary with higher powers of pressure.

Most commercial plasma reactors operate at 13.56 MHz, a U.S. Federal Communications Commission approved frequency, or below 1 MHz. The choices have been made more by accident than technical design. However changing excitation frequency can alter key plasma characteristics that have important effects on chemical processing.[4] Below the ion transit frequency (~2 MHz) potentials and (≥0.1 Torr) ion energies are high, up to ~1000V even at high pressure. Of course at any pressure, high ion energy favors damage-induced ion-enhanced etching. Other transition frequencies are expected, but associated chemical effects have scarcely been explored.[4]

Power increases the density of radicals and ions as well as the ion energy. In every case when power density is low (≤ 0.5 W cm^{-2}) raising power will increase etch rates. However excessive power causes detrimental substrate heating and surface damage from intense ion bombardment. Etch rates, generally follow an Arrhenius-type dependence on substrate temperature, etch rate $\propto e^{-E_a/kT}$. The effective activation energy E_a is material dependent so selectivity of one material over another varies exponentially with temperature. Increasing temperature usually accelerates chemical etching more than ion enhanced etching, often to the point that it competes with ion damage enhanced etching and reduces anisotropy.

ETCHING GASES

All etching feed gases are a source of atoms that form volatile materials by combining with the substrate. Commercial etching of silicon and its compounds is almost exclusively done by forming the fluoride or chloride.[3] Safety, convenience and economics usually dictate the selection. Hence non-toxic and commercially available halogen carriers such as freons and SF_6 are preferred over the pure halogens F_2, Cl_2, Br_2, ClF_3 and other hazardous gases. However toxic materials are used where the benefits are great enough and safety precautions are manageable- for instance Cl_2 and NF_3 can be handled safely with commercial equipment and offer process benefits that justify the expense. By contrast, reactive oxidizers such as F_2, ClF_3 and the interhalogens are usually considered too dangerous, and a similar view is beginning to apply to carcinogens such as CCl_4.[5]

The feed gases can supply elements to volatilize substrate material in many ways. It is usual for halogen atoms formed in the gas phase to adsorb on surfaces and react, as is the case for most silicon etching in chlorine or fluorine based plasmas. Common feed gases, etching species and etching mechanisms are illustrated in Table I:

ETCHING SPECIES	SOURCE GAS	ADDITIVE	MATERIALS	MECHANISM	SELECTIVE OVER
F	CF_4 SF_6 NF_3	O_2 O_2 None	Si	Chemical	SiO_2 Resist
CF_x-film	CF_4 C_2F_6	H_2 H_2	SiO_2/Si_3N_4	Ion-energetic	Si
Cl	Cl_2	None C_2F_6	undoped Si n-type Si	Ion-energetic Ion-Inhibitor	SiO_2
Cl	Cl_2	BCl_3 CCl_4	Al	Ion-inhibitor	Resist

Table 1. Examples of plasma etching gases and radicals.

A variety of additives are listed in Table I. Additives generally fall into three classes. *Oxidants* are used to increase etchant concentration or suppress polymer. The addition of O_2 to the fluorine bearing gases is an example. *Radical scavengers* such as hydrogen increase the concentration of film formers. This is done to promote selective SiO_2 etching by the mechanisms discussed below. Finally the plasma decomposition products of CCl_4 and C_2F_6 play a dual role as native oxide etchants and sidewall film-formers. Since Cl atoms do not etch the thin oxide films that form on Si, oxides must be removed and the surfaces kept clear to allow halogen atom attack. Finally, *inert gases* (usually Ar or He which are not shown in the table) are often added to help stabilize a plasma, enhance anisotropy or to control the etching rate by dilution.

ETCHING OF SILICON

Almost all silicon plasma etching today is done by converting Si into the final volatile products SiF_4 or $SiCl_4$. These products are thought to be produced by reactions between silicon and the atomic or molecular halogens. The elementary interaction of halogen with silicon is inherently complicated by temperature effects, morphology and crystallographic effects, trace impurities, ion bombardment, doping effects and, in all likelihood, other factors which have not yet been identified. However, according to our understanding, most components other than the halogens, are added for their side effects, discussed above, rather than to influence the rate of etching. Ion bombardment, considered as an "ingredient, is perhaps the exception since it causes the etching of undoped Si by Cl.

Etching silicon with F-atoms

When clean silicon is exposed to atomic fluorine, it quickly acquires a fluorinated "crust," that extends about 5 monolayers into the bulk. Evidence suggests that that F atoms penetrate the top of this layer and attack subsurface bonds Si-Si bonds, liberating two gaseous desorption products- the free radical SiF_2 and the stable end product SiF_4.[6] These two reaction channels have precisely the same activation energy, possibly because there is a common activated state that undergoes dissociation to form SiF_2 or stabilization. The stabilized surface species are fluorinated further to form SiF_4. SiF_2 can be separately followed since it reacts with F or F_2 to form an excited state of SiF_3 which emits a broad visible chemiluminescence, peaking around 500nm:

$$F + F-Si- \rightarrow SiF_2$$

$$SiF_2 + F(F_2) \rightarrow SiF_3^*(+ F)$$

$$SiF_3^* \rightarrow SiF_3 + h\nu_{continuum}$$

Studies show most product leaves the surface as SiF_4, with SiF_2 probably amounting to 5 - 30 percent.

The etch rates for fluorine atoms etching Si and SiO_2 follow an Arrhenius expression, as shown in Table II:

FILM	A	E_A (kcal/mole)	RATE $\dot{A}$/min (298K, $n_F = 3 \times 10^{15} cm^{-3}$)
Si	2.86×10^{-12}	2.48	2250
SiO_2	0.614×10^{-12}	3.76	55

Table 2. Preexponential factors, activation energy and room temperature F-atom etching rate for Si and SiO_2. The rate equation is $ER(\dot{A}/min) = An_F \ T^{1/2} e^{-E_A/RT}$.

These equations show the room temperature selectivity for F-atom etching of Si over SiO_2 is about 40:1. However if the plasma is allowed to heat the substrate the selectivity will fall. Obviously, the wafer surface temperature is an essential variable for process control.

It has been suggested many times that silicon etching by F_2, XeF_2 and other "plasmaless etchants" goes by the same basic process as F-atom etching. But studies show the kinetics and product distribution in XeF_2 etching are remarkably different[7], while F_2 etching proceeds with distinct kinetics[8] and a much less (approximately a monolayer) fluorinated surface.[9]

Many plasma feed gas mixtures produce F-atoms as the dominant etching species. These include F_2, CF_4, CF_4/O_2, SiF_4/O_2, SF_6, SF_6/O_2, NF_3 and ClF_3[3]. In all cases high selectivity over SiO_2 and Si_3N_4 can be achieved and measured activation energies are close to the ideal value in line with the notion that the basic mechanism and etchant are the same.

CF_4 and SF_6 feeds are preferred over pure F_2 because of their low toxicity; however they form unsaturated species (oligomers derived from CF_2 and fluorosulfur radicals (S_xF_y) in the discharge which can react with free F atoms, and sometimes form polymeric residues. Oxygen is frequently added to these plasmas and it has at least two different effects. First, in accordance with the etchant-unsaturate model discussed below, O atoms react with unsaturated CF_x radicals to promote F-atom formation and suppress oligomeric species and polymerization. This enhances the silicon etch rate, and gives better selectivity than pure CF_4, because unsaturated species selectively etch SiO_2 in the presence of ion bombardment. Second, when enough O_2 is present in the feed, O chemisorbs on the silicon surface making it more "oxide-like" and slowing etching.

These effects can be seen in Figure 3 where the etch rate of both Si and SiO_2 increase dramatically as oxygen is added to the feed mixture.

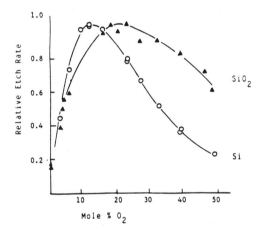

Fig. 3. Effect of oxygen concentration on the etch rates of Si and SiO_2 in a CF_4/O_2 discharge (after ref. 10).

This is the result of increasing F-atoms concentrations. Superficially, oxygen can be considered to "burn" fluorocarbon radicals

$$O + CF_x \rightarrow \begin{cases} COF_2 \\ CO \\ CO_2 \end{cases} + F, F_2$$

With large oxygen additions the rates fall because of dilution by oxygen. It interesting to note that the detailed chemistry of CF_4/O_2 discharges has been modelled and the results are in close agreement with experimental data.

Note that the peak etching rates of Si and SiO_2 in Figure 3 are different. With small oxygen additions both Si and SiO_2 etching are proportional to F atom concentration, but with larger oxygen additions the Si etching lags the the SiO_2 rate because of competition between F and O for surface sites. In effect the Si surface becomes more like an oxide.

Anisotropic etching of silicon in fluorine-containing plasmas is practically impossible under most plasma etching conditions because of the rapid spontaneous chemical reaction. Low pressure plasmas with high substrate "bias" plasmas in CF_4 and SF_6 are an apparent exception, but in this regime the gas phase concentration of F-atoms is lower relative to the adsorbed halocarbon species and the ion bombardment flux. Hence it is quite possible that adsorbed fluorocarbon species are the etchant in this regime, and the mechanisms are unclear. Adsorbed halocarbons and halocarbon radicals are an effective etchant for SiO_2 and, interestingly, the selectivity for Si over SiO_2 is well below 10:1 for these anisotropic processes (compared with 40:1 at room temperature.

Plasma dissociation of feed mixtures forms reactive halogen atoms and unsaturated species. These species tend to recombine with each other. A general model of these reactions:

$$\text{e + Halocarbon} \rightarrow \frac{\text{Saturated}}{\text{Species}} + \frac{\text{Unsaturated}}{\text{Species}} + \text{Atoms}$$

$$\left.\begin{array}{l}\text{Reactive Atoms}\\\text{Reactive Molecules}\end{array}\right\} + \text{Unsaturates} \rightarrow \text{Saturates}$$

$$\text{Unsaturated Species} \rightarrow \text{Oligomers} \rightarrow \text{Polymers}$$

$$\text{Atoms + Surfaces} \rightarrow \left\{\begin{array}{l}\text{Chemisorbed Layer}\\\text{Volatile Products}\end{array}\right.$$

$$\text{Unsaturates + Surfaces}\left[+\frac{\text{Initiating}}{\text{Radicals}}\right] \rightarrow \text{Films}$$

provides a useful basis for formulating fluorocarbon/oxidant feed gas mixtures.[11] Unsaturated species such as CF_2, are formed through electron-impact dissociation of the halocarbon. There is usually a small steady-state concentration of F atoms because they combine with CF_2 radicals and unsaturated species such as C_2F_4. The most reactive atoms (F,O) do not coexist with appreciable concentrations of unsaturates; either these atoms or the unsaturates are depleted. When atoms that can gasify a substrate predominate, etching takes place. Film formation is observed when the unsaturates are present in excess and adsorb on surfaces where polymerization proceeds. Alternatively, the unsaturated radicals may only form sidewall films that result in anisotropic etching in the presence of ion bombardment. Polymerization of unsaturates is inhibited by surfaces that react with fluorocarbon radicals to form entirely volatile products (e.g. SiO_2 surfaces).

Oxidant additions to a plasma alter the balance between halogen atoms and unsaturates. More reactive oxidants will be preferentially consumed by unsaturates, tending to increase the relative concentration of less reactive halogen atoms, and while doing so they will suppress polymer formation. The effect may be illustrated by the addition of oxygen to a CF_4 plasma, discussed above. Oxygen consumes unsaturated species and increases the F concentration.

By contrast, the addition of hydrogen, unsaturated halocarbon feed, hydrocarbons or etchable material (loading effect) removes free fluorine. For instance hydrogen added to a CF_4 plasma removes F by the rapid reaction

$$F + H_2 \rightarrow HF + H$$

making it richer in unsaturated species.

Etching silicon with chlorine

In contrast to fluorine atom plasmas, which are used for rapid chemical etching of Si, chlorine plasmas are usually used as anisotropic Si etchants.[3] Cl atoms, do not etch SiO_2 at normal processing temperatures ($<300°C$). Chlorine based silicon etching can be understood in terms of a few basic facts. First, Cl and Cl_2 etch undoped silicon very slowly (~100Å/min below 100°C at 0.1 torr), or not at all (depending on the crystallographic orientation). However Cl will etch pure silicon in the presence of energetic ion bombardment ("damage" ion-enhanced etching). Second, heavily n-type doped silicon and polysilicon are rapidly and spontaneously etched by Cl atoms. Cl_2 also slowly attacks the n-doped material. This means that inhibitor chemistry must be used for anisotropic chlorine etching of heavily n-doped silicon, Virtually the same remarks apply to bromine plasma etching of silicon.

The steps in silicon etching by chlorine are typical of many other etching systems:

etchant formation

$$e + Cl_2 \rightarrow 2\, Cl + e$$

adsorption of etchant on the substrate

$$\begin{cases} Cl \\ Cl_2 \end{cases} \rightarrow Si_{surf} - n\, Cl$$

either chemical reaction to form product

$$Si - nCl \rightarrow SiCl_{x(ads)}$$

and/or ion-assisted reaction to form product

$$Si - nCl \overset{(ions)}{\rightarrow} SiCl_{x(ads)}$$

and, finally, product desorption

$$SiCl_{x(ads)} \rightarrow SiCl_{x(gas)}$$

Any one of these steps can be rate-limiting. The relative importance of the spontaneous (3) and ion-assisted (4) reactions varies with the etchant and substrate. Purely chemical reaction between undoped silicon and Cl atoms is slow at ordinary temperatures so anisotropic etching can be promoted by damage-induced anisotropy. On the other hand heavily doped silicon, a conductor in integrated circuits, reacts rapidly with Cl and anisotropic patterning of this material requires an inhibitor chemistry. In the same vein, the F atom silicon etching reaction similar to (5) (in CF_4/O_2 plasmas, for instance) is so rapid that F atom etching is isotropic.

Figure 4 shows the effect of frequency on the applied voltage and etch rate of undoped silicon etching in a pure Cl_2 plasma at constant power.[12] All else equal, the applied voltage reflects sheath potential and the sharp increase in etch rate and voltage with decreasing frequency is symptomatic of energy-driven ion enhanced etching.

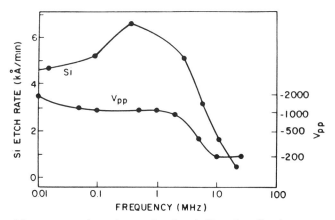

Fig. 4. Effect of frequency on the etch rate of undoped silicon in a Cl_2 plasma at 0.3 Torr (from ref. 12).

The reason why chlorine etch rates depend on doping levels is a subject of current study. However one explanation, which seems consistent with data, is that n-type doping raises the Fermi level and reduces the energy barrier for charge transfer to chemisorbed chlorine. Chlorine (or bromine) atoms are tightly bound to specific sites on undoped silicon surface and steric hindrance impedes the etchant from reaching subsurface Si-Si bonds. With charge transfer, the silicon-halogen bond becomes more ionic in character, allowing bound Cl access to more configurations. Impinging Cl then can chemisorb more easily, and penetrate the surface layer to react.

An example of inhibitor based chlorine plasma etching is shown in Figure 5. Cl_2 supplies the etchant while C_2F_6 is the source of sidewall-protecting inhibitor species.

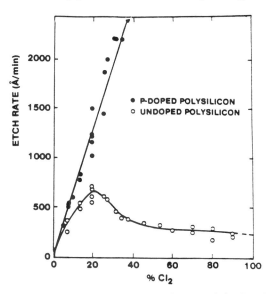

Fig. 5. Etch rates of n-doped and undoped polysilicon in Cl_2/C_2F_6 mixtures at 0.35 torr (after ref. 11).

As chlorine is added to a pure C_2F_6 feed, undoped and undoped polysilicon etched rates initially increase. The heavily doped polysilicon etch rate then shows a rapid, almost linear rise with increasing amounts of added chlorine, while the etch rate for undoped polysilicon remains at a relatively low value. This difference reflects the fact that chlorine atoms easily etch doped polysilicon. Significantly, the doped polysilicon etching is almost completely anisotropic until 10-15% chlorine is in the feed. After this point there is insufficient inhibitor-forming material to protect the sidewalls of etched features. A short plasma exposure to pure C_2F_6 attacks the native silicon oxide so etching can start.

ETCHING SiO_2

SiO_2 can be etched in fluorine atom generating feed gases or in mixtures that are rich in unsaturated fluorocarbon radicals. In silicon microelectronic processing, it is usually necessary to etch SiO_2 preferentially over silicon. Since F-atoms etch Si faster (see above), the unsaturated gas feeds are mostly used. However, since the group V fluorides are involatile, F-atom plasmas *can* be used to etch oxide layers on III-V semiconductors, with the advantage that there is no carbonaceous film residue. At higher pressures and excitation frequency (≥ 0.1 Torr, ≥ 5 MHz) where ion bombardment energy is low, F-atom etching of SiO_2 is chemical, and therefore isotropic. However at low frequency (≤ 1 MHz) and/or low pressure, damage driven ion enhanced attack dominates. Oxide etching in unsaturated plasmas can *only* go by the anisotropic damage driven mechanism. Unfortunately this means there is no way known to selectively plasma etch SiO_2 over Si with an isotropic profile. These alternatives are summarized in Table III.

Additive	Etchant	Conditions	Mechanism	Selectivity	Rate
O_2	F	High Pressure High Frequency	Isotropic (Chemical)	Over III-V's For Silicon	Low
		High Pressure Low Frequency	Energetic Ion-Enhanced		Moderate
H_2	C_xF_y	Low Frequency Low Pressure	Energetic Ion-Enhanced	Over Silicon	Moderate

Table 3. Plasma etching SiO_2 under various conditions.

Halogen atoms are transferred to the substrate through an intermediate when SiO_2 is etched in unsaturate-forming "fluorine-deficient" plasmas. Thin (~ 30Å) fluorocarbon films form at the oxide interface and persist during etching. Ion damage produces dangling bonds and radical groups at the fluorocarbon-substrate interface where reactions between the CF_x-film and the oxygen in the SiO_2 lattice form volatile products like CO, CO_2, COF_2, and SiF_4. For example the following reactions have been proposed:

$$\begin{array}{c} R \quad / \\ \diagdown / \\ Si \\ / \diagdown \\ O \quad O \end{array} + CF_3 \rightarrow \begin{array}{c} R \quad F \\ \diagdown / \\ Si \\ / \diagdown \\ O \quad O \end{array} + CF_2$$

$$\begin{array}{c} R \quad O \\ \diagdown / \\ Si \\ / \diagdown \\ O \quad O \end{array} + \begin{array}{c} F \quad F \\ \diagdown / \\ C = C \\ / \diagdown \\ F \quad F \end{array} \rightarrow \begin{array}{c} R \quad F \\ \diagdown / \\ Si \\ / \diagdown \\ O \quad O \end{array} + CF_2 + CF_2O$$

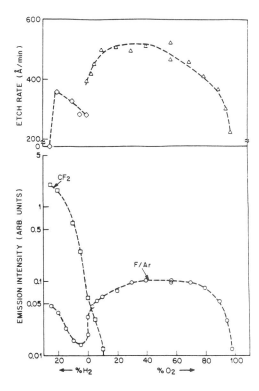

where R denotes F, O, or bound fluorocarbon. Studies show that these fluorocarbon films also form on exposed Si areas during oxide etching where there is no route to gaseous products so that carbonaceous layer blocks attack and improves selectivity.

Fig. 6. Effect of O_2 and H_2 additions on the etch rate of SiO_2 (upper) and optical emission from 100 KHz CF_4 plasmas (lower). The lower electrode was covered with Si wafers. (after ref. 14).

Starting with CF_4, adding either O_2 or H_2 increases the etch rate of SiO_2 under anisotropic ion energy enhanced etching conditions- low frequency and high pressure, or low pressure. However the chemistry under these two conditions is drastically different, following the scheme in Table III. Figure 6 shows the etch rate and emissions from F-atoms and CF_2 in the discharge versus additives. When oxygen is added F atoms are the etchant, while films derived from CF_2 react at the surface with H_2 is additions. This is reflected by in the emission spectrum (Fig. 6). Note that excessive hydrogen additions cause gross plasma polymerization and all etching stops beyond the leftmost side of composition curve..

The remaining condition in Table III, oxygen additions at high frequency, results in slow purely chemical SiO_2 etching by F, discussed above, which is isotropic but not selective (over Si).

REFERENCES

1. L. C. Parillo, Ch. 11, pps. 445-505 in *VLSI Technology*, S. M. Sze, ed. (McGraw Hill, NY, 1983).

2. L. F. Thompson and M. J. Bowden, "The Lithographic Process," Ch. 2, pps. 16-85 in *Introduction to Microlithography*, L. F. Thompson, C. G. Willson, M. J. Bowden, eds., ACS Symp. Ser. 219, (Amer. Chem. Soc, Wash., 1983).

3. D. L. Flamm, V. M. Donnelly and D. E. Ibbotson, in *VLSI Electronics: Microstructure Science*, Vol. 8, eds. N. G. Einspruch and D. M. Brown, Academic Press, NY (1983), Chapter 8; D. L. Flamm and J. A. Mucha, in *Chemistry of the Semiconductor Industry*, eds. A. Ledwith and S. J. Moss, Blackie and Son, Glasgow (1987) Chapter 15; D. L. Flamm and G. K. Herb, *Plasma-Materials Interactions*, Vol. 1, eds. D. L. Flamm and D. M. Manos, Academic Press, Orlando, in press (1987) Chapter 1.

4. D. L. Flamm, *J. Vac. Sci. Technol. D. L. Flamm*, J. Vac. Sci. Technol. A4, 729(1986).

5. G. K. Herb, "Safety, Health and Engineering Considerations for Plasma Processing," in *Plasma-Materials Interactions*, Vol. 1, eds. D. L. Flamm and D. M. Manos, Academic Press, Orlando, in press (1987).

6. D. L. Flamm, V. M . Donnelly and J. A. Mucha, *J. Appl. Phys.*, 52, 3633 (1981).

7. D. E. Ibbotson, D. L. Flamm, J. A. Mucha, and V. M. Donnelly, *Appl. Phys. Lett.*, 44, 1129 (1984): M. J. Mitchell, M. Suto, L. C. Lee and T. J. Chuang, *J. Chem. Phys.*, in press (1987).

8. J. A. Mucha, V. M. Donnelly D. L. Flamm and L. M. Webb, *J. Phys. Chem.*, 85, 3529 (1981).

9. C. D. Stinespring, A. Freedman, *Appl. Phys. Lett.*, 48, 718 (1986).

10. C. J. Mogab, A. C. Adams and D. L. Flamm, *J. Appl. Phys.*, 49, 3796 (1979).

11. D. L. Flamm, *Plasma Chem. Plasma Proc.*, 1, 37 (1981).

12. V. M. Donnelly, D. L. Flamm, and R. H. Bruce, *J. Appl. Phys.*, 58, 2135 (1985).

13. C. J. Mogab and H. J. Levenstein, *J. Vac. Sci. Technol*, 17, 1721 (1980).

14. D. L. Flamm, V. M. Donnelly and D. E. Ibbotson, *J. Vac. Sci. Technol. B1*, 23 (1983).

Chapter 37

Mechansims of silicon CVD

Pauline Ho, William G.Breiland and Michael E. Coltrin — Sandia National Laboratories, Albuquerque, NM 87185.

ABSTRACT

The microelectronics and solar-cell industries rely heavily on silicon thin films. Such films are generally produced via chemical vapor deposition (CVD) from silanes or chlorosilanes. We describe a coordinated program of experimental and theoretical research into the fundamental mechanisms of CVD. The experimental program uses laser spectroscopic techniques for *in situ* measurements during CVD. The theoretical program consists of a computer model that contains a detailed description of the coupled fluid mechanics and gas-phase chemical kinetics in silane CVD.

INTRODUCTION

Chemical vapor deposition (CVD) is widely used for the production of thin films for electronics, protective coatings and other materials processing/production applications. CVD involves flowing a reactive gas (often diluted in a carrier gas) through a reactor containing a heated substrate. Chemical reactions occur that deposit a solid film on the substrate. Gaseous reaction products and any unreacted starting materials flow out of the reactor. In particular, CVD is heavily used in microelectronic and photovoltaic technologies which are predominantly silicon-based. This paper includes a brief review of some of these applications and processing techniques for silicon materials, many of which involve silane or substituted silanes. A review of our work [1-4] on the detailed mechanisms of silane chemical vapor deposition (CVD) is included as an example. Other techniques are reviewed to illustrate the range of silicon chemistry involved in these technologies.

The fabrication of microelectronic devices starts with substrates cut from a single crystal of silicon. The large single crystals (boules) are generally grown from melted, purified silicon. A typical process for producing purified silicon involves formation of a chlorosilane from a lower grade of silicon, purification of the chlorosilane via distillation, then decomposition of the purified gas back to a polycrystalline solid.

Among the other processing steps used in production of microelectronic devices are the deposition of epitaxial (single crystal) or polycrystalline thin films of silicon. Epitaxial silicon is superior to polysilicon (polycrystalline silicon) because the grain boundaries in polysilicon degrade the electrical properties of the material. However, polysilicon films can be deposited on insulators, whereas epitaxial films are generally only grown on single crystal silicon. Polysilicon is therefore used on upper layers for interconnects, resistors and capacitors. A great deal of current research is devoted to the deposition of epitaxial silicon on insulating materials (silicon-on-insulator research).

Epitaxial films are generally deposited via the hydrogen reduction of chlorosilanes at temperatures of 1000-1100°C. In contrast, polysilicon films are usually deposited from silane at temperatures of $\sim700°$C and low total pressures. The combination of lower temperatures and the large batches used in low-pressure processes makes polysilicon deposition cheaper than epitaxial silicon deposition. Low deposition temperatures also make polysilicon deposition better suited to the latter stages of device processing, when exposure of the wafers to high temperatures must be minimized to avoid interdiffusion of dopants.

Single crystal, polycrystalline and amorphous silicon have all been used in photovoltaics, although single crystal and amorphous hydrogenated silicon (a-Si:H) have received the most attention. Single crystal silicon solar cells have higher efficiencies than amorphous silicon devices, but are considerably more expensive to produce. The cost of growing a single crystal of silicon, sawing it into wafers, then polishing and processing the wafers is considerably higher than that for depositing an amorphous silicon film. Sunlight is also absorbed more efficiently by a-Si:H than by a silicon single crystal. Although amorphous films have been produced by sputtering techniques, the material is primarily produced by silane CVD. Several variations of silane CVD are used, including plasma-enhanced CVD, homogeneous CVD (HOMOCVD), photochemical CVD and deposition from SiH_4/SiF_4 mixtures.

EXPERIMENTAL AND THEORETICAL STUDIES OF SILANE CVD

This section describes a detailed study of the fundamental mechanisms of the silane CVD system. Although the silane system has important applications to polycrystalline and amorphous silicon films, we have chosen this system primarily because it is relatively simple and can serve as a model for systems with more complex chemistries.

The fundamental chemistry and physics of CVD can be divided into gas-phase and surface processes. Gas phase processes include chemistry and fluid mechanics, both of which are well-established fields. The field

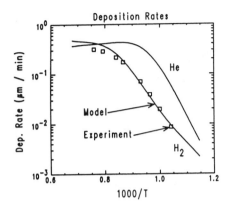

Fig. 1. Deposition rates predicted by the model as a function of temperature and carrier gas. Experimental deposition rates in hydrogen carrier gas taken from Ref. 8.

of surface chemistry is not as mature as gas-phase chemical kinetics, so our research has focused on the events that occur in the gas phase during deposition.

The theoretical part of the program [1,2] consists of a computer model that contains a detailed description of the coupled fluid mechanics and gas-phase chemical kinetics in silane CVD. The model uses finite-difference methods to numerically solve the boundary layer equations of fluid flow and chemical rate expressions that describe the deposition of silicon from silane. It is a steady-state, two-dimensional model. Initially, a mechanism of 120 reversible, elementary reactions was considered to describe the thermal decomposition of silane and the subsequent reactions of intermediate species. Thermochemical and kinetic data for these reactions were obtained from the literature whenever possible. Otherwise, thermochemical data were obtained from *ab initio* electronic structure calculations [5,6], and kinetic parameters estimated using techniques described by Benson [7]. A sensitivity analysis was used to reduce this mechanism to the 27 reactions shown in Table 1. The initial step in the silane decomposition mechanism is production of SiH_2 (Reaction 1 in Table 1).

Formation of SiH_3 by loss of H from silane is slow, so silicon species containing an odd number of hydrogens are predicted to be unimportant in silane CVD [1,2] although they are included in the model for completeness. Surface reactions are included in the model in the form of reactive sticking coefficients (RSCs). The RSC for silane on a hot silicon surface is relatively low (10^{-4} to 10^{-6}) and was derived from literature data (see Ref. 2 for details). In the absence of any experimental information, the RSC for intermediate species such as SiH_2, Si_2H_2, Si_2, etc., were set to one. The model predicts gas-phase temperature and velocity fields, and density fields for the 17 chemical species in the kinetics mechanism, deposition rates and deposition uniformity as a function of control variables such as susceptor temperature, flow velocity, total pressure, reactant gas composition, carrier gas, and cell dimension.

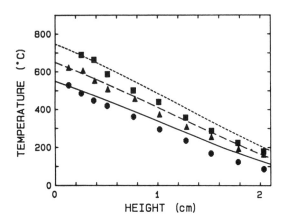

Fig. 2. Profiles of gas-phase temperatures obtained from the rotational Raman spectrum (data points) and model predictions (curves) for three susceptor temperature: 550°C, octagons and solid curve; 650°C, triangles and dashed curve; 750°C, squares and dotted curve.

Figure 1 shows an Arrhenius-type plot of silicon deposition rates predicted by the model as a function of susceptor temperature for atmospheric-pressure CVD (APCVD) in He and H_2 carrier gases, along with experimental deposition rates in H_2 taken from the literature [8]. In addition to reproducing the experimental data, the model also predicts several well-known features of silane CVD: the transition between high-temperature behavior (weak temperature dependence) and low-temperature behavior (strong temperature dependence) and the higher deposition rates in He carrier gas. The carrier-gas-dependence results from the fact that H_2 is a product of the silane decomposition reaction and many of the other reactions. Gas-phase reactions convert relatively unreactive silane into species such as SiH_2 and Si_2H_4 which are more reactive with the silicon surface than is silane. An atmosphere of H_2 suppresses formation of these intermediates, leading to slower deposition rates at a given temperature.

Although correct predictions of deposition rates are a necessary test of a model for CVD, deposition rate data give little information on the details of the chemistry. The experimental part of the program [3,4] provides such information through the use of laser spectroscopic techniques for *in situ* measurements during CVD. The remainder of this section summarizes some of our previously published comparisons between the model predictions and the *in situ* measurements.

The experimental apparatus consists of a research CVD reactor with laminar gas flow and windows for optical access. Fluorescence or scattered light is collected at right angles to the laser beam, which is focused above the heated surface. This geometry provides excellent spatial resolution. Varying the measurement position relative to the heated susceptor was done by translating the cell relative to the laser beam and collection optics. Pulsed-laser Raman spectroscopy and laser-excited fluorescence spectroscopy were used to measure spatial profiles of gas-phase temper-

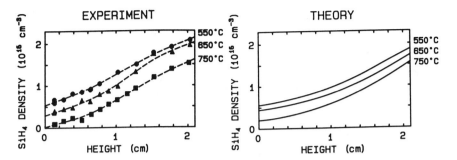

Fig. 3. Silane density (in units of 10^{16} cm^{-3}) versus height above the susceptor for three susceptor temperatures.

atures and species densities. Gas temperatures were obtained using rotational Raman spectroscopy [3]. Laser-excited fluorescence was used to obtain relative density profiles of Si_2 and Si atoms [3,4].

Temperature profiles above the heated susceptor are important in determining the extent to which chemical reactions occur in the gas phase. In Fig. 2 we compare the measured temperature as a function of height above the susceptor with the model predictions for three susceptor temperatures. The calculated and experimental temperature profiles agree that steep temperature gradients exist in the gas. Although the agreement is not exact, it shows that the 2-D boundary layer model provides a sufficiently accurate description of the temperature field (and fluid flow) in the experimental reactor to yield realistic chemical predictions.

Figure 3 compares experimental silane densities as a function of height above the susceptor with predictions of the model for three susceptor temperatures. Both the theory and experiment show a substantial decrease in the silane density between 2 cm and 0 cm, reflecting the large temperature gradients seen in Fig. 2. There is somewhat more depletion of the silane in the 650°C and 750°C experimental curves than is predicted by the model. These results, however, show that the model gives a satisfactory description of the silane density curves.

Analysis of the model shows that three effects contribute to the shape of the silane density profiles. The first effect is ideal gas expansion; the temperature is higher near the susceptor, so the density of the gas decreases. Gas-phase decomposition is the second effect leading to a decrease in silane density close to the susceptor. The third effect is thermal diffusion, the separation of two components in a gas mixture of different mass and/or size in a temperature gradient. In the SiH_4/He system, thermal diffusion decreases the silane density near the susceptor.

Compared with silane, chemical species formed as intermediates in gas-phase reactions should have qualitatively different profiles as a function of height above the susceptor. The rates of both creation and destruction of such species are controlled by the gas-temperature field. Experimental and predicted profiles of the intermediate species Si_2 are given in Fig. 4. The shapes of the profiles are in qualitative agreement. Both

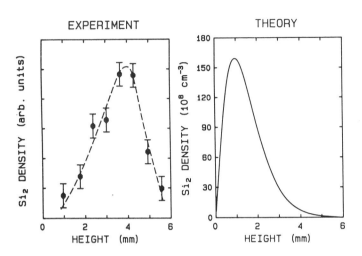

Fig. 4. Si$_2$ density versus height above the susceptor. Experimental data give relative densities only. Susceptor temperature 740°C.

have a maximum in the gas phase a few mm above the susceptor, although the one predicted by the model is closer to the susceptor. The fact that the Si$_2$ profile has a maximum in the gas phase confirms that this species is formed via gas-phase reactions rather than surface reactions.

Density profiles of gas-phase Si atoms predicted by the model and experimental measurements are given in Fig. 5. The shape of the Si atom profiles are similar to the Si$_2$ profiles above. The model accurately predicts the position of the maximum in the density profiles, as well as the tendency for the maximum to occur closer to the susceptor as the temperature is decreased. However, the model predicts that the maximum Si atom density increases somewhat more rapidly with increasing susceptor temperature than is seen experimentally.

The model predicts that the Si atom density (as well as the Si$_2$ density) should decrease dramatically as H$_2$ is added to the He carrier gas. Figure 6 shows the predicted and measured effects of adding H$_2$ on the maximum Si atom density. The experimental measurements show a strong suppression of Si atoms by H$_2$. However, the model predicts a substantially stronger suppression of Si atom density with the addition of the first few Torr of H$_2$, followed by a weaker decrease as more H$_2$ is added. This qualitative disagreement (and a similar result for the Si$_2$ dependence on H$_2$) represents a significant breakdown in the model. A kinetic analysis showed that reactions 1 and -9 are primarily responsible for producing Si atoms, while reaction 8 (and -22 above 1000°C) is the primary destruction channel. This analysis also showed, however, that the disagreement between the model and experiment could not be resolved by varying the rate constants for these reactions [4].

We believe that the disagreement in the predicted H$_2$ effect may be due to the omission from the model of a mechanism for gas-phase nucle-

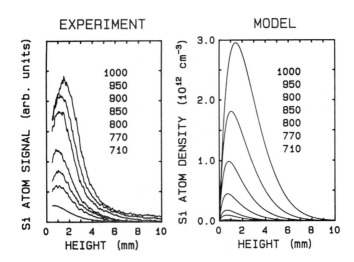

Fig. 5. Si atom density versus height above the susceptor for several susceptor temperatures (listed on the figure), with the highest temperature corresponding to the largest profile and decreasing monotonically. Experimental data give relative densities only.

ation of particulates, which we observe under some conditions in silane CVD. The formation of silicon particles and clusters provides an additional removal pathway for silicon in the gas phase, so Si atom densities predicted by the model could be artificially high in the absence of this mechanism. The overestimation would be more severe at high temperatures, where particle formation occurs readily, but less severe in the presence of a few Torr of H_2, where particle production is suppressed. The conditions under which particles are observed are similar to those for which the model predicts the largest amounts of the intermediates Si, Si_2, and Si_3. It is likely that these species are related to the formation of particulates, and that reactions leading to formation of Si_2 and Si_3 (reactions 1, 3, 17, 18, -20 and -22 in Table 1) are important in particle formation.

Taken together, the results of the theoretical modeling and the laser diagnostic measurements clearly show the importance of gas-phase chemistry in silane CVD. However, as the chemical environment in a CVD reactor changes with operating conditions, so do the relative importance of gas-phase chemistry, surface chemistry and fluid mechanics in determining film deposition rates. When chemical reactions are very fast, mass transfer (diffusion) rates limit deposition rates. When gas-phase reactions are very slow, surface-reaction rates can determine deposition rates. Despite this variability, many features of silane CVD can be understood in terms of gas-phase chemical reaction kinetics and fluid mechanics. The theoretical framework exists for treating these aspects of CVD in detail. Treating the surface reactions in comparable detail, however, is not feasible at this time.

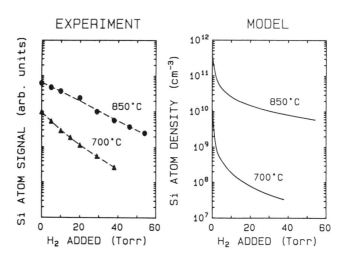

Fig. 6. The maximum in the Si atom density profile as a function of partial pressure of H_2 added to the helium carrier gas. Experimental data give relative densities only.

OTHER CVD SYSTEMS

Although silane CVD is technologically important, it is only one of many materials processing techniques. This section presents several other processing techniques involving silicon chemistry. The work presented above on silane CVD illustrates the level of detail to which we can understand a CVD process. A comparably detailed description of other CVD systems would require substantial chemical information about the system of interest.

The hydrogen reduction of chlorosilanes is used extensively in CVD, particularly for depositing epitaxial silicon. $SiCl_4$, $SiCl_3H$ and $SiCl_2H_2$ are used, although $SiCl_4$ is the least expensive starting material. Chlorosilane CVD has been the subject of a number of equilibrium calculations, so generally-accepted values for thermochemical data are available for most of these species [5]. Although an equilibrium analysis of the gas phase chemistry in chlorosilane CVD may be a useful guide at the high temperatures (1000-1100°C) now used for epitaxial silicon deposition, current work in this field emphasizes reducing processing temperatures. Understanding chlorosilane CVD at lower temperatures will probably require a reaction kinetics analysis. At present, however, few kinetics data are available for reactions of SiH_nCl_m molecules [10].

Silicon dioxide rivals silicon in importance for the fabrication of microelectronic devices. It is used for protective layers and is the primary masking material used to produce the localized regions of doped silicon that form junctions. It is used as an insulating material to separate active devices on a wafer, as well as being a part of devices, i.e. gate oxides in field-effect transistors. Several techniques are used for producing SiO_2 films. Direct oxidation of silicon can be done either with oxygen or water vapor. "Dry" oxidation produces denser films than "wet" oxidation, but has a much lower oxidation rate. CVD is generally used to deposit SiO_2 films on non-silicon surfaces. Depending on film-property requirements, SiO_2 can be deposited from mixtures of silane or a chlorosilane and N_2O, CO_2, or O_2, or from $Si(OC_2H_5)_4$ (TEOS). SiO_2 films can also be deposited by plasma-enhanced CVD (PECVD).

Silicon nitride and silicon oxynitride are also used as insulators in microelectronic devices. Again, these materials are used for separating electrically-active regions, for passivating (protective) layers, and within devices. Si_3N_4 is deposited via CVD from NH_3 and either SiH_4 or $SiCl_2H_2$. Si_2N_2O is deposited from mixtures of SiH_4 or $SiCl_2H_2$ and N_2O and NH_3. PECVD techniques are also used to produce nitride films. PECVD adds a new dimension of complexity to the problem of understanding the fundamental mechanisms of a process. The plasma adds electron and ion dynamics, electric fields, and reactions of charged species (ions and electrons) to the problem. Data on such processes are not usually available.

Other silicon-containing materials involved in microelectronics include metal silicides (i.e. WSi_2, $MoSi_2$, or $TaSi_2$) which are used for interconnects. Silicate glasses containing phosphorus or boron are used as sources for diffusion-doping of silicon. Doping can also be done during silicon deposition by adding small amounts of PH_3, AsH_3, or B_2H_6 to the reacting gas, although this alters deposition rates. Silicon itself is also used as a dopant in GaAs technology.

ACKNOWLEDGEMENT

We thank Michael P. Youngman for providing expert technical assistance in the experimental work, and Robert J. Kee and James A. Miller for their support of the modeling described here. This work was performed at Sandia National Laboratories, supported by the U.S. Department of Energy, Office of Basic Energy Sciences under contract No.DE-AC04-76DP00789.

REFERENCES

1. M. E. Coltrin, R. J. Kee, and J. A. Miller, *J. Electrochem. Soc.* **131**, 425 (1984).
2. M. E. Coltrin, R. J. Kee, and J. A. Miller, *J. Electrochem. Soc.* **133**, 1206 (1986).
3. W. G. Breiland, M. E. Coltrin, and P. Ho, *J. Appl. Phys.* **59**, 3267 (1986).
4. W. G. Breiland, P. Ho, and M. E. Coltrin, *J. Appl. Phys.* **60**, 1505 (1986).
5. P. Ho, M. E. Coltrin, J. S. Binkley and C. F. Melius, *J. Phys. Chem.* **89**, 4647 (1985).
6. P. Ho, M. E. Coltrin, J. S. Binkley and C. F. Melius, *J. Phys. Chem.* **90**, 3399 (1986).
7. S. W. Benson, *Thermochemical Kinetics*, Wiley and Sons, New York, 1976.
8. C. H. J. van den Brekel, PhD. Thesis, University of Nijmegen, The Netherlands (1978).

9. C. G. Newman, H. E. O'Neal, M. A. Ring, F. Leska, and N. Shipley, *Int. J. Chem. Kin.* **11**, 1167 (1979).

10. P. Ho, W. G. Breiland, and R. W. Carr, *Chem. Phys. Lett.* **132**, 422 (1986).

Table 1. Reaction mechanism.

	Reaction	A^a	β^a	E^a
R1	$SiH_4 \longleftrightarrow SiH_2 + H_2$	9.81×10^{37}	-7.79	61.92
R2	$SiH_4 \longleftrightarrow SiH_3 + H$	3.69×10^{15}	0	93
R3	$SiH_4 + SiH_2 \longleftrightarrow Si_2H_6$	5.01×10^{12}	0	1.29
R4	$SiH_4 + H \longleftrightarrow SiH_3 + H_2$	1.04×10^{14}	0	2.5
R5	$SiH_4 + SiH_3 \longleftrightarrow Si_2H_5 + H_2$	1.77×10^{12}	0	4.4
R6	$SiH_4 + SiH \longleftrightarrow Si_2H_3 + H_2$	1.45×10^{12}	0	2
R7	$SiH_4 + SiH \longleftrightarrow Si_2H_5$	1.43×10^{13}	0	2
R8	$SiH_4 + Si \longleftrightarrow SiH_2 + SiH_2$	9.31×10^{12}	0	2
R9	$Si + H_2 \longleftrightarrow SiH_2$	1.15×10^{14}	0	2
R10	$H_2SiSiH_2 \longleftrightarrow SiH_2 + SiH_2$	1.00×10^{16}	0	59
R11	$SiH_2 + H \longleftrightarrow SiH + H_2$	1.39×10^{13}	0	2
R12	$SiH_2 + H \longleftrightarrow SiH_3$	3.81×10^{13}	0	2
R13	$SiH_2 + SiH_3 \longleftrightarrow Si_2H_5$	6.58×10^{12}	0	2
R14	$SiH_2 + Si_2 \longleftrightarrow Si_3 + H_2$	3.55×10^{11}	0	2
R15	$SiH_2 + Si_3 \longleftrightarrow Si_2H_2 + Si_2$	1.43×10^{11}	0	18.8
R16	$H_2SiSiH_2 \longleftrightarrow Si_2H_2 + H_2$	3.16×10^{14}	0	53
R17	$Si_2H_6 \longleftrightarrow H_2SiSiH_2 + H_2$	2.51×10^{14}	0	52.2
R18	$H_2 + SiH \longleftrightarrow SiH_3$	3.45×10^{13}	0	2
R19	$H_2 + Si_2 \longleftrightarrow Si_2H_2$	1.54×10^{13}	0	2
R20	$H_2 + Si_2 \longleftrightarrow SiH + SiH$	1.54×10^{13}	0	40
R21	$H_2 + Si_3 \longleftrightarrow Si + Si_2H_2$	9.79×10^{12}	0	42.6
R22	$Si_2H_5 \longleftrightarrow Si_2H_3 + H_2$	3.16×10^{14}	0	53
R23	$Si_2H_2 + H \longleftrightarrow Si_2H_3$	8.63×10^{14}	0	2
R24	$H + Si_2 \longleftrightarrow SiH + Si$	5.15×10^{13}	0	5.3
R25	$Si_3H_8 \longleftrightarrow Si_2H_6 + SiH_2$	4.90×10^{15}	0	52.99
R26	$Si_3H_8 \longleftrightarrow H_3SiSiH + SiH_4$	4.79×10^{14}	0	49.24
R27	$H_3SiSiH \longleftrightarrow H_2SiSiH_2$	6.31×10^{12}	0	29.2

^{a} Arrhenius parameters for the rate constants in the forward direction written in the form $k = AT^\beta exp(-E / RT)$. The units of A depend on the reaction order, but are given in terms of mols, cubic centimeters, and seconds. E is in kcal/mol. Rate constants for the back reactions can be calculated from the reaction thermochemistry.

Chapter 38

Applications of methyldichlorosilane in the preparation of silicon-containing ceramics

Dietmar Seyferth*, Gary H. Wiseman, Yuan-Fu Yu, Tom S. Targos, Christine A. Sobon, Timothy G. Wood and Gundrun E. Koppetsch – Department of Chemistry, Massachusetts Institute of Technology, Cambridge, Massachusetts 02139.

ABSTRACT

The hydrolysis, ammonolysis, sodium condensation, and ethynylenation products of methyldichlorosilane can be converted to useful preceramic polymers. The chemistry involved in these conversions is summarized.

INTRODUCTION

The Rochow-synthesis of methylchlorosilanes by the reaction of gaseous methyl chloride with a solid mass of silicon/copper alloy has been practiced since the 1940's.[1] The major product of this reaction as practiced industrially is dimethyldichlorosilane, $(CH_3)_2SiCl_2$, the workhorse of the silicones industry, but other products are formed as well, albeit in much lower yield. Among these are CH_3SiCl_3, $(CH_3)_3SiCl$ and CH_3SiHCl_2. The latter, methyldichlorosilane, has some commercial applications.[2,3] It presents interesting options for further chemical conversion: in addition to its two very reactive Si-Cl bonds, it has a reactive Si-H bond.[2,4] (The Si-CH$_3$ bond, on the other hand, is kinetically quite stable). In some cases, exclusive reactions of the Si-Cl bonds are possible. In other cases, exclusive reactions of the Si-H bond can be effected. Toward some reagents, both the Si-Cl and Si-H

bonds are reactive, but usually at different rates. Thus the
chemistry of methyldichlorosilane is potentially rich and
variable. We have found it to be an excellent precursor for a
number of different preceramic polymer systems which we shall
discuss below.

THE HYDROLYSIS OF METHYLDICHLOROSILANE

The hydrolysis of methyldichlorosilane can be carried out in
such a manner that a high yield of cyclic oligomers, $\underline{1}$, is
obtained, in addition to linear polysiloxanes, $\underline{2}$.[5]

$\underline{1}$, n = 1, 2, 3 . . . $\underline{2}$, n > 1

For instance, hydrolysis of CH_3SiHCl_2 in dichloromethane solution at room temperature by slow addition of bulk water gave
the distribution of cyclic oligomers shown in Table I.[5b]
Since the Si atoms contain CH_3 and H substituents, cis and
trans isomers of the cyclic oligomers will be possible. Mostly
linear polysiloxanes, $\underline{2}$, also can be prepared. One such
polymeric product, a liquid of molecular weight 2000-5000
(vendor data), is sold under the designation PS-122 by the
Silanes and Silicones Group of Dynamit Nobel (formerly Petrarch
Systems). Such linear polysiloxanes, as well as polysiloxanes
obtained by cohydrolysis of CH_3SiHCl_2 and $(CH_3)_2SiCl_2$, have
found industrial application.[2]

The chemistry of $[CH_3Si(H)O]$-containing polysiloxanes, both
the cyclic oligomers and the linear polymers, has been investigated. The reactions studied have involved in the main
hydrosilylation,[4] the catalyzed additions of their Si-H bonds
to C=C bonds of diverse olefinic substrates,, although other
Si-H reactions are possible. Thus, for instance, conversion of
the Si-H linkages of the linear polymer to reactive Si-Cl bonds
has been carried out using allyl or benzyl chloride with Pd/C
catalyst as the chlorination reagent.[6] Other reactions of
$[CH_3Si(H)O]_n$ cyclic oligomers have been used to introduce
metal functionality. For instance, reactions of $[CH_3Si(H)O]_n$

Table 1. Composition of cyclo-$[CH_3Si(H)O]_n$ mixtures formed in the hydrolysis of CH_3SiHCl_2.

n	mol %	n	mol %
3	trace[a]	11	1.57
4	36.97	12	1.34
5	26.43	13	1.21
6	13.80	14	1.04
7	6.95	15	0.87
8	3.72	16	0.68
9	2.46	17	0.51
10	1.88	18	0.38
		19	0.22
		20-22	trace

[a] Some of the cyclic trimer may have been lost on concentration of the solution.

cyclics with $Co_2(CO)_8$ and $[\eta-C_5H_5Fe(CO)_2]_2$ gave interesting
cobalt- and iron- containing cyclosiloxanes with Si-Co and
Si-Fe bonds.[7]

In our own research we have used reactions of both the cyclic
$[CH_3Si(H)O]_n$ oligomers and the linear polysiloxanes containing
this repeat unit to prepare polymers containing metal alkoxide
side groups. The impetus for preparing such materials was
given by a need for coating materials for carbon/carbon composites that would serve to protect them from high temperature
oxidative degradation. Current thinking suggested that hafnium
or zirconium containing materials might be suitable in this
application, so our focus was on the preparation of polymeric
systems containing these elements.

The chemistry of zirconium and hafnium alkoxides is well
developed[8], and we chose to approach this goal via metal
alkoxide chemistry, using the reactivity of the $[CH_3Si(H)O]_n$
oligomers and polymers to prepare the desired metal alkoxide-
containing polymers.

Metal alkoxides containing alkenoxy groups, $CH_2=CH(CH_2)_nO$
$n = 0, 1, 2 . . .$, can be prepared in different ways. For our
purposes we required two alkenoxy substituents on the metal for
cross-linking hydrosilylation reactions with $[CH_3Si(H)O]_n$.
The simplest approach to such alkoxides of hafnium and zirconium involves the reaction of two molar equivalents each of a
saturated alcohol and an unsaturated alcohol with one equivalent of the hafnium or zirconium tetrachloride, using ammonia
as HCl acceptor. This would be expected to give a mixture of
species, $(RO)_nM(OCH_2CH=CH_2)_{4-n}$ (when allyl alcohol was
the unsaturated alcohol used), with n values of 0-4 possible,
but with the average composition "$(RO)_2M(OCH_2CH=CH_2)_2$". An
alternate designed synthesis of such a metal alkoxide uses the
preformed $(RO)_2MCl_2$ (a known, stable compound type) in a
reaction with an unsaturated alkoxide reagent. The reaction of
$(iPrO)_2TiCl_2$ with two molar equivalents of CH_2CHOLi (prepared
by the fragmentation reaction of tetrahydrofuran by n-butyllithium[9]) is an example of this approach. The reaction of alkoxide materials of average composition "$(iPrO)_2M(OCH_2CH=CH_2)_2$"
(M = Hf and Zr) with commercial $[CH_3Si(H)O]_n$ PS-122 polysiloxane, using various reactant ratios and using as hydrosilylation catalyst $H_2PtCl_6 \cdot 6H_2O$, in toluene at ~110°C gave glassy
solids which were initially soluble in organic solvents such as
benzene, toluene and THF. However, once these products were
isolated from solution they tended to become insoluble on
storage under nitrogen at room temperature. The cross-linking
process must have been effective in forming a network structure
since the pyrolysis of these materials (to 1000°C in a stream
of argon) left a residue in 80 weight % yield (ceramic yield).
That the cross-linking involves hydrosilylation of allyloxy
groups to build $-CH_2CH_2CH_2OM(OPr^1)_2OCH_2CH_2CH_2-$ bridges

between siloxane chains was demonstrated by the proton NMR spectra of the products. Since these products are initially soluble, their solutions may be used in vacuum dip-coating of carbon/carbon composite substrates. A testing program of the oxidation resistance of carbon/carbon composites treated with these zirconium- and hafnium-containing products is in progress. It is clear that polysiloxane-anchored metal alkoxides of many other metals can be made by this procedure and we are examining further possibilities.

THE AMMONOLYSIS OF METHYLDICHLOROSILANE

The reaction of ammonia with methyldichlorosilane was reported first by Brewer and Haber in 1948.[10] They obtained a viscous oil which decomposed on attempted vacuum distillation with release of ammonia. An insoluble gel remained as residue. Sememova, Zhinkin and Andrianov[11] studied the same reaction and reported obtaining by vacuum distillation a 25% yield of $[CH_3SiHNH]_4$ and a 10% yield of a bicyclic compound formulated as $(CH_3SiHNH)_3[CH_3SiH]_{1.5}N]_2$. In our hands, the ammonolysis of CH_3SiHCl_2 proceeded smoothly in diethyl ether or THF at 0° C. Very little insoluble polysilazane precipitated with the ammonium chloride. Removal of solvent in vacuum from the filtrate left a mobile oil in high yield Its proton NMR spectrum was in agreement with the formulation $[CH_3SiHNH]_n$, and a cryoscopic molecular weight determination gave values of n between 4.7 and 5.4. This ammonolysis product, probably a mixture of cyclic oligomers and, perhaps, also some linear species was not a useful silicon carbonitride precursor. Its pyrolysis to 1000°C gave a ceramic yield of only 20%. Application of the ammonium salt-induced thermal polymerization of Rochow and Krüger to the CH_3SiHCl_2 ammonolysis resulted in evolution of ammonia and formation of a very viscous polysilazane oil ($SiCH_3:NH:SiH = 3:0.55:1$ vs $3:1:1$ for $[CH_3SiHNH]_n$). Pyrolysis of this material, however, gave a ceramic yield of only 36%.

In a search for better ways of converting the CH_3SiHCl_2 ammonolysis product to a useful silicon carbonitride precursor, we applied the dehydrocyclodimerization reaction which had been reported by Fink in 1965 (eq. 1)[13].

In each cyclic [CH$_3$SiHNH]$_n$ oligomer there are "n" sets of
adjacent Si-H and N-H bonds. If the process shown in eq. 1
could be induced, then a polymer consisting in good part of
fused cyclosilazane rings would result. (No doubt, the process
would be more complicated, since base-induced intermolecular
Si-H/ N-H reactions (to give an Si-N bond and H$_2$) also were
known.

Since the potassium in eq. 1 would quickly be converted to KH,
we used KH as the basic catalyst. The hoped-for dehydrocyclopolymerization did indeed take place and we obtained organic-
soluble low polymers (MW between 800 and ca. 2000, depending on
experimental condition).[14] These still retained unreacted N-K$^+$
functionality and, therefore, had to be treated with a suitable
electrophile to obtain the "neutral" polymer. (Usually, CH$_3$I or
(CH$_3$)$_2$SiHCl was the electrophile used). Sufficient Si(H)-N(H)
functionality was present in the neutral polymer (a typical
formulation, after a CH$_3$I quench, was (CH$_3$SiHNH)$_{0.39}$(CH$_3$-
SiN)$_{0.57}$(CH$_3$SiHNCH$_3$)$_{0.04}$, so that on pyrolysis to 1000°C further
efficient thermal cross-linking occurred and the ceramic yields
obtained are high: almost 85%. The volatiles released during
the pyrolysis consist of H$_2$, CH$_4$ and a trace NH$_3$. The silicon
carbonitride ceramic residue, based on elemental analysis
contains about 65% by weight of Si$_3$N$_4$ equivalent, 27% by weight
of SiC equivalent and 5% by weight of excess carbon. This
material is amorphous and crystallizes only above 1400°C. At
this point, distinct β-Si$_3$N$_4$ and α-SiC phases can be observed
by x-ray diffraction. Our studies of this useful system and
its applications (production of ceramic fibers and coatings,
used as a binder for ceramic powders and in the "up-grading" of
Si-H containing polymers which by themselves are poor or only
moderately good preceramic materials[15]) are continuing.

THE SODIUM CONDENSATION OF METHYLDICHLOROSILANE

Linear polysilylenes (or polysilanes), [RR'Si]$_n$, whose backbone is a chain of silicon atoms, have received much attention
in recent years. As such[16] and as precursors to polycarbo-
silanes,[17] they have attained considerable importance in the
materials chemistry of silicon. Most of the polysilanes
prepared to date bear two organic substituents on the silicon
atoms. They generally are prepared by the action of an alkali
metal on the respective diorganodichlorosilane.[16]

The action of alkali metals on methyldichlorosilane has been
examined in these laboratories[18] and also by Brown-Wensley
and Sinclair at 3M.[19] In this reaction, attack at the Si-Cl
linkages is preferred, but reaction can also occur at the Si-H
bond. To what extent such attack at Si-H occurs is very dependent on the reaction conditions used. For instance, a reaction
of CH$_3$SiHCl$_2$ with an excess of sodium carried out in 7:1
by volume) hexane/THF gives liquid products of composition

(by NMR) $[(CH_3SiH)_x(CH_3Si)_y]_n$, with $x = 0.75-0.9$ and
$y = 0.25-0.1$ and $n = 14-16$. Thus, between 10 and 25% of the
Si-H bonds have reacted. This leads to some cross-linking
since hydrogen loss generates trifunctional silicon atoms.
These polysilanes are not good ceramic precursors. On pyrolysis to 1000°C the ceramic yield obtained from them ranged
between 12 and 27% and the composition of the ceramic product
(on the basis of elemental analysis was 1.0 mole SiC + 0.42 g
atom Si. A better product, at least in terms of ceramic yield
on pyrolysis, was obtained when the sodium condensation of
CH_3SiHCl_2 was carried out in THF alone. As might be
expected, the extent of reaction of the Si-H bond was considerably greater and the composition of the product was, on the
average, $[(CH_3SiH)_{0.4}(CH_3Si)_{0.6}]_n$. That is, 60% of the Si-H
bonds had reacted. As a result, the polymeric product was much
less soluble in hydrocarbon solvents, but it was soluble in
THF. The greater cross-linking also had as a useful consequence
that the ceramic yield on pyrolysis to 1000°C was increased
to 60%. However, the problem of the elemental composition of
the ceramic residue remained, the composition being (on the
basis of elemental analysis) 1.0 SiC + 0.49 Si. Such an excess
of silicon (mp 1414°C) would be expected to compromise the
high temperature applications of this ceramic material.

The $[(CH_3SiH)_x(CH_3Si)_y]_n$ compositions contain reactive Si-H
and Si-Si functionality which should provide the basis for
further chemical conversions which might serve to convert the
at first sight unpromising polymers to useful preceramic
materials. The hydrosilylation reaction has already been
mentioned and we have used this reaction to good advantage
in the present instance.[20] Reactions of $[(CH_3SiH)_x(CH_3Si)_y]_n$
(both of 7 hexane/1 THF and THF preparations) with various
compounds and polymers containing two or more vinyl groups were

examined. Best results were obtained with the ammonolysis
product of $CH_3(CH_2=CH)SiCl_2$. In this ammonolysis reaction
mostly cyclic species are formed, $[CH_3(CH_2=CH)SiNH]_n$, and in
our hydrosilylation experiments we used the cyclic trimer
$[CH_3(CH_2=CH)SiNH]_3$ which was separated by distillation.
Although the hydrosilylation reaction can be catalyzed by free
radical initiators as well as by low valent, coordinatively
unsaturated transition metal species,[4] we chose to use as
catalyst a member of the former class, azobisisobutyronitrile,
$(CH_3)_2(CN)CN=NC(CN)(CH_3)_2$ (AIBN). The thermal
decomposition of this compound is relatively rapid at 80°C
and provides the initiating radicals, $(CH_3)_2(CN)C\cdot$, and
molecular nitrogen. In one such experiment, a mixture of
$[(CH_3SiH)_{0.91}(CH_3Si)_{0.09}]_n$ and $[CH_3(CH_2=CH)SiNH]_3$
(Si-H:Si-Vi ratio ~6) in benzene was heated briefly in the
presence of a catalytic amount of AIBN. The soluble product
(MW 2100) had the composition (NMR and analysis)
$[(CH_3SiH)_{0.73}(CH_3Si)_{0.1}(CH_3SiCH_2CH_2Si(CH_3)(CH=CH_2-NH)_{0.17}]$.

Its pyrolysis (to $1000^\circ C$) under argon gave a black ceramic product in 77% yield. Elemental analysis of the ceramic allowed the calculation of the composition as 1.0 SiC + 0.03 SI_3N_4 + 0.04 C, which is equivalent to 1 weight % of free carbon.

The hydrosilylation reaction of $[(CH_3SiH)_{0.91}(CH_3Si)_{0.09}]_n$ with the trifunctional $[CH_3(CH_2=CH)SiNH]_3$ thus appears to have produced a network polymer. This is indicated by the good pyrolysis yield of ceramic residue. Furthermore, the chemical composition of the ceramic produced is now quite satisfactory. This improved preceramic polymer was found to be applicable to the preparation of ceramic fibers and ceramic bars and served well as a binder for β-SiC powder.

The hydrosilylation of unsaturated metal alkoxides, for instance, "$Hf(OPr^1)_2(OCH_2CH=CH_2)_2$" with $[(CH_3SiH)_{0.8}(CH_3Si)_{0.2}]_n$ using $H_2PtCl_6 \cdot 6H_2O$ as catalyst (SiH/SiVi molar ratio = 5) in toluene also was examined. A white, glassy solid was produced (MW ~1630) whose pyrolysis to $1000^\circ C$ left a black ceramic residue in 86% yield. It would appear that these systems also may be applicable to the protection of carbon/carbon composites toward oxidation.

It is also possible to convert the $[(CH_3SiH)_x-(CH_3Si)_y]_n$ polymers to useful hybrid polymers[15a] by reaction with major amounts of the $[(CH_3SiHNH)_a]_n$ species which is the product of the KH-induced dehydrocyclopolymerization of the CH_3SiHCl_2 ammonolysis product before quenching with an electrophile[14b] (see previous section). After a CH_3I quench, a new polymer can be isolated which on pyrolysis gives a ceramic yield in the 75-85% range. Furthermore, the excess silicon which usually results in the pyrolysis of the $[(CH_3SiH)_x(CH_3Si)_y]_n$ polysilazane/polysilane ratio is used.

THE ALKYNYLENATION OF METHYLDICHLORSILANE

We have studied the reactions of CH_3SiHCl_2 and other dichlorosilanes with magnesium acetylide. The latter reagent, easily prepared in THF medium by reaction of commercial di-butylmagnesium with gaseous acetylene, reacts readily with chlorosilanes. Its reaction with CH_3SiHCl_2 produces the expected $[CH_3Si(H)C\equiv C]_n$, but the molecular weight is low (~900). Nevertheless, pyrolysis of this material (under argon to $1000^\circ C$) results in an 82% yield of a black ceramic residue. Its composition (by elemental analysis) is 1.0 SiC + 0.5C, which   is   an   unacceptably high carbon content. Similar low polymer products were obtained using $(CH_3)_2SiCl_2$ and $(CH_3)(CH_2=CH)SiCl_2$ as starting materials: $[(CH_3)_2SiC\equiv C]_n$, a poorly soluble powder, and

$[(CH_3)(CH_2=CH)SiC\equiv C]_n$, a soluble wax, MW 1100. The latter on pyrolysis under argon to $1000^{\circ}C$ gave 1.0 SiC + 1.0C (83% ceramic yield). However, hydrosilylation of this product with $[(CH_3SiH)_{0.4}(CH_3Si)_{0.6}]_n$ resulted in a new preceramic polymer whose pyrolysis also gave a good ceramic yield If appropriate stoichiometry was used, the excess C and excess Si of the respective starting materials were balanced out and elemental analysis showed a nearly 1:1 Si/C ratio.

CONCLUSIONS

Methyldichlorosilane has proved to be a very useful starting material for the preparation of preceramic polymer systems. Its Si-Cl bonds provide the means for forming the initial oligomers or polymers which often in themselves are not useful in the preparation of ceramic materials. However, the reactivity of the Si-H bond then can be brought into play to convert these polymers to useful preceramic materials. Its commercial availability and relatively low cost make it a very attractive starting material for silicon-based materials.

ACKNOWLEDGEMENTS

This research was made possible through generous support from the Air Force Office of Scientific Research and the Office of Naval Research.

REFERENCES

1. (a) Rochow, E.G., "An Introduction to the Chemistry of the Silicones", 2nd edition, 1951, Wiley: New York, Chapter 2.

    (b) Zuckerman, J.J. Advan. Inorg. Chem. Radio Chem., 6 (1964) 383.

2. Noll, W. "Chemistry and Technology of Silicones", Academic Press: New York, 1968, pp. 589-591.

3. Fox, H.W.; Taylor, P.W.; Zisman, W.A. Ind. Eng. Chem. (Industr.), 39 (1947) 1401.

4. Lukevics, E.; Belyakova, Z.V.; Pomerantseva, M.G., Voronkov, M.G. J. Organomet. Chem. Library, 5 (1977) 1.

5. (a) Sokolov, N.N.; Andrianov, K.A.; Akomova, S.M., J. Gen. Chem. USSR, 26 (1956) 933.

    (b) Seyferth, D.; Prud'homme, C.; Wiseman, G.H., Inorg. Chem., 22 (1983) 2163.

6. Pai, Y.-M.; Servis, K.L.; Weber, W.P., Organometallics, 5 (1986) 683.

7. Harrod, J.F.; Pelletier, E., <u>Organometallics</u>, <u>3</u> (1984)
    1064.

8. Bradley, D.C.; Mehrotra, R.C.; Gaur, D.P.   "Metal Alk-
    oxides", Academic Press, New York, 1978.

9. Bates, R.B.; Kroposki, L.M.; Potter, D.E., <u>J. Org. Chem.</u>
    <u>37</u> (1972) 560.

10. Brewer, S.D.; Haber, C.P., <u>J. Am. Chem. Soc.</u>, <u>70</u> (1948)
    3888.

11. Semenova, E.A.; Zhinkin, D.Ya.; Andrianov, K.A., <u>Izv.</u>
    <u>Akad. Nauk SSSR, Otd. Khim. Nauk.</u> (1962) 2036.

12. Krüger, C.R.; Rochow, E.G., <u>J. Polym. Sci.</u> Part A, <u>2</u>
    (1964) 3179.

13. Monsanto Co., Neth. Appl. 6, 507,996 (23 Dec. 1965); <u>Chem.</u>
    <u>Abstract.</u>, <u>64</u> (1966) 19677d.

14. (a) Seyferth, D.; Wiseman, G.H., <u>J. Amer. Ceram. Soc.</u>, <u>67</u>
    (1984) C-132.

    (b) Seyferth, D.; Wiseman, G.H., U.S. Patent 4,482,699
    (1984).

    (c) Seyferth, D.; Wiseman, G.H., Chapter 38 in "Science
    of Ceramic Chemical Processing," L.L. Hench and D.R.
    Ulrich, editors, Wiley: New York, 1986.

15. (a) Seyferth, D.; Yu, Y-F., Proceedings of the Fourth
    Annual IUCCP Symposium, "Design of New Materials," Texas
    A&M University, March 25, 1986.

    (b) Seyferth, D.; Yu, Y-F., U.S. Patent 4,645,807 (1987).

    (c) Seyferth, D.; Yu, Y-F., U.S. Patent 4,650,837 (1987).

16. West, R., <u>J. Organomet. Chem.</u>, <u>300</u> (1886) 327.

17. Yajima. S., <u>Am. Ceram. Soc. Bull.</u>, <u>62</u> (1983) 893.

18. Wood, T. G., Ph.D. Dissertation, M.I.T., 1984.

19. Brown-Wensley, K.A.; Sinclair, R.A., U.S. Patent 4,537,942
    (1985).

20. Seyferth, D.; Yu, Y-F., U.S. Patent 4,639,501 (1987).

PART IX

PHYSICAL CHEMISTRY, THEORETICAL STUDIES AND SPECTROSCOPY

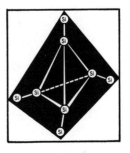

Chapter 39

Recent advances in
1) mechanism and kinetics of
alkylsilylene reactions & in
2) silicon hydride kinetics
and thydride molecules

M.A. Ring and H.E. O'Neal – Department of Chemistry, San Diego State University, San Diego, CA 92182-0328.

PART 1. CONCERNING THE KINETICS OF ALKYLSILYLENE INTRAMOLECULAR ISOMERIZATIONS AND DECOMPOSITIONS

Alkylsilylenes with R > CH_3 are known to decompose mainly to olefins and smaller silylenes.[1] However, a number of other reaction pathways, such as isomerizations to vinylsilanes and isomerizations to silacyclic compounds are also observed. The nature of the products of any given RSiH reaction depends in a striking fashion on the reaction conditions. For example, in the decomposition of n-PrSiH (single pulse shock tube (SPST)[2] reaction conditions, 1045 – 1240 K) the products were propylene and ethylene, roughly in a 3/2 ratio, while the only product of the analogous flash vacuum pyrolysis (FVP)[3] of n-PrSiD at 980 K was propylene. Similarly, in the SPST, 1150 – 1250 K[4] decomposition of n-BuSiH the products were ethylene, propylene, 1-butene, cis and trans butene, and silacyclopentane, (in respective yields of about 0.17, 0.12, 0.27, 0.13, and 0.30) while for the analogous decomposition of n-BuSiMe (FVP, at 980 K)[3], the only products were butenes.

The accepted mechanism for these pyrolyses involves silacyclic intermediates produced via intramolecular insertions into C–H bonds. This is illustrated below in Scheme I for the n-butylsilylene decomposition.

Scheme 1: Mechanisms of n-butylsilylene decomposition.

$$n-Bu\ddot{S}iH \xrightarrow{1} CH_3CH_2CH-\underset{\underset{SiH_2}{\diagdown}}{CH_2} \xrightarrow{2} CH_3CH_2CH=CH_2+:SiH_2$$

$$\downarrow 3$$

$$CH_3CH_2\underset{\underset{SiH_3}{|}}{CHCH_3}$$

$$\downarrow 4$$

$$CH_3CH-\underset{\underset{\ddot{S}iH_2}{\diagdown}}{CHCH_3} \xrightarrow{5} c,t-CH_3CH=CHCH_3+:SiH_2$$

$$CH_3\underset{\underset{SiH_2}{|}}{CH}-\underset{\underset{CH_2}{|}}{CH_2} \xrightarrow{7} CH_3CH=CH_2 + SiH_2=CH_2$$

$$n\text{-}Bu\ddot{S}iH \xrightarrow{6}$$

$$\xrightarrow{8} CH_2=CH_2 + CH_3CH=SiH_2$$

$$CH_3CH_2CH_2CH_2\ddot{S}iH \xrightarrow{\quad 9 \quad} \overline{CH_2CH_2CH_2CH_2}\dot{S}iH_2$$

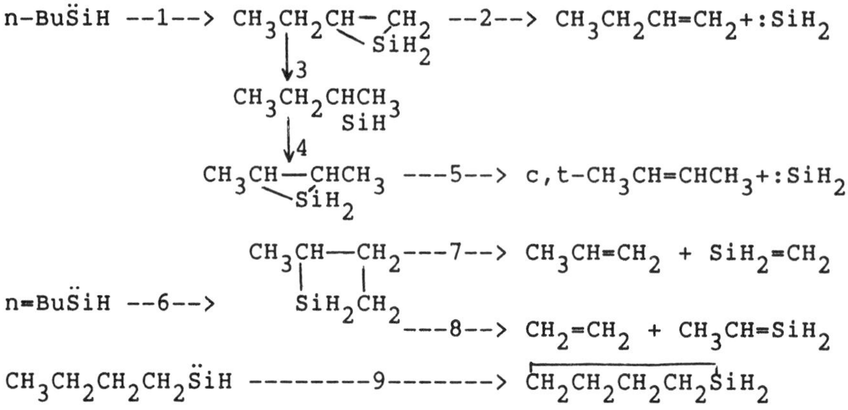

Perhaps the best evidence for this kind of mechanism
comes from observations on the reverse reactions
(i.e. silylene additions to olefins). Thus when SiH_2
was reacted with ethylene[5] the main products were:
ethylsilane (static reactor conditions, 'low' T and
'high' silane concentrations); ethylsilane and
vinylsilane in comparable yields (stirred flow
conditions, 'moderate' T and 'low' silane conc.);
vinylsilane (SPST conditions, 'high' T and 'moderate'
silane conc.). This reaction system, with its
increasing importance of vinylsilane with increasing
temperatures and/or decreasing silane concentrations,
is most reasonably explained via the silacyclopropane
intermediate mechanism shown in Scheme II since the
1,2 H-migration ring opening reactions of silacyclo-
propanes are clear analogues of the well known
cyclopropane to proplylene isomerization.

Scheme 2: Partial mechanism of the reaction of silylene with ethylene in the presence
of silane.

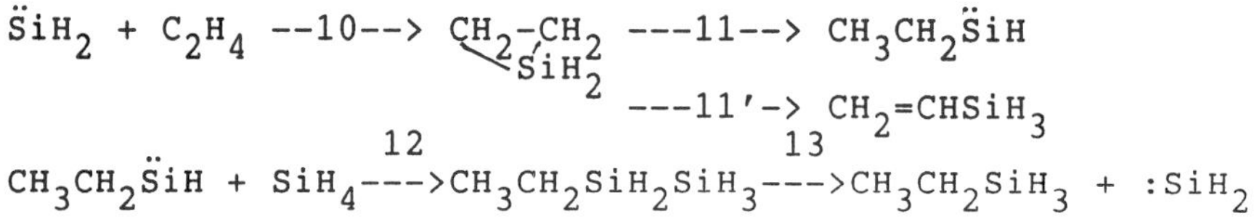

$$\ddot{S}iH_2 + C_2H_4 \xrightarrow{10} \underset{\underset{\ddot{S}iH_2}{\diagdown}}{CH_2-CH_2} \xrightarrow{11} CH_3CH_2\ddot{S}iH$$

$$\xrightarrow{11'} CH_2=CHSiH_3$$

$$\underset{12}{CH_3CH_2\ddot{S}iH} + SiH_4 \xrightarrow{\quad} CH_3CH_2SiH_2SiH_3 \xrightarrow{\quad} \underset{13}{CH_3CH_2SiH_3} + :SiH_2$$

On a qualitative level, alkysilylene decompositions
appear to be reasonably well understood. However,
there are disturbing problems on the kinetic level.
Why do the products of these reactions change so
markedly with such relatively modest condition
changes? Why are the products of silacyclopropane
type intermediates so much more important than those

of larger silacyclic rings --or, started another way, why should increasing temperature favor the production of the larger silacyclic rings? (One would expect, because of the possible development of ring strain in the transition states of the cyclization processes, that activation energies should decrease with increasing size of the cyclic ring product. This should favor large ring formation at low T and small ring formation at high T.)

Until recently no hard data on the kinetics of alkylsilylene reactions (i.e. decompositions or isomerizations) have been available, although crude estimates of their activation energies and rate constants have been made. Thus considerations of the endothermicities of the decompositions of propyl-silylene and ethylsilylene (rxn 14,15), coupled with observations of 'no trapping' under butadiene inhibition at SPST reaction conditions, have placed 'apparent' upper and lower limits of about 30 ± 4 Kcal on the activation energies of these decompositions.[6]

$$C_2H_5\ddot{S}iH \xrightarrow{14} \underset{\underset{SiH_2}{\diagdown}}{CH_2-CH_2} \xrightarrow{15} C_2H_4 + :SiH_2 \quad \underline{Units}$$

$$\Delta H_f: \quad 46.1 \qquad\qquad (?) \qquad\qquad 12.5 \quad 64.3 \quad Kcal/mol$$

However, a much lower decomposition activation energy follows from unpublished data of Davidson[7] on the decomposition of n-PrSiMe under butadiene inhibition at temperatures around 700 K. Thus the observed 60/1 ratio of the silacyclopentene trapping products of MeSiH (the product of the PrSiMe decomposition) and of reactant n-PrSiMe, respectively, coupled with $[C_4H_6] \approx 10^{-5}$ M, and the most recent value for the rate constant of silylene-butadiene trapping ($k_T \approx 3\times10^{10}$ $M^{-1}sec^{-1}$)[8] gives a rate constant of n-PrSiMe decomposition of $k_d \approx 1.8 \times 10^7$ sec^{-1}. With an estimated $A_d \approx 10^{13}$ sec^{-1}, this results in an activation energy of $E_d \approx 18.4$ Kcal, a value actually 9 Kcal lower than the estimated 27 Kcal rxn enthalpy (see later discussion of this curious result).

Estimates of activation energies for alkylsilylene intramolecluar C-H insertion reactions have been made by analogy with the activation energies of C-H insertions of RSiR' species (R and R' = H or Me) with methane. These, in turn, have been calculated from the experimental kinetic parameters[10] for methane elimination from alkylmonosilanes. Thus the inter-molecular C-H insertions with CH_4 range from about 17 Kcal (for $:SiH_2$) to about 25 (for $Me_2Si:$), and show an

increase of about 4 Kcal/mol for each Me for H
substitution. Activation energies for alkysilylene
intramolecular insertions (into C-H bonds), then, are
expected to have activation energies comparable to
their intermolecular reaction counterparts but to be
higher by any ring strain developed in the cyclic
transition states, and lower by some amount equal to
or less than the difference inbound strenghts of the
reacting C-H bonds (i.e. 104 Kcal in methane, 98 Kcal
in methyl, 95 Kcal in the methylene, etc.)[11]

We have recently completed a study of the kinetics of
reactions induced by the trapping of silylene by 1-
butene (static system, 643-689 K)[12] and results of
this study provide the first direct information on
the kinetics and mechanisms of alkylsilylene
decomposition and isomerization reactions. The
proposed mechanism is shown below in Scheme III.

Scheme 3: Silylene rxns with silane and 1-butene (static system reaction conditions).

SiH_4 + (M) ---16--> $:SiH_2$ + H_2 + (M)
SiH_2 + SiH_4 --17--> Si_2H_6
SiH_2 + $1-C_4H_8$ -18--> $CH_2=CHCH_2CH_3$--19--> $CH_3CH_2CH_2CH_2\ddot{S}iH$
$\quad\quad\quad\quad\quad\quad\quad\quad\quad\quad\underset{SiH_2}{}$
$\quad\quad\quad\quad\quad\quad\quad\quad\quad\quad\quad\quad$--20--> $CH_3\underset{:SiH}{C}HCH_2CH_3$

$CH_3\underset{:SiH}{C}HCH_2CH_3$--21-> $CH_3CH-\underset{SiH_2}{C}HCH_3$--22-> $CH_3CH=CHCH_3$ + $:SiH_2$

$sC_4H_9\ddot{S}iH$ + SiH_4--23--> $sBuSiH_2SiH_3$--24-->$sBuSiH_3$ + $:SiH_2$
$nC_4H_9\ddot{S}iH$ + SiH_4--25--> $nBuSiH_2SiH_3$--26-->$nBuSiH_3$ + $:SiH_2$

Products of the reaction are disilane, n-butylsilane,
sec-butylsilane, and cis and trans 2-butene.
Relative product yields of the butylsilanes and
2-butenes proved to be independent of percent
conversion, total pressure, and temperature
($n-BuSiH_3$/$s-BuSiH_3$ $\approx$ 0.24; c,2-butene/t,2-butane~
1.65). Since 1-butene and silane compete in the
trapping of silylene, it was possible to obtain a
value of $k_{18}=10^{11.1}$ $M^{-1}sec^{-1}$ for the rate constant
of the former process (from the product yields of
disilane and butylsilanes and $k_{17}=6.6x10^{10}$ $M^{-1}s^{-1}$).[13]
More significantly, by studying the yield variations
of $s-BuSiH_3$/2-butene with changes in [SiH_4] at
various temperatures, it was possible to obtain rate
constant parameters for the s-butylsilylene isomerization to dimethylsilacyclopropane, and for its
decomposition to 2-butene and SiH_2. The pertinent rate
equation is shown below, where $f_{23}=k_{24}/(k_{24}+k_{-23})$.

$$Y(sBuSiH_3)/Y(2-C_4H_8)=k_{23}f_{23}(1+k_{-21}/k_{22})[SiH_4]k_{21} \quad I$$

An Arrhenius treatment of the rate constants derived from the data and equ I gave log A_{exp} (M^{-1})=-1.64 +0.5 and E_{exp}=-10.7+1.6 Kcal (with errors of +2σ). Since the competing reactions are trapping of sec-butyl-silylene by silane (which may have an activation energy of about 1.7 Kcal)[14] and its decomposition to olefin, the results indicate a decomposition activation energy significantly lower than the rxn enthalpy of about 23 Kcal[9]. This result, which parallels the Davidson finding, seems impossible (and would be impossible if the butylsilylene decomposition occurred in a single step, concerted process). However, by Scheme III, the decomposition is a consecutive step process, and this mechanism can accommodate the present findings. This is illustrated in the reaction coordinate diagrams of Fig 1. The activation energy diagram shows an apparent minimum barrier to the decomposition of 23 Kcal (i.e. the reaction energy, ΔE^o), however, the free energy of activation diagram shows that the real barrier to reaction occurs during the silacyclopropane formation reaction (where the E_a barrier is only about 10.5[9] Kcal). Analysis of the kinetic and thermodynamic data for the decomposition reaction has resulted in

the following: $k_{21}=10^{11.8}xe^{-10500}$ $_{cal/RT}$ sec^{-1} (C-H

insertion); $k_{-21}=10^{13.5}xe^{-11600/RT}$ sec^{-1} (silacyclo-

propane ring opening to s-butylsilylene), $k_{22} =$

$10^{17.9}xe^{-23100cal/RT}$ (ring opening and decomposition

to olefin), and k_{-22} k_{18} $10^{11.1}$ $M^{-1}sec^{-1}$ (2-butene

trapping of silylene). Errors in these Arrhenius parameters arise from uncertainties in the thermodynamics as well as from the kinetics and could easily be as large as 3-4 Kcal in E and a corresponding uncertainty in A (i.e. a factor of 10 - 30). However, even with these errors, the trends are clear and lead to a number of interesting conclusions. First, the high A-factor for rxn 22 means that silacyclopropane decompositions (and their reverse silylene addition to olefin reactions) proceed through 'loose' transition states; a consecutive step mechanism, possibly involving biradical formation, is inferred. Second, the very low activation energy of ≈ 10.5 Kcal for the intramolecular C-H insertion rxn 21 (when compared to its intermolecular RSiH + CH_4 reaction counterpart of 21 Kcal) implies little or no

development of ring strain energy in the 3 membered
cyclic transition state. Further, the entire 9 Kcal
difference in reacting C-H bond strengths (104 Kcal
in CH_4 and 95 Kcal in $-CH_2-$) is reflected in the
transition state energetics. While surprising, this
result is quite consistent with the observations that
3 membered ring C-H insertions are the favored mode
of reaction for alkylsilylenes at all temperatures.

The data of Davidson on the nPrSiMe decomposition now
also makes sense and is, in fact, consistent with our
results. Thus the nPrSiMe decomposition rate
constant can be estimated from the s-BuSiH rate
constant by raising the activation energy of the
latter by 4 Kcal (i.e. Davidson's rxn involves an
RSiR' silylene as opposed to our RSiH silylene).

Thus, $k_{PrSiMe} \approx k_{21}/(1 + k_{-21}/k_{22})xe^{-4000/RT}$. This

has a value at 700 K of 1.6×10^7 sec^{-1}, and is in good

agreement with the value deduced previously.

It is now possible to give a plausible answer to why
3 membered ring formation is favored over larger ring
production in the intramolecular cyclization
processes of long chain alkylsilylenes. The answer
simply stated is that three membered ring formations
are concerted processes, with low activation energies
and entropies, while larger ring formations (and the
formation of products which 'appear' to follow from
their formations) are consecutive step processes of
relatively high activation energy and entropy. Thus
larger ring products are only observed at high
temperatures. The above follows from the fact that
1) there is no ring strain developed in the
transition states of 3 membered ring C-H insertion
reactions, and 2) A factors of concerted
intramolecular insertion reactions must decrease by
about a factor of 10 with each one unit increase in
product ring size (due to internal rotational
restriction). Thus 3 membered ring formation (when
possible) should always be the favored concerted mode
of reaction for alkylsilylenes. A consecutive step
mechanism which fits the energetic and entropic
requirements for larger ring formations is shown in
Scheme IV. The activation energy for this process
is about 25 Kcal (i.e. no activation energy for the
biradical back reaction to silylene or for its ring
closing). This is 14 Kcal higher than 3 membered ring
formation and should therefore be favored at high T.

Scheme 4: Biradical mechanism of alkylsilylene intramolecular cyclization.

$$CH_3CH_2CH_2CH_2\ddot{S}iH--\dot{C}H_2CH_2CH_2CH_2\dot{S}iH_2--\overline{\dot{C}H_2CH_2CH_2CH_2\dot{S}iH_2}$$

ΔH_f^o: 36.2   61.5 $E_a \sim$ 25 (in kcal)

Part 2. Silylene Kinetics and Thermochemistry:

Recent laser absorption[15] and laser excited fluoroescence[16] time resolved measurements of silylene loss during insertion and addition reactions with various substrates (most importantly H_2, SiH_4, and Si_2H_6), along with theoretical calculations of reaction profiles[17] and heats of formation of silicon hydride species (most importantly (SiH_2, SiH_3SiH, and SiH_2SiH_2)[18-23] have shaken the foundations of long standing beliefs about silylene reactivities and silylenethermochemistries. Thus rate constants for the reactions of silylene have been found to be much faster (by one or more orders of magnitude) than previously believed, and heats of formation of silylenes have been calculated to be 5 -10 Kcal/mol higher than previously believed. For comparison, 'new' and 'old' kinetic and thermochemical results on silylene are shown in Table I.

Most of the 'old' kinetic and thermochemical values for silylene are based on one pivotal reaction: the decomposition of disilane[23]. For example, the reaction kinetics of silylene insertion into SiH_4 (i.e. the reverse of the decomposition reaction) follow from the Arrhenius parameters and overall thermochemistry of this process. The assumption made is that the activation energy for insertion is zero $\pm$ 2 Kcal. The same assumption leads to the 'old' silylene heat of formation of 59 +2 Kcal/mol. All other kinetic results come from silylene competitive rate trapping studies referenced to the SiH_2 + SiH_4 reaction as standard. Prior to the 'new' results, there was no reason to question existing values as the disilane decomposition appeared to be an extremely well studied and well understood reaction. Thus its Arrhenius parameters acceptably correlated static and shock tube rate constant data over a 400 K temperature range[24], and few reactions have been that thoroughly studied. Nevertheless, as evidence mounted in support of the 'new' results, there was a need to reexamine the earlier data and assumptions used in their analysis. The question to be answered was "how could the old values, derived by tried and true classical kinetic techniques, be so much in error, if indeed they were in error?"

Addressing this question, Frey, Walsh and Watts
(FW&W)[25] pointed out that the experimental rate
constant data and Arrhenius parameters for the
disilane decomposition under single pulse shock tube
conditions suggested a considerably larger degree of
pressure fall-off than could be accounted for by the
accepted Arrhenius parameters (i.e. $A = 10^{14.5}$ sec^{-1}
and $E = 49.3$ Kcal, from Bowrey and Purnell (BP)). On
the assumption that the Arrhenius parameters of the
decomposition might be significantly higher, FW&W
found that a better fit to the shock tube data could
be realized with $A_\infty = 10^{16.0}$ sec^{-1} and $E_\infty = 53.2$ Kcal.
These parameters produced $\Delta H_f(SiH_2) \approx 65$ Kcal/mol,
which is within the errors of the 'newer' values.

The FW&W suggestion prompted our reexamination of the
disilane decomposition[21] under conditions similar to
those of B&P. The pressure dependence of the rate
constant increase of only 38% (B&P reported a 22%
rise for a pressure increase from 23 to 101 torr
at 572 K), and the Arrhenius parameters of the
decomposition, evaluated at 3 pressures from 10 to
500 torr, reflect only minor pressure dependencies.
However, the significant finding was that the
Arrhenius parameters were much higher than those of
B&P. Good fits of the experimental pressure effects
on rate constants and Arrhenius parameters were
achieved via RRKM calculations with $A_\infty = 10^{15.75}$ sec^{-1}
and $E_\infty = 52.2$ Kcal. The results are shown in Table II.
With the assumption of zero activation energy for
the reverse insertion reaction (M std state), these
results give $\Delta H_f(SiH_2) = 64.5$ Kcal/mol, quite close
to the FW&W recommended value.

It is interesting to note that very similar heat of
formation values for silylene are obtained from
existing activation energies for silylene formation[26]
from the pyrolyses of methyldisilane and trisilane
$\Delta H_f(SiH_2) = 64.1$ and 63.9 Kcal/mol, respectively. In
retrospect it is surprising that the kinetics of
these reactions were not given more consideration in
the earlier adoption of a silylene heat of formation
value since, because of reactant size, neither of
these reactions under the conditions studied can be
troubled by pressure fall-off effects.

The most glaring kinetic difference between the 'old'
and 'new' values is that of the activation energy of
SiH$_2$ insertion into H$_2$[27,17]: (8 vs 0 Kcal). This
problem is resolved by a combination of our revised
Arrhenius parameters for the disilane decomposition
and our revised RRKM calculations on the silane
decomposition (see below). Coupled with the
experimental relative rate constant measurement of
$k_{(SiH2+H2)}/k_{(SiH2+SiHr)} \approx 0.02$ (200 torr, 670 K)[27], and

with the reaction entropies of these processes, these revisions given an activation difference for silylene insertions into H_2 and SiH_4 of only 0.6 Kcal (in agreement with the theoretical calculations).

One of the factors responsible for the confusion summarized in Table I concerns the validity of RRKM calculations for reactions in their pressure fall-off regimes. Silane is well into its pressure fall-off under all conditions thus far employed, and fall-off in the disilane reaction can be significant at pressures below 10 torr (static reaction conditions), and at higher temperatures at much higher pressures (as in SPST reactions, T > 850 K, P ≈ 2500 torr). The major problem with RRKM calculations is the proper assignment of collision efficiencies for the bath gas molecules, and is particularly severe in calculations at high temperatures because the exact behavior of collision efficiencies with increasing temperature is not really known at this time (although equations for this effect which should be reasonably valid have been developed), and because bath gases of low collision efficiency are often involved (e.g. in SPST and stirred flow studies the bath gas is usually argon). In addition, while β_c's for any given bath gas would be expected to be similar from one reaction system to another, their transferrability has not been sufficiently demonstrated; important differences may exist. Consequently β_c's should be experimentally evaluated at least at one temperature for each system. Earlier RRKM calculations on silane and disilane were made prior to, or roughly at the same time, as these problems were being debated and resolved on a theoretical level. Thus the first calculations on silane were made assuming that β_c(argon) was temperature independent and equal to the value determined in another reaction system (the CH_3NC isomerization[28]). The results were $E_\infty = 59.5$ Kcal[28], and this value was subsequently adopted for the SiH_4 reaction. The latest calculations, using the Troe temperature dependence relation for β_c[29] (but still assuming the transferrability of β_c for Ar), give $E_\infty = 57.9$ Kcal[27]. This should be a much better number, but it is still based on an assumption which needs experimental verification. Acceptance of RRKM fall-off results at this time, therefore, must be made with caution.

It is now clear that an unfortuante combination of factors were responsibly for the 'old', and now clearly erroneous, kinetic and thermodynamic values on silylene and its reactions. The new time resolved laser results and theoretical calculations have forced us to confront these factors and correct them.

Fortunately, correction has been successful and reasonably painless, proving that classical kinetic data is still not only a "tried" but also a "true" pathway to valid kinetic and thermodynamic results.

Table 1. Comparisons of old and new kinetic and thermodynamic data on silylene and its reactions.

Rxn:	SiH_2+H_2	SiH_2+SiH_4	$SiH_2+Si_2H_6$	$\Delta H_f(SiH_2)$*	
New:	logk=7.78	logk=10.82	logk=11.54	68.1	63.4
				65.6	64.5
				69+3	
Old:	logA=10.8	logA=9.5	logA=9.6	59.2+2	
	E=8.3+2.0	E=1.2+2.0	E=0+2.0		

Units of k and A are $M^{-1}sec^{-1}$; units of E are Kcal; and the new k values are for the 298 K condition.
* References for $\Delta H_f(SiH_2)$ are respectively, 18-23.

Table 2. Arrenius parameters of the disilane decomposition as a function of total pressure: experimental and calculated.

Press.**	LogAexp*	Eexp(cal)*	logAcalc	Ecalc(cal)
500	15.62+0.19	51,810+470	15.57	51,703
150	15.38+0.05	51,300+300	15.41	51,349
10	15.13+0.27	50,922+330	14.80	50,103

** Pressure in torr.
* The errors show correspond to the +2 or 95% confidence limit; 150 and 500 torr runs were made on propane-disilane mixtures while 10 torr runs were made one pure disilane.

REFERENCES

1. M.A. Ring, H.E. O'Neal, S.F. Rickborn and B.A. Sawrey, Organometallics, 2, 1891 (1983).
2. B.A. Sawrey, H.E. O'Neal, M.A. Ring, D. Coffey, Int. J. Chem. Kinet., 16, 801 (1984).
3. N. Tillman and T.J. Barton, to be published.
4. B.A. Sawrey, H.E. O'Neal and M.A. Ring, Organometallics, 6, 720 (1987).
5. D.S. Rogers, K.L. Walker, M.A. Ring and H.E. O'Neal, Organometallics, in press.
6. S.F. Rickborn, M.A. Ring and H.E. O'Neal, Int. J. Chem. Kinet., 16, 31 (1984).
7. I.M.T. Davidson, private communication.

8. J.O. Chu, D.B. Beach and J.M. Jasinski, to be published.
9. G. Licciardi, H.E. O'Neal and M.A. Ring, to be published.
10. I.M.T. Davidson and M.A. Ring, J. Chem. Soc., Faraday Trans. 1, 76, 1520 (1980); B.A. Sawrey, H.E. O'Neal, M.A. Ring, D. Coffey, Int. J. Chem. Kinet., 16, 31 (1984); P.S. Neudorfl and O.P. Strausz, J. Phys. Chem., 82, 241 (1978); S.F. Rickborn, D.S. Rogers, M.A. Ring and H.E. O'Neal, J. Phys. Chem., 90, 408 (1986).
11. S.W. Benson, Thermochemical Kinetics, 2nd edition, Wiley, New York (1976).
12. A. Dickinson, K.E. Nares, M.A. Ring and H.E. O'Neal, submitted to Organometallics.
13. G. Inoue and M. Suzuki, Chem. Phys. Lett., 122, 361 (1985).
14. R. Walsh, private communication.
15. J.J. Jasinski, J. Phys. Chem., 90, 555 (1986).
16. G. Inoue and M. Suzuki, Chem. Phys. Lett., 122, 361 (1985).
17. M.S. Gordon, T.N. Truong and E.K. Bonderson, J. Am. Chem. Soc., 108, 1421 (1986); M.S. Gordon and D.R. Gano, J. Am. Chem. Soc., 106, 5421 (1984).
18. P. Ho, M.E. Coltrin, J.S. Binkley and C.F. Melius, J. Phys. Chem., 89, 4647 (1985).
19. J.A. Pople, B.T. Luke, M.J. Frisch and J.S. Binkley, J. Phys. Chem., 89, 2198 (1985).
20. J. Berkowitz, J.P. Greene, H. Cho and B. Ruscic, J. Chem. Phys., 86, 1235 (1987).
21. J.G. Martin, M.A. Ring and H.E. O'Neal, Int. J. Chem. Kinet., in press.
22. S.K. Shin and J.L. Beauchamp, J. Phys. Chem., 90, 1507 (1986).
23. M. Bowrey and J.H. Purnell, Proc. Roy. Soc. Ser. A, 321 341 (1971).
24. J. Dzarnoski, S.F. Rickborn, H.E. O'Neal and M.A. Ring, Organometallics, 1, 1217 (1982).
25. H.M. Frey, R. Walsh and I.M. Watts, J. Chem. Soc., Chem. Comm. 1189 (1986).
26. A.J. Vanderwielen, M.A. Ring and H.E. O'Neal, J. Am. Chem. Soc., 97, 993 (1975).
27. J.W. Erwin, M.A. Ring and H.E. O'Neal, Int. J. Chem. Kinet., 17, 1067 (1985); R.T. White, R.L. Espinos-Rios, D.S. Rogers, M.A. Ring and H.E. O'Neal, Int. J. Chem. Kinet., 17 1029 (1985).
28. C.G. Newman, H.E. O'Neal, M.A. Ring, F. Leska, and N. Shipley, Int. J. Chem. Kinet., 11, 1167 (1970).
29. J. Troe, J. Chem. Phys., 66, 4745, 4758 (1977).

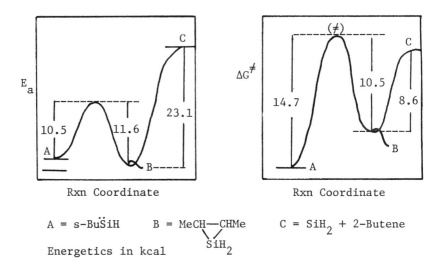

A = s-Bu$\ddot{S}$iH   B = MeCH—CHMe   C = SiH$_2$ + 2-Butene

Energetics in kcal $\overset{\diagdown \diagup}{SiH_2}$

Fig. 1. Reaction coordinate diagrams of S-Bu$\ddot{S}$iH decomposition.

Chapter 40

Absolute rate constants
for silylene reactions in the gas phase

J.M. Jasinski and J.O. Chu – IBM Thomas J. Watson Research Center, Yorktown Heights, NY 10598, USA.

ABSTRACT

Absolute rate constants for the reaction of silylene with a wide range of reaction partners have been measured using laser resonance absorption flash kinetic spectroscopy. Rate constants for these reactions are reported at 5 Torr total pressure and 298K. The results show that silylene reacts essentially gas kinetically with silanes, olefins, acetylene and 1,3-butadiene, is unreactive with saturated hydrocarbons and exhibits a range of reactivity with diatomics.

INTRODUCTION

Silylenes, divalent silicon species, are well established as important transient species in organosilicon and silicon hydride chemistry. However, almost all of our information about the intermediacy and reactivity of these species comes from classical mechanistic and kinetic studies. These studies provide information on silylenes only indirectly, through observation of stable trapping products. As is always the case when transient species are postulated in chemical mechanisms, direct spectroscopic observation of these elusive entities is highly desirable. This is especially true for silylenes, since organosilicon and silicon hydride chemistry is frequently complex, often involving multiple reaction pathways including formation of solid polymeric materials or inorganic thin films.

This paper describes recent efforts to selectively generate and directly detect the simplest silylene, SiH_2, in the gas phase. Silylene is generated by excimer laser photodissociation of organic and inorganic silanes and detected by time-resolved tunable dye laser absorption spectroscopy. The goal of the work is to provide reliable fundamental photochemical, thermochemical and kinetic information which can be compared to the results of indirect studies, compared to the predictions of ab initio calculations and used to assess quantitatively the role of SiH_2 in complex chemical processes such as the chemical vapor deposition of silicon containing thin films for microelectronics applications.

BACKGROUND

Silylene has a singlet electronic configuration in the ground state and, consequently, its chemistry is characterized by insertion reactions into single bonds and addition reactions with olefins, acetylenes and conjugated dienes. The addition reactions are frequently followed by rearrangement of the initial adduct. The potential energy surfaces for these reactions are characterized, qualitatively, by electrophilic attack of the empty 3p orbital of silylene, followed by nucleophilic attack of the lone electron pair. There have been numerous classical mechanistic studies of SiH_2 reactions, typically employing pyrolysis, shock tube, infrared laser or nuclear recoil methods for silylene generation and gas chromatography or mass spectrometry for the analysis of stable products. Kinetic data has most often been obtained by examining competitive trapping of silylene by two reactants. Much of this work has been systematically reviewed [1,2]

While most of the kinetic data available for silylene reactions is in the form of relative rates or differential Arrhenius parameters, there have been several attempts to derive absolute rate parameters using the methods of thermochemical kinetics. John and Purnell [3] estimated Arrhenius parameters for the reaction of SiH_2 with H_2, SiH_4 and Si_2H_6 and also extracted a heat of formation for silylene of 58 kcal/mole from their data. More recently, Rogers et al. [4] have estimated Arrhenius parameters for a number of silylene reactions, including addition to acetylene and 1,3-butadiene which are frequently used as silylene traps in competitive rate experiments.

There have also been a number of ab initio theoretical studies of silylene reactivity and thermochemistry. The most recent of these calculations are in poor agreement with the experimental results discussed above. Gordon et al. [5] have calculated a much smaller barrier for insertion of silylene into molecular hydrogen ($\sim$1 kcal/mole) than that estimated by John and Purnell (5.5 kcal/mole) or that estimated by White et al. [6] ($\sim$10 kcal/mole) and Ho et al. [7] obtain a heat of formation for silylene of 68 kcal/mole.

Most recently, two direct absolute rate studies, a laser induced fluorescence experiment by Inoue and Suzuki [8] and our laser resonance absorption flash kinetic study [9], and two relative rate studies [10,11], all based on photochemical generation of silylene from phenylsilane have appeared. The results of these studies do not agree well with previous experimental work, but do support the most recent ab initio results. In particular, our measurement [9] of the rate constant for reaction of silylene with D_2, which can be used to approximate the high pressure limiting rate constant for reaction with H_2 (vide infra) provided the first experimental support for both a small barrier to reaction and a higher silylene heat of formation. Additional support for the smaller reaction barrier comes from the laser induced fluorescence studies and from scavenging studies by Frey et al. [10]. Additional support for the higher heat of formation is now available from ion cyclotron resonance studies [12], guided ion beam studies [13] and photoionization studies [14].

LRAFKS STUDIES

In order to make direct measurements of silylene reaction rate constants, we use laser resonance absorption flash kinetic spectroscopy (LRAFKS) to generate and directly detect silylene in real time [9]. A schematic of the LRAFKS apparatus is shown in Figure 1. Briefly, silylene is generated in a slowly flowing gas cell in the presence of 1-100 torr of buffer gas (typically He) by excimer laser photodissociation of phenylsilane or disilane at 193 nm or photodissociation of iodosilane at 248 nm. The different precursors are necessary because not all of the reaction partners we wish to study are suitably transparent at 193 nm, because some reaction partners are chemically incompatible with one or more precursor, and because measurements of the same reaction rates using different sources of silylene provide an experimental check for possible artifacts arising from internally or translationally non-thermal populations of silylene. The transient SiH_2 population is created during the 10-20 nsec excimer laser pulse and detected with a narrow bandwidth tunable dye laser tuned to individual rotational lines of the well known A ← X electronic transition of silylene. Scanning the dye laser and integrating the transient absorption signal provides a Doppler limited transient absorption spectrum, which allows unambiguous identification of the signals as arising from silylene (Figure 2). For kinetic studies, reactant gases are added to the cell at a constant total pressure, the dye laser is tuned to a strong SiH_2 line, and the transient decay of the silylene absorption signal is measured (on a microsecond time scale) as a function of reactant gas partial pressure under pseudo-first order conditions. A plot of SiH_2 decay rate vs reactant gas partial pressure gives a straight line with a slope corresponding to the bimolecular rate constant for silylene removal by the added reactant.

Table I presents a summary of the reactions which we have thus far studies using LRAFKS. Each rate constant has been determined at room temperature and a total pressure of 5 torr. The total system pressure is an important parameter in SiH_2 studies because the reactions are three body associations. Silylene inserts into a single bond or adds to a double bond to form a chemically activated adduct which must be collisionally stabilized or decompose by some channel other than loss of SiH_2 in order to contribute to silylene removal. As a result, especially with small molecules and in cases where there are no other low energy decomposition routes available to the adduct, silylene removal rate constants will be pressure dependent. This is particularly true for the case of reaction with H_2. Figure 3 shows the dependence of the rate constant on total pressure over the range 1 to 100 torr along with the results of an RRKM calculation. The RRKM analysis of the pressure dependence gives a limiting high pressure rate constant of 3×10^{-12} cm^3 molecule^{-1} s^{-1} This type of effect is also responsible for the rate constants for reaction with D_2 and HD being much greater than that for reaction with H_2 at 5 torr. In the isotopic cases, the chemically activated silylene can decompose to SiD_2 or SiHD as well as SiH_2. Since the isotopic silylenes have different absorption spectra than SiH_2, their production contributes to the loss of SiH_2 signal in the LRAFKS experiment [9].

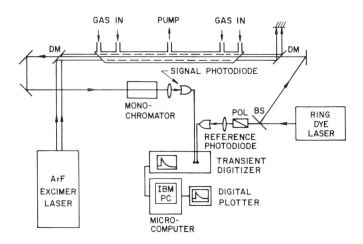

Fig. 1. Schematic of the LRAFKS apparatus.

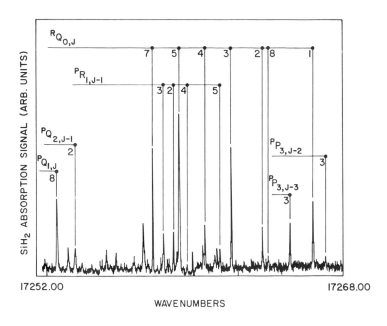

Fig. 2. Silylene transient absorption spectrum near the center of the
$^1B_1(0,2,0)\leftarrow{}^1A_1(0,0,0)$ band.

Table 1. Absolute rate constants for silylene removal at 5 torr and 298K.

Reactant	Rate Constant (cm^3 molecule^{-1} s^{-1})	Silylene Precursor	Photolysis Wavelength (nm)
H_2	$4.4 \pm 0.8 \times 10^{-13}$ $4.9 \pm 0.9 \times 10^{-13}$	ϕSiH_3 Si_2H_6	193 193
HD	$2.0 \pm 0.4 \times 10^{-12}$	ϕSiH_3	193
D_2	$2.1 \pm 0.4 \times 10^{-12}$	ϕSiH_3	193
HCl	$6.8 \pm 1.0 \times 10^{-12}$	ϕSiH_3	193
Cl_2	$1.4 \pm 0.2 \times 10^{-10}$	ϕSiH_3	193
NO	$1.7 \pm 0.2 \times 10^{-11}$	SiH_3I	248
O_2	$7.6 \pm 1.0 \times 10^{-12}$ $7.8 \pm 1.0 \times 10^{-12}$	ϕSiH_3 SiH_3I	193 248
CO	$8.0 \pm 2.0 \times 10^{-14}$	ϕSiH_3	193
SiH_4	$1.4 \pm 0.2 \times 10^{-10}$	ϕSiH_3	193
Si_2H_6	$2.8 \pm 0.3 \times 10^{-10}$ $2.1 \pm 0.2 \times 10^{-10}$	ϕSiH_3 SiH_3I	193 248
CH_4	$2.5 \pm 0.5 \times 10^{-14}$	ϕSiH_3	193
C_2H_6	$1.2 \pm 0.5 \times 10^{-14}$	ϕSiH_3	193
C_2H_4	$5.2 \pm 0.5 \times 10^{-11}$ $5.5 \pm 0.5 \times 10^{-11}$	SiH_3I ϕSiH_3	248 193
C_3H_6	$1.2 \pm 0.1 \times 10^{-10}$	ϕSiH_3	193
C_4H_6	$1.9 \pm 0.2 \times 10^{-10}$	SiH_3I	248
C_2H_2	$9.8 \pm 1.2 \times 10^{-11}$	SiH_3I	248

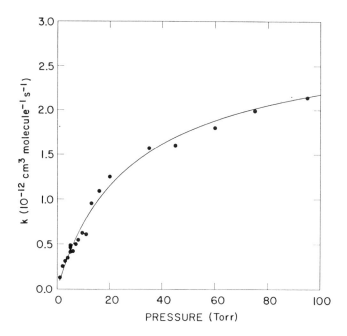

Fig. 3. Pressure dependence of the rate constant for reaction of silylene with hydrogen. Line is the result of an RRKM calculation.

Space limitations precluded a detailed case by case discussion of the rate constants in Table I. Such an analysis is presented elsewhere [9,17,18]. Qualitatively, several trends are apparent. Silylene reacts at essentially gas kinetic rates with silanes, olefins, acetylene and 1,3-butadiene, is unreactive with saturated hydrocarbons, and exhibits a range of reactivity with diatomics. The rate constants for reaction with methane and ethane are upper limits only, limited by the purity of the hydrocarbon gases [16]. The actual rate constants could be orders of magnitude slower. The reaction of SiH_2 with NO and O_2 are interesting. Although the product channels have not yet been determined, the final products are likely to be SiO and NH_2 or H_2O respectively. Formation of these sets of products is quite exothermic and the reactions may in fact produce chemiluminescence from SiO or NH_2. The reaction with NO is also interesting because NO has frequently been used as an SiH_3 scavenger, occasionally with the additional assumption that it does not scavenge SiH_2 [15,16].

Quantitatively, the data in Table I are extremely interesting and somewhat disturbing when compared with estimates of absolute silylene reactivity discussed in the previous section. The rate constants for reaction with H_2, SiH_4 and Si_2H_6 are 10^4, 10^2 and 10 times faster, respectively, than values calculated at room temperature from John and Purnell's Arrhenius parameters [3]. The rate constants for reaction with acetylene and butadiene are $\sim 10^5$ times faster than the values calculated from the Arrhenius parameters suggested by Rogers et al. [4] Agreement between our results, the laser induced fluorescence experiments [8] and the scavenging experiments of Walsh's group [10,11] are considerably better, often within experimental error or at worst differing by a factor of 2-3. The one exception is in the case of methane. Our results are in agreement with those of Eley et al. [11]. Methane is unreactive with silylene. This is in significant disagreement with the laser induced fluorescence measurements [8] which give a rate constant of ~1 x 10^{-12}. A possible cause for this apparently rapid rate constant has been suggested [10].

CONCLUSION

We have used LRAFKS to make direct measurements of silylene removal rates by a number of reaction partners. In general the results disagree badly with previous indirect kinetic studies, but agree well with recent direct and indirect kinetic studies in which silylene is generated photochemically. The results strongly support recent high level ab initio calculations of silylene potential energy surfaces and thermochemistry. The major discrepancy to be resolved is why our direct time-resolved silylene removal studies produce rate constants that are orders of magnitude faster than values deduced from various product trapping studies. The next logical step in resolving these issues is to perform experiments which directly detect removal of silylene and formation of product species in real time. Such experiments may well be possible, at least for a few simple cases, if a more general detection scheme such as infrared absorption spectroscopy using tunable diode lasers can be applied to silylene reaction kinetics.

ACKNOWLEDGEMENT

The authors would like to thank Dr. David Beach for preparing samples of iodosilane and for helpful discussions and Mr. Richard Estes for technical support.

REFERENCES

1. Gaspar, P. P. *Reactive Intermed.* 1985, **3**, 333, and references therein.

2. Davidson, I. M. *Ann. Rep. Prog. Chem.* 1985, **Sect. C82**, 47, and references therein.

3. John, P.; Purnell, J. H. *J. Chem. Soc., Faraday Trans. 1* 1973, **69**, 1455.

4. Rogers, D. S.; O'Neal, H. E.; Ring, M. A. *Organometallics* 1986, **5**, 1467.

5. Gordon, M. S.; Gano, D. R.; Binkley, J. S.; Frisch, M. J. *J. Am. Chem. Soc.* 1986, **108**, 2191.

6. White, R. T.; Espino-Rios, R. L.; Rogers, D. S.; Ring, M. A.; O'Neal, H. E. *Intern. J. Chem. Kinet.* 1985, **17**, 1029.

7. Ho, P.; Coltrin, M. E.; Binkley, J. S.; Melius, C. F.; *J. Phys. Chem.* 1985, **89**, 4647.

8.   Inoue, G.; Suzuki, M. *Chem. Phys. Lett.* 1985, **122**, 361.

9.   Jasinski, J. M. *J. Phys. Chem.* 1986, **90**, 555.

10.   Frey, H. M.; Walsh, R.; Watts, I. M. *J. Chem. Soc. Chem. Commun.* **1986**,1189.

11.   Eley, C. D.; Rowe, M. A.; Walsh, R. *Chem. Phys. Lett.* 1986, **126**, 153.

12.   Shin, S. K.; Beauchamp, J. L. *J. Phys. Chem.* 1986, **90**, 1507.

13.   Boo, B. H.; Armentrout, P. B. *J. Am. Chem. Soc.* 1987, in press.

14.   Berkowitz, J.; Greene, J. P.; Cho, H.; Ruscic, B. *J. Chem. Phys.* 1987, **86**, 1235.

15.   Longeway, P. A.; Weakliem, H. A.; Estes, R. D. *J. Phys. Chem.* 1984, **88**, 3282.

16.   Longeway, P. A.; Estes, R. D.; Weakliem, H. A. *J. Phys. Chem.* 1984, **88**, 73.

17.   Chu, J. O.; Beach, D. B.; Jasinski, J. M. *J. Phys. Chem* 1987, submitted.

18.   Chu, J. O.; Jasinski, J. M., to be published.

Chapter 41

Thermochemistry and reactivity of silylenes

R. Walsh – Department of Chemistry, University of Reading, Whiteknights, PO Box 224, Reading RG6 2AD, UK.

ABSTRACT

This article describes the impact made on our knowledge of the reactivity and thermochemistry of silylenes, through the recent application of time-resolved techniques to their detection and monitoring. Recently obtained rate constants for SiH_2, $SiMe_2$ and $SiCl_2$ are reviewed and discussed. The mechanism of the silylene Si-H insertion reaction is considered. A re-examination of the methylated disilane pyrolyses leads to new values for the enthalpies of formation of the silylenes.

INTRODUCTION

In the field of gaseous organosilicon chemistry, silylenes have the same importance as intermediates as have alkyl radicals in hydrocarbon chemistry. The mechanisms of high temperature decompositions of alkanes typically proceed via free radical chain mechanisms, whereas the pyrolyses of many alkylsilanes (and alkyldisilanes) proceed via mechanisms involving silylenes. A considerable body of knowledge has been accumulated concerning such mechanisms, much of it summarised in a series of excellent review articles by Gaspar [1]. Until recently most experimental information has come from high temperature thermal decomposition studies, with mechanistic inferences drawn from a combination of stable product detection and kinetic analysis. As illustrative of this one may cite selections of the work of Barton [2], Davidson [3] Ring and O'Neal [4] and Gaspar [5].

A picture has emerged that silylenes, such as SiH_2 and $SiMe_2$, are highly reactive species. What has been lacking, however, until recently, has been experimental work involving the direct monitoring if these species in order to answer quantitatively the question "just how reactive are silylenes ?" Undoubtedly, the application of time-resolved techniques like Flash Photolysis has been hampered to some extent by the lack of available and convenient photolytic sources of silylenes, as well as the demanding question of how to monitor the silylenes. These problems are now being overcome in several laboratories and gas-phase rate data have been obtained recently by Jasinski [6] and Inoue and Suzuki [7] for SiH_2, by Strausz and colleagues for $SiCl_2$ [8,9], by Stanton et al for SiF_2 [10] and ourselves for $SiMe_2$ [11]. The availability of this new data, inevitably, is already causing an alteration of perspective on the question of reactivity. Since rate studies are intimately tied up with thermochemistry of intermediates, some changes in accepted thermochemistry have come about. This may be described as a modest unheaval rather than a revolution.

It is the purpose of this article to review our work in this area together with some of its implications. Mention should be made of similar time-resolved studies of $SiMe_2$ and other silylenes in solution by the Gaspar group [12,13] although lack of space precludes a detailed discussion. It is our feeling that while gas phase studies are likely to offer the more direct index of reactivity, solution studies will show whether solvation effects are important for silylenes as they undoubtedly are for the analogous carbenes.

TIME-RESOLVED STUDIES OF SiH_2

Inoue and Suzuki [7] were the first to obtain kinetic data for SiH_2, using the 193 nm photolysis of $C_6H_5SiH_3$ as the source and laser induced fluorescence to detect SiH_2. This was followed by a study of the reaction of SiH_2 with D_2 by Jasinski [6] using the same source but laser resonance absorption at $\lambda=579.4$ nm to monitor SiH_2. The rate data are collected in Table 1 which also shows some further figures based on relative rate measurements by ourselves [14] but using the rate constant for SiH_2 with C_2H_4 [7] as a reference value. All the rate constants were obtained at room temperature. To date there have been no temperature dependent studies, and therefore measurements of activation energies. However, the rate constants for the Si-H insertion and π-type addition reactions are all close to collision rates and so little, if any barrier should be expected in these cases. It should be added that these rates are all considerably faster than some estimates based on relative rate studies extrapolated from high temperatures [15]. The rate constant for SiH_2 with D_2 is ca 26 times greater than the value for H_2 (not quoted in the table) obtained by Inoue and Suzuki [7] . This apparent discrepancy may be rationalised by the observation that the recombination reaction

$$SiH_2 + H_2 \rightarrow SiH_4$$

is in a pressure dependent (non-second order) region [6].   We

Table 1. Rate constants for reaction of $SiH_2(^1A_1)$ at 298 K.

Reactant	$k/10^{10} M^{-1} s^{-1}$	Ref.
SiH_4	6.6	[7]
Si_2H_6	34	[7]
$C_6H_5SiH_3$	6.6	[14]
Me_3SiH	3.1	[14]
C_2H_4	5.8	[7]
C_2H_2	6.0	[14]
$MeC\equiv CMe$	6.6	[14]
O_2	0.072	[14]
D_2	0.16	[6]

have carried out relative rate studies and RRKM calculations
[16] which reconcile these differences.   Jasinski's study of
SiH_2 with D_2 represents the true second order rate constant to
a good approximation because any breakdown of the intermediate
vibrationally excited SiH_2D_2 is more likely to yield $SiHD + HD$,
or $SiD_2 + H_2$ than regenerate $SiH_2 + D_2$.

Further rate measurements on SiH_2 by Jasinski [17] are discussed
in conjunction with $SiMe_2$ studies in the next section.

TIME-RESOLVED STUDIES OF SiMe₂

We recently reported [11] the first gas-phase time-resolved
kinetic data for $SiMe_2$ using $Me_3SiSiMe_2H$ (PMDS) and Si_3Me_8
photolyses (at 193 nm) as the sources and Ar^+ laser resonance
absorption to monitor $SiMe_2$ via its now well-established visible
absorption spectrum [18].   The variety of available laser
lines in the wavelength range 457-518 nm allowed us to confirm
the identity of the $SiMe_2$ spectrum.   Kinetic data were obtained
for a variety of Si-H insertion processes by study of the
signal decay characteristics at 458 nm in the presence of
various and varying quantities of the reactive substrate
molecules.   The decay traces obtained were found to correspond
to clean exponentials in all cases, thus indicating the straight
forward pseudo-first order decay expected for scavenging of
$SiMe_2$ in the presence of excess substrate.   It was further
demonstrated that the decay constants were linear functions of
substrate concentration.   Experiments were carried out in
excess of inert gas (5 Torr of SF_6) to eliminate the possibility

of hot molecule reactions. The results are shown in Table 2, which includes the data for more recent studies with unsaturates [19] as well as earlier studies with the methylated

Table 2. Rate constants for reactions of $SiMe_2$ ($^1A_1$) at 298 K.

Reactant	$k/10^9 M^{-1} s^{-1}$
SiH_4	0.12 ± 0.02
$MeSiH_3$	1.1 ± 0.1
Me_2SiH_2	3.3 ± 0.3
Me_3SiH	2.7 ± 0.3
Me_4Si	$\leqslant 0.03$
PMDS	25 ± 1
C_2H_4	14 ± 1
$1,3-C_4H_6$	45 ± 3
CH_3OH	0.60 ± 0.06

silanes [11]. In the latter case the experimental low limit of reaction of $SiMe_2$ with $SiMe_4$ confirmed existing belief [1] that insertion into C-H or Si-C bonds is very inefficient if it occurs at all. For the other silanes, end product analyses confirmed that the reaction occurred *via* Si-H bond insertion.

Comparison of the data of Tables 1 and 2 shows that not only are the Si-H insertion rate constants of $SiMe_2$ slower than those of SiH_2 but also that there is a significant variation in reactivity of $SiMe_2$. Dimethyl silylene is in fact both a less reactive and more discriminating species than silylene. The comparison of reactivity with SiH_4 is the most striking where $SiMe_2$ is nearly 10^3 less reactive than SiH_2. The variation of rate constants for the $SiMe_2$ within the series of methyl silanes indicates a strong enhancing effect of methyl group substitution. Whether this is due to an A factor or activation energy effect is not yet known but we suspect the latter, since generally fast reactions usually have fairly similar A factors. However, the *origin* of this trend is most probably the change in bond polarity of the Si-H bonds with methyl substitution since bond dissociation energy measurements [20] indicate little if any variation in Si-H bond strengths across this series. (The explanation for the high insertion rate constants for PMDS may, however, involve a bond strength effect since recent studies by Griller et al [21] have shown that $D(Me_3SiSiMe_2-H)$ is ca 85 kcal mol^{-1} some 5 kcal mol^{-1} less than the typical value of 90 kcal mol^{-1} for $D(Si-H)$ for the methylated monosilanes). An explanation in which methyl

groups on either the reacting silylene or the substrate silane
act as weakly electronegative substituents would appear to fit
in with current theoretical ideas of the nature of the insertion
reaction [22,23]. These ideas are represented in figure 1.

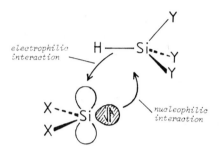

Fig. 1. Representation of electronic interactions occurring during silylene insertion into
as Si-H bond.

The insertion process may be viewed as comprising an
"electrophilic" stage followed by a "nucleophilic" stage.
Electronegative substituents (X) on the silylene will cause
orbital contraction thus disfavouring both the acceptor and
donor character of the silylene. Electronegative substituents
(Y) on the silane enhance the bond polarity $Si^{\delta+}-H^{\delta-}$, by making
the silicon more positively charged and thereby favouring the
nucleophilic stage of the process.

As far as previous data on these insertion processes are
concerned, our absolute rate constants are not in very good
agreement with relative rates obtained from either solution
studies [24] or higher temperature gas phase studies [25].

Sufficient data has now accumulated to make a worthwhile
comparison of rate constants for the addition processes. Table
3 shows a comparison for SiH_2, $SiMe_2$ and $SiCl_2$ for several
substrate molecules. The figures indicate that the trends of
reactivity with silylene substituent observed for the Si-H
insertion reactions, are also present for the π-type addition
process, viz. SiH_2 is more reactive than $SiMe_2$ which, in turn,
is considerably more reactive than $SiCl_2$. The difference
between $SiMe_2$ and SiH_2 is, however, not very great and the high
values for the rate constants for both species suggest
processes with little if any activation barrier. The lower
rate constants for $SiCl_2$ probably imply an activation energy.
For SiH_2 and $SiMe_2$ it is, furthermore, plausible that addition
rates to 1,3-butadiene should be ca 2-3 times greater than to
ethylene. The rate constant for SiH_2 with C_2H_4 in Table 3,
obtained by Jasinski [17] is slightly lower than that of
Table 1.   There is no necessary disagreement here since the

Table 3. Comparison of rate constants[a] for silylene addition reactions.

Reactant	Silylene		
	SiH_2[b]	$SiMe_2$[c]	$SiCl_2$[d]
C_2H_4	$\geqslant 3.3 \times 10^{10}$	1.4×10^{10}	7.8×10^7
C_3H_6	7.2×10^{10}	-	2.3×10^8
$1,3-C_4H_6$	1.1×10^{11}	4.5×10^{10}	-
C_2H_2	5.9×10^{10}	-	4.3×10^7

a. k in $M^{-1}s^{-1}$ b. ref. [17]   c. ref. [19]   d. ref.[8]

rates were measured at different pressures and Jasinski has
found the reaction to be pressure dependent. This may be
indicative of incomplete stabilisation of the initially formed
silacyclopropane. Some of the rate constants of Tables 3 and 1
are in poor agreement with relative rate studies, for instance
that of SiH_2 with SiH_4 and $1,3-C_4H_6$ [26]. A note of caution
should be sounded, however, in comparing kinetic studies based
on reactant disappearance with those from product detection.
If all reaction channels have not been identified, different
answers will inevitably result. These π-type addition
reactions of silylenes are likely to be particularly prone to
this difficulty since while it is reasonably well-established
from mechanistic studies [1] that initial interaction produces
silycyclopropanes (or silacyclopropenes) these products are
known to be extremely labile and difficult to handle. To our
knowledge they have not yet been detected in any of the time-
resolved studies [8,17,19]. In relative rate studies, 1,3-
butadiene has been a favourite substrate [15,26,27] because
stable silacyclopentenes are formed at high temperatures.
However, it is known that these products result from isomerisa-
tion of initially formed vinylsilacyclopropanes [28] and if
these intermediates have alternative fates than ring expansion,
such as stabilisation (or analytical loss) at low temperatures,
or reversible decomposition or other channels at high
temperatures [29] then product ratios must be interpreted with
caution.

The gas phase data on $SiMe_2$ [11,19] suggest not only greater
reactivity but also a different pattern of reactivity from
solution studies [13]. Space limitation precludes a discussion
of this.

THERMOCHEMICAL IMPLICATIONS

The rate constants obtained in these studies offer the
opportunity to refine estimates of some important thermochemical quantities, viz. the heats of formation of silylene
and the methylsilylenes. This is possible since there is a
considerable amount of published information on the decomposition of disilane and the methyldisilanes which are the reverse
processes of silylene, Si-H insertion reactions.

In principle for any such decomposition,

$$A \underset{-1}{\overset{1}{\rightleftarrows}} B + C \qquad\qquad (1,-1)$$

the enthalpy change is given by

$$\Delta H^{o}_{1,-1} = E_1 - E_{-1}$$

where E_1 and E_{-1} are forward and reverse activation energies.
For several decompositions, values of E_1 are available.
Values of E_{-1} representing silylene insertion activation
energies are not yet known (as discussed) but may be estimated.
This approach has been used in the past to obtain values of
$\Delta H_{1,-1}$ and thereby the silylene heats of formation [20, 30].
A more reliable approach, however, involves estimation of the
entropy change ΔS^o for reaction (1,-1) and then the use of the
relationship,

$$\log_{10}(A_1/s^{-1}) = \log_{10}(A_{-1}/M^{-1}s^{-1}) + [\Delta S^o/R - (1 + \ln R'T)]/\ln 10$$

to obtain A_1. A check can then be made on the measured
Arrhenius A factor for decomposition, and if necessary A_1 and
E_1 can be adjusted prior to estimation of $\Delta H^{o}_{1,-1}$. Whilst
this procedure involves knowing A_{-1} (the A factor for silylene
insertion) it turns out that, because the silylene rate
constants are so high, there is relatively little error in
$\Delta H_{1,-1}$ introduced by the uncertainty in A_{-1}.

To obtain ΔH_f^{o} for the silylene from information on $\Delta H^{o}_{1,-1}$ it
is further necessary to know enthalpies of formation of
methyl substituted mono- and di-silanes. The former are well
established [31] and the latter may be reasonably reliably
estimated [32]. We have used this approach previously for the
particular dissociations,

$$Si_2H_6 \underset{}{\overset{}{\rightleftarrows}} SiH_4 + SiH_2$$

$$Me_3SiSiMe_2H \underset{}{\overset{}{\rightleftarrows}} Me_3SiH + SiMe_2$$

to obtain $\Delta H_f^{o}(SiH_2)$ [16] and $\Delta H_f^{o}(SiMe_2)$ [11,33].

In recent work we have extended this to include further

examples [32].   The results are shown (in part) in tables 4 and
5.   These tables show the A factor estimates for the $SiMe_2$ and

Table 4. Entropy changes and A factors for some methyldisilane decompositions.

Reaction	ΔS^a	$\min^b$	$\log(A/s^{-1})$ expt.	prefd.[b]
$Me_3SiSiMe_2H \rightarrow Me_3SiH + SiMe_2$	34.2	14.75	12.9 ± 0.3^c	15.3
$Me_2SiHSiMe_2H \rightarrow Me_2SiH_2 + SiMe_2$	32.1	14.4	–	15.2
$MeSiH_2SiMe_2H \rightarrow MeSiH_3 + SiMe_2$	29.5	13.3	12.6 ± 2.0^c	14.8
$H_3SiSiMe_2H \rightarrow SiH_4 + SiMe_2$	29.9	12.5	–	15.0

a.   cal $K^{-1}mol^{-1}$ at 650 K   b. See text   c. ref. [38]

Table 5. Entropy changes and A factors for some methyldisilane decompositions.

Reaction	ΔS^a	$\min^b$	$\log(A/s^{-1})$ expt.	prefd.[b]
$Me_3SiSiH_3 \rightarrow Me_3SiH + SiH_2$	36.1	16.2	14.48 ± 0.3^c	16.2
$Me_2SiHSiH_3 \rightarrow Me_2SiH_2 + SiH_2$	32.5	15.7	–	15.7
$MeSiH_2SiH_3 \rightarrow MeSiH_3 + SiH_2$	31.1	15.4	15.28 ± 0.15^d	15.6
$Si_2H_6 \rightarrow SiH_4 + SiH_2$	32.7	15.8	14.52^e	16.1

a.   cal K^{-1} mol^{-1} at 650 K   b.   See text   c.   ref. [39]
d.   ref. [40]   e.   ref. [36]

SiH_2 producing reactions, in terms of the minimum values (set
equal to the measured rate constants) and preferred values,
based on constant per-H insertion A factors (A_{-1}) for the
silylenes.   These calculations suggest that measured A factors
are in many cases somewhat low.   Reinvestigation of the
pyrolysis of Si_2H_6 [34,35] has shown the prediction to have
been correct and the new Arrhenius parameters are significantly
higher than the original values [36].   A new study of the
pyrolysis of $Me_3SiSiMe_2H$ by Davidson et al [37]   however
appears to support his earlier work [38] and there remains,
therefore, a substantial discrepancy in this case, as in

several of the other disilane pyrolyses.

The silylene heats of formation calculated using this approach are shown in Table 6. $\Delta H_f^{\,o}(SiH_2)$ is some 7 kcal mol^{-1} higher

Table 6. Enthalpies of formation of silylene and Si_2H_4.

Species	$\Delta H_f^{\,o}$/kcal mol^{-1}	Reference
SiH_2	65.3(±1.5)	[16]
SiHMe	43.9(±3)	[32]
$SiMe_2$	26.1(±2)	[11,32]
$SiHSiH_3$	74.6(±2)	[43]
$H_2Si=SiH_2$	≤ 63.1	[43]

than the previously accepted value derived by the simpler thermochemical arguments [30], although it is in good agreement with more recent measurements and theoretical calculations (as detailed in ref. [16]). $\Delta H_f^{\,o}(SiMe_2)$ has been more uncertain until recently. The present value [11,32] is some 4 kcal mol^{-1} higher than our own earlier estimate [33,41] based on an erroneous solution rate constant [42] for the reaction of $SiMe_2$ with Et_3SiH. The stabilisation energies corresponding to these heats of formation form part of a consistent correlation with silylene substituent electronegativity [41]. Also shown in Table 6 are the heats of formation of the two Si_2H_4 isomers recently derived by us from a revised mechanistic modelling analysis of the H atom/SiH_4 reaction [43]. These figures also represent an upward revision over previous values.

CONCLUSION

The advent of time-resolved techniques to the study of silylenes is leading to new insight both to their reactivity and stability. Although the first direct kinetic data have now been obtained for several silylenes this work is clearly at an early stage, and we can look forward to substantial progress in our understanding of the mechanisms of reaction of organo-silanes in the next few years. In particular the availability of more reliable elementary rate constants for silicon-containing intermediates should lead to improved modelling of the bewildering variety of high temperature organosilane rearrangements.

ACKNOWLEDGEMENTS

The author gratefully acknowledges helpful discussions with

colleagues at Reading and pre-publication manuscripts and
valuable comments from Bob Carr, Iain Davidson, Peter Gaspar,
David Griller, Joe Jasinski, Ed. O'Neal and Morey Ring.

REFERENCES

1. P.P. Gaspar in Reactive Intermediates, Eds, M. Jones Jr.
   and R.A. Moss, Wiley, New York, 1, 229 (1978); 2, 335
   (1981); 3, 333 (1985).
2. For example; W.D. Wulff, W.F. Goure and T.J. Barton, J. Am.
   Chem. Soc., 100,6236 (1978).
3. For example; I.M.T. Davidson and R.J. Scampton, J. Organometallic Chem., 271, 249 (1984).
4. For example; J.W. Erwin, M.A.Ring and H.E. O'Neal, Int. J.
   Chem. Kinet., 17, 1067 (1986).
5. For example; B.H. Boo and P.P. Gaspar, Organometallics, 5,
   698 (1986).
6. J. Jasinski, J. Phys. Chem., 90, 555 (1986).
7. G. Inoue and M. Suzuki, Chem. Phys. Lett., 122, 361 (1985).
8. I. Safarik, B.P. Ruzsicska, A. Jodhan, P.O. Strausz and
   T.N. Bell, Chem. Phys. Lett., 113, 71 (1985).
9. V. Sandhu, A. Jodhan, I. Safarik and O.P. Strausz, Chem.
   Phys. Lett., 135, 260 (1987).
10. A.C. Stanton, A. Freedman, J. Wormhoudt and P.P. Gaspar,
    Chem. Phys. Lett., 122, 190 (1985).
11. J.E. Baggott, M.A. Blitz, H.M. Frey, P.D. Lightfoot and
    R. Walsh, Chem. Phys. Lett., 135, 39 (1987).
12. P.P. Gaspar, B.H. Boo, S. Chari, A.K. Ghosh, D. Holten,
    C. Kirmaier and S. Konieczny, Chem. Phys. Lett., 105, 153
    (1984).
13. P.P. Gaspar, D. Holton, S. Konieczny and J.Y. Corey, Acc.
    Chem. Res., in press (1987).
14. C.D. Eley, M.C.A. Rowe and R. Walsh, Chem. Phys. Lett.,
    126, 153 (1986).
15. D.S.Rogers, H.E. O'Neal and M.A. Ring, Organometallics,
    5, 1467 (1986).
16. H.M.Frey, R. Walsh and I.M. Watts, J. Chem. Soc.,Chem.
    Commun., 1189 (1986).
17. J.O. Chu, D.B. Beach and J.M. Jasinski, personal
    communication.
18. H. Vancik, G. Raabe, M.J. Michalczyk, R. West and J. Michl,
    J. Am. Chem. Soc., 107, 4097 (1985).
19. J.E. Baggott, M.A.Blitz, H.M. Frey, P.D. Lightfoot and
    R. Walsh, to be published.
20. R. Walsh, Acc. Chem. Res., 14, 246 (1981).
21. J.M. Kanabus-Kaminska, J.A. Hawari, D. Griller and
    C. Chatgilialoglu, personal communication.
22. M.S. Gordon and D.R. Gano, J. Am. Chem. Soc., 106, 5421
    (1984) and refs. therein.
23. A. Sax and G. Olbrich, J. Am. Chem. Soc., 107, 4868 (1985)
    and refs. therein.
24. T.Y. Gu and W.P. Weber, J.Organometallic Chem.,195, 29 (1980).

25. I.M.T. Davidson and N.A. Ostah, J. Organometallic Chem., 206, 149 (1981).
26. P.P. Gaspar, S. Konieczny and S.H. Mo, J. Am. Chem. Soc., 106, 424 (1984).
27. J. Dzarnoski, S.F. Rickborn, H.E. O'Neal and M. A. Ring, Organometallics, 1, 1217 (1982).
28. D. Lei, R-J. Hwang and P.P. Gaspar, J. Organometallic Chem., 271, 1 (1984).
29. P.P. Gaspar and D. Lei, Organometallics, 5, 1276 (1986).
30. P. John and J.H. Purnell, J. Chem. Soc., Faraday Trans 1, 69, 1455 (1973).
31. A.M. Doncaster and R. Walsh, J. Chem. Soc., Faraday Trans.2, 82, 707 (1986).
32. R. Walsh, submitted for publication.
33. R. Walsh, J. Phys. Chem., 90, 389 (1986).
34. J.G. Martin, M.A.Ring and H.E. O'Neal, Int. J. Chem. Kinet., in press (1987), personal communication.
35. K.F. Roenigk, K.F. Jensen and R.W. Carr, J. Phys. Chem., in press (1987), personal communication.
36. M. Bowrey and J.H. Purnell, Proc. Roy. Soc. A, 321, 341 (1971).
37. I.M.T. Davidson, K.J. Hughes and S. Ijadi-Maghsoodi, Organometallics, 6, 639 (1987).
38. I.M.T. Davidson, J.I. Matthews, J. Chem. Soc., Faraday Trans. 1, 72, 1403 (1976).
39. M.A Ring and D.P. Paquin, J. Am. Chem. Soc., 99, 1793 (1977).
40. A.J. Vanderwielen, M.A.Ring and H.E. O'Neal, J. Am. Chem. Soc., 97, 993 (1975).
41. R. Walsh, Pure and Appl. Chem., 59, 69 (1987).
42. A.S. Nazran, J.A. Hawari, D. Griller, I.S. Alnaimi and W.P. Weber, J. Am. Chem. Soc., 108, 5041 (1986); see also ibid, 106, 7267 (1984).
43. R. Becerra and R. Walsh, submitted for publication.

Chapter 42

Theoretical studies
of organosilicon chemistry

Mark S. Gordon, Kim K. Baldridge, Jerry A. Boatz, Shiro Kosecki, and Michael W. Schmidt – Department of Chemistry, North Dakota State University, Fargo, North Dakota 58105.

Five years ago, Schaefer published a paper entitled "The Sili-con-Carbon Double Bond: A Healthy Rivalry Between Theory and Experiment" [1]. This was a landmark paper, not only because it summarized what were at that time major experimental vs. theoretical discrepancies in the molecular structure and stability of silaethylene, but because it focussed our atten-tion on the extraordinary potential for experimental and theoretical chemists to collaborate in the development of an emerging field of chemistry. Indeed, if Schaefer were writing that paper today, he might call it "Organosilicon Chemistry: A Successful Interaction Among Experimentalists and Theoreti-cians". Because we have benefited greatly from such inter-actions, the latter title represents the theme of this paper. The topics to be considered here are: silylenes, aromatic metals, strained rings, anions and radicals, and penta-coordinated silicon.

SILYLENES

These reactive intermediates play as important a role in organo-silicon chemistry as do their carbon counterparts, carbenes, in organic chemistry. As in carbene chemistry, a major inter-est in silylene chemistry is how the electronic state of the species (i.e., singlet or triplet) affects the reactivity. Potential energy profiles, as a function of the H-Si-H angle,

are shown in Figure 1 for the lowest three states of silylene
[2]. For each point on the surface, the bond length has been
optimized using full valence (FV) FORS MCSCF wave function [3]
with the 6-31G(d,p) basis set [4]. Energies obtained by aug-
menting the MCSCF wavefunctions with a second order configuration interaction (SOCI) are also shown in the figure. The two
sets of curves are nearly parallel and predict a closed shell
($^1A_1$) ground state, about 18 kcal/mol below the $^3B_1$ state.

An MCSCF energy contour diagram for the triplet state is shown
in Figure 2. The dotted line in this figure is the seam of
common energy for the triplet and closed shell singlet states.
The plus sign indicates the minimum for the triplet state,
while the X is the lowest energy point along the seam. The
key point here is that the distance (in energy) between these
two points is only 0.04 eV and the structural difference is
slight. Since the zero point vibrational energy of the triplet
is 0.32 eV, the upper state will easily intersystem cross to
the ground state if the spin-orbit coupling matrix elements are
significant. Thus, not only is the ground state of silylene a
singlet, but the triplet may well be difficult to access.

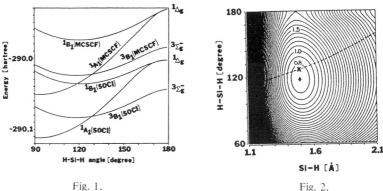

Fig. 1.   Fig. 2.

In view of the foregoing, attention naturally turns to identifying substituents which might preferentially stabilize the triplet state. From the analogous carbene chemistry, one suspects
that lone-pair-containing electronegative substituents will
preferentially stabilize the singlet state [5], and this is
indeed verified by both experiment [6] and theory [7]. This
suggests that the introduction of electropositive substituents
might stabilize the triplet more than the closed shell singlet.
Alternatively, if bulky enough substituents are introduced, the
R-Si-R angle will open. In the limit of a 180° angle, the
triplet must become the ground state (see Figure 1).

Dimethyl substitution actually slightly increases the singlet-
triplet splitting of silylene, even though the angle about
silicon undergoes a small increase [8]. Di-t-butyl substitution has a much larger geometric effect, opening the singlet

and triplet angles by 13.5 and 6.5°, relative to the dimethyl compound. Even so, the ground state of di-t-butyl silylene is still predicted to be the closed shell singlet [8], and experimental evidence [9] supports this prediction. It should be noted, however, that Gaspar and co-workers have experimental evidence that di-adamantyl substitution does indeed lead to a silylene with a triplet ground state [10].

The highly electropositive substituent lithium has a dramatic effect on the ordering of the states of silylene. The energies of the four lowest lying states of Li_2Si as a function of the Li-Si-Li angle are illustrated in Figure 3. The Li-Si bond length was optimized at each angle for each state at the SCF/6-31G(d) [4] computational level, and the energies were calculated with MP2 [11] /6-31G(d) wave functions. The two lowest states are triplets: $^3B_1$ and $^3A_2$. The former is the analog of the well-known low-lying triplet in carbenes and silylenes with a linear structure, while the latter has a very small (65.5°) bond angle and a very long (2.625A) bond length. The $^3A_2$ state, which is predicted to be the ground state at the highest level of theory [8], may best be described as a weakly bound complex between atomic silicon and diatomic lithium. A similar description may be applied to the singlet analog.

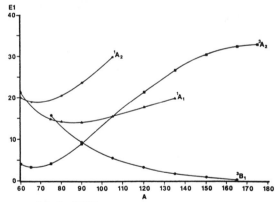

Fig. 3. Angular potentials for SILI2.

It should be noted that the A_2 states have also been predicted for the parent SiH_2 [12], but there these states are much higher in energy than $^1A_1$ and $^3B_1$. Mono-lithiated silylene also has a triplet ground state [7b,13] and exhibits low-lying states analogous to the A_2 states of Li_2Si.

Since lithium may not be the most desirable substituent from an experimental point of view, disilylsilylene and its isomers has also been investigated [14]. It is predicted that the electropositive silyl substituent does indeed lower the singlet-triplet splitting relative to the parent silylene; however, the singlet is still estimated to be lower by about 6 kcal/mol. The lowest

energy isomer on the singlet potential energy surface is tri-
silacyclopropane. This compound is predicted to be 10 kcal/mol
below trisilapropene and 24 kcal/mol below disilylsilylene.
This is in contrast with the hydrocarbon analogs, for which the
global minimum is propene. Calculations are under way in this
laboratory on the low-lying states of Li-Si-SiH$_3$. The ground
state of FSiLi has recently been predicted to be a triplet [15],
and replacing F with SiH$_3$ is not expected to alter this.

A particularly interesting silylene, decamethylsilicocene - the
silicon analog of ferrocene - , has recently been isolated by
Jutzi and co-workers [16]. The crystal reveals two structures
(Figure 4), one (structure A) with coplanar rings and D$_{5d}$ symmetry and the other (B) in which the two rings are tilted
relative to each other. The geometry of the hydrocarbon parent
of structure A has been optimized with SCF/3-21G* wave functions
[17]. Figure 5 illustrates an electron density difference diagram, in which the (approximately unperturbed) electron density
0.7 bohr above the upper ring is subtracted from the electron
density at an equal distance below the same ring. The increase
in (pi) electron density (solid contours) near the silicon and
the depletion of electron density (dotted contours) in the
vicinity of the ring CC bonds are apparent. A harmonic force
field carried out on this molecule has a degenerate pair of
negative eigenvalues (frequency = 98 cm^{-1}), suggesting that the
structure is not a minimum on the potential energy surface.
The normal modes (Figure 6) corresponding to the imaginary
frequencies lead to structure B of Figure 4.

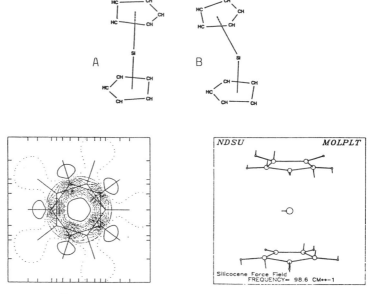

Fig. 5.   Fig. 6.

Aromaticity, the stabilization of a cyclic compound due to delocalization of pi electrons, has been defined in a variety of ways. The most convenient .approach for electronic structure theory calculations is to use chemical reactions (termed iso-desmic reactions) [18] in which the number of formal bonds of each type is conserved. Such an approach minimizes the change in the correlation error from reactants to products and there-fore minimizes the error in the calculation at the SCF level. Two types of homodesmic reactions, bond separation reactions [18] and superhomodesmic reactions [19], are illustrated in Scheme I for six-membered rings. The bond separation reactions relate the parent ring to completely isolated singly and doubly bound fragments, while the superhomodesmic reactions relate the ring to its closest acyclic analogs. One expects the energy differences for the latter reactions to be smaller and more accurate than those for the former, but the trends are expected to be similar.

Scheme 1: Bond separation reactions.

$+5CH_4+XH_4 \rightarrow 2CH_3-CH_3+CH_3-XH_3+2CH_2=CH_2+CH_2=XH_2$

$+5CH_4+XH_3 \rightarrow 2CH_3-CH_3+CH_3-XH_2+2CH_2=CH_2+CH_2=XH$

Superhomodesmic Reactions

$\left\{ \begin{array}{c} 2XH_2=CH-CH=CH_2 \\ + \\ CH_2=CH-CH=CH_2 \end{array} \right\} \rightarrow \left\{ \begin{array}{c} 2XH_2=CH-CH=CH-CH=CH_2 \\ + \\ CH_2=CH-XH=CH-CH=CH \end{array} \right\}$

$\left\{ \begin{array}{c} 2XH-CH-CH=CH_2 \\ + \\ CH_2=CH-CH=CH_2 \end{array} \right\} \rightarrow \left\{ \begin{array}{c} 2XH=CH-CH=CH-CH=CH_2 \\ + \\ CH_2=CH-X=CH-CH=CH_2 \end{array} \right\}$

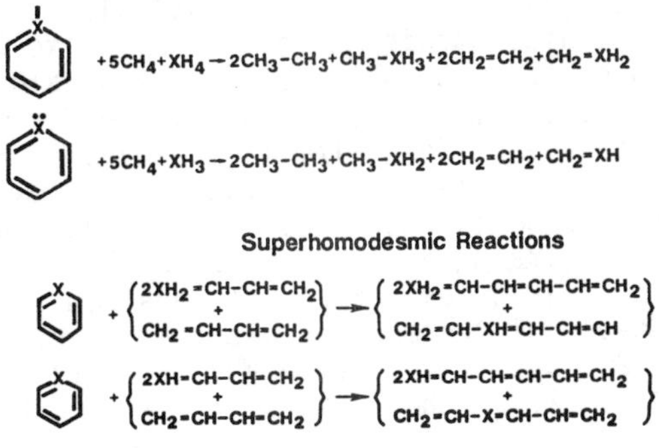

The results of SCF/3-21G* calculations for both types of reactions, with X taken from groups IV or V, are summarized in Figure 7. As expected, the trends for the two types of reac-tions are the same. For group IV elements there is a sharp drop in aromaticity from the second to the third period (C to Si), but Si and Ge are predicted to have nearly the same aro-maticity. The bond separation reactions predict another sharp drop from Ge to Sn. The superhomodesmic calculations on Sn are still in progress. According to the bond separation reactions, there is much less variation in the aromatic character of the group V elements. The calculations available to date on the superhomodesmic reactions are in agreement with this.

STRAINED RINGS

Strain energies of small rings may be calculated using reactions similar to those discussed above. The strain energies for the

Fig. 7. 3-21G* aromaticities.

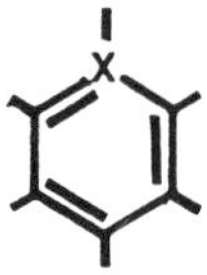

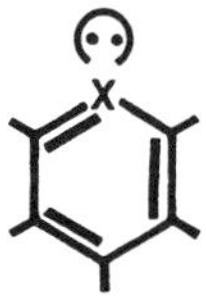

x	Bond Separation	Super homodesmic
C	60.18	25.99
Si	46.44	17.97
Ge	45.57	16.75
Sn	30.55	---
N	64.19	25.78
P	56.56	23.30
As	53.49	---
Sb	50.49	---

cycloalkanes and monosilacycloalkanes, calculated from the appropriate homodesmic reactions [20] with MP2/6-31G(d) wave functions and including zero point vibrational corrections, are summarized in Table 1. For the cycloalkanes, the calculated and experimental strain energies are in excellent agreement with each other. The silacycloalkane strain energies are less well known experimentally; however, the estimated value for silacyclobutane is 24 kcal/mol [21], again in excellent agreement with the theoretical prediction. Note that the strain in silacyclopropane is much greater than that in cyclopropane, whereas the strain in the two four-membered rings is almost the same. This may be explained, at least in part, by recalling that, because of the much greater difference in the size of 3s ($<r>$ = 2.22 bohr) vs. 3p ($<r>$ = 2.75 bohr) orbitals relative to 2s ($<r>$ = 1.59 bohr) vs. 2p ($<r>$ = 1.69 bohr), the n=3

Table 1. Strain energies (kcal/mol).

Molecule	Enthalpy (0 K) Theory	Experiment
Cyclopropane	27.3	25.8
Cyclobutane	25.8	25.7
Cyclopentane	6.1	5.3
Cyclohexane	-0.1	-0.4
Silacyclopropane	41.4	
Silacyclobutane	24.7	
Silacyclopentane	4.5	
Silacyclohexane	4.0	

orbitals are much less prone to hybridize. Thus, silacyclobutane is quite able to accomodate a structure which has a 100° CCC angle, an 80° CSiC angle, and two 90° CCSi angles, whereas cyclobutane must accomodate four CCC angles which are slightly less than 90°.

The same homodesmic reactions used to estimate strain energies may also be used to predict heats of formation of the strained rings [22]. The results of such calculations are summarized in Table 2. The calculated heats of formation for the cyclo-

Table 2. Calculated and experimental heats of formation (kcal/mol).

Molecule	Theory	Experiment
Cyclopropane	18.3	16.8
Cyclobutane	12.8	12.6
Cyclopentane	-10.3	-11.1
Cyclohexane	-19.8	-20.2
Silacyclopropane	37.0	
Silacyclobutane	19.0	
Silacyclopentane	-1.8	
Silacyclohexane	-9.1	

alkanes are in excellent agreement with experiment. This lends some confidence that the values predicted for the silacyclo-alkanes are reasonably accurate as well.

RADICALS AND ANIONS

Methylsilane is a useful prototype molecule with which to probe the ease of removing a hydrogen atom or a proton from carbon vs. silicon. For the neutral species [23], using MP4 [24]/ 6-31G(d,p)//6-31G(d,p) wave functions, it is found that CH_3SiH_2 is nearly 11 kcal/mol lower in energy than its isomer SiH_3CH_2. This is in good agreement with the experimental value of 9 kcal/mol and may be explained largely in terms of the relative strengths of CH vs. SiH bonds.

The analogous isomerization energies are not so obvious for the anions, since removal of a proton from silicon will leave that (very electropositive) silicon with a substantial negative charge. On the other hand, the gas phase acidity of methane is nearly 2 eV greater than that of silane, so in those reference compounds it is 40 kcal/mol more difficult to remove the proton from carbon. For methylsilane, the relative ease with which a proton can be removed from the carbon is reduced by a factor of two (relative to methane vs. silane). However, the silicon-centered anion is still more stable by nearly 20 kcal/mol at the MP4/6-31++G(d,P) [25] computational level, despite the negative charge on the silicon. This result is particularly intriguing, since removal of the proton by NH_2^- is found experimentally to be easier at the carbon end [26]. This suggests the existence of a larger barrier for the reaction at the silicon end. One can imagine a variety of transition states for the overall reaction

$$NH_2^- + CH_3SiH_3 \rightarrow NH_3 + CSiH_5^-,$$

including pentacoordinated silicon anions. As a first step in probing the potential energy surface for the reaction, we have initiated a systematic investigation of such species.

PENTACOORDINATED SILICON ANIONS

The structures and energetics of pentacoordinated silicon anions have been calculated at the RHF/6-31G(d) and MP2/6-31++G(d,p), respectively. Sample structures are shown in Figure 8. For $FSiH_4^-$, the axial Si-F and Si-H bonds are somewhat longer than their tetracoordinated analogs, as expected. In contrast, when F is replaced by Cl, the axial Si-Cl bond is more than 1.3A longer than that in SiH_3Cl [27]. This is a typical contrast between second and third period nucleophilic substituents. For monosubstitutions, the latter appear to be more like ion-dipole complexes between silane and the ion, rather than true penta-coordinated compounds. This is less true when a third period substituent is balanced against another third period substituent, as in SiH_3Cl_2 with both chlorines axial. Then, the Si-Cl bond stretches by only 0.3A.

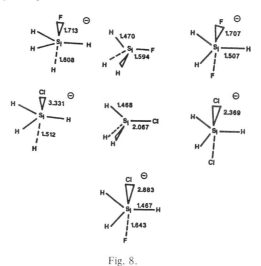

Fig. 8.

The foregoing structural considerations are supported by the calculations of the enthalpies for the reactions

$$SiH_3XY^- \rightarrow SiH_3X + Y^-.$$

The results of these calculations are summarized in Tables 3 and 4. For the mono-substituted compounds, Y = H yields a stabilization energy of nearly 17 kcal/mol, whereas if Y is a group from the second period the stabilization is about 30

Table 3. $SiH_3XY^- \to SiH_3X + Y^-$ (kcal/mol).   MP2/6–31++G(d,p)//SCF/6–31G(d).

X	Y	ΔH_0	X	Y	ΔH_0
H	H	16.7			
H	F	29.1	F	H	42.2
H	OH	29.9	OH	H	31.6
H	CH_3	30.7	CH_3	H	18.2
H	Cl	4.5	Cl	H	58.2
H	SH	4.0	SH	H	46.6
H	SiH_3	-0.2	SiH_3	H	28.3

kcal/mol. In contrast, third period substituents are only marginally stable, as suggested by the structures discussed above.

For the disubstituted compounds, third period substituents are

Table 4. $SiH_3XY^- \to SiH_3X + Y^-$ (kcal/mol). MP2/6–31++G(d,p)//SCF/6–31G(d).

X	Y	ΔH_0	X	Y	ΔH_0
F	F	49.4			
Cl	Cl	21.0			
Cl	F	57.7	F	Cl	15.6
OH	OH	41.3			
OH	F	39.5	F	OH	51.2
OH	Cl	9.1	Cl	OH	62.8
CH_3	CH_3	32.4			
SiH_3	SiH_3	6.4			
NH_2	NH_2	42.1			
SH	SH	17.6			
CH_3	F	29.0	F	CH_3	54.1

most stabilized when they are opposite another substituent from the third period. For example the stabilization of Cl^- is 21 kcal/mol when it is opposite Cl, whereas it is only stabilized by 15.6 and 9.1 kcal/mol when it is opposite F or OH, respectively. This balancing of bonding has been termed "sacrificial bonding" [28].

ACKNOWLEDGEMENTS

The authors have benefitted from collaborations with several colleagues, including Professors Philip Boudjouk, Robert West, Robert Damrauer, Peter Gaspar, Tom Barton, Iain Davidson, Robin Walsh, Major Larry Davis, Lt. Col. Larry Burggraf, and Dr. Richard Hilderbrandt. The research discussed was supported by grants from the Petroleum Research Fund, administered by the American Chemical Society, the National Science Foundation (NSF), and the Air Force Office of Scientific Research (AFOSR). Computer time for the calculations was made available by a variety of sources, including the San Diego Supercomputer Center (via a grant from the NSF), the North Dakota State University Computer Center, the NSF (through a grant for the purchase of Celerity C1260D and microVAX II computers) and the AFOSR (through a grant for a VAX 11/750 computer).

REFERENCES

1. H.F. Schaefer III, Accts. Chem. Res., $\underline{15}$, 283 (1982).
2. S. Koseki and M.S. Gordon, J. Mol. Spectrosc., $\underline{123}$, 392 (1987).
3. K. Ruedenberg, M.W. Schmidt, M.M. Gilbert, and S.T. Elbert, Chem. Phys., $\underline{71}$, 41, 61, 65 (1982).
4. (a) P.C. Hariharan and J.A. Pople, Theor. Chim. Acta, $\underline{28}$, 213 (1973);
    (b) M.S. Gordon, Chem. Phys. Lett., $\underline{76}$, 163 (1980).
5. (a) D. Feller, W.T. Borden, and e.R. Davidson, Chem. Phys. Lett., $\underline{71}$, 22 (1980);
    (b) N.C. Baird and K.F. Taylor, J. Am. Chem. Soc., $\underline{100}$, 1333 (1978);
    (c) V. Staemmler, Theor. Chim. Acta, $\underline{35}$, 309 (1974).
6. T. Suzuki, K. Hakuta, S. Saito, and E. Hirota, J. Chem. Phys., $\underline{82}$, 3580 (1985).
7. (a) M.E. Colvin, R.S. Grev, H.F. Schaefer III, and J. Bicerano, Chem. Phys. Lett., $\underline{99}$, 399 (1977);
    (b) B.T. Luke, J.A. Pople, K. Krogh-Jespersen, Y. Apeloig, M. Karni, J. Chandrasekhar, and P.v.R. Schleyer, J. Am. Chem. Soc., $\underline{108}$, 270 (1986).
8. M.S. Gordon and M.W. Schmidt, Chem. Phys. Lett., $\underline{132}$, 294 (1986).
9. (a) P. Boudjouk, private communication;
    (b) R. West, private communication.
10. M. Xiao and P. Gaspar, Eighth International Symposium on Organosilicon Chemistry, St. Louis, 1987, paper PAB15.
11. R. Krishnan and J.A. Pople, Int. J. Quantum Chem., $\underline{14}$, 91 (1978).
12. (a) B. Wirsam, Chem. Phys. Lett., $\underline{14}$, 214 (1972);
    (b) J.E. Rice and N.C. Handy, Chem. Phys. Lett., $\underline{107}$, 365 (1984).
13. M.E. Colvin, J. Breulet, and H.F. Schaefer III, Tetrahedron, $\underline{41}$, 1429 (1985).

14. M.S. Gordon and D. Bartol, J. Am. Chem. Soc., in press.
15. M.E. Colvin, H.F. Schaefer III, and J. Bicerano, J. Chem. Phys., 83, 4581 (1985).
16. P. Jutzi, D. Kanne, and C. Kruger, Angew. Chem., 98, 163 (1986).
17. K.K. Baldridge and M.S. Gordon, Angew. Chem. Int Eng. Ed., submitted.
18. W.J. Hehre, R. Ditchfield, L. Radom, and J.A. Pople, J. Am. Chem. Soc., 92, 4796 (1970).
19. P. George, M. Trachtman, C.W. Bock, and A.M. Brett, Theor. Chim. Acta, 38, 121 (1975).
20. J.A. Boatz, M.S. Gordon, and R.L. Hilderbrandt, J. Am. Chem. Soc., submitted.
21. R. Walsh, private communication.
22. R.L. Disch, J.M. Schulman, and M.L. Sabio, J. Am. Chem. Soc., 107, 1904 (1985).
23. T.J. Barton, A. Revis, I.M.T. Davidson, S. Ijadi-Maghsoodi, K.J. Hughes, and M.S. Gordon, J. Am. Chem. Soc., 108, 4022 (1986).
24. R. Krishnan, M.J. Frisch, and J.A. Pople, J. Chem. Phys., 72, 4244 (1980).
25. M.J. Frisch, J.A. Pople, and J.S. Binkley, J. Chem. Phys., 80, 3265 (1984).
26. R. Damrauer, C.H. Depuy, and S. Kass, in preparation.
27. M.S. Gordon, L.P. Davis, L.W. Burggraf, and R. Damrauer, J. Am. Chem. Soc., 108, 7889 (1986).
28. L.W. Burggraf and L.P. Davis, Eighth International Organosilicon Symposium, St. Louis, 1987, paper B20.

Chapter 43

Molecular mechanics calculations for predictions of organosilane structures and reactivities

F.K. Cartledge, S. Profeta, Jr.†, S. Cho and R. Unwalla – Department of Chemistry, Louisiana State University, Baton Rouge, LA 70803-1804, USA. †Allergan Pharmaceuticals, 2525 Dupont Drive, Irvine, CA 92715, USA.

INTRODUCTION

This paper will briefly review a few basic things about molecular mechanics calculations and give an overview of the kinds of silicon systems for which parameters exist and are available for the general user public. After doing that, I want to present some results from our own recent work. Molecular mechanics has been slow to be accepted as a useful tool in the arsenal of theoretical organic chemistry. I believe that two reasons are principally to blame. First, the method is not based on quantum mechanics, which has dominated theoretical considerations for half a century. Secondly, at a time in which the mainstream of theoretical thought is going away from a dependance on empirical short-cuts, molecular mechanics is essentially an empirically-based method. However, molecular mechanics has clearly carved out a niche into which other theoretical methods are not likely to intrude soon. In the 1950s organic and biochemists began to realize the importance of conformational analysis. The analysis of steric and stereoelectronic effects on reactivity, for instance, requires detailed knowledge of conformational preferences. Molecular mechanics can give a

great deal of accurate information about very large
systems in a very rapid calculation. So-called
"molecular modelling" has become a recognized field
of research in recent years, supported by hardware
and software of increasing sophistication. A
principal component of the software underpinning
these methods is molecular mechanics.

Readily available programs which are easy to use
without a great deal of theoretical sophistication
allow the following: energy minimization as a
function of structure; partitioning of energy into
several sources, e.g., stretching, bending, van der
Waals interactions, dipolar effects, etc.; the
capability of finding minima other than the global
minimum; some ability to trace energy changes as a
function of conformational change and thus predict
conformational barriers; and, depending upon para-
meterization, calculation of dipole moments and heats
of formation. The calculations are fast enough that
they can be performed on a microcomputer.

The method requires the availability of a
significant amount of empirical data, enough to
define a so-called "force field" familiar to vibrational spectroscopists for generations. The data
required before a class of compounds can be treated
includes stretching and bending force constants and
sufficient information about rotational barriers to
define torsional parameters. Silicon chemists are
fortunate that there have been vibrational spectroscopists interested in the area for a long time.
Furthermore, two of the major contributors to the
development of molecular mechanics in the first
place, Allinger and Mislow, were interested in
treating silicon systems. Some parameters for Si
systems have been available from the beginning of
widespread use of the method.

Good parameters are available for calculations
on a wide variety of compounds containing Si-C, Si-H
and Si-Si bonds, and more tentative parameters are
available for Si-O and Si-Cl containing compounds [1-
3]. Highly strained molecules generally require some
modification of parameters, notably the so-called
stretch-bend parameters in MM2. Small modifications
are necessary to do a good job, even with four-
membered rings. Silicon-containing ring systems much
more strained than that have not been treated, to the
best of our knowledge. We are most commonly using
the MM2 program, which is available from QCPE. That
issue contains effective Si parameters, but there is
an updated set in the dissertation of M. S. Frierson
[2]. Information required to define torsional parameters is that most often likely to be missing for
systems of interest. When direct experimental data

about rotational preferences is not available,
quantum mechanical calculations are possible sources
of the required information. In that regard, Si
chemists are also lucky. Enough ab initio calculational work has been done with Si systems to allow a
good assessment of what basis sets are likely to be
adequate to give good structure and relative energy
information. For instance, we have recently defined
torsional parameters to fit the MM2 program for the
systems: Si-C-C-Si and Si-C-C=C, using the Gaussian
82 set of programs and a 3-21G(*) basis set [4] to
provide confirmation that the MM2 parameters give an
adequate treatment of energy changes with bond
rotation.

PARAMETERS FOR THE Si-C-C-Si FRAGMENT

It is common that those of us who learned our
conformational analysis as organic chemists are
surprised by the behavior of Si analogs. I want to
trace our work related to the Si-C-C-Si fragment,
because it will give you a good idea of how parameters are developed, and it will also highlight some
surprising conformational results. Parameter
development generally starts with a search for
available experimental data in the form of crystal
structures, electron diffraction or infrared or
microwave spectroscopic data. In the case of the Si-
C-C-Si fragment, only one Raman study [5] has been
done on the parent molecule with H substitution, and
only three crystal structures, for very complex
molecules, exist. Consequently, we considered it
best to start from scratch with ab initio calculations. Table 1 gives some results of calculations of
relative energies of various conformations of 1,4-
disilabutane by several methods. They are in general

Table 1. Relative energies (kcal/mol) of 1,4-Di-silabutane.

Angle	MM2	STO-3G*/ MM2	3-21G*// MM2	MP2// 3-21G*	MP3// 3-21G*
0°	4.84	5.97	6.25	5.60	4.79
60°	1.34	0.88	1.50		
70°*	1.20	1.00	1.47	1.05	
90°	1.84	1.70			
120°	3.35	2.70	2.45		
180°	0	0	0	0	0

*Gauche minimum in the optimized structures.

agreement in a variety of respects, particularly the
structure of the minimum energy conformer and the
energy values for various barriers. Figure 1 shows
torsional energy changes for four related molecules,
butane, propylsilane, ethylmethylsilane and 1,4-
disilabutane. Butane, propylsilane and disilabutane

Fig. 1. Torsional energies of 1,4-disilabutane ($\bigcirc$), ethylmethylsilane ($\triangle$), propylsilane ($\square$) and butane ($\diamondsuit$).

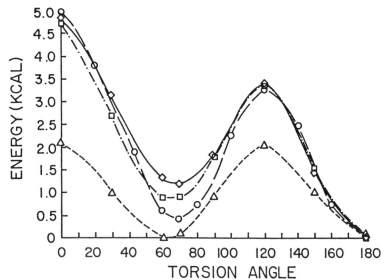

are quite similar in terms of rotational barriers in
relation to the lowest energy, anti form. Ethylmethylsilane is quite different in having much lower
barriers. The principal cause for these lower
barriers appears to be the attractive-dominant van
der Waals interactions arising from longer C-Si bonds
in ethylmethylsilane compared to butane. On the
other hand, disilabutane behaves like butane, even
though the longer bonds are present in disilabutane
as well. Population analysis on the Si-C-C-Si
systems shows the terminal Si atoms to be quite
positive, and we suggest that the attractive van der
Waals interactions are opposed by a repulsive dipolar
effect, resulting in overall behavior closer to that
of butane than that of ethylmethylsilane.

A second feature of interest in the calculations
is the quite long C-C bond between two Si centers.
This feature shows up in all the ab initio calculations and also in the x-ray structures of two of the
three complex analogs. We suggest that in MM2
calculations, a zero energy C-C bond length of 1.563
Å is appropriate when the C-C bond is flanked by two
Si atoms. That is an increase of 0.04 Å over the

value in the standard MM2 force field. The bond
lengthening leads to some very interesting results in
two larger molecules containing the Si-C-C-Si unit.

DISILYLCYCLOHEXANES

For a number of years we have been interested,
both experimentally and theoretically, in cyclic
systems containing either Si in the ring or in
substituents. As a result of correspondence with
Professor William Kitching, we were interested in
conformational effects in silyl-substituted cyclohexanes. Prof. Kitching has determined experimentally
an A value for the trimethylsilyl group (2.5
kcal/mol) [6] and has also reported NMR studies of
bis(trimethylsilyl)cyclohexanes [7]. The latter
systems contain the Si-C-C-Si unit in the 1,2-
disubstituted derivatives. They are also Si analogs
of t-butyl-substituted cyclohexanes, and the latter
have been of major interest to organic chemists for
years, since they represent the possibility to "lock"
in a particular cyclohexane conformation. However,
considerable interest has attended the possibility
that in the process of demanding equatorial place-
ment, the t-butyl group may also distort the cyclohexane ring. Indeed, force field methods applied to
1,2-di-t-butylcyclohexanes show that the trans isomer
can be expected to exist as a mixture of a diaxial
chair and a twist conformation of the cyclohexane
ring [8]. This latter result agrees with conclusions
based on a variety of spectroscopic data. A homo-
morph of trans-1,2-di-t-butylcyclohexane actually
exists in a diaxial chair conformation in the crystal
[9].
The MM2 calculated energies for the conformers
of 1.2-bis(trimethylsilyl)cyclohexane (BTMSC) are
(kcal/mol): trans (ax, ax chair) 0; trans (eq, eq
chair) 1.20; cis (eq, ax chair) 1.96; trans (twist)
2.34. The corresponding calculated energies for 1,2-
di-t-butylcyclohexane (DTBC) are: 0; 5.91, 4.46 and
0.39. Calculations were made on several twist forms,
the most stable of which is listed above, and is 2.3
kcal/mol lower in energy than the next most stable
twist form. These findings can be rationalized in
the following way. In the diequatorial trans-BTMSC,
as well as in the cis-isomer, there are not only
reduced non-bonded repulsive interactions, compared
to t-butyl groups, but also substantial attraction
between the Me$_3$Si groups. These attractive interactions arise from longer Si-C bond lengths and Me/Me
distances. The relatively long C-C bond between the
Si atoms also acts to place the silyl groups farther
away from each other. By contrast, in the die-

quatorial and equatorial/axial DTBC cases repulsive
non-bonded interactions dominate. A torque is
applied around the C(exo)-C(ring)-C(ring)-C(exo)
dihedral, resulting in a flattening of the ring. A
similar, but much less pronounced, flattening is
calculated for BTMSC.

Our calculational results are relevant to NMR
structural assignments made earlier by Eaborn and
Kitching. Eaborn et al. [10] prepared BTMSC, as well
as the 1,3- and 1,4-isomers, by Li/EtNH$_2$ reduction of
the appropriate disilylbenzene and succeeded in
separating cis- and trans-BTMSC by GC. A temperature-dependent NMR spectrum was observed for one
isomer, assigned the cis geometry, but no temperature
dependance was observed for the other isomer in the
range 25 to -80 °C. The latter, trans-isomer was
assumed to be a conformationally fixed diequatorial
derivative. By contrast, our work indicates that the
most stable form of trans-BTMSC is the axial/axial
conformer and that the equatorial/equatorial conformer is 1.2 kcal/mol higher in energy. The two
conformers can reasonably be expected to exist in
rapid equilibrium with one another at room temperature, but with an equilibrium composition of approximately 90% axial/axial. Our rationale for the
lack of temperature dependance in the NMR spectrum is
simply that throughout the temperature range, one is
always observing essentially the spectrum of the
diaxial conformer.

Kitching has argued from NMR data that he has in
hand the pure cis-3,4-bis(trimethylsilyl)cyclohexene
(BTMSCE), which can be converted to cis-BTMSC by
hydrogenation [7]. The argument hinges on the
magnitude of the gamma-effect of the SiMe$_3$ attached
to C4 on the ^{13}C chemical shift of C6. The magnitude
of the gamma-effect implies that the C4-SiMe$_3$ has
some axial character. Our calculations of the most
stable structures for both cis- and trans-BTMSCE
indicate significant axial character for the C4-SiMe$_3$
in both isomers. The trend observed in the silyl-
substituted cyclohexene isomers is the same as with
the cyclohexanes; namely, the axial/axial trans-
isomer is the lowest energy conformer, followed by
the axial/equatorial cis-isomer (3.9 kcal/mol) and
the equatorial,equatorial trans-isomer (4.1
kcal/mol). Thus, while we believe that all of the
structural assignments made by Eaborn and Kitching
are correct, we disagree with the original bases for
making the assignments.

As probably could have been expected, Me$_3$Si
groups do not have the major effects on cyclohexane
geometries that t-Bu groups do. Nevertheless, the
effects are still substantial, leading to conforma-

tional preferences that are very different from those assumed in the prior literature.

1,2-DISILACYCLOBUTANES

A second system containing the Si-C-C-Si unit attracted our attention; namely, the head-to-head silene dimers obtained after photolysis of acyltris-(trimethylsilyl)silanes [11]. An x-ray crystal structure of one of these dimers showed an essential-ly planar ring and a C-C bond length in the ring that is one of the longest C-C bonds ever observed. Brook has suggested that both of these structural features are probably due to steric effects in these highly encumbered four-membered rings. We considered this to be a system that would present a challenge to force field methods, but which would also represent a case in which molecular mechanics, if successful, could make a real contribution to understanding the origins of some unusual structural features.

We have done full geometry optimizations using Gaussian 82 and the 3-21G(*) basis set for the parent 1,2-disilacyclobutane and a number of methyl-sub-stituted derivatives. We have done MM2 optimizations of these same molecules, as well as some with t-butyl substituents. The MM2 parameters are those of Frierson, with the exception of the longer zero energy C-C bond between two Si atoms, as discussed above, and several bending parameters for Si-Si-X units in the four-membered ring. In particular, the modified four-membered ring stretch-bend parameters of Frierson appear to do a better job than the original MM2 set. Table 2 shows several calculated structural features for the parent molecule, as well as the experimental x-ray data for the highly hindered version, 1,1,2,2-tetrakis(trimethylsilyl)-3,4-di-t-butyl-3,4-bis(trimethylsiloxy)-1,2-dis-ilacyclobutane, 1 [11]. The C-C bond is indeed a long one in the calculated structures, but not as long as in 1. The structures for these simple derivatives are all puckered, not flat. Increasing the number of Me substituents decreases the ring dihedral angle, but it actually increases the ring inversion barrier. These effects appear to be a result of a complex set of both attractive and repulsive non-bonded interactions.

Clearly, however, the Me substituents do not mimic what is going on in 1, which is much more sterically encumbered. Consequently, we have done MM2 calculations on 1,2-disilacyclobutanes with t-butyl substituents, and some of those results are

Table 2. Structural features of 1,2-disilacyclobutane and $\underline{1}$ (lengths in Å, angles in degrees, energies in kcal/mol).

Feature	MM2	3-21G(*)	Expt. ($\underline{1}$)
Si-Si	2.314	2.327	2.37
Si-C	1.904	1.915	2.00
C-C	1.595	1.593	1.66
Si-Si-C	77.3	77.4	
Si-C-C	98.3	98.8	
H-Si-H (av)	113.9	107.7	
Si-C-C-Si	-26.9	-25.1	0
C-C-Si-Si	22.2	20.7	0
C-Si-Si-C	-18.7	-17.4	0
barrier	0.90	0.98	

shown in Table 3. Now we do in fact reproduce the effects observed in $\underline{1}$. Increasing $\underline{t}$-butyl substitu-

Table 3. Bond lengths, ring dihedrals and inversion barriers for 1,2-disilacyclobutane derivatives.

Molecule	C-Si (av)	C-C	Si-C-C-Si	Barrier
c-1,2-diMe	1.900	1.597	27.30	1.15
t-1,2-diMe	1.898	1.597	27.12	0.68
octaMe	1.911	1.623	22.40	2.63
c-1,2-ditBu	1.900	1.600	30.14	3.09
t-1,2-ditBu	1.897	1.598	27.67	0.66
c-1,3-ditBu	1.904	1.599	26.34	1.28
t-1,3-ditBu	1.903	1.603	22.44	0.13
c-1,2,3,4-tetratBu	1.914	1.636	24.60	7.62
t-1,2,3,4-tetratBu	1.910	1.620	3.49	0.08

tion increases the ring C-C bond length. With four $\underline{t}$-butyl groups the values approach the 1.66 Å value observed in $\underline{1}$. Also, as $\underline{t}$-butyl groups are added, the ring dihedral angles vary in an understandable way, assuming that steric repulsions between $\underline{t}$-butyl groups are responsible for a torque applied to the ring angle. When $\underline{t}$-butyl groups are cis to one another on the ring, they can occupy pseudo-equatorial positions on a puckered ring and minimize interactions. However, a trans substitution pattern does not provide that possibility, and the molecule with a t-butyl group on each ring atom, all trans, is calculated to have an essentially planar ring.

As a result of our successes thus far, we are encouraged to believe that molecular mechanics can be a very important contributor to studies of conformational effects in organosilicon chemistry. Stereoelectronic effects are at least as important in Si chemistry as they are in carbon chemistry. Molecular mechanics calculations can define expected geometries in silanes which might be proposed as probes for conformational effects and hence will have very wide use in coming years. We are particularly interested in systems which have leaving groups on Si or on carbons α or β to Si, and we are currently working on parameter sets to treat Cl in those positions [3].

REFERENCES

1. Burkert, U.; Allinger, N.L. Molecular Mechanics, American Chemical Society, Washington, D.C., 1982.

2. Frierson, M., Ph.D. Dissertation, University of Georgia, 1984.

3. Cho, S. and Cartledge, F.K., unpublished studies.

4. Pietro, W.J.; Francl, M.M.; Hehre, W.J.; DeFrees, D.J.; Pople, J.A.; Binkley, J.S. J. Am. Chem. Soc. 1982, 104, 5039-50.

5. Petelenz, B.U.; Shurvell, H.F.; Phibbs, M.K. J. Mol. Struct. 1980, 64, 183-92.

6. Wickham, G.; Kitching, W. Organometallics 1983, 2, 541-7.

7. Kitching, W.; Olszowy, A.H.; Drew, M.G.; Adcock, W. J. Org. Chem. 1982, 47, 5153-6.

8. van de Graaf, B.; Baas, J.M.; Wepster, B.M.; Recl. Trav. chim. Pays-Bas 1978, 97, 268-73; van de Graaf, B.; Baas, J.M.; Widya, H.A. ibid., 1981, 100, 59-65.

9. van Koningsveld, H. Acta Cryst. 1973, B29, 1214.

10. Eaborn, C.; Jackson, R.A.; Pearce, R. J. Chem. Soc., Perkin Trans. I 1975, 475-7.

11. Brook, A.G.; Nyburg, S.C.; Reynolds, W.F.; Poon, Y.C.; Chang, Y-M.; Lee, J-S.; Picard, J-P. J. Am. Chem. Soc. 1979, 101, 6750-2.

Chapter 44

The acidity of silica surfaces

Dr. Michael L. Hair – Manager, Synthesis & Exploratory Research, Xerox Research Centre of Canada, 2660 Speakman Drive, Mississauga, Ontario, Canada L5K 2L1.

ABSTRACT

The acidity of a silica surface can be obtained either by titrating the surface from an aqueous suspension or by ir investigation of frequency shifts of surface silanol groups measured in gas-solid systems. The latter case measures only the acidity of the surface hydroxyl groups but both methods give similar results.Incorporation of alumina and boria into the surface introduces a new type of acid site - the Lewis acid site - which profoundly affects the acidity at the molecular level.

THE ACIDITY OF SILICA SURFACES

It is often convenient for chemists to classify compounds as either acidic or basic and to interpret interactions between them in these general terms. It is universally accepted that the surface of a silica material is acidic and thus it is able to interact with bases. Primary evidence for the acidic nature of silica surfaces came originally from titration experiments.. The silica particle is considered to be a solid acid which has hydroxyl groups on its surface. In solution these hydroxyl (or silanol) groups ionize to give a negative particle and a proton. The proton can undergo ion exchange and thus the particle surface can be titrated and the intrinsic acidity constant (pK_a) of the surface can be determined from the titration. Typical values are reported for instance by Schindler and Kamber[1] who determined a value of 6.8 ± 0.2 for the surface acid association constant. Similar values can be inferred from electro kinetic measurements, though it should be noted that these measurements refer to the slippage plane in the fluid and are thus indirect[2].

Such measurements are obviously of interest but they shed little light on the molecular nature of the silica surface or the processes which are likely to be involved in carrying out a chemical reaction with the surface. In this respect, quantitative infrared spectroscopy has proven to be an invaluable tool in identifying surface species and in delineating the pathways and kinetics of surface processes [3-6]. More recently nmr has proven to be very useful in identifying co-ordination around the surface silicon atoms and confirming earlier ir interpretations[7]. Such studies have clearly identified the presence of hydroxyl groups on the silica surface and identified their chemical reactivity.

When a silica has been prepared from aqueous solution (or allowed to stand in a humid environment for a long period of time) the surface of the silica is usually completely hydroxylated and contains a layer of adsorbed water. The adsorbed water may be removed by evacuation and heating to about 120°C. At this point, the silica surface contains about 4.6 OH groups per nm^2 which corresponds to 1 OH per surface silicon atom. On raising the temperature, some of these hydroxyl groups condense with the elimination of water and the formation of siloxane groups on the silica surface (Fig.1).

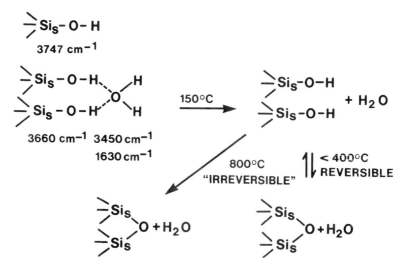

Fig. 1. Schematic showing hydration — rehydration of silica surface. Redrawn from ref. (3).

Up to a temperature of about 450°C this dehydration appears to be reversible and re-exposure to moisture recreates the original surface. Above 450°C, however, this reaction becomes less and less reversible and, on further heating, the water which is lost during the dehydration cannot be readsorbed. The cause of the irreversibility is not known, but is presumed to be due to a restructuring of the surface at the elevated temperatures. Studies of the infrared spectra show that *two* types of hydroxyl group occur on the silica surface: a so-called freely vibrating silanol group which occurs at 3747cm^{-1} and hydroxyl groups which are hydrogen bonded to each other and which exhibit an absorption at 3650 cm^{-1}. It is the hydrogen bonded hydroxyl groups which are eliminated during the dehydration and, above 800°C only the freely vibrating SiOH group is left on the surface. At this point, the silica surface will not readsorb water and is often referred to as being hydrophobic. A spectrum of the SiO$_2$ surface heated to 500°C is shown in Fig. 2.

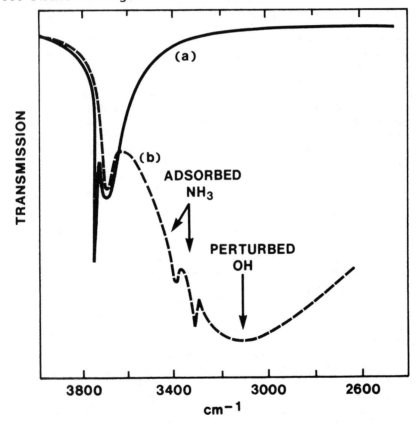

Fig. 2. Infra red spectrum of Cabosil surface (a) heated in vacuum to about 500°C and (b) after adsorbtion of NH$_3$. Redrawn from ref. (15).

This spectrum clearly shows the free hydroxyl groups at 3747 cm^{-1} and the hydrogen bonded silanol groups at 3650 cm^{-1}. It is well established that the hydroxyl groups at 3747 cm^{-1} are the site for the specific adsorption of molecules containing lone-pair electrons. This is exemplified by the adsorption of ammonia onto the silica surface is (Fig. 2). Interaction of the ammonia molecule with the surface silanol is evidenced by the disappearance of the band at 3747cm^{-1} and physically adsorbed ammonia is revealed by the two absorptions at 3400 and 3320 cm^{-1} attributed to the asymmetric and symmetric stretching vibrations of the physically adsorbed ammonia species. Evacuation causes desorption of the ammonia and the spectrum returns to its original form. It should be noted that the band at 3650 cm^{-1} is not affected by this adsorption process. For the purposes of this discussion, however, the important feature is to note that the narrow absorption at 3747 cm^{-1} is replaced by a broad band centering about 3100 cm^{-1} which is attributed to the formation of the Si$_s$OH----NH$_3$ hydrogen bond. The OH stretching frequency has been shifted by about 670 cm^{-1}. This type of interaction has been well studied and surface chemists have attempted to relate the spectroscopic data with heats of adsorption, etc. Specifically, Knözinger[8] has reviewed this hydrogen bonding interaction and lists 95 adsorbed molecules for which spectral data is available. Table 1 lists the frequency shifts observed for the adsorption of about a dozen of these molecules. The most significant feature of this table is that it can be immediately noted that there is a correlation between the frequency shift reported for the hydrogen bonding interaction with the silica surface and the basicity of the donor molecule. A method is thus suggested for the determining of the acid dissociation constant of the freely vibrating surface silanol group at low surface coverage in the absence of solvent.

The nature of the hydrogen bonding interaction between organic solvents and dissolved molecules which contain hydroxyl groups (i.e., alcohols, phenols) has been the subject of many studies and results obtained prior to 1959 have been well summarized by Pimentel and McLellan[9]. It is particularly noted that when a hydroxyl containing compound is dissolved in solution that the band due to the OH vibration is moved to lower frequencies due to the hydrogen bonding interaction. Bellamy and Williams[10] have shown that the perturbation of the OH band increases systematically with the increasing basicity of the solvent. Moreover, they have shown that the frequency shifts of an unsubstituted phenol in a series of solvents can be plotted against the corresponding values of methanol in the same solvents to give a straight line. Bellamy, Hallam and Williams[11] have extended the solution data to show that $\Delta\gamma_{OH}$ is a direct function of the acidity of the hydroxyl group and the basicity of the solvent, and this relationship has been shown to be correct over a range of pK$_a$ values from −0.3 (trifluoroacedic acid) to + 19 (dibutynol)[13]. In order to measure the acidity of the surface silanol groups Hair and Hertl[12] carried out an analgous investigation with a series of

Table 1. Hydroxylband frequency shifts (in cm^{-1}) in solution and on surface groups.

DONOR	SILICA	METHANOL	PHENOL	TRICHLOROPHENOL
Hexane	30	0	---	---
CCl4	45	0	---	---
CS2	60	6	---	---
Benzene	120	32	---	---
Xylene	155	40	62	40
Acetaldehyde	280	---	---	---
Acetone	395	112	131,236	140,290
Trimethylmethoxysilane	477	134	---	---
Diethyl ether	460	142	280	345
Ammonia	675	---	---	---
Pyridine	765	280	460	600,1010
Diethylamine	ca 900	220,400	ca 650	ca 830,1030
Triethylamine	975	251,446	ca 700	910-1030

hydroxylated surfaces and donor molecules. CCl_4 solutions of methanol, phenol and trichlorophenol (pK$_a$'s = 15.1, 9.9, and 5.5 respectively) were used as calibration standards. This data is also shown in Table 1. A plot of the observed alcohol hydroxyl band frequency shifts in the presence of the various bases is plotted against the frequency shift observed when the same donors are adsorbed on silanol groups (Fig. 3).

The slopes of these lines are then plotted against the known pK$_a$ values (Fig. 4) and by interpolation, the pK$_a$ of the Si$_s$OH surface grouping is determined to be 7.1.

This is in surprisingly good agreement with the value obtained by the original titration experiments.

The infrared method has also allowed determination of the acidity of surface hydroxyl groups attached to magnesium (15.5), boron (8.8), phosphorous (−0.4) and silica-alumina (7.1). The values agree well with those which might be anticipated based upon electronegativity principles. The result for silica-alumina, however, is somewhat surprising. Silica-alumina compounds are widely used in the catalyst industry for their high surface acidity and thus the observation that the SiOH group on silica-alumina surface has an acidity approximating that on pure silica surface is an apparent contradiction. This has been explained by Rouxhet' and Semples[13] following a very careful analysis of the spectrum of the perturbed OH. Before adsorption the spectra of the SiO$_2$-Al$_2$O$_3$ samples examined by Rouxhet and Semples exhibited a sharp

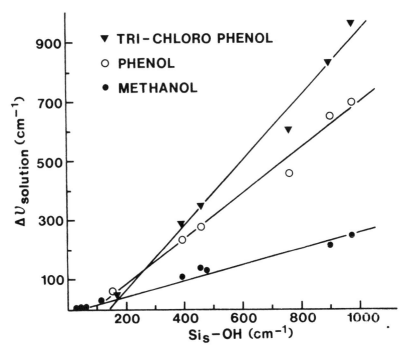

Fig. 3 Hydroxyl frequency shifts measured for trichlorophenol-, phenol-, methanol-base interactions in CCl_4 plotted against data for same bases physically adsorbed on SiO_2. Redrawn from ref (12).

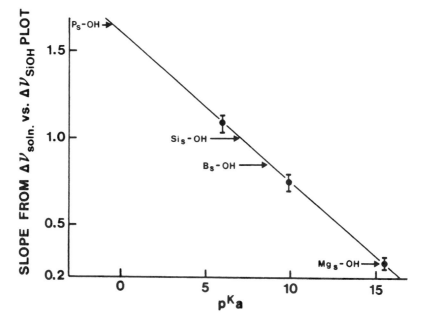

Fig. 4. Slopes from Fig. 3. plotted against known pK_a values. Redrawn from ref. (12).

band at about 3750 cm^{-1} characteristic of free surface hydroxyl groups. On adsorbing acetonitrile, however, they observed that the perturbed band had considerable asymmetry. This could be resolved into a major component which exactly followed the perturbations of the "pure" surface silanol group and second component which could be attributed to a group of much higher acidity. Indeed, measured frequency shifts suggested a surface pK_a of between –4 and –8 for this second silanol grouping. Thus, with adequate care, the method can be used to identify highly acidic hydroxyl groups which are present in small concentrations on the surface.

The nature of the acidic groups on silica-alumina surfaces has been the subject of study for several decades and the results of these studies are very relevant when considering molecular reactions at silica surfaces. The presence of alumina in the silica structure has two main effects.

(i)   The introduction of an Al^{3+} ions into the tetrahedral environment of the silica surface leads to an unsaturation or Lewis acid site.

(ii) This unsaturation can have an inductive effect on a neighbouring silanol group rendering it much more acidic. The effects are shown diagrammatically in Fig. 5.

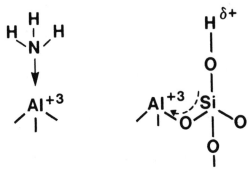

Fig. 5. Proposed structure of surface Lewis acid site and its inductive effect on adjacent silanol.

The first effect gives rise to a very different type of adsorbtion site. This is indeed a strongly acidic site and it is revealed, for instance during adsorption of NH_3, by the formation of a strongly covalently bonded species. Thus in the adsorbtion of NH_3, on such a surface the infrared spectrum shows the presence of four absorbtions. Two of these are due to the physically adsorbed NH_3 and two additional vibrations appear due to formation of the new surface complex. Thermal studies show that this surface complex is stable up to 250°C in keeping with the much stronger bond formation.

The second effect is of great importance when considering chemical reactions with surface silanol groups. The presence of Lewis acid impurities affects the proton acidity of such groups and this in turn affects the kinetics of the chemical reaction. This has been convincingly demonstrated by Hertl and Hair in a study of the reaction of various silanes with a silica surface in which boron has been purposely introduced[14]. Not only do the reactions with the surface silanol groups occur at much lower temperatures, but the order of the reaction is also often affected.

In conclusion it can be said that our present state of knowledge of the acid nature of silica containing surface is in general very good: There is agreement between titration, electrokinetic and spectroscopic experiments. However, as in any surface study, reactions at surfaces must invoke a molecular approach and emphasis must focus on small numbers of active centres. Interpretations based on bulk approaches are likely to be misleading.

REFERENCES

1)   Schindler P. and Kamber H. R., *Helv. Chim. Acta,* **51,** 1781 (1968).

2)   Lyklema J., *Faraday Soc. Disc.,* **52,** 318 (1971).

3)   Hair, M. L., *Infrared Spectroscopy in Surface Chemistry.* Dekker, New York, 1967.

4)   Little, L. H., *Infrared Spectra of Adsorbed Species.* Academic Press, New York, 1966.

5)   Hair, M. L., J. Colloid and Interface Science, **60,** 154 (1977).

6)   Hair, M. L. and Hertl, W., J. Phys. Chem., **73,** 2372 (1969).

7)   Maciel, G. E., Science, **226,** 282-288, (1984).

8)   Knözinger, H., *Hydrogen Bonds in Systems of Adsorbed Molecules,* Ch. 27 of *The Hydrogen Bond,* Volume 3, Edited by P. Schuster, G. Zundel and C. Sandorfy, North-Holland, Amsterdam (1976) pp. 1263-1364.

9)   Pimentel, G. C. and McClellan, A. L., *The Hydrogen Bond,* Freeman and Co. San Francisco, Calif., 1959.

10)   Bellamy, L. J. and Williams, R. L., *Proc. Roy. Soc.,* A254, 119, (1960).

11)   Bellamy, L. J., Hallam, H. E. and Williams, R. L., *Trans. Faraday Soc.,* 1958, **54,** 1120.

12)　Hair, M. L. and Hertl, W., J. Phys. Chem., **74,** 91 (1970).

13)　Rouxhet, P. G. and Semples, R. E., J. Chem. Soc. Faraday Trans., 1 **70,** 2021 (1974).

14)　Hair, M. L. and Hertl, W., J. Phys. Chem., **77,** 1965 (1973).

15)　Hair, M. L., J. Non-Crystalline Solids., **19,** 299 (1975).

Chapter 45

Polysilane photochemistry
and laser desorption mass spectrometry

Thomas F. Magnera, V. Balaji and Josef Michl – Center for Structure and Reactivity, Department of
Chemistry, The University of Texas at Austin, Austin, TX 2

The recent synthesis of soluble polysilanes of high molecular
weight, poly-(RR'Si), has elicited considerable scientific and
technological interest [1]. Laser ablation of these polymers is
of potential utility for optical recording and for self-
developing photoresists [2-4]. A knowledge of its mechanism would
be useful if progress is to be made in minimizing the amount of
debris that settles on the unexposed areas and interferes with
clean image production.

The solution photochemistry of polysilanes is quite complex and is
poorly understood [5]. In view of the possible involvement of
thermal processes in photoablation, its mechanism might not only
be at least equally complex but could also be quite different. To
our knowledge, the nature of the gaseous products from the
irradiation of neat polysilanes has only been investigated once
before and reported in the proceedings of a conference [3]. The
conclusion derived from electron-impact (EI) mass spectra of the
volatilized material was that the main constituent in the absence
of air was the monomeric silylene RR'Si:, and that smaller amounts
of larger chain fragments up to pentamers were also present. It
was proposed that the main chain-degrading process was (1):

$$-SiRR'-SiRR'-SiRR'\bullet \quad \rightarrow \quad -SiRR'-SiRR'\bullet + RR'Si: \quad (1)$$

It seemed to us that the published spectra of the copolymer of

dimethylsilylene and methylcyclohexylsilylene [3] did not warrant
the conclusion that RR'Si: represents the bulk of the ejected
material since ions of larger mass appeared more abundant than
RR'Si$^{\bullet+}$ and smaller ions. Moreover, some of the latter could be
due to ion fragmentation. Although reaction (1) appears
plausible, it has never been documented, and it is questionable
whether it could compete successfully with others. At room
temperature, the Me$_3$SiSiMe$_2\bullet$ radicals disproportionate and
recombine about evenly, and do not fragment according to (1) [6],
but of course, in the irradiated neat polymer, the temperature is
surely higher, possibly promoting (1). However, if straightforward chain degradation of the polymer occurs, it seemed to us
that reaction (2) should be at least competitive with (1),
considering the strength of the disilene π bond [7]:

$$-SiRR'-SiRR'-SiRR'\bullet \quad \rightarrow \quad -SiRR'\bullet + RR'Si=SiRR' \quad (2)$$

This type of process has not been documented in the literature
either, but is of course well known in carbon chemistry. The
closest known analog probably is the fragmentation of
Me$_3$SiMe$_2$SiCH$_2\bullet$ to Me$_3$Si$\bullet$ and Me$_2$Si=CH$_2$, believed to occur
at about 300°C [8].

Finally, the observation [3] of ions ascribable to the oligomers
of a silanone, (RR'SiO)$_n$, in the EI mass spectrum of material
photoablated in air, does not provide a definitive proof of the
presence of RR'Si: either, since other silicon-based reactive
intermediates could give similar products.

Because of these doubts, we have reexamined the issue. Our EI
mass spectra of the material photoablated with 308 nm light are
qualitatively similar to those published for related polysilanes
[3], but our conclusions are very different.

Quite apart from the specifics of polysilane chemistry, the
mechanistic aspects of laser photoablation of polymers have been
of considerable interest since its original discovery [9], and the
relative merits of the photochemical and photothermal mechanisms
have been argued extensively [10]. Our results agree with those
of ref. 4 and suggest that the relative importance of the
photothermal component in polysilanes has been underestimated [3].

EXPERIMENTAL PART

Poly(dimethylsilane) [poly-(Me$_2$Si)], poly(methylcyclohexyl-
silane) [poly-(MeChSi)], poly(methylphenylsilane) [poly-
(MePhSi)], poly(methyl-n-propylsilane) [poly-(MePrSi)], poly(di-
n-butylsilane) [poly-(poly-Bu$_2$Si)], poly(di-n-pentylsilane) [poly-
(Pn$_2$Si)], poly(di-n-hexylsilane) [poly-(Hx$_2$Si)], and its
isotopomers fully deuterated on α carbons or β carbons and fully
^{13}C-labelled on α carbons were purchased or synthesized by
published procedures [11] from the appropriate monomers. The

measurements were done on a home-built triple-quadrupole mass spectrometer [12] whose transmission was calibrated using perfluoro-tri-n-butylamine. A 308 nm line of a Lumonics HyperEX-400 excimer laser was fired at 10 Hz and focused to a 1×2 mm spot on the polysilane target, which was slowly tracked. The fluence per pulse was varied from 90 to 250 mJ/cm^2, without affecting relative ion intensities in the mass spectra. Electron-impact ionization (50 V) of the ablated vapor at two different locations and multiphoton ionization (MPI) about 1 mm above the sample surface were used. In some experiments, the ablated material was collected in excess argon on a cold plate for examination of its UV-visible absorption, or in a cold trap for GC-MS analysis of volatile components.

RESULTS AND DISCUSSION

The laser desorption EI mass spectra (Figure 1) are similar in part to the spectra of saturated acyclic low-molecular weight silanes and oligosilanes, known [13-18] to suffer extensive fragmentation and rearrangement after ionization. However, they also contain additional more highly unsaturated ions and in this sense resemble the spectra of silenes [19]. This difference could also be caused by cyclic structures. However, the spectra of the isocyclic permethylated cyclosilanes are distinctly different in that they have intense parent ion peaks [20].

The salient features of our EI spectra are:

(i) *Closed-shell* (even-electron, odd-mass) *silicon-containing ions* dominate, particularly in the high-mass region. The alkylated polysilanes poly-(R_2Si) yield four ion series: $Si_nH_kR_{2n-k+1}^+$, $Si_nH_{k-2}R_{2n-k+1}^+$ (k ≤ 2n), $SiC_mH_{2m+3}^+$, and $SiC_mH_{2m+1}^+$. Intensity drops from n=1 to n=2 and strongly thereafter, but for the smaller R's, ions with n up to 5 are detectable. Ions with m=1 are particularly intense, except for $HMeChSi^+$ from poly-(MeChSi). The last two series are absent in the case of poly-(Me_2Si) and poly-(MePhSi), and weak in poly-(MeChSi). Isotopic labelling on poly-(Hx_2Si) demonstrated extensive hydrogen scrambling and shifts of Si atoms to carbons to which they were not originally bonded.

(ii) *Open shell* (odd-electron, even mass) *silicon-containing ions* are present. $RR'Si^{\bullet+}$ is always less intense than $RR'HSi^+$, except in the case of poly-(MeChSi). The loss of ethylene from propyl and even more so, from larger alkyls is important. Isotopic labelling on poly-(Hx_2Si) showed that the α and β carbons are lost in the process. The $(RR'Si)_2^{\bullet+}$ ion is also observed when the alkyls are small. The $(RR'Si)_3^{\bullet+}$ ion is weak.

The conclusions from observations (i) and (ii) are:

The relative intensities of ions whose mass exceeds that of $RR'Si^{\bullet+}$ are high, particularly when R and R' are small, and $RR'Si$:

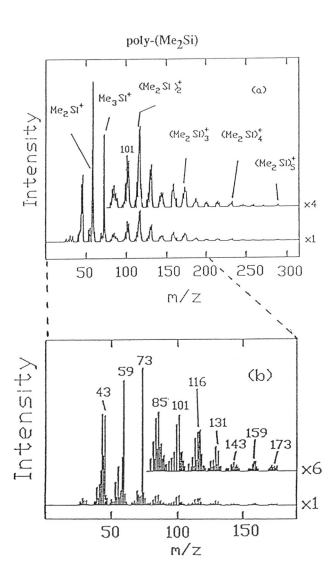

Fig. 1. Laser desorption impact mass spectrum of poly-(Me$_2$Si) under low (a) and high (b) resolution.

cannot possibly represent the bulk of the ablated material. Indeed, it may not be present at all, and reaction (1) need not be postulated.

The relatively low intensity of the $(RR'Si)_2^{\bullet +}$ peak argues against the presence of much disilene $(RR'Si)_2$ but does not disprove it. Reaction (2) need not be postulated.

The very low intensity of the higher $(RR'Si)_n^{\bullet +}$ peaks argues strongly against the presence of significant amounts of the isocyclic cyclosilanes $(RR'Si)_n$. However, cyclic carbosilane structures may be present and would account for the results of ^{13}C isotopic labelling experiment.

The high abundance of ions such as H_2MeSi^+ and HR_2Si^+ is most easily understood if Si-H bonds are already present in the neutral precursor. Isotopic labelling on poly-(Hx_2Si) shows that these H atoms originate not only in the α and β methylene groups, but predominantly in the more distant ones.

The mass spectra do not permit an actual structural assignment of the silicon-containing ablated material. The fundamental difficulty is the prevalent uncertainty as to which fragmentation processes occur already during the ablation and which ones results from the subsequent ionization (e.g., the loss of the α and β carbons from an alkyl, presumably in the form of ethylene).

(iii) *Ions derived from hydrocarbons* represent a significant fraction of the total intensity for all of the polysilanes except poly-(Me_2Si). The hydrocarbons are terminal alkenes from the alkylated polysilanes, cyclohexene from poly-(MeChSi), and benzene and toluene from poly-(MePhSi). These ions are readily detected by their characteristic patterns in the mass spectra and comparison with authentic samples. We cannot exclude the possibility that poly-(Me_2Si) forms some methane or ethane. On three of the polymers, we have demonstrated unequivocally that the hydrocarbons are formed already in the photoablation process by trapping them and identifying them in GC MS. Isotopic labelling on poly-$(HxSi_2)$ demonstrated that 1-hexene is formed by scission of the Si-C_α bond and removal of one of the β hydrogens.

This previously unnoticed hydrocarbon formation provides a key to further progress. The observation represents the first time that a structure of a polysilane photoablation product has been proven unambiguously and contrasts with the only vague notions derived above under (i) and (ii). Below, we shall use it to develop a tentative mechanistic scheme for the photoablation process.

The laser desorption MPI mass spectra of poly-(Me_2Si) (Figure 2) are strikingly similar to its laser desorption EI spectra and do not change much as the wavelength of the ionizing radiation is varied. The ionization efficiency curve has a large peak near

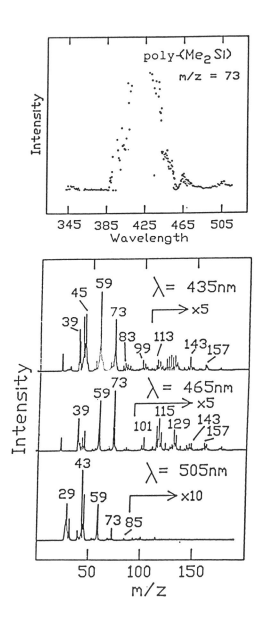

Fig. 2. Laser desorption multiphoton ionization mass spectrum of poly-(Me_2Si) at three wavelengths (bottom) and the wavelength dependence of the abundance of the Me_3Si^+ peak (top).

425 nm and small peaks near 465 and 505 nm.

The location of the ionization efficiency peaks is right for the resonant ionization of Me_2Si: but not for $(Me_2Si)_2$. In glassy solution, the former has an absorption maximum at 450 nm [21,22] and the latter at 345 nm [22,23]. However, the 460 nm peak in Figure 2 is very small, although a one-photon resonance should be highly efficient. Even more important, at all three wavelengths, ions larger than $Me_2Si^{\bullet+}$ are formed and this $Me_2Si^{\bullet+}$ ion is not even observed, arguing convincingly against the resonant ionization of Me_2Si: Except at 505 nm, ions even larger than $(Me_2Si)_2^{\bullet+}$ are also observed. Finally, an experiment in which the material ablated from poly-(Me_2Si) was trapped on a cold window in excess argon revealed no absorption maxima in this region, only end absorption below 300 nm. All this agrees with the conclusion reached from the EI spectra: there are only negligible amounts of Me_2Si: and $(Me_2Si)_2$ in the material ablated from poly-(Me_2Si).

We feel that a likely origin of the structure observed in the MPI efficiency curve are two-photon resonances with the trisilane, tetrasilane and possibly pentasilane chromophores which absorb near 215, 235, and 250 nm, respectively [24]. Other possibilities cannot be excluded. For instance, the 505 nm peak could be easily due to a two-photon resonance with the silene chromophore, whose absorption band lies near 260 nm [22,25], and this would be better compatible with the absence of high-mass ions in the mass spectrum obtained at this wavelength (Figure 2).

Taken together, the EI and MPI spectra not only exclude RR'Si: and $(RR'Si)_2$ as significant constituents of the ablated material, but also suggest quite strongly that most if not all photochemical and/or pyrolytic damage has healed in the ablated fragments, so that they should be thought of preferably as ordinary fairly stable linear, branched or cyclic organosilanes, and except for a possible presence of a C=Si bond, not as reactive intermediates. If we think of the ablation region as analogous to a miniature pyrolytic oven, we appear to be interrogating the effluent far downstream and not the fleeting reactive intermediates in the hot zone. We have made several attempts to vary the time between the ablating laser pulse and the time of MPI observation but have seen no significant differences in the spectra down to 1 μs delay.

Photochemical versus photothermal ablation. The usual argument in favor of a purely photochemical mechanism, with a negligible contribution from photothermal effects and ground-state reactions, is based on the observation of clean and steep crater walls around the ablated region and on the high activation energies for polymer degradation by conventional pyrolysis. It is argued that thermal processes lead to melting and ill-defined crater walls, and that the temperatures and time-scales associated with the ablation process are insufficient to permit significant thermal degradation of the polymer chain [10a].

Since C-Si bond scission in a polysilane has never been observed during irradiation in room-temperature solution, it seems clear that it is due to a thermal rather than a photochemical process. Our results therefore provide strong independent support for the claim [4] that the relative importance of the photothermal component in the photoablation process has been underestimated.

This need not be in conflict with the arguments based on the shape of the crater walls and on activation energies for chain degradation. In the sharply defined irradiation region, primary photochemical damage will occur and may well open thermal reactive channels with much lower activation energies than those available elsewhere in the intact polymer. The irradiated region may thus suddenly become thermally very labile and almost all of the bond breaking that degrades the polymer chains may occur in the ground electronic state.

The mechanism of polysilane photoablation. Since we have unambiguously identified only one of the ablation products, we cannot propose a mechanism with any degree of confidence, and shall make only a tentative proposal.

To our knowledge, only one class of alkylated silicon compounds cleaves the Si-C bond readily with the removal of a β hydrogen and formation of a terminal alkene. These are the alkylsilylenes [8,26] and the activation energy is about 30 kcal/mol [27,28]. The reaction may proceed directly or by silylene insertion into the β C-H bond (3):

$$-\overset{..}{\underset{..}{Si}}-CH_2-CH_2- \quad \rightarrow \quad -\overset{\overline{\quad\quad\quad}}{SiH}-CH_2-\overset{}{CH}- \quad \rightarrow \quad -\overset{..}{\underset{..}{Si}}H \; + \; CH_2=CH- \quad (3)$$

There are at least two well documented ways in which the silylene precursor could arise in the irradiated polymer, in addition to the hypothetical but possible processes (1) and (2) followed by a disilene-to-silylsilylene rearrangement. One is the homolytic photodissociation of an Si-Si bond followed by radical disproportionation [6,22,29] and subsequent silylsilene to silylmethylsilylene rearrangement [30], which leads to the structure $-SiR_2-CHR''-Si-R$. There is some evidence that the first two of these steps play a role in solution photochemistry of polysilanes [5]. The second possibility is a purely thermal process of 1,1-elimination on silicon, which dominates the pyrolysis of disilanes [31] and trisilanes [32]. This leads to the structure $-SiR_2-Si-R$.

Once we accept the hypothesis that the terminal alkene is formed by elimination from an alkylsilylene, rather than by some other facile but as yet undiscovered reaction (Occam's razor), the general features of the remainder of the mechanistic picture follow automatically. Silylene insertion into a C-H bond has an activation energy of about 20 kcal/mol in the absence of strain

[33]. This is substantially less than the 30 kcal/mol needed for (3), presumably because of steric strain in the β-insertion transition state, and we are therefore forced to conclude that quite a few ring-forming and possibly some cross-linking C-H insertions occur, both by silylenes that have not yet eliminated an alkene and those that have. Five-membered and perhaps also six-membered rings should be formed preferentially.

The formation of silacyclobutane rings could also be favorable relative to the alkene elimination (ring strain in silacyclobutane is 13 kcal/mol [33], in silacyclopropane 20 [33] or 23 [34] kcal/mol). Even the insertion that leads to a disilacyclopropane may well be competitive.

The "healing" of the damage induced by irradiation should therefore produce heterocycles containing one or more carbon and one or more silicon atoms, one or more of which carry a hydrogen. There is good precedent for this general type of ring formation in pyrolytic processes involving silylsilylene intermediates [22,32,35].

A second photochemical or thermal chain rupture, possibly nearly concurrent with the first one, is required to produce a low-molecular weight fragment, unless the initial scission happened near a chain end. The second damage site could again repair in an analogous fashion, producing a bicyclic heterocycle.

Some of the smaller rings thus formed may be quite unstable at the ablation temperatures [22]. The thermal transformations of disilacyclopropanes may lead to 1,3-disilacyclobutanes and 1,2,4-trisilacyclopentanes, silacyclobutanes will fragment to alkenes and silenes (activation energy about 60 kcal/mol [36]), etc. Moreover, silylenes can isomerize reversibly to silenes (activation energy about 40 kcal/mol [37]) and silylsilylenes to disilenes, so that a variety of rearrangements are possible in principle [22]. It would be useless at the present time to speculate which of the various paths will be most important and we only note that our proposal that section (3) plays a key role and that the ablated material therefore should consist primarily of an alkene and of alkylated polysilacycloalkanes and polysilabicyclo-alkanes, possibly with some unsaturation, is compatible with the observed mass spectra. Ring-forming silylene insertions provide a natural explanation for the migration of hydrogen atoms from α, β, and more distant positions of the alkyl chain onto silicon, and for the transfer of the Si atoms to carbon atoms other than α.

CONCLUSIONS

An examination of the laser desorption mass spectra of seven representative polysilane homopolymers revealed the presence of a significant photothermal component in the polymer degradation process, involving a Si-C bond scission, suggesting that this behavior is general.   A tentative mechanistic scheme consistent with the available data is proposed.   The spectra are characteristic for each polymer and could be easily used for their analysis.

In constrast with previous *in vacuo* studies [3], we find under our conditions that the monomeric silylenes do not constitute the bulk of the material volatilized.

Acknowledgment.   This project was supported by the U.S. National Science Foundation.   We are grateful to Dr. K. Klingensmith for performing the argon matrix isolation experiment.

REFERENCES

1. West, R. J. Organomet. Chem. 1986, 300, 329.
2. Hofer, D. C.; Jain, K.; Miller, R. D. IBM Tech. Discl. Bull. 1984, 26, 5683.
3. Zeigler, J. M.; Harrah, L. A.; Johnson, A. W. Proc. of SPIE, Advances in Resist Technology and Processing II, 1985, 539, 166.
4. Marinero, E. E.; Miller, R. D. Appl. Phys. Lett., in press (1987).
5. Trefonas, III, P.; West, R.; Miller, R. D. J. Am. Chem. Soc. 1985, 107, 2737; Michl, J.; Downing, J. W.; Karatsu, T.; Klingensmith, K. A.; Wallraff, G. M.; and Miller, R. D., in "Inorganic and Organometallic Polymers", ACS Symposium Series, Zeldin, M.; Wynne, K. eds., American Chemical Society, Washington, D.C., in press.
6. Hawari, J. A.; Griller, D.; Weber, W. P.; Gaspar, P. P., preprint (1987).
7. Michalczyk, M. J.; West, R.; Michl, J. Organometallics 1985, 4, 826.
8. Gusel'nikov, L. E.; Polyakov, Yu. P.; Volnina, E. A.; Nametkin, N. S. J. Organomet. Chem. 1985, 292, 189.
9. Srinivasan, R.; Mayne-Bauton, V. Appl. Phys. Lett. 1982, 41, 576; Srinivasan, R.; Leigh, W. J. J. Am. Chem. Soc. 1982, 104, 6784.
10. (a) Srinivasan, R. Science, 1986, 234, 559
    (b) Srinivasan, R.; Braren, B.; Dreyfus, R. W. J. Appl. Phys. 1987, 61, 372; Gorodetsky, G.; Kazyaka, T.; Melcher, R. L.; Srinivasan, R. Appl. Phys. Lett. 1985 46, 828; Brannan, J. H.; Lankard, J. R.; Baise, A. I.; Burns, F.; Kaufman, J. Appl. Phys. 1985, 58, 2036; Larciprete, R.; Stuke, M. Appl. Phys. B 1987, 42, 181.
11. Miller, R. D.; Hofer, D.; McKean, D. R.; Willson, C. G.; West, R.; Trefonas, III, P. in "Materials for Microlithography", ACS Symposium Series No. 266, Thompson, L. F.; Willson, C. G.; Frichet, J. M. J. Eds., American Chemical Society, Washington, D.C., 1984, 293.
12. Magnera, T. F.; David, D. E.; Orth R.; Stulik, D.; Jonkman, H. T.; Michl, J., submitted for publication.

13. Schwarz, H., in "The Chemistry of Organosilicon Compounds", Patai, S.; Rappoport, Z., eds., Wiley, New York, in press.
14. Nakadaira, Y.; Kobayashi, Y.; Sakurai, H. J. Organomet. Chem. 1973, 63, 79. 15. Brook, A. G.; Harrison, A. G.; Kallury, R. K. M. R. Org. Mass Spetrom. 1982, 17, 360.
16. Aitken, C.; Harrod, J. F.; Gill, U. S. Can. J. Chem., in press (1987).
17. Gaidis, J. M.; Briggs, P. R.; Shannon, T. W. J. Phys. Chem. 1971, 75, 974.
18. Polivanov, A. N.; Bernadskii, A. A.; Zhun, V. I.; Bochkarev, V. N. Zh. Obshch. Khim. 1978, 48, 2703.
19. Pope, K. R.; Jones, P. R. Organometallics 1984, 3, 354.
20. Kinstle, T. H.; Haiduc, I.; Gilman, H. Inorg. Chim. Acta 1969, 3, 373.
21. Drahnak, T. J.; Michl, J.; West, R. J. Am. Chem. Soc. 1979, 101, 5427.
22. Raabe, G.; Michl, J. Chem. Rev. 1985, 85, 419.
23. West, R.; Fink, M. J.; Michl, J. Science (Washington, D.C.) 1981, 214, 1343.
24. R. West, in "Comprehensive Organometallic Chemistry", Wilkinson, G.; Stone, F. G. A.; Abel, E. W. eds., Pergamon Press, Oxford, 1982, Vol. 2, 365.
25. Drahnak, T. J.; Michl, J.; West, R. J. Am. Chem. Soc. 1981, 103, 1845.
26. Barton, T. J.; Burns, G. T. Organometallics 1983, 2, 1.
27. Rickborn, S. F.; Ring, M. A.; O'Neal, H. E. Int. J. Chem. Kinetics 1984, 16, 1371.
28. Sawrey, B. A.; O'Neal, H. E.; Ring, M. A.; Coffey, Jr., D. Int. J. Chem. Kinetics 1984, 16, 801.
29. Gammie, L.; Safarik, I.; Strausz, O. P.; Roberge, R.; Sandorfy, C. J. Am. Chem. Soc. 1980, 102, 378; Tokach, S. K.; Koob, R. D. J. Am. Chem. Soc., 1980, 102, 376; Cornett, B. J.; Choo, K. Y.; Gaspar, P. P. J. Am. Chem. Soc. 1980, 102, 377; Doyle, D. J.; Tokach, S. K.; Gordon, M. S.; Koob, R. D. J. Phys. Chem. 1982, 86, 3626.
30. Barton, T. J.; Burns, S. A.; Burns, G. T. Organometallics 1982, 1, 210.
31. Davidson, I. M. T.; Howard, A. V. J. Chem. Soc. Faraday Trans. 1 1975, 71, 69.
32. Chen, Y. S.; Cohen, B. H.; Gaspar, P. P. J. Organomet. Chem. 1980, 195, C1.
33. Davidson, I. M. T.; Scampton, R. J. J. Organomet. Chem. 1984, 271, 249.
34. Walsh, R. J. Phys. Chem. 1986, 90, 389.
35. Davidson, I. M. T.; Hughes, K. J.; Scampton, R. J. J. Organomet. Chem. 1984, 272, 11.
36. Conlin, R. T.; Kwak, Y.-W. J. Am. Chem. Soc. 1986, 108, 834.
37. Schaefer III, H. F. Acc. Chem. Res. 1982, 15, 283; Nakase, S.; Kudo, T. J. Chem., Soc. Chem. Commun., 1984, 141; Davidson, I. M. T.; Ijada-Maghsoodi, S.; Barton, T. J.; Tillman, N. J. Chem. Soc., Chem. Commun. 1984, 478.

Chapter 46

The spectroscopy and photochemistry of some silicon halide and silicon hydride molecules

O.P. Strausz, V. Sandhu, B. Ruzsicska, I. Safarik and T.N. Bell* – Department of Chemistry, University of Alberta, Edmonton, Alberta, Canada, T6G 2G2. *Department of Chemistry, Simon Fraser University, Burnaby, B.C., Canada, V5A 1S6.

Apart from SiF_2 [1], little has been known until recently about the UV spectrum or chemistry of silicon dihalides. Although the first claim for the UV spectrum of $SiCl_2$ was made 50 years ago [2], that claim was shown to be erroneous some 40 years later [3]. The UV spectrum of neither $SiBr_2$ nor SiI_2 has been reported.

In 1985 we published a study [4] of the flash photolysis of Si_2Cl_6 in the vapour phase. The principal end product was $SiCl_4$, detected and identified by gc and ms analysis. The intermediacy of Cl atoms in the reaction was shown by scavenging experiments using $t-C_4H_8-2$ or $i-C_4H_{10}$ which led to the production of copious amounts of HCl. In the scavenging experiments with olefins it was also noted that the yield of the $SiCl_4$ product was not significantly affected by the added olefin. This observation showed that $SiCl_4$ was formed molecularly in the primary step. Had $SiCl_4$ been formed from free radical precursors in reactions such as $SiCl_3 + Cl + M \rightarrow SiCl_4 + M$ or $2SiCl_3 \rightarrow SiCl_4 + SiCl_2$, the presence of an olefin would have had a suppressing effect on the $SiCl_4$ yield via the steps $Cl + t-C_4H_8 \rightarrow \cdot C_4H_8Cl$ and $SiCl_3 + t-C_4H_8 \rightarrow Cl_3SiC_4H_8$. Employing the kinetic spectroscopic method [4], three transient spectra were detected following flash photolysis, two of them appeared in absorption and one in emission.

The banded emission spectrum in the 286–307 nm range and the banded absorption spectrum in the 274–297 nm range have been identified as belonging to the $\tilde{B}^2\Sigma^+$-$\tilde{X}\Pi_{1/2;3/2}$ and $\tilde{B}'^2\Delta$-$\tilde{X}^2\Pi_r$ transitions of the SiCl radical, Tables 1 and 2.

Table 1. Absorption spectrum of SiCl($X^2\pi_r$).

Assigned Wavelength (nm)	Transition[a]		Ovcharenko et al. 1962,1963 [5]
296.93	A(0–1)	Q_1	297.00
294.21		$^OP_{12}$	294.21
	A(0–0)		
294.13		P_2	294.16
292.40		P_1	292.43
	A(0–0)		
292.32		Q_1	292.38
286.54		P_1	286.58
	A(1–0)		
286.49		Q_1	286.52
285.17	A(2–1)	Q_1	285.22
282.62		$^OP_1$	282.66
	A(2–0)		
282.58		P_2	282.60
282.32	B(0–0)	P_2	282.35
280.99		P_1	281.02
	A(2–0)		
280.93		Q_1	280.96
280.68	B(0–0)	Q_1	280.96
279.77	A(3–1)	Q_1	279.77
277.26	A(3–0)	P_2	277.30
275.68		P_1	275.72
	A(3–0)		
275.63		Q_1	275.67
274.54	A(4–1)	Q_1	274.58

[a] $A=\tilde{B}^2\Sigma^+-\tilde{X}^2\Pi_r$; $B=\tilde{B}'^2\Delta-\tilde{X}^2\Pi_r$.

The second absorption spectrum, Figure 1, featured a broad envelope dominated by a single vibrational progression centered at 317.4 nm. The two plausible carriers of this spectrum are the SiCl2 and SiCl3 radicals. Since, however, it has been established by chemical methods that SiCl4 is the principal stable primary product, the likely carrier of the spectrum is

Table 2. Emission spectrum of $SiCl(\tilde{B}^2\Sigma^+)$.

Wavelength (nm)	Band Assignment $\tilde{B}^2\Sigma^+-\tilde{X}^2\Pi_r$
286.5	$1-0(^2\Pi_{1/2})$
288.2	$1-0(^2\Pi_{3/2})$
292.2	$0-0(^2\Pi_{1/2})$
294.0	$0-0(^2\Pi_{3/2})$
295.3	$1-2(^2\Pi_{1/2})$
296.9	$0-1(^2\Pi_{1/2})$
297.2	$1-2(^2\Pi_{3/2})$
298.6	$0-1(^2\Pi_{3/2})$
300.0	$1-3(^2\Pi_{1/2})$
301.6	$0-2(^2\Pi_{3/2})$
301.9	$1-3(^2\Pi_{3/2})$
303.5	$0-2(^2\Pi_{3/2})$
304.8	$1-4(^2\Pi_{1/2})$
306.5	$0-3(^2\Pi_{1/2})$
306.7	$1-4(^2\Pi_{3/2})$

the $SiCl_2$ radical formed in the primary step:

$$Si_2Cl_6 + h\nu(\lambda\approx170\text{-}220 \text{ nm}) \rightarrow SiCl_4 + SiCl_2 \quad \Delta H\lesssim49.2 \text{ kcal mol}^{-1}$$

Additional primary steps of potential importance are:

$$Si_2Cl_6 + h\nu \rightarrow SiCl(\tilde{X}^2\Pi_r) + Cl + SiCl_4 \quad \Delta H\lesssim136.3 \text{ kcal mol}^{-1}$$
$$Si_2Cl_2 + h\nu \rightarrow SiCl(\tilde{X}^2\Pi_r) + Cl \quad \Delta H\lesssim106.2 \text{ kcal mol}^{-1}$$
$$SiCl(\tilde{X}^2\Pi_r) + h\nu \rightarrow SiCl(\tilde{B}^2\Sigma^+) \quad T_{oo} = 97.2 \text{ kcal mol}^{-1}$$

The step:

$$SiCl_2 + h\nu \rightarrow SiCl(\tilde{B}^2\Sigma^+) + Cl \quad \Delta H\lesssim203.4 \text{ kcal mol}^{-1}$$

would be endoergic and is of no significance.

In order to gain independent confirmation of SiCl2 being the carrier of the new spectrum we selected a source compound which, on flash photolysis, could yield only SiCl2 but no

SiCl3 radical, namely, 1,1-dichlorosilacyclobutane, ⟨SiCl₂.

The spectrum obtained from this source is indeed identical to that obtained from the flash photolysis of Si2Cl6, Figure 1c, rendering the assignment unambiguous.

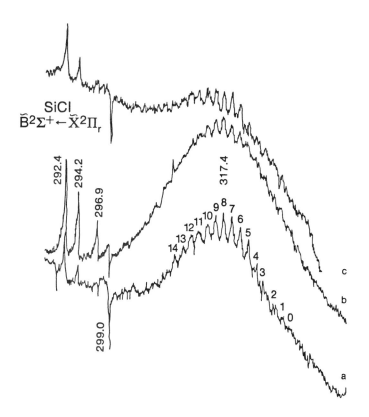

Fig. 1. UV spectra of $SiCl_2(\tilde{A}^1B_1)\leftarrow(\tilde{X}^1A_1)$: a) 0.2 torr Si_2Cl_6/50 torr Ar, 40 μsec delay, quartz reaction cell; b) 0.2 torr Si_2Cl_6/50 torr Ar, 30 μsec delay, suprasil reaction cell; c) 4 torr $SiCl_2$/46 torr Ar, 30 μsec delay, suprasil reaction cell. Reproduced from Reference 4 with permission by Elsevier Science Publishers B.V., The Netherlands.

The mean vibrational spacing as determined from the 15 lowest vibrational transitions had a value of 148 cm^{-1} which appears to be within the range expected for ν'_2, the bending vibration in the upper electronic state. For the $^1A_1$ ground electronic state of $SiCl_2$ the ν_2'' mode had a value of 202 cm^{-1}, as determined from IR spectroscopy [6]. This lends further support to our assignment of the spectrum because the known UV spectra of the Group IV atom dihalides [7] are dominated by such a ν_2' progression, owing to the larger bond angle in the upper than in the lower electronic state of the transition.

Moreover, the scavenging experiments with t–C4H8–2 yielded two addition products, neither of which showed detectable parent ions but both gave two characteristic fragment ions corresponding to C4H7$^+$ and SiCl2H$^+$, suggesting the occurrence of the following reactions:

Evidence has also been obtained, from the ms analysis, for the addition reaction of SiCl to the olefin

as was indicated by the appearance of an m/e 238 peak which corresponds to the mass of the dimeric adduct.

Thus, after nearly half a century of controversy, the UV spectrum of the SiCl2 radical has been detected. Asundi, Karim and Samuel [2] first described in 1938 a structured emission spectrum from flowing SiCl4 vapour discharges and assigned it to two band systems of the SiCl2 radical, originating from excited states lying at 29952 and 28925 cm^{-1} above the ground state. The higher lying of these states was later suggested to be the $\tilde{A}^1B_1$ state and the transition giving rise to one of the band systems, to be $\tilde{A}^1B_1 \rightarrow \tilde{X}^1A_1$ [8]. A continuous absorption spectrum with an intensity maximum around 315.0 nm was observed by Wieland and Heise [9] in 1951 from the reaction of Si with SiCl4 at 800–900°C and was attributed to the SiCl2 radical. No experimental details were given, nor was the spectrum shown or described. In 1968, while studying the IR spectrum of Ar matrix isolated SiCl2 at liquid helium temperature from the UV photolysis of H2SiCl2 and D2SiCl2, Milligan and Jacox [6] reported the observation of a broad unstructured absorption in the photolyzates near 315 nm. In a more recent development, Cornet and Dubois [3] proposed that the two emission band systems discovered by Asundi et al. [2] and assigned by them to SiCl2, are in fact emitted by other radicals. Asundi et al. allowed the SiCl4 vapour to flow over phosphorous pentoxide for the removal of moisture, not knowing that SiCl4 reacts with P2O5 [10] to yield OPCl3 which, in a condensed electrical discharge, is known to lead to the formation of the PO and P2 molecules [3]. Thus, Cornet and Dubois were able to account for each of the several dozen bands reported by Asundi et al. in terms of the known transitions of the PO and P2 molecules.

According to recent ab initio type MO calculations [11] the lowest excited state of SiCl2 is $^1B_1$ with a computed $\tilde{A}^1B_1$–$\tilde{X}^1A_1$ excitation energy of 30677 cm^{-1}, compared to the experimentally estimated value of 30319 cm^{-1}. The computation also predicts an ∿18% increase in the bond angle in the $^1B_1$, relative to the $^1A_1$ state, which provides a plausible rationale for the appearance of the long ν_2' progression in the spectrum.

In subsequent developments Washida *et al.* discovered an emission continuum from the reactions

SiH2Cl2, SiHCl3, SiCl4 + hν (VUV) →

with λmax ∿340 nm, which they postulated to be due to the same transition of SiCl2 reported before by Ruzsicska *et al.* [4]. Later, Sameith *et al.* [13] subjected flowing SiCl4 vapour, in the absence and presence of silicon, to an rf discharge and observed two emission bands emanating from the reactions taking place. The shorter wavelength band with λmax ∿330 nm showed 52 vibrational bands. From the analysis of the band progression they concluded that the spectrum was identical to the absorption spectrum reported by Ruzsicska *et al.* [4] and that the vibrational assignment of the absorption spectrum was correct. More recently, Washida *et al.* [14] carried out a more detailed study of the SiCl2 spectrum using laser-induced fluorescence at room temperature, and in a supersonic jet at low temperature. From the high resolution fluorescence excitation spectrum they concluded that the vibrational assignment of Ruzsicska *et al.* [4] was two vibrational levels shifted and that the 317.4 nm band was due to the $v_2' = 10 - v_2'' = 0$ transition. This correction in the assignment led to a value of 30003.6 cm^{-1} for the electronic excitation energy of the $\tilde{A}^1B_1$ state of SiCl2.

Time resolved flash spectroscopic studies [15] of the decay of SiCl2 showed that the decay is first order in SiCl2. In the presence of an added olefin, acetylene, oxygen or nitric oxide, the decay accelerated but remained first order in SiCl2 concentration and was also first order in the concentration of the added reagent. The bimolecular rate constant values measured to date are given in Tables 3 and 4 for hydrocarbons and oxygen, respectively.

Table 3. Rate constants for the reactions of SiCl2 with olefins and acetylenes.

Substrate	$k(M^{-1}s^{-1})$
C2H2	$(4.3\pm1.1)\times10^7$
C2H4	$(7.8\pm2.0)\times10^7$
C3H6	$(2.3\pm0.6)\times10^8$
t-C4H8	$(3.1\pm0.8)\times10^8$
1-C4H6	$(7.9\pm1.9)\times10^8$

From the rate constant values for hydrocarbon reactions it is evident that SiCl2 is less reactive than SiH2 and that it exhibits an electrophilic character. With O2 the reactivity appears to be slightly higher than that of SiH2 and the reaction mechanism may be different.

Table 4. Rate constants for the reactions of SiCl2 with O2 and NO.

Reaction	$k(M^{-1}s^{-1})$	Reference
SiF2 + O2	1.2×10^4	[16]
SiMePh + O2	3×10^8	[17]
SiH2 + O2	7.2×10^8	[18]
SiCl2 + O2	$(3.4 \pm 0.7) \times 10^9$	[19]
SiCl2 + NO	$(2.0 \pm 0.31) \times 10^9$	[20]

In efforts to generate the hitherto unknown SiBr2 radical we have explored the flash photolysis ($\lambda \sim 170$-225 nm) of SiBr4 [21]. The most important primary step considered is SiBr4 + hν → SiBr2 + Br2 ΔH = 82 kcal mol-1 followed by SiBr2 + hν → SiBr + B̆r ΔH = 83.7 kcal mol-1 and SiBr + hν → SiBr($\tilde{B}^2\Sigma$) T_{oo} = 97.2 kcal mol-1.

Three transient spectra have been detected, two in absorption and one in emission. The emission spectrum featured 18 vibrational bands in the 294.3-319.2 nm range and one of the absorption spectra, 21 vibrational bands in the 273.3-304.8 nm range. Both of these spectra are due to the SiBr radical, involving the $\tilde{B}^2\Sigma$-$\tilde{X}^2\Pi_{1/2}$ and $\tilde{B}^2\Sigma$-$\tilde{X}^2\Pi_{3/2}$ transitions. The wavenumber of all the observed bands can be calculated from the formulae:

$$^2\Sigma\text{-}^2\Pi_{1/2} : \nu_{head}(cm^{-1}) = 33571 + (571.2u' - 2.4u'^2) - (425.3u'' - 1.5u''^2)$$

$$^2\Sigma\text{-}^2\Pi_{3/2} : \nu_{head}(cm^{-1}) = 33153 + (571.2u' - 2.4u'^2) - (425.3u'' - 1.5u''^2)$$

The second absorption band detected is a continuum extending from 340 to 400 nm with λmax ~ 362 nm, Figure 2. This spectrum is due to either the SiBr3 or SiBr2 radical. We believe that the carrier of the spectrum is the SiBr2 radical for the following reason:

a) the primary step in the direct photolysis of most saturated carbon and silicon compounds is a molecular mode of elimination (SiBr4 → SiBr2 + Br2);

b) the spectrum is at about the right location for SiBr2, judging from the position of the SiCl2 spectrum;

c) the available data on the Group IV atom dihalides [22] would appear to suggest that λmax for SiBr2 should be around 360 nm.

The decay of both SiBr and SiBr2 was found to be first order and if the reactions removing these radicals are insertion into the undecomposed substrate,

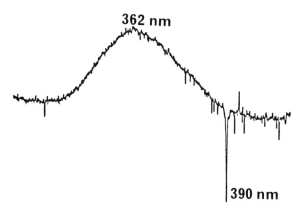

Fig. 2. Absorption spectrum of $SiBr_2(\tilde{A}\,^1B_1 - \tilde{X}\,^1A_1)$.

SiBr + SiBr4 → Si2Br5

SiBr2 + SiBr4 → Si2Br6

then the estimated rate constants are $\sim 1.0 \times 10^{10}$ and 3.3×10^8 M^{-1} s^{-1}, respectively.

   The final topic we would like to briefly touch upon is the disproportionation reaction of the trimethylsilyl radical, which has been investigated by several research groups reporting kd/kc values in the range 0.03-0.48 [23].

   In most studies the kd/kc ratios were determined from the yields of the combination product hexamethyldisilane and the methanol-scavenged disproportionation product, dimethylsilaethylene, (CH3)2Si=CH2 + ROH → (CH3)2ROSi-CH3. In a recent careful study of this latter reaction we found that upon standing in a Pyrex vessel, mixtures of (CH3)3SiH - the usual precursor of the (CH3)3Si radical - and CH3OH or C2H5OH, a dark reaction takes place giving H2, (CH3)3SiSi(CH3)3 and (CH3)3SiOR. In a detailed study we were able to minimize this interfering dark reaction on the one hand, and derive a fairly precise estimate for its contribution, on the other.

   Using the triplet mercury photosensitization of (CH3)3SiH as a source of the (CH3)3Si radical in the presence of 10-20 torr CH3OH or 10 torr C2H5OH, Figure 3, in this way it became possible to determine the value of kd/kc as being 0.10±0.01. We believe that this is the correct value for the gas phase disproportionation to combination ratio of the trimethylsilyl radical at room temperature.

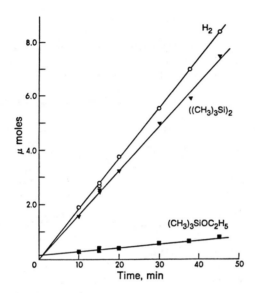

Fig. 3. Product yields from the Hg(3Pl) photosensitized decomposition of 150 torr (CH3) 3SiH in the presence of 10 torr C2H5OH as a function of exposure time.

ACKNOWLEDGEMENT

We thank the Natural Sciences and Engineering Research Council of Canada for financial support.

REFERENCES

1. H. Burger and R. Eujen, Top. Curr. Chem., 50 (1974) 7.
2. R.K. Asundi, M. Karim and R. Samuel, Proc. Phys. Soc., (London) 50 (1938) 581.
3. R. Cornet and I. Dubois, J. Phys., B10 (1977) L69.
4. B.P. Ruzsicska, A. Jodhan, I. Safarik, O.P. Strausz and T.N. Bell, Chem. Phys. Lett., 113 (1985) 67.
5. L. Ovcharenko, Y.Y. Kuzyakov and V. Tatevskii, Opt. Spectrosc. Suppl. 2 (1963) 6.
6. D.E. Milligan and M.E. Jacox, J. Chem. Phys., 49 (1968) 1938.
7. J.W. Hastie, R. Hauge and J.L. Margrave, J. Mol. Spectrosc., 29 (1969) 152.
8. Reference 1, p.1.
9. K. Wieland and M. Meise, Angew. Chem., 63 (1951) 438.
10. H. Remy, Treatise on Inorganic Chemistry, Elsevier, Amsterdam, 1956.
11. R.K. Gosavi and O.P. Strausz, Chem. Phys. Lett., 123 (1986) 65.

12. N. Washida, Y. Matsumi, T. Hayashi, T. Ibuki, A. Hiraya and K. Shobatake, J. Chem. Phys., 83 (1985) 2769.

13. D. Sameith, J.P. Monch, H.J. Tiller and K. Schade, Chem. Phys. Lett., 128 (1986) 483.

14. M. Suzuki, N. Washida and G. Inoue, Chem. Phys. Lett., 131 (1986) 24.

15. I. Safarik, B.P. Ruzsicska, A. Jodhan, O.P. Strausz and T.N. Bell, Chem. Phys. Lett., 113 (1985) 71.

16. A.C. Stanton, A. Freedman, J. Wormhoudt and P.P. Gaspar, Chem. Phys. Lett., 122 (1985) 122.

17. P.P. Gaspar, in: Proc. 7th Intl. Symp. Organosilicon Chem., Ed., H. Sakurai, Horwood, Chichester, U.K., 1985, p.87.

18. C.D. Eley, M.C.A. Rowe and R. Walsh, Chem. Phys. Lett., 126 (1986) 153.

19. V. Sandhu, A. Jodhan, I. Safarik, O.P. Strausz and T.N. Bell, Chem. Phys. Lett., 135 (1987) 260.

20. V. Sandhu, to be published.

21. B.P. Ruzsicska, A. Jodhan, I. Safarik, O.P. Strausz and T.N. Bell, Chem. Phys. Lett., in press.

22. B.P. Ruzsicska, Ph.D. Thesis, University of Alberta, Edmonton, Canada, 1983.

23. M.A. Nay, G.N. Woodall, O.P. Strausz and H.E. Gunning, J. Am. Chem. Soc., 87 (1965) 179; S.K. Tokach and R.D. Koob, J. Phys. Chem., 83 (1979) 774 and J. Am. Chem. Soc., 102 (1980) 376; B.J. Cornett, K.Y. Choo and P. Gaspar, J. Am. Chem. Soc., 102 (1980) 377; L. Gammie, I. Safarik, O.P. Strausz, R. Roberge and C. Sandorfy, J. Am. Chem. Soc., 102 (1980) 378; E. Bastian, P. Potzinger, A. Ritter, H.P. Schuchmann, C. von Sonntag and G. Weddle, Ber. Bunsenges. Phys. Chem., 84 (1980) 58; D.J. Doyle, S.K. Tokach, M.S. Gordon and R.D. Koob, J. Phys. Chem., 86 (1982) 3626.

Chapter 47

Mass spectrometry and ion-molecule reactions in silanes

F.W. Lampe – Department of Chemistry, The Pennsylvania State University.

Although formally similar in structure, even the simplest silanes and paraffin hydrocarbons exhibit very different behavior when subjected to gas-phase ionization such as occurs in the ionization chamber of a mass spectrometer. Both the low-pressure mass spectra (10^{-6} torr) and the high-pressure mass spectra (0.5 torr) of CH_4 [2] and SiH_4 [3] exhibit striking differences as may be seen in Table I.

Table 1. Comparison of low-pressure mass spectra of CH_4 and SIH_4.

Molecule	M^+	MH^+	MH_2^+	MH_3^+	MH_4^+
CH_4	3	5	10	86	100
SiH_4	27	30	100	78	0

Table 2. Comparison of high-pressure mass spectra of CH_4 and SiH_4.

Molecule	MH_5^+	$M_2H_3^+$	$M_2H_4^+$	$M_2H_5^+$	$M_2H_7^+$	$M_3H_8^+$	$M_3H_9^+$	$M_4H_{11}^+$
CH_4	100	3	5	85	0	0	0	0
SiH_4	0	15	8	0	77	46	100	54

EXISTENCE OF SiH_4^+

The low-pressure, electron–impact mass spectrum of SiH_4 at 70 eV (Table I) shows no intensity that can be attributed to the parent molecular ion SiH_4^+, although the corresponding ion is the major one in the mass spectrum of CH_4. Even when the energy of the impacting electrons is reduced to below the lowest appearance potential of fragment ions in SiH_4 no evidence of SiH_4^+ has been found. Indeed, Gordon [4] has reported quantum mechanical calcuations that indicate the energy barrier for decomposition of SiH_4^+ to SiH_3^+ and H to be only 0.07 eV (1.6 kcal./mole).

The existence of small amounts of SiH_4^+ in the photoionization of SiH_4 using an argon resonance lamp as a light souce (emission lines at 11.62 eV and 11.83 eV) has been reported [5]. However, other recent photoionization studies [6,7] of SiH_4 have not observed SiH_4^+ and we must at this stage consider the existence to be possible but somewhat doubtful.

A very striking difference in the high-pressure mass spectrum of SiH_4 as compared with CH_4 is the absence of SiH_5^+ whereas at 0.5 torr CH_5^+ is the major ion present. The most obvious reason for this difference may well be the absence of SiH_4^+, because, as will become apparent later, none of the reactions of the other primary ions, i.e. SiH^+, SiH_2^+, and SiH_3^+, with SiH_4 to form SiH_5^+ is energetically feasible and CH_5^+ does arise from reaction of CH_4^+ with CH_4.

APPEARANCE POTENTIALS AND ENTHALPIES OF FORMATION OF SiH_x^+ IONS

The appearance potentials of the primary ions from SiH_4 have now been well studied by both electron impact [9–14] and photoionization [7,8,15–17] and the formation and disappearance of these ions by ion-molecule reactions [18–20] has been extensively investigated. The results have led to a reliable self-consistent set of standard enthalpies of formation of the SiH_x^+ (x = 0–3) ions. The recommended values at 298K in kcal/mole are as follows:

$$\Delta H_f^\circ(SiH_3^+) = 238 \pm 1$$

$$\Delta H_f^\circ(SiH_2^+) = 278 \pm 1$$

$$\Delta H_f^\circ(SiH^+) = 274 \pm 1$$

$$\Delta H_f^\circ(Si^+) = 297 \pm 1$$

EXISTENCE OF SiH_5^+ AND THE PROTON AFFINITY OF SiH_4

It has already been pointed out that a striking difference between the high-pressure mass spectra of SiH_4 and CH_4 is the total absence of SiH_5^+ and the dominance of CH_5^+. The ion SiH_5^+ does exist, however, and is formed readily [21,22] in reaction of NH_2^+, $C_2H_5^+$, $C_2H_6^+$, and CH_5^+ with SiH_4, viz.

$$RH^+ + SiH_4 \rightarrow R + SiH_5^+ \ (R = NH, C_2H_2, C_2H_5, \text{ and } CH_4) \quad (1)$$

All reactions (1) were observed to be exothermic. On the other hand, formation of SiH_5^+ in collisions of $C_3H_8^+$ ions with SiH_4 was found, on the basis of the dependence of reaction cross-sections on relative kinetic energy, to be endothermic [21]. Since the standard enthalpies of formation of all species in (1) except SiH_5^+ are known [23], it is possible to bracket the standard enthalpy of formation of SiH_5^+ as lying in the range of 217-223 kcal./mole. Equivalently, this may be expressed in terms of Bronsted base strength by stating that the proton affinity of SiH_4, i.e. $\Delta H°$ of (2),

$$SiH_5^+ \rightarrow SiH_4 + H^+ \quad (2)$$

is in the range of 150-156 kcal./mole. This result indicates that SiH_4 is a stronger base than CH_4 by 35 kcal./mole in the gas-phase. Similar experiments [24] have shown that GeH_4 has a proton affinity in the gas-phase in the range of 162-164 kcal./mole and is thus a stronger base than SiH_4.

Theoretical calculations [25-29] relative to the energies of the protonated forms of SiH_4 and GeH_4 have confirmed the experimental results and have produced optimized geometries for these interesting species which are shown in Figure 1.

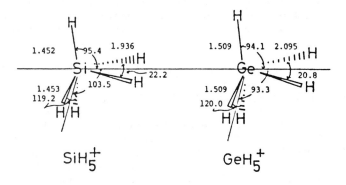

Fig. 1. Comparison of optimized geometry of GeH_5^+ with those of SiH_5^+

ION-MOLECULE REACTIONS IN PURE SILANES

It is clear from Table II that when pure SiH_4 is ionized a gas-phase polymerization to higher silicon hydride ions, particularly the protonated higher silanes, i.e. $Si_2H_7^+$, $Si_3H_9^+$, $Si_4H_{11}^+$, etc. takes place. A similar occurence is observed in the ionization of pure Si_2H_6 [30]. This gas-phase ionic polymerization in SiH_4 and Si_2H_6 is in sharp contrast to the behavior of ionized CH_4. In this latter gas, the high-pressure mass spectrum [2] is dominated by only two ions, namely CH_5^+ and $C_2H_5^+$, the relative intensities of which become independent of pressure above about 1 torr. This high abundance and pressure-independence forms the basis of methane chemical ionization mass spectrometry.

The major primary ions in the mass spectrum of SiH_4 are SiH_2^+ and SiH_3^+. As the pressure in the ion-source is increased permitting collisions of these ions with SiH_4, the SiH_2^+ is converted efficiently to SiH_3^+ by the hydride transfer reaction (3). As the pressure is increased still further SiH_3^+

$$SiH_2^+ + SiH_4 \rightarrow SiH_3^+ + SiH_3 \tag{3}$$

undergoes a third-order association reaction with SiH_4 to form $Si_2H_7^+$, as shown by (4), a process that gradually removes SiH_3^+ from the spectrum.

$$SiH_3^+ + 2SiH_4 \rightarrow Si_2H_7^+ + SiH_4 \tag{4}$$

Further increases in pressure consume the protonated disilane by (5), viz.

$$Si_2H_7^+ + SiH_4 \rightarrow Si_3H_9^+ + H_2 \tag{5}$$

We have observed (5) by tandem mass spectrometry [19] and we assume that analogous processes such as (6) will occur with the protonated forms of the higher silanes leading to the observed polymerization. Analogous reactions are observed in ionized Si_2H_6 [30].

$$Si_3H_9^+ + SiH_4 \rightarrow Si_4H_{11}^+ + H_2 \tag{6}$$

ION-MOLECULE REACTIONS IN SILANE MIXTURES

The major bimolecular ion-molecule reaction in pure silanes is that of hydride ion transfer to an attacking positive ion. The dominance of hydride ion transfer is even more apparent when silanes are subjected to attack by positive ions not containing silicon. Thus, all of the primary ions in the mass spectra of CH_4 and even CH_5^+ react with SiH_4 predominantly by removal of a hydride ion [31]. The predominance of this reaction, which probably reflects the $Si^{\delta+}-H^{\delta-}$ polarization of the Si-H bond,

can be judged from the data in Table III. These data are
relative reaction cross-sections at ion energies of 1-1.5 eV
(LAB) for formation o the various product ions when the primary
ions of CH_4 collide with SiH_4. The hydride abstraction process
is represented by (7).

$$M^+ + SiH_4 \rightarrow SiH_3^+ + MH \text{ (or } M + H_2) \tag{7}$$

Table 3. Secondary ion formation in the CH_4-SiH_4 System

Primary Ion	Relative Cross Section for Formation of								
	Si^+	SiH^+	SiH_2^+	SiH_3^+	$SiCH^+$	$SiCH_3^+$	$SiCH_3^+$	$SiCH_4^+$	$SiCH_5^+$
C^+	120	180	300	1260	118	308	--	--	--
CH^+	140	228	310	980	--	320	230	--	--
CH_2^+	50	150	50	986	--	42	246	--	--
CH_3^+	--	--	--	1210	--	--	31	--	28
CH_4^+	--	--	252	810	--	--	10	17	25
CH_5^+	--	--	--	1350	--	--	--	--	--

Other ions that have been observed to act as M^+ are CF_3^+, OH^+,
H_2O^+, H_3O^+, $C_2H_2^+$, $C_2H_3^+$ and $C_2H_4^+$. Indeed any ion that can
energetically undergo hydride transfer with SiH_4 appears to do
so. The only ion we [32] have found that SiH_3^+ will abstract
hydride ions from is GeH_4, undoubtedly reflecting the greater
basicity of GeH_4 as compared with SiH_4.

Since all the primary and major secondary ions derived from CH_4
react with SiH_4 by hydride ion removal to yield SiH_3^+ and since
the major ion-molecule reaction in pure SiH_4 is also hydride
ion transfer, it follows that even in a mixture that is rich in
CH_4, increasing the total pressure should result mainly in the
formation and subsequent reaction of SiH_3^+. This has been
observed in our laboratory for a mixture that is 90% CH_4 and
10% SiH_4 and similar results have been obtained [33] for a
mixtures even richer in CH_4.

This dominant feature of hydride abstraction from Si-H bonds
has also been observed [34-36] in the attack of CH_3^+, $C_2H_5^+$ and
$C_3H_7^+$ ions on CH_3SiH_3, $(CH_3)_2SiH_2$, $(CH_3)_3SiH$, $C_2H_5SiH_3$,
$(C_2H_5)_2SiH_2$ and $(C_2H_5)_3SiH$.

In addition to hydride ion abstraction from silanes, alkyl
anion transfer has also been observed to be a major process
[34-36]; that is, reaction such as (8) occur with large cross-
section.

$$C_2H_5^+ + (CH_3)_2SiH_2 \rightarrow C_3H_8 + CH_3SiH_2^+ \tag{8}$$

While hydride ion transfers to attacking positive ions are
predominant ion-molecule reactions of silanes, silicon-

containing positive ions react with halogen-containing
molecules principally by halide abstraction. Thus in a
CF_4-SiH_4 mixture [37,38] the predominant reactions are (9) and
(10).

$$CF_3^+ + SiH_4 \rightarrow CF_3H + SiH_3^+ \tag{9}$$

$$SiH_3^+ + CF_4 \rightarrow SiH_3F + CF_3^+ \tag{10}$$

The first of these reactions proceeds via a collision complex
that is of suffcient lifetime for a process involving the
breaking and forming of six bonds to occur. The process was
observed in a tandem mass spectrometer under single collision
conditions. Since CH_3^+ was detected as a product and the
reaction was observed to be exothermic, the process must be as
shown in (11).

$$CF_3^+ + SiH_4 \rightarrow CH_3^+ + SiF_3H \ (\text{or } SiF_2 + HF) \tag{11}$$

This reaction which involves breakage of 6 bonds and formation
of 6 bonds must result from a single collision. A mechanism
describing this rather remarkable process is shown below.

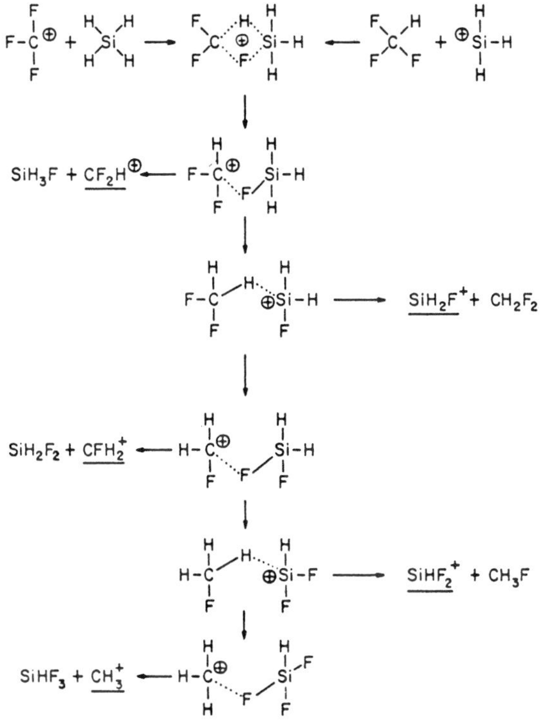

Fig. 2. Mechanism of reaction of CF_3^+ with SiH_4.

DYNAMICS OF HYDRIDE TRANSFER REACTIONS

Direct processes

While the dynamics of most of the hydride transfer reactions observed in silane and silane mixtures have not been investigated, detailed studies of the dynamics of the reactions

$$SiH_2^+ + SiH_4 \rightarrow SiH_3^+ + SiH_3 \tag{12}$$

$$CH_3SiH^+ + CH_3SiH_3 \rightarrow CH_3SiH_2 + CH_3SiH_2^+ \tag{13}$$

$$Si^+ + CH_3SiH_3 \rightarrow CH_3Si^+ + SiH_3 \tag{14}$$

have been carried out [39,40]. It has been found that (12) and (13) occur by both hydride ion and hydrogen atom transfer and predominantly by spectator stripping processes in which only the moiety abstracted receives any momentum from the attacking positive ion. Thus in describing ion-beam studies of these processes, the reactions are more correctly written

$$\overrightarrow{SiH_2}^+ + SiH_4 \rightarrow \overrightarrow{SiH_3}^+ + SiH_3 \tag{12a}$$

$$\overrightarrow{SiH_2}^+ + SiH_4 \rightarrow SiH_3^+ + \overrightarrow{SiH_3} \tag{12b}$$

$$\overrightarrow{CH_3SiH}^+ + CH_3SiH_3 \rightarrow \overrightarrow{CH_3SiH_2}^+ + CH_3SiH_2 \tag{13a}$$

$$\overrightarrow{CH_3SiH}^+ + CH_3SiH_3 \rightarrow CH_3SiH_2^+ + \overrightarrow{CH_3SiH_2} \tag{13b}$$

In (12a)-(13b) the species without the arrow have only thermal kinetic energy while those with the arrow have an energy of 1-2 eV in the forward direction of the incident ion beam. Thus the major reactions observed in such collisions do not involve what is normally thought of as an intermediate complex. In both (12) and (13) the attacking positive ion captures either a hydrogen atom or hydride ion without the SiH_3, $^+SiH_3$, CH_3SiH_2 and $CH_3SiH_2^+$ moieties receiving any momentum or energy.

Reaction (14) is still under investigation in our laboratory and appears to be an example of a gas-phase inversion reaction, that may be written

$$\overrightarrow{Si}^+ + CH_3SiH_3 \rightarrow \left[\begin{array}{c} H \\ | \\ \overrightarrow{Si} \cdots \overset{|}{\underset{H}{C}} \cdots \overrightarrow{SiH_3} \\ H \end{array} \right] \rightarrow {}^+Si-CH_3 + \overrightarrow{SiH_3} \tag{14}$$

Because of the similarity in mass of Si^+ and SiH_3^+, the product ion $^+SiCH_3$ has essentially only thermal kinetic energy.

Long-lived intermediate complexes

When a beam of SiH_3^+ ions collides with C_2H_4 the major reaction
[41-44] is to form $SiC_2H_5^+$ and H_2. In addition to these
products, it is found that for kinetic energies of SiH_3^+ below
2.6 eV an intermediate complex, $SiC_2H_7^+$, is also registered
directly at the detector of a tandem mass spectrometer.
Moreover, this complex is found to represent a significant
fraction of the relative product yield even in the low pressure
limit when no collisional stabilization occurs. Thus the
reaction is not a direct one as for hydride transfer but
proceeds as written in (15a,b)

$$SiH_3^+ + C_2H_4 \rightarrow SiC_2H_7^{+*} \qquad (15a)$$

$$SiC_2H_7^{+*} \rightarrow SiC_2H_5^+ + H_2 \qquad (15b)$$

The fact that $SiC_2H_7^+$ is registered at the detector in the low-
pressure limit indicates that the mean lifetime of $SiC_2H_7^{+*}$ is
at least 10^{-5} seconds. This is a very long lifetime for such a
complex of this degree of complexity. Indeed, no analogous
complex is observed under any conditions when SiH_3^+ is replaced
by CH_3^+, yet all other indications relative to the CH_3^+/C_2H_4
reaction suggest that it too proceeds by an intermediate
complex.

Detailed studies [44,45] of the SiH_3^+/C_2H_4 and CH_3^+/C_2H_4
reactions have shown that the explanation for the different
behavior with respect to lifetime of collision complexes is due
solely to the nature of the potential energy profiles of the
two reactions. In the SiH_3^+/C_2H_4 reaction, the energy barrier
separating $SiC_2H_7^+$ from the products $SiC_2H_5^+$ and H_2 is higher
than the barrier leading back to reactants [44]. The result is
that under conditions of low-energy reactant ions, at least 98%
of total reaction is back to reactants and the complexes
$SiC_2H_7^{+*}$ have a very long lifetime. In the CH_3^+/C_2H_4 reaction,
the barrier separating $C_3H_7^{+*}$ from products is considerably
lower than that leading to reactants [45]. The result is that
less than 1% of the total reaction is backward to reactants.
Moreover, no matter how low the initial energy of the
reactants, the total energy of the system is always appreciably
above the barrier to products and the reaction of the
intermediate complex to products is always very fast, leading,
of course, to a very short lifetime of the complex. The
potential energy profiles are sketched in Figure 3. Similar
behavior is observed in the SiH_3^+/C_6H_6 and CH_3^+/C_6H_6 systems
[46].

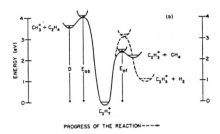

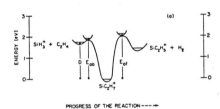

Fig. 3. Approxiamte potential-energy diagrams (a) for the reaction of silyl ion with ethylene and (b) for the reaction of methyl ion with ethylene.

REFERENCES

1. U.S. Department of Energy Document No. DE-AS02-76ER03416.
2. F. H. Field and M. S. B. Munson, J. Am. Chem. Soc. 87, 3289 (1965).
3. T.-Y. Yu, T. M. H. Cheng, V. Kempter, and F. W. Lampe, J. Phys. Chem. 76, 3321 (1972).
4. M. S. Gordon, Chem. Phys. Lett. 59, 410 (1978).
5. W. Genuit, A. H. J. Boerboom, and T. R. Govers, Int. J. Mass Spectrom. Ion Proc. 62, 341 (1984).
6. A. Ding, R. A. Cassidy, L. S. Cordis, and F. W. Lampe, J. Chem. Phys. 83, 3426 (1985).
7. K. Borlin, T. Heinis, and M. Jungen, Chem. Phys. 103, 93 (1986).
8. P. Haal and A. Rahbee, Chem. Phys. Lett. 114, 571 (1985).
9. H. Neuert and H. Clasen, Z. Naturforsch. 7A, 410 (1952).
10. F. E. Saalfeld and H. J. Svec, Inorg. Chem. 2, 46 (1963).
11. W. C. Steele, L. D. Nichols, and F. G. A. Stone, J. Am. Chem. Soc. 84, 4441 (1962).
12. P. Potzinger and F. W. Lampe, J. Phys. Chem. 73, 3912 (1969).
13. J. D. Morrison and J. C. Traeger, Int. J. Mass Spectrom. Ion Phys. 11, 289 (1973).
14. H. Chatham, D. Hils, R. Robertson, and A. Gallagher, J. Chem. Phys. 81, 1770 (1984).
15. B. P. Pullen, T. A. Carlson, W. E. Moddeman, G. K. Schweitzer, W. E. Bull and F. A. Grim, J. Chem. Phys. 53, 768 (1970).

16. A. W. Potts and W. C. Price, Proc. Roy. Soc. (London) A326, 165 (1972).
17. J. M. Dyke, N. Jonathan, A. Morris, A. Ridha, and M. J. Weister, Chem. Phys. 81, 481 (1983).
18. J. M. S. Henis, G. W. Stewart, M. K. Tripodi, and P. P. Gaspar, J. Chem. Phys. 57, 389 (1972).
19. T.-Y. Yu, T. M. H. Cheng, V. Kempter, and F. W. Lampe, J. Phys. Chem. 76, 3321 (1972).
20. B. H. Boo and P. B. Armentrout, J. Am. Chem. Soc. 109, xxxx (1987) (In Press).
21. T. M. H. Cheng and F. W. Lampe, Chem. Phys. Lett. 19, 532 (1973).
22. M. D. Sefcik, J. M. S. Henis, and P. P. Gaspar, J. Chem. Phys. 61, 4329 (1974).
23. H. M. Rosenstock, K. Draxl, B. W. Steiner, and J. T. Heron, J. Phys. Chem. Ref. Data 6, (1977), Suppl. No. 1.
24. S. N. Senzer, R. N. Abernathy, and F. W. Lampe, J. Phys. Chem. 84, 3066 (1980).
25. H. Hartmann, L. Papula, and W. Strehl, Theor. Chim. Acta 17, 131 (1970).
26. P. von R. Schleyer, Y. Apeloig, D. Arad, B. Luke, and J. Pople, Chem. Phys. Lett. 95, 477 (1983).
27. H. Hartmann, L. Popula, and W. Strehl, Theor. Chim. Acta 19, 155 (1970).
28. S. Kohde-Sudoh, S. Ikuta, O. Nomura, S. Katagiri, and M. Imamura, J. Phys. B. 16, L529 (1983).
29. S. Kohda-Sudoh, S. Ikuta, O. Nomura, S. Katagiri, and M. Imamura, Rep. Inst. Phys. Chem. Res., Japan 59, 152 1983).
30. T. M. H. Cheng, T.-Y. Yu, and F. W. Lampe, J. Phys. Chem. 78, 1184 (1974).
31. T. M. H. Cheng, T.-Y. Yu, and F. W. Lampe, J. Phys. Chem. 77, 2587 (1973).
32. K. P. Lim, G. Johnson and F. W. Lampe, Unpublished Results.
33. J. R. Krause and P. Potzinger, Int. J. Mass Spectrom. Ion Phys. 18, 303 (1975).
34. G. W. Goodloe, E. R. Austin, and F. W. Lampe, J. Am. Chem. Soc. 101, 3472 (1979).
35. G. W. Goodloe and F. W. Lampe, J. Am. Chem. Soc. 101, 5649 (1979).
36. G. W. Goodloe and F. W. Lampe, J. Am. Chem. Soc. 101, 6028 (1979).
37. J. R. Krause and F. W. Lampe, J. Am. Chem. Soc. 98, 7826 (1976).
38. J. R. Krause and F. W. Lampe, J. Phys. Chem. 81, 281 (1979).
39. T. M. Mayer and F. W. Lampe, J. Phys. Chem. 78, 2195 (1974).
40. T. M. Mayer and F. W. Lampe, J. Phys. Chem. 78, 2429 (1974).

41. T. M. Mayer and F. W. Lampe, J. Phys. Chem. 78, 2433
     (1974).
42. R. N. Abernathy and F. W. Lampe, Int. J. Mass Spectrom.
     Ion Phys. 51, 3 (1983).
43. W. N. Allen and F. W. Lampe, J. Am. Chem. Soc. 99, 6816
     (1977).
44. R. N. Abernathy and F. W. Lampe, J. Am. Chem. Soc. 103,
     2573 (1981).
45. R. N. Abernathy and F. W. Lampe, Int. J. Mass Spetrom.
     Ion Phys. 41, 7 (1981).
46. W. N. Allen and F. W. Lampe, J. Am. Chem. Soc. 99
     2943 (1977).

Chapter 48

^{29}Si NMR spectroscopy
in organic chemistry

J. Schraml – Institute of Chemical Process Fundamentals, Czechoslovak Academy of Sciences, 165 02
Prague, Czechoslovakia.

In recent years ^{29}Si NMR spectroscopy has become an indispensable tool for structural studies of silicon-containing inorganic materials. High-resolution ^{29}Si NMR spectra provided detailed structural information about such important materials as zeolites, clays, glasses and others. In several instances the results of ^{29}Si NMR investigations have led to revisions of traditional structural concepts. In view of the spectacular success of ^{29}Si NMR in the field of inorganic chemistry, the applications to organic materials appear to have been neglected. It is the aim of this lecture to show that ^{29}Si NMR is also a convenient and useful analytical tool for studies of polyfunctional organic materials like polyols, saccharides, wood constituents or petrol fractions which all pose problems for other methods of structure determination.

In our discussion we shall be primarily concerned with silicon-29 chemical shifts and to a much lesser extent also with long-range silicon-proton coupling constants. Other spectral parameters (isotopic effects, spin-lattice and spin-spin relaxation times, nuclear Overhauser effect) will be left out as their interpretation in terms of molecular structure is not so straightforward.

Already the first measurements of ^{29}Si NMR spect-
ra (for the review of the early work see |1|) esta-
blished a non-monotonous dependence of the ^{29}Si chemical shift on the number of electronegative substituents. Attempts to explain this U-shaped (or sag-
ging pattern) dependence led to several different experiments that were all designed to clarify the role
which various structural factors play and thus to de-
cide between theoretical alternatives (for a review
of ^{29}Si chemical shift theories see |2|). The well
known example of such experiment is the study of chemical shift dependence on Hammett substituent constant, $\sigma(Y)$, in substituted phenylsilanes $X_3SiC_6H_4Y$.
Ernst et al. |3| have shown that the sign of the slope
changes with the nature of the substituent X.

In our laboratory we have studied remote substituent effects in other systems. We were mainly concerned with compounds of general structure

$$X_n(CH_3)_{3-n}Si(CH_2)_mY$$

in which the effect of the remote substituent Y could
be varied by increasing its separation from the silicon atom. The substituent effects have shown the dependence on the chain length m and on the nature of
directly bonded substituents X |2,4|. The results were
consistent with the concept of $(p - d)_\pi$ bonding
and its assumed influence on the silicon chemical
shift. According to this concept the remote substituent effect should be amplified if oxygen (or other
atom capable of $(p - d)_\pi$ bonding) is inserted between
the silicon atom and the alkyl chain bearing the remote substituent Y. Measurements of a series of compounds with such inserted oxygen atom

$$X_n(CH_3)_{3-n}Si-O-(CH_2)_mY$$

confirmed this expectation for the case of n = 0
(i.e., in trimethylsilyl derivatives, in other cases
with n > 0 the situation was more complex) |5|. The
amplifying effect of the oxygen link could be quantitatively evaluated by comparison of Hammett-type de-
pendences of silicon chemical shifts in substituted
trimethylphenylsilanes and trimethylphenoxysilanes.
We found |6| that the silicon-29 chemical shift was
twice as sensitive to substituent effects when the
trimethylsilyl group was connected to the rest of the
molecule via an oxygen link than when it was connected directly (despite the shorter distance or the
lower number of intervening bonds between the silicon
atom and the substituent Y). Later, similar amplifying effects were found for nitrogen and sulphur bridges

by other authors |7|.

The high sensitivity of the silicon-29 chemical shift to substituent effects when the effects are transmitted through the above links is the basis of all the analytical applications of ^{29}Si NMR spectroscopy, including the mentioned applications to inorganic materials and the analysis of silicones.

For the applications to organic materials it is equally important that many functional groups (X-H) can be silylated, i.e. converted to trimethylsilyl derivatives by a simple reaction:

$$R-X-H + (CH_3)_3SiY \longrightarrow R-X-Si(CH_3)_3 + YH$$

(for a review of silylating reactions including technical details see |8|). Trimethylsilylation is often used for isolation of polyfunctional natural compounds, for gas-chromatographic separation, for reversible blocking of functional groups and for other purposes. The compounds are then usually identified by different spectroscopic methods including, ^{1}H and ^{13}C NMR which are excellent methods for establishing carbon skeleton but still rather weak in determining the type and number of functional groups present in the parent or original molecule. (Several important functional groups do not contain carbon atoms at all or their carbon-13 and proton chemical shifts are not sufficiently characteristic, being dependent on concentration, pH, temperature and are affected by exchange processes). ^{1}H and ^{13}C NMR spectra of the trimethylsilylated derivatives permit determination of the functional groups and increase its precision by introducing 9 equivalent protons or 3 equivalent carbon atoms into each functional group but the spread of chemical shifts of these groups is very narrow. Hence, it was suggested to measure instead ^{29}Si NMR spectra of the trimethylsilyl derivatives of polyfunctional compounds |4,5,9,10|.

The ^{29}Si NMR spectrum of the trimethylsilylated derivative exhibits as many lines as there are functional groups originally present in the polyfunctional compound. Of course, ^{29}Si NMR spectroscopy can be combined not only with the total silylation but also with partial or selective trimethylsilylation which produces derivatives with only particular types of functional groups silylated.

The method (total silylation combined with ^{29}Si NMR) was successfully tested on a number of compounds of different classes like polyols |5|, oligosaccharides (including annomeric mixtures of trisaccharides) |11-13|, amino- and hydroxyacids |14|, lignin constituents |15,16| etc. In all cases we could see the correct number of ^{29}Si NMR lines that corresponded to the number of nonequivalent functional groups. In the case of symmetry equivalent groups, naturally, the lines coincide but such a case can be easily discerned according to the intensity of the line. Accidental overlap is rather rare in ^{29}Si NMR spectroscopy of molecules of moderate size as the silicon-29 lines are very narrow (the linewidth is often determined by the efficiency of proton decoupling).

Because of inherently low sensitivity of ^{29}Si NMR, the described method of functional group analysis could be applied to problems only when large amount of material was available (a few hundred mg). This limitation was eliminated after general polarization transfer schemes INEPT [17] and DEPT [18] were invented (for a review see [19]). Owing to particular features of the trimethylsilyl derivatives these pulse sequences can be used routinely [20] for measurement of ^{29}Si NMR spectra. Provided that certain precautions are taken, the spectra yield even quantitative data with maximum error in the concentration estimate less than 20 % (relative) [21].

Obviously, with the number of different functional groups determined by the described procedure, one inquires after the nature of the groups present. Or, in other words, what other information can be deduced from the silicon chemical shift values.

So far, we have studied in detail factors affecting the silicon-29 chemical shift in trimethylsiloxy groups, i.e., in the derivatives with $O-Si(CH_3)_3$ grouping. There, the silicon chemical shifts are influenced by polar effects (the chemical shift varies linearly with polar Taft σ^* constant of substituent R in Me_3SiOR derivatives [22]) and by steric effects. For example, in trimethylsilylated dihydroxyadamantanes, the interactions of proximate oxygen atoms produce diamagnetic shifts up to 2.3 ppm (in the $2^A,4^A$-isomer) [23]. Analogous steric effects are, however, paramagnetic in ortho substituted benzenes [24] due to different spatial relationship between the interacting groups. In trimethylsilylated cyclic enols, the silicon chemical shift can be used as a sensitive measure of ring size (for small rings up to cyclodecene) [25]; in steroids trimethylsilylation of OH group in position 3 followed by ^{29}Si NMR spectral measurement can be used as an indicator of cis or trans arrangement of the rings A/B [26] etc.

In order to be able to utilize the described relations between the silicon chemical shift and molecular structure, the lines in the spectrum of a polyfunctional compound must be first assigned.

Similarly as in other branches of NMR spectroscopy, the simplest approach is to collect sufficient data for model compounds and to assign the lines in unknown compounds or their mixtures by comparison with the collected data. This is, perhaps, the only feasible approach for such complicated mixtures as lignins and petrol fractions. Next, the slightly more advanced stage, would be to deduce an empirical relationship from the model data and then extend their validity to the investigated compound. Two such relationships were found for monosaccharides; silicon chemical shifts were always found in the order δ(Si-2) > δ(Si-4) > δ(Si-3) in trimethylsilylated methyl β-D-xylopyranoside derivatives |27| and the silicon shifts followed a Hammett-type dependence in 1,6-anhydro-β-D-glucopyranoside derivatives |28|. Usefulness of such empirical relationships is problematic as the extrapolation from the model data to the compound under study cannot be always justified.

Generally applicable are, however, experimental methods based on spin-spin coupling between the silicon-29 and some other nucleus. The first experiments of this type employed selective deuteration |29| but for obvious reason the experiments were limited to model compounds. More useful are off-resonance selective decoupling experiment |9| which do not require any special chemical treatment of the sample and can be performed on any NMR spectrometer. As a result of such experiments, trimethylsiloxy groups that are bound to primary, secondary, and tertiary carbon atoms are differentiated. The off-resonance decoupling experiments can be combined with INEPT or DEPT techniques |30| with considerable sensitivity benefit.

When the lines of skeletal protons or carbons are assigned (at least partially), the silicon chemical shifts can be assigned through different correlation experiments. The most sensitive are the experiments based on a modified INDOR technique |31| which requires a CW spectrometer. Now popular are heteronuclear 2D chemical shift correlation experiments that can also be adopted to Si - H chemical shift correlations |32|. Of course, such experiments are time consuming but we have demonstrated that using a 200 MHz spectrometer, the spectra of trisaccharides can be assigned by this method routinely |12|.

The most powerful but also the most demanding are the experiments based on ^{29}Si - ^{13}C coupling. Two such experimental techniques were tried: measurement of ^{13}C NMR spectra with selective silicon-29 decoupling |34| and the selective heteronuclear INADEQUATE experiment |35|. These experiments require a modification of the existing spectrometers to allow irradiation of the sample with four different r.f. frequencies (lock, proton decoupling, other nucleus decoupling and observation). However, because of the large spectral dispersion of ^{13}C NMR spectra the results are promising.

With the capacity to assign the lines in the spectrum of a polyfunctional compound, the method surpasses other methods for functional group analysis like those based on the use of ^{19}F NMR spectroscopy. Good quantitative data, good sensitivity and established relationships to molecular structural parameters speak also for the combination of ^{29}Si NMR spectroscopy with trimethylsilylation.

REFERENCES

1. Williams E. A., Cargioli J. D.: "Silicon-29 NMR Spectroscopy", in Annual Reports on NMR Spectroscopy (Webb G. A., ed.) Vol. 9, p. 221. Academic Press, London 1979.
2. Schraml J.: "NMR Spectroscopy in the Investigation and Analysis of Carbon-Functional Organosilicon Compounds", in Carbon-Functional Organosilicon Compounds (Chvalovský V. and Bellama J. M., eds.) p. 121. Plenum Press, New York 1984.
3. Ernst C. R., Spialter L., Buell G. R., and Wilhite D. L.: J. Am. Chem. Soc. 96, 5375 (1974).
4. Schraml J. and Bellama J. M.: "^{29}Si Nuclear Magnetic Resonance", in Determination of Organic Structures by Physical Methods (Nachod F. C., Zuckerman J. J., and Randall E. W., eds.) Vol. 6, p. 203. Academic Press, New York 1976.
5. Schraml J., Pola J., Jancke H., Engelhardt G., Černý M., and Chvalovský V.: Coll. Czech. Chem. Commun. 41, 360 (1976).
6. Schraml J., Koehler P., Licht K., and Engelhardt G.: J. Organometal. Chem. 121, C1 (1976).
7. Pestunovich V. A., Larin M. F., Pestunovich A. E., and Voronkov M. G.: Izv. Akad. Nauk SSSR, Ser. Khim. 1455 (1977).
8. Pierce A. E.: "Handbook and General Catalog", 1979-80, p. 173. Pierce Company, Rockford, Ill. 1978.
9. Gale D. J. Haines A. H., and Harris R. K.: Org. Magn. Reson. 7, 635 (1975).
10. Bayer E., Jung G., Breitmaier E., Hunziker P., Koenig W., Voelter W.: in the book "Peptides 1971", Proc. 11th Eur. Pept. Symp., Wien 1971 (Nesvadba H., ed.), p. 265. North-Holland, Amsterdam 1973.
11. Petráková E., Schraml J., Hirsch J., Kvíčalová M., Zelený J., and Chvalovský V.: Coll. Czech.

Chem. Commun., in press.
12. Schraml J., Petráková E., and Hirsch J.: Magn.
    Reson. Chem. 25, 75 (1987).
13. Schraml J., Petráková E., Hirsch J., Čermák J.,
    Chvalovský V., Teeäär R., Lippmaa E.: Coll.
    Czech. Chem. Commun., in press.
14. Schraml J., Pola J., Chvalovský V., Marsmann H.
    C., and Bláha K.: Coll. Czech. Chem. Commun. 42,
    1165 (1977).
15. Brežný R., Schraml J., Kvíčalová M., Zelený J.,
    and Chvalovský V.: Holtzforschung 39, 297 (1985).
16. Brežný R. and Schraml J.: Holtzforschung, in
    press.
17. Morris G. A. and Freeman R.: J. Am. Chem. Soc.
    101, 760 (1979).
18. Doddrell D. M., Pegg D. T., and Bendall M. R.:
    J. Magn. Reson. 48, 323 (1982).
19. Blinka T. A., Helmer B. J., and West R.: Adv.
    Organometal. Chem. 23, 193 (1984).
20. Schraml J.: Coll. Czech. Chem. Commun. 48, 3402
    (1983).
21. Schraml J., Blechta V., Kvíčalová M., Nondek L.,
    and Chvalovský V.: Anal. Chem. 58, 1892 (1986).
22. Schraml J., Chvalovský V., Mägi M., and Lippmaa
    E.: Coll. Czech. Chem. Commun. 46, 377 (1981).
23. Schraml J., Včelák J., Chvalovský V., Engelhardt
    G., Jancke H., Vodička L., and Hlavatý J.: Coll.
    Czech. Chem. Commun. 43, 3179 (1978).
24. Schraml J., Chvalovský V., Jancke H. and Engel-
    hardt G.: Org. Magn. Reson. 9, 237 (1977).
25. Schraml J., Sraga J., and Hrnčiar P.: Org. Magn.
    Reson. 21, 73 (1983).
26. Schraml J., to be published.
27. Schraml J., Petráková E., Pihar O., Hirsch J.,
    and Chvalovský V.: Org. Magn. Reson. 21, 666
    (1983).
28. Schraml J., Včelák J., Černý M., and Chvalovský
    V.: Coll. Czech. Chem. Commun. 48, 2503 (1983).
29. Haines A. H., Harris R. K., and Rao R. C.: Org.
    Magn. Reson. 9, 432 (1977).
30. Schraml J.: J. Magn. Reson. 59, 515 (1984).
31. Schraml J., Larin M. F., Pestunovich V. A.:
    Coll. Czech. Chem. Commun. 50, 343 (1985).
32. Schraml J.: 3rd International Symposium on NMR
    Spectroscopy, Abstract p. 49, Tábor 1982.
33. Schraml J., Petráková E., Pelnař J., Kvíčalová
    M., and Chvalovský V.: J. Carbohydr. Chem. 4,
    393 (1985).
34. Past J., Puskar J., Schraml J., and Lippmaa E.:
    Coll. Czech. Chem. Commun. 50, 2060 (1985).
35. Past J., Puskar J., Alla M., Lippmaa E., and
    Schraml J.: Magn. Reson. Chem. 23, 1076 (1985).

Chapter 49

Unstable intermediates in pyrolysis and alkali metal vapors dehalogenation of organosilicon compounds

L.E. Gusel'nikov – Topchiev Institute of Petrochemical Synthesis of the USSR Academy of Sciences, Moscow, USSR.

INTRODUCTION

The object of our kinetic and mechanistic studies of gas phase thermal decomposition processes is (1) to relate the thermal stability to the structure of organosilicon compounds, (2) to find out pathways of formation of products, and (3) to characterize the short-lived intermediates containing elements in an unusual coordination. The mechanisms of hydrocarbons and silahydrocarbon thermal decomposition processes have certain resemblance in that both may involve intermediates having tri- or divalent Group IVB elements: radicals, carbenes, their silicon analogs. The main difference is due to the fact that unlike unsaturated hydrocarbons compounds with multiple $p_\pi - p_\pi$ bonded silicon are not the end reaction products but unstable intermediates.

Thus, in studying the pyrolysis mechanism of organosilicon compounds, account must be taken of the specific group of low-coordination compounds of tetravalent silicon, which are thermodynamically, but not kinetically, stable molecules and exhibit excessively high reactivity.

Twenty years ago we described the first silaalkene - 1,1-dimethyl-1-silaethylene - which was obtained and characterized as unstable intermediate in studying the gas phase thermal decomposition of 1,1-dimethyl-1-silacyclobutane [1]. Now the problem of

existence of compounds with different types of multiple p_{π} - p_{π} bonds involving silicon has been solved
[2-4]. A large number of multiply bonded intermediates as well as a few stable compounds containing
Si=C, Si=Si, Si=N, and Si=P double bonds have been
described [2-7]. The latter having sterically hindered bulky substituents are rather exceptions among
the huge abundance of unstable intermediates.
   Here we present some results of our studies on
thermal decomposition of cyclic organosilicon compounds as well as on dehalogenation reactions of
organochlorosilanes using alkali metal vapors.

KINETIC AND MECHANISTIC STUDIES OF THERMAL DECOMPOSITION OF HETEROCYCLIC COMPOUNDS OF SILICON

   With the aim to study the effect of ring size,
nature of heteroatom, and the character of substituents on the direction and rate of thermal decomposition, we studied the pyrolysis of silacyclo-
alkanes and thiasilacycloalkanes.
   Kinetic studies were made in a special device
using the pulse pyrolysis gas chromatography technique. The intermediates were identified by the low-
temperature matrix IR-spectroscopy method [8].

1,1-dimethyl-silacycloalkanes

   Monosilacyclobutanes having no functional groups
(Si-H, Si-Cl, etc.) upon pyrolysis are known to undergo clean [4 → 2 + 2]-cyclodecomposition with the
formation of unstable silaalkenes which dimerize
yielding 1,3-disilacyclobutanes.
   On going from 1,1-dimethyl-1-silacyclobutane
(DMSCB) to a 5-membered heterocycle - 1,1-dimethyl-
1-silacyclopentane (DMSCP) - the decomposition temperature increases by about 200°C and the selectivity decreases. If in pyrolysis of DMSCB ethylene is
the only gaseous product, then in the case of DMSCP
hydrogen, methane, acetylene, and propene are formed.
Correspondingly, the Arrhenius parameters of the
thermal decomposition reaction vary: the activation
energy increases and the factor A decreases (DMSCB:
log A = 15.3 s^{-1}, E = 61.6 kcal/mole; DMSCP: log A =
14.5 s^{-1}, E = 71.4 kcal/mole). The study of DMSCP
pyrolysis by the low-temperature matrix IR-spectroscopy method has revealed no ϑ_{Si-H} absorption, but
the $\vartheta_{C=C}$ 1590 cm^{-1} band of vinyldimethylsilane and
the 1640 cm^{-1} band of allyltrimethylsilane as well
as weak absorptions of methyl (613 cm^{-1}) and allyl
(808 cm^{-1}) radicals were observed. Similarly, the
formation [9]   of   two   alkenylsilanes
upon   pyrolysis   of   DMSCP   is

apparently caused by isomerization of 1,5-biradicals
and is due to the cleavage of different C-C bonds of
the 5-membered ring. However, the yield of volatile
organosilicon reaction products is low.

4-silaspiroalkanes (SSA). Intramolecular rearrangements of 1-methylene-1-silacycloalkanes

Earlier it has been shown that the difference in
the direction and selectivity of thermal decomposition of 4-sila[3,3]spiroheptane (SSH) [10] - of the
molecule in which one and the same silicon atom
participates in the formation of two monosilacyclobutane ring - and its homologs containing only one
4-membered ring: 4-sila[3,4]spirooctane (SSO) and
4-sila[3,5]spirononane (SSN) [11] is due to the
properties of 1-methylene-1-silacycloalkanes - unstable intermediates formed upon [4 → 2+ 2]-cyclo-
decomposition of spirocompounds. In particular, a
striking difference in the tendency of MSCA to
cyclodimerize was noticed in moving from n = 3 to
n = 4 and 5. Pyrolysis   in a flow reactor
in a current of inert gas yielded mainly nonvolatile
oily products, whereas SSO and SSN gave a good yield
of MSCA cyclodimerization products. Such a variation
in the composition of SSH and SSO pyrolysis products
is due to the occurrence of intramolecular transformations of MSCA, which are most pronounced in the
case of 1-methylene-1-silacyclobutane and manifest
themselves in the case of 1-methylene-1-silacyclo-
pentane.
To reveal intramolecular transformations of
MSCA, we studied pyrolysis of SSO at low pressures,
i.e. under conditions when the bimolecular reactions of the intermediates are suppressed. Earlier,
by this method, it was ascertained that in pyrolysis
of SSH occured the intramolecular transformation of
1-methylene-1-silacyclobutane into silacyclopent -
3-ene [10]. This transformation, as thought in [12,
13], takes place via the formation of silacyclopentylidene (silene-silylene rearrangement). The SSO
low pressure pyrolysis products (6.10^{-2} Torr,
730°C) were found to have composition substantially
different from those obtained under conventional
pyrolysis conditions. In the latter case, at a 60-70
percent conversion, only ethylene is formed, whereas
in the former case, other hydrocarbons to the extent of up to 40% of the yield of ethylene are also
obtained. Besides propylene, buta-1,3-diene, penta-
1,3-diene and 1-silacyclohexene have been formed.
Such a change in the direction of thermal transformation of SSO is due to (1) low pressure in the reaction system and (2) much higher than   conven-

tional pyrolysis temperature. Below is given the
proposed scheme for thermal decomposition of SSO in
a gaseous phase.

$$\longrightarrow \quad H_2 + C_2H_4 + C_3H_6 + C_4H_6 + C_5H_8 + H_2Si \quad C_5H_8$$

Thus, the selectivity of formation of products
in SSO pyrolysis can be controlled by varying the
pressure and temperature, and creating thereby the
conditions that would favour either cyclodimerization or silene-silylene rearrangement of 1-methy-
lene-1-silacyclopentane resulting in the expansion
of the ring to the six-membered one.

A comparison of Arrhenius parameters of these
processes (SSH: log A = 13.7 s^{-1}, E = 52.9 kcal/mole;
SSO: log A = 15.1 s^{-1}, E = 60.4 kcal/mole) with
those for DMSCB is indicative of a somewhat higher
rate of thermal decomposition of silaspiroalkanes.

1,1,3-trimethyl-1-silacyclobutane (1,1,3-TMSCB) and
1,1,2-trimethyl-1-silacyclobutane (1,1,2-TMSCB)

Earlier [14,15] we have found that the presence
of methyl group at position 3 of the ring permits of
isomerization of 1,1,3-TMSCB into allyltrimethylsilane (log A = 15.6 s^{-1}, E = 63.2 kcal/mole) which
accompany its [4 → 2+2]-thermodecomposition (log A =
16.3 s^{-1}, E = 63.2 kcal/mole). Isomerization proceeds via a six-membered cyclic transition state
necessary for 1,5-H migration in the biradical formed upon cleavage of the C-C bond of the ring.

In 1975 Barton and co-workers [16] have shown
that [4 → 2+2]-cyclodecomposition takes place in two
ways in pyrolysis of 1,1,2-TMSCB, one being accompanied also by isomerization. We have studied the
kinetics of these processes. The Arrhenius parra-
meters of the total process (log A = 15.7 s^{-1}, E =
61.4 kcal/mole) reflect somewhat lesser thermal stability of 1,1,2-TMSCB compared to DMSCB. The activation energies of formation of ethylene and propylene as well as of vinyldimethylethylsilane which is
the isomerization product, were found to be alike
within the experimental error. The different rates
of alternative [4 → 2+2]-cyclodecomposition processes
and isomerization are entirely defined by the entropy factor (log A) which for the reaction of for-

mation of propylene and isomerization are respectively 0.7 and 1.4 s^{-1} less than for the formation of
ethylene. Below are given the entropies of activation.

Such a change in the reactivity is indicative of
a more "tight" transition state of the second and
third reactions compared to the first one.

1-vinyl-1-methyl-1-silacyclobutane (VMSCB)

The Arrhenius parameters of thermal decomposition
of this monosilacyclobutane (log A = 15.3 s^{-1}, E =
61.4 kcal/mole) are that of DMSCB, the
selectivity being very high in both cases.
Ethylene (yield 100 moles per 100 M of the transformed VMSCB) and 1,3-dimethyl-1,3-divinyl-1,3-di-
silacyclobutane (DMDVDSCB) - a cyclodimer of 2-
methyl-2-silabuta-1,3-diene - were the only reaction
products [12]. According to [17,18] the selectivity
of VMSCB decomposition is less. Besides ethylene and
DMDVDSCB, a complex mixture of methylsilanes and
hydrocarbons, and also allen were detected. Our kinetic data correlate well with those of low-pressure pyrolysis of MVSCB [18], the "allyl" stabilization of the silylvinyl radical being small to
exert such an influence on the activation energy as
noticed in vinylcyclobutane pyrolysis [19].

1-methyl-1-silacyclobutane (MSCB)

The hydrogen atom bonded to the silicon of the
monosilacyclobutane ring has a pronounced effect on
the thermal decomposition process because of the
silaethylene-silylene rearrangement [20-28].
Dependence of the yield of main products on the
conversion of MSCB is shown in Fig. 1.

Fig. 1.

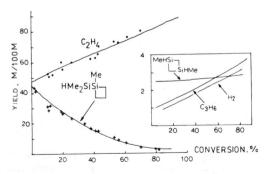

At low conversions the yield of ethylene equals 50M/
100M. At a higher conversion due to thermal decomposition
of the dimethylsilylene to MSCB adduct whose maximum

yield 50M/100M is observed at the very early stage of pyrolysis. Minor products are hydrogen and propylene (about 1M/100M), their yields being greater by 3-4-folds at higher temperatures.   of 1,3-dimethyl-1,3-disilacyclobutane were formed on converting 100 moles of MSCB, the selectivity being independent of conversion.

The Arrhenius parameters of MSCB total decomposition (log A = 13.2 s^{-1}, E = 52.4 kcal/mole) are close to those reported in Ref. 24 (log A = 14.0 s^{-1}, E = 52.6 kcal/mole).

When MSCB pyrolysis was carried out in the presence of 10-fold excess buta-1,3-diene the Arrhenius parameters became usual for [4→2+2]-thermocyclo-decomposition (log A = 15.3 s^{-1} and E = 60.4 kcal/mole). In this case, the formation of ethylene was described by log A = 16.0 s^{-1} and E = 63.5 kcal/mole. According to [20], low-pressure pyrolysis of MSCB carried out at 5 Torr and in the presence of 5-fold excess butadiene yielded log A = 14.9 s^{-1} and E = 59.1 kcal/mole.

The activation energy of the propylene formation reaction was found to be 15 kcal/mole higher than of ethylene, the value of log A being more by 2.5 s^{-1}. This seems to be in contradiction with the data reported in Refs. 26 and 27.

Because the selectivity of formation of 1-methyl-1-(dimethylsilyl)-1-silacyclobutane is high at low conversions, one cannot make use of MSCB pyrolysis as preparative method of obtaining this compound. Its yield can be raised by copyrolysis of MSCB with excess pentamethylmethoxydisilane under conditions when MSCB does not decompose, but disilane provides for the desired amount of dimethylsilylene.

Fig. 2.

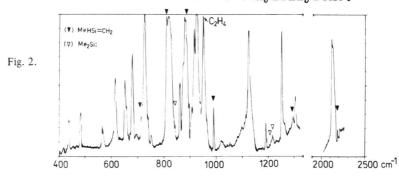

Attempts to detect silaalkanes by the low-temperature matrix IR-spectroscopy method in pyrolysis of 1-silacyclobutane were not successful [29,30]. We succeeded in matrix isolating 1-methylsilaethylene during pyrolysis of MSCB (670°C, 6.10^{-2} Torr). The 2181, 991, 880, 811, and 712 cm^{-1} infrared absorptions of 1-methyl-1-silaethylene [31-34] were observed (Fig. 2).

3,3-dimethyl-3-silathietane (DMST) [35,36]

Pyrolysis of DMST containing both silicon and sulfur atoms in the 4-membered ring yields ethylene and tetramethylcyclodisilthiane (TMCDST). According to the proposed scheme [below] ethylene and dimethylsilanthione (precursor of TMCDST) are formed in [4→2+2]-cyclodecomposition of the transient 2-silathietane which appears to be the product of DMST [4→2+2]-cyclodecomposition - [2+2]- cycloaddition sequence.

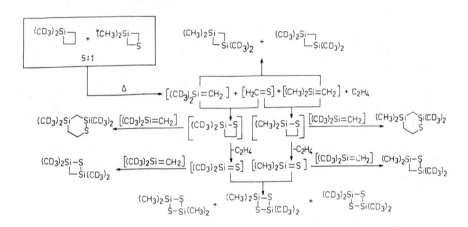

The absorption bands of DMSE and thioformaldehyde as well as of ethylene and TMCDST were identified in the matrix IR-spectra of DMST pyrolysis products. Compared to DMSCB, the presence of sulfur atom at position 3 of the monosilacyclobutane ring somewhat increases the rate of DMST thermal decomposition (log A = 15.1 s^{-1}, E = 58.9 kcal/mole). Chemical evidence of the intermediacy of DMSE, thioformaldehyde, dimethylsilanthione, and 2-silathietane derived from copyrolysis of DMST with excess DMSCB-d$_6$ [37]. [2+2]-Cycloaddition of DMSE to dimethylsilanthione was employed to obtain 2,2,4,4-tetramethyl-2,4-disilathietane (TMDST) via copyrolysis of DMSCB and thietane. In this reaction are also formed cyclocarbosilthianes - the insertion products of both DMSE and dimethylsila Si-S bond of TMDST [39,39].

DEHALOGENATION OF CHLOROMETHYLCHLORO- AND DICHLOROSILACYCLOALKANES USING ALKALI METAL VAPORS

Alkali metal vapors dehalogenation reaction appeared to be a convenient method of generating silyl radicals, silylenes, and silaalkenes [40]. We have applied this reaction for obtaining methylenesilacycloalkanes and cyclic silylenes: 1-methylene-1-silacyclopentane, silacyclopentylidene and silacyclohexylidene - the intermediates mentioned above in discussing the thermal decomposition of 4-silaspiroalkanes.

Dehalogenation of 1-chloromethyl-1-chloro-1-silacyclopentane yields mainly a cyclic dimer of 1-methylene-1-silacyclopentane (MSCP). No products indicating intramolecular rearrangements of MSCP were found.

Thus, under dehalogenation conditions, isomerization of MSCP into silacyclohexylidene probably does not take place. We attempted at obtaining silacyclohexylidene by dehalogenating 1,1-dichloro-1-silacyclohexane. In so doing, traces of penta-1,3-diene and larger amounts of penta-2-ene were detected. Decomposition is typical also for other cyclic silylenes: silacyclopentylidene and silacyclobutylidene. In the former are obtained cis- and trans- butenes, in the latter, ethylene and propylene. The possible pathways of their formation are shown in the Scheme:

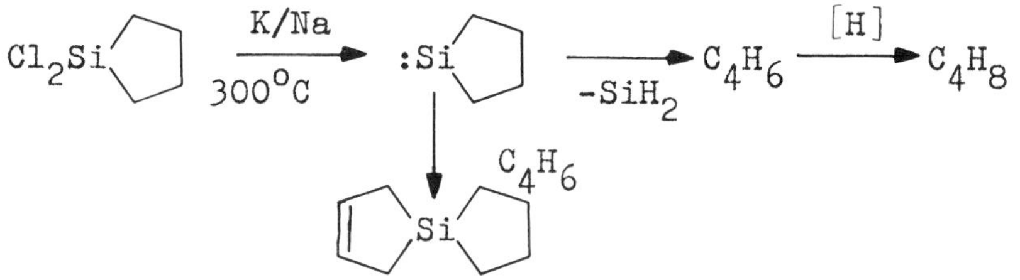

$$Cl_2Si \overset{K/Na}{\underset{300^\circ C}{\longrightarrow}} :Si \overset{}{\underset{-SiH_2}{\longrightarrow}} C_4H_6 \overset{[H]}{\longrightarrow} C_4H_8$$

$$\downarrow C_4H_6$$

The intermediacy of silacycloalkylidenes is proved by the formation of adducts when 1,1-dichloro-1-silacycloalkanes are dehalogenated in the presence of 20-fold excess butadiene. Thus, the decomposition of cyclic silylenes results in the formation of conjugate dienes, whereas noncyclic silylenes [40,41] decompose yielding corresponding alkenes. Presently we are engaged in a more comprehensive study of these quite unexpected decomposition reactions of cyclic silylenes.

ACKNOWLEDGEMENTS

The results discussed in this chapter have been

obtained with the untiring efforts of my co-workers:
V.V. Volkova, P.E. Ivanov, E.A. Volnina, and Yu.P.
Polyakov whose contribution is gratefully acknowl-
edged.

REFERENCES

1. L.E. Gusel'nikov and N.S. Nametkin, in "Advances in Organosilicon Chemistry", Mir Publishers, Moscow, 1985, p.69.
2. L.E. Gusel'nikov and N.S. Nametkin, Chem. Rev.,
   79, 529 (1979).
3. G. Raabe and J. Michl, Chem. Rev., 85, 419(1985).
4. A.G. Brook, Adv. Organomet. Chem., 25, 1 (1986).
5. N. Wiberg, K. Schulz, and G. Fischer, Angew.
   Chem., 97, 1058 (1985).
6. C.N. Smith, F.M. Lock, and F. Bickelhaupt,
   Tetrahedron Lett., 25, 3011 (1984).
7. V.D. Romanenko, A.V. Ruban, A.B. Drapailo, and
   L.N. Markovskii, Zh. Obsch. Khim., 55,2793(1985).
8. L.E.Gusel'nikov, L.V. Shevelkova, V.V. Volkova,
   L.M. Vedeneeva, P.E. Ivanov, G. Zimmermann, G.
   Bach, U. Ziegler, B. Ondruschka, and F.-D.
   Kopinke, in "Fifth Symposium on Petrochemistry
   of the Socialist Countries". Proceedings, p.297,
   Burgas, September 16-21, 1986.
9. L.E. Gusel'nikov, V.V. Volkova, U. Ziegler, G.
   Zimmermann, B. Ondruschka, P.E. Ivanov, and L.V.
   Schevelkova, Izv. Akad. Nauk SSSR, Ser. Khim.
   1986, 2152.
10. N.S. Nametkin, L.E. Gusel'nikov, V.Yu. Orlov, R.
    L. Ushakova, O.V. Kuzmin, and V.M. Vdovin, Dokl.
    Akad. Nauk SSSR, 211, 106 (1973).
11. N.S. Nametkin, L.E. Gusel'nikov, V.Yu. Orlov, N.
    N. Dolgopolov, P.L. Grinberg, and V.M. Vdovin,
    Zh. Obsch. Khim., 45, 69 (1975).
12. R.L. Ushakova, Author's Abstract of the Candi-
    date of Sci. Thesis (Chem.),Moscow, INKhS AN
    SSSR, 1975.
13. T.J. Barton, G.T. Burns, and D.Gschneidner, Or-
    ganometallics, 2, 8 (1983).
14. N.S. Nametkin, R.L. Ushakova, L.E. Gusel'nikov,
    E.D. Babich, and V.M. Vdovin, Izv. Akad. Nauk
    SSSR, Ser. Khim., 1970, 1676.
15. L.E. Gusel'nikov, N.S. Nametkin, and N.N. Dolgo-
    polov, J. Organomet. Chem., 169, 165 (1979).
16. T.C. Barton, G. Marquardt, and J.A. Kilgour, J.
    Organomet. Chem., 85, 317 (1975).
17. N. Auner and J. Grobe, J. Organomet. Chem., 197,
    13 (1980).
18. I.M.T. Davidson, A.M. Fenton, P. Jackson, and F.
    T.Lawrence, J.Chem.Soc.,Chem. Commun.,1982, 806.
19. H.M. Frey, R. Pottinger, J. Chem. Soc., Faraday

I, 74, 1827 (1978).

20. R.T. Conlin and D.L. Wood, J. Amer. Chem. Soc.,
    103, 1843 (1981).
21. R.T. Conlin and Y.W. Kwak, Organometallics, 3,
    918 (1984).
22. R.T. Conlin and Y.W. Kwak, J. Amer. Chem. Soc.,
    108, 834 (1986).
23. T.J. Barton, S.A. Burns, and G.T. Burns, Organo-
    metallics, 1, 210 (1982).
24. I.M.T.Davidson, S.Iyadi-Maghsoodi, T.J.Barton and
    N. Tillman, J. Chem. Soc., Chem.Commun., 1984,478.
25. I.M.T. Davidson and R.J. Scampton, J. Organomet.
    Chem., 271, 249 (1984).
26. I.M.T. Davidson, A. Fenton, S. Iyadi-Maghsoodi,
    R.J. Scampton, N. Auner, J. Grobe, N.Tillman, and
    T.J. Barton, Organometallics,3, 1593 (1984).
27. I.M.T. Davidson, in "Organosilicon and Bioorgano-
    silicon Chemistry (H. Sakurai, Ed.), Ellis
    Horwood Ltd., Chichester, 1985, p. 75.
28. R.Walsh, J. Chem. Soc., Chem. Commun.,1982, 1415.
29. A.K. Mal'tsev, V.N. Khabashesku, and O.M. Ne-
    fedov, Dokl. Akad. Nauk SSSR, 247, 383 (1979).
30. G. Maier, G. Michm, and H.P. Reisenauer, Chem.
    Ber., 117, 2351 (1984).
31. T.J. Drahnak, J. Mihl, and R. West, J. Amer.
    Chem. Soc., 101, 5427 (1979).
32. C.A. Arrington, K.A.Klingensmith, R. West, and
    J. Mihl, J. Amer. Chem. Soc., 106, 525 (1984).
33. G. Raabe, H. Vancik, R. West, and J. Michl, J.
    Amer. Chem. Soc., 108, 671 (1986).
34. G. Maier, G. Mihm, H.P. Reisenauer, and D. Lit-
    tmann, Chem. Ber., 117, 2369 (1984).
35. L.E. Gusel'nikov, V.V. Volkova, V.G. Avakyan, N.
    S.Nametkin, M.G. Voronov, S.V.Kirpichenko, and E.
    N.Suslova, Dokl. Akad. Nauk SSSR, 272,892(1983).
36. L.E. Gusel'nikov, V.V. Volkova, V.G. Avakyan, N.
    S. Nametkin, M.G.Voronov, S.V.Kirpichenko, and
    E.N.Suslova, J. Organomet. Chem., 254, 173(1983).
37. V.V.Volkova, Author's Abstract of the Cand. of
    Sci. Thesis (Chem.), Moscow, INKhS AN SSSR,1987.
38. L.E. Gusel'nikov, V.V. Volkova, E.A. Volnina, V.
    G. Avakyan, and N.S. Nametkin, Dokl. Akad. Nauk
    SSSR, 274, 1106 (1984).
39. L.E. Gusel'nikov, V.V. Volkova, V.G. Avakyan, E.
    A. Volnina, V.G. Zaikin, N.S. Nametkin, A.A.
    Polyakova, and M.I. Tokarev, J. Organomet. Chem.,
    271, 191 (1984).
40. L.E.Gusel'nikov and N.S.Nametkin, in "Organosil-
    icon and Bioorganosilicon Chemistry",(H.Sakurai,
    Ed.), Ellis Horwood Ltd.,Chichester, 1985,p.115.
41. L.E. Gusel'nikov, E. Lopatnikova, Yu.P. Polyakov,
    and N.S. Nametkin, Dokl. Akad. Nauk SSSR, 253,
    1387 (1980).

Index